VOLUME SIXTY SEVEN

ADVANCES IN ECOLOGICAL RESEARCH

Functional Microbiomes

ADVANCES IN ECOLOGICAL RESEARCH

Series Editors

DAVID A. BOHAN
Agroécologie, AgroSup Dijon, INRAE,
Université de Bourgogne Franche-Comté,
Dijon, France

ALEX J. DUMBRELL
School of Life Sciences,
University of Essex,
Wivenhoe Park, Colchester,
Essex, United Kingdom

VOLUME SIXTY SEVEN

ADVANCES IN ECOLOGICAL RESEARCH

Functional Microbiomes

Edited by

DAVID A. BOHAN

Agroécologie, AgroSup Dijon, INRAE,
Université de Bourgogne Franche-Comté,
Dijon, France

ALEX DUMBRELL

School of Life Sciences,
University of Essex,
Wivenhoe Park, Colchester,
Essex, United Kingdom

ELSEVIER

ACADEMIC PRESS
An imprint of Elsevier

Academic Press is an imprint of Elsevier
125 London Wall, London, EC2Y 5AS, United Kingdom
The Boulevard, Langford Lane, Kidlington, Oxford OX5 1GB, United Kingdom
525 B Street, Suite 1650, San Diego, CA 92101, United States
50 Hampshire Street, 5th Floor, Cambridge, MA 02139, United States

First edition 2022

ISBN: 978-0-323-98593-2
ISSN: 0065-2504

For information on all Academic Press publications
visit our website at https://www.elsevier.com/books-and-journals

Publisher: Zoe Kruze
Acquisitions Editor: Jason Mitchell
Developmental Editor: Federico Paulo Mendoza
Production Project Manager: Abdulla Sait
Cover Designer: Miles Hitchen

Typeset by STRAIVE, India

Contents

Contributors *ix*
Preface: Functional Microbiomes *xiii*

1. Linking microbial body size to community co-occurrences and stability at multiple geographical scales in agricultural soils **1**

Pengfa Li, Alex J. Dumbrell, Muhammad Saleem, Lu Kuang, Ting Li, Lu Luan, Weitao Li, Guilong Li, Meng Wu, Baozhan Wang, Jiandong Jiang, Ming Liu, and Zhongpei Li

1. Introduction 2
2. Methods 5
3. Results 11
4. Discussion 19
5. Data accessibility statement 23

Acknowledgements 23
References 23

2. The functional microbiome of grapevine throughout plant evolutionary history and lifetime **27**

Paola Fournier, Lucile Pellan, Didac Barroso-Bergadà, David A. Bohan, Thierry Candresse, François Delmotte, Marie-Cécile Dufour, Virginie Lauvergeat, Claire Le Marrec, Armelle Marais, Guilherme Martins, Isabelle Masneuf-Pomarède, Patrice Rey, David Sherman, Patrice This, Clémence Frioux, Simon Labarthe, and Corinne Vacher

1. Introduction 28
2. The grapevine functional microbiome throughout evolutionary history 29
3. The grapevine functional microbiome throughout plant lifetime 39
4. Let's make the grapevine microbiome more functional 58
5. Conclusion and perspectives 71

Acknowledgements 74
References 74

3. Compendium of analytical methods for sampling, characterization and quantification of bioaerosols **101**

Corinne Whitby, Robert M.W. Ferguson, Ian Colbeck, Alex J. Dumbrell, Zaheer A. Nasir, Emma Marczylo, Rob Kinnersley, Philippa Douglas, Gill Drew, Kam Bhui, Mark Lemon, Simon Jackson, Sean Tyrrel, and Frederic Coulon

1. Introduction 104
2. Air sampling methods 109
3. Bioaerosol analysis methods 149
4. Sampling and analysis methods to assess exposure and risk 177
5. Conclusions and future perspectives 202

Acknowledgements 206
List of contributors 206
References 207

4. A microbial solution to oil sand pollution: Understanding the microbiomes, metabolic pathways and mechanisms involved in naphthenic acid (NA) biodegradation **231**

Corinne Whitby

1. Introduction 232
2. Natural vs anthropogenic naphthenic acid (NA) exposed microbiomes 237
3. Anthropogenic naphthenic acid (NA) contaminated microbiomes 247
4. Biodegradation of naphthenic acids (NAs) 263
5. Conclusions 273
6. Future perspectives 274

References 275

5. The gut microbiome in health and disease: Inflammatory bowel diseases **289**

Salma El-Sahhar and Patrick Varga-Weisz

1. Introduction 290
2. Genetic analysis highlights the role of the gut microbiota in IBD 295
3. How do changes in the microbiome influence IBD? 303
4. How does the environment contribute to gut microbiome changes? 309
5. Can microbial-based treatments be the answer? 314
6. Conclusion and outlook 318

References 320

6. Mind the gaps: What do we know about how multiple chemical stressors impact freshwater aquatic microbiomes? 331

Alessia Bani, Kate C. Randall, Dave R. Clark, Benjamin H. Gregson, Drew K. Henderson, Erin C. Losty, and Robert M.W. Ferguson

1. Introduction 332
2. The freshwater aquatic chemical exposome 335
3. Selection of chemicals and literature search 345
4. Responses of freshwater microbial communities to eight important chemicals: What do we know so far? 350
5. Embracing complexity: Multiple stressors in freshwaters 355
6. Everything is connected: Indirect effects of chemical stressors 357
7. Antimicrobial resistance (AMR): Are freshwaters the key link in the one health approach? 358
8. Nutrient pollution: Does nothing else matter? 360
9. Beyond taxonomy: A new paradigm for biomonitoring of freshwater ecosystems 362
10. Bridging the gap 366
11. Summary and future directions 367

Acknowledgements 368
References 368

Contributors

Alessia Bani
School of Life Sciences, University of Essex, Wivenhoe Park, Colchester, Essex, United Kingdom

Didac Barroso-Bergadà
INRAE, Université Bourgogne Franche-Comté, Agroécologie, Dijon, France

Kam Bhui
University of Oxford, World Psychiatric Association Collaborating Centre, Oxford, United Kingdom

David A. Bohan
INRAE, Université Bourgogne Franche-Comté, Agroécologie, Dijon, France

Thierry Candresse
Univ. Bordeaux, INRAE, UMR BFP, Villenave d'Ornon, France

Dave R. Clark
School of Life Sciences; Institute for Analytics and Data Science, University of Essex, Wivenhoe Park, Colchester, Essex, United Kingdom

Ian Colbeck
School of Life Sciences, University of Essex, Colchester, Essex, United Kingdom

Frederic Coulon
School of Water, Energy and Environment, Cranfield University, Cranfield, United Kingdom

François Delmotte
INRAE, Bordeaux Sciences Agro, ISVV, SAVE, Villenave-d'Ornon, France

Philippa Douglas
UK Health Security Agency, Radiation, Chemical and Environmental Hazards Directorate, Oxfordshire; NIHR Health Protection Research Unit in Environmental Exposures and Health at the University of Leicester, Leicester, United Kingdom

Gill Drew
School of Water, Energy and Environment, Cranfield University, Cranfield, United Kingdom

Marie-Cécile Dufour
INRAE, Bordeaux Sciences Agro, ISVV, SAVE, Villenave-d'Ornon, France

Alex J. Dumbrell
School of Life Sciences, University of Essex, Colchester, Essex, United Kingdom

Salma El-Sahhar
School of Life Sciences, University of Essex, Colchester, United Kingdom

Robert M.W. Ferguson
School of Life Sciences, University of Essex, Wivenhoe Park, Colchester, Essex, United Kingdom

Paola Fournier
INRAE, Univ. Bordeaux, BIOGECO, Pessac; INRAE, Bordeaux Sciences Agro, ISVV, SAVE, Villenave-d'Ornon, France

Clémence Frioux
Inria, Univ. Bordeaux, INRAE, Talence, France

Benjamin H. Gregson
School of Life Sciences, University of Essex, Wivenhoe Park, Colchester, Essex, United Kingdom

Drew K. Henderson
School of Life Sciences, University of Essex, Wivenhoe Park, Colchester, Essex, United Kingdom

Simon Jackson
University of Plymouth, Plymouth, United Kingdom

Jiandong Jiang
Department of Microbiology, Key Lab of Microbiology for Agricultural Environment, Ministry of Agriculture, College of Life Sciences, Nanjing Agricultural University, Nanjing, China

Rob Kinnersley
Environment Agency, Bristol, United Kingdom

Lu Kuang
Department of Microbiology, Key Lab of Microbiology for Agricultural Environment, Ministry of Agriculture, College of Life Sciences, Nanjing Agricultural University, Nanjing, China

Simon Labarthe
INRAE, Univ. Bordeaux, BIOGECO, Pessac; Inria, Univ. Bordeaux, INRAE, Talence, France

Virginie Lauvergeat
EGFV, Univ. Bordeaux, Bordeaux Sciences Agro, INRAE, ISVV, Villenave d'Ornon, France

Claire Le Marrec
Univ. Bordeaux, UMR oenologie, INRAE, Bx INP, Bordeaux Sciences Agro, ISVV, Villenave d'Ornon, France

Mark Lemon
De Montfort University, Leicester, United Kingdom

Guilong Li
State Key Laboratory of Soil and Sustainable Agriculture, Institute of Soil Science, Chinese Academy of Sciences, Nanjing, China

Pengfa Li
Department of Microbiology, Key Lab of Microbiology for Agricultural Environment, Ministry of Agriculture, College of Life Sciences, Nanjing Agricultural University; State Key Laboratory of Soil and Sustainable Agriculture, Institute of Soil Science, Chinese Academy of Sciences, Nanjing, China

Ting Li
Department of Microbiology, Key Lab of Microbiology for Agricultural Environment, Ministry of Agriculture, College of Life Sciences, Nanjing Agricultural University, Nanjing, China

Weitao Li
CAS Key Laboratory of Tropical Forest Ecology, Xishuangbanna Tropical Botanical Garden, Chinese Academy of Sciences, Menglun, Mengla, Yunnan, China

Zhongpei Li
State Key Laboratory of Soil and Sustainable Agriculture, Institute of Soil Science, Chinese Academy of Sciences, Nanjing, China

Ming Liu
State Key Laboratory of Soil and Sustainable Agriculture, Institute of Soil Science, Chinese Academy of Sciences, Nanjing, China

Erin C. Losty
School of Life Sciences, University of Essex, Wivenhoe Park, Colchester, Essex, United Kingdom

Lu Luan
State Key Laboratory of Soil and Sustainable Agriculture, Institute of Soil Science, Chinese Academy of Sciences, Nanjing, China

Armelle Marais
Univ. Bordeaux, INRAE, UMR BFP, Villenave d'Ornon, France

Emma Marczylo
UK Health Security Agency, Radiation, Chemical and Environmental Hazards Directorate, Oxfordshire; NIHR Health Protection Research Unit in Environmental Exposures and Health at the University of Leicester, Leicester, United Kingdom

Guilherme Martins
Univ. Bordeaux, UMR oenologie, INRAE, Bx INP, Bordeaux Sciences Agro, ISVV, Villenave d'Ornon, France

Isabelle Masneuf-Pomarède
Univ. Bordeaux, UMR oenologie, INRAE, Bx INP, Bordeaux Sciences Agro, ISVV, Villenave d'Ornon, France

Zaheer A. Nasir
School of Water, Energy and Environment, Cranfield University, Cranfield, United Kingdom

Lucile Pellan
INRAE, Bordeaux Sciences Agro, ISVV, SAVE, Villenave-d'Ornon, France

Kate C. Randall
School of Life Sciences, University of Essex, Wivenhoe Park, Colchester, Essex; Applied Sciences, University of Northumbria, Newcastle Upon Tyne, United Kingdom

Patrice Rey
Université de Pau et des Pays de l'Adour, E2S UPPA, CNRS, Institut des Sciences Analytiques et de Physicochimie pour l'Environnement et les Matériaux—UMR 5254, IBEAS, Pau, France

Muhammad Saleem
Department of Biological Sciences, Alabama State University, Montgomery, Alabama, United States

David Sherman
Inria, Univ. Bordeaux, INRAE, Talence, France

Patrice This
UMR AGAP Institut, Univ Montpellier, CIRAD, INRAE, Institut Agro, Montpellier, France

Sean Tyrrel
School of Water, Energy and Environment, Cranfield University, Cranfield, United Kingdom

Corinne Vacher
INRAE, Univ. Bordeaux, BIOGECO, Pessac, France

Patrick Varga-Weisz
School of Life Sciences, University of Essex, Colchester, United Kingdom; São Paulo Excellence Chair, Department of Genetics and Evolution, Microbiology and Immunology, Institute of Biology, University of Campinas, Campinas, SP, Brazil

Baozhan Wang
Department of Microbiology, Key Lab of Microbiology for Agricultural Environment, Ministry of Agriculture, College of Life Sciences, Nanjing Agricultural University, Nanjing, China

Corinne Whitby
School of Life Sciences, University of Essex, Colchester, Essex, United Kingdom

Meng Wu
State Key Laboratory of Soil and Sustainable Agriculture, Institute of Soil Science, Chinese Academy of Sciences, Nanjing, China

Preface: Functional Microbiomes

This first volume of a two-volume issue on functional microbiomes presents a snapshot of opinion and work at the interface of research on microbial biology and ecological communities, networks and functions. Our aim in developing these volumes was to move beyond microbiomes as a simple listing of the microbial species, OTUs or ASVs present, and towards microbiomes that have dynamics, structure and function, as an attempt to deliver on the promise of microbiome research of better understanding, management, health and sustainability in the systems where microbes are believed to play an ecologically regulatory role.

Volume I presents six chapters that cover a spectrum of levels of organisation, from local to regional and continental scales, and from the bowels of an individual human host to individual plants, and the distinct environmental biomes of the soil, freshwater and air. In "Linking microbial body size to community co-occurrences and stability at multiple geographical scales in agricultural soils," Li et al. argue that body size structures microbial communities and networks. This rather unexpected finding mirrors that in macrobial systems, where the body size of predators and associated allometries such as mouth gape size determine the structuring of trophic ecological networks. Fournier et al. explore the evolution of the microbiome of the grapevine with domestication, in "The functional microbiome of grapevine throughout plant evolutionary history and lifetime," with a view to identifying future avenues of research for this economically and culturally important system. The methods necessary to sample, characterize and quantify the microbiomes of aerosols are discussed by Whitby et al. in "Compendium of analytical methods for sampling, characterisation and quantification of bioaerosols." The microbiomes of bioaerosols have recently received much scientific and societal attention due to the emergence and global spread of COVID-19 and the threat of bioterrorism, necessitating these techniques. The fourth chapter of the volume, "A microbial solution to oil sand pollution: Understanding the microbiomes, metabolic pathways and mechanisms involved in naphthenic acid (NA) biodegradation" by Corinne Whitby, seeks to unravel and understand the functions of naphthenic acid biodegradation in natural and anthropogenic bitumen-rich environments, including the genes, enzymes and mechanisms of the microbiome. The role of the microbiome in inflammatory bowel diseases that affect a significant proportion of human populations, particularly in the developed world, is the subject of "The gut microbiome in health and disease: Inflammatory bowel diseases." The authors, El-Sahhar and Varga-Weisz, argue that the evidence supports a significant role for microbiomes in the development of pathology and that future long-term, longitudinal studies will propel the field from *cataloguing correlations to understanding causation.* The final chapter of the volume is "Mind the gaps: What do we know about how multiple chemical stressors impact freshwater aquatic microbiomes?" by Bani et al. In it, the authors argue that, as many of the services we derive from freshwaters are based upon microbial processes, it is critical that we understand how freshwater microbiomes are affected by stressors, therefore necessitating the inclusion of taxonomic and trait-based information in biomonitoring.

This selection of content demonstrates that a functional understanding of microbiomes has the potential for a major economic and societal impact but, as all the authors of this volume note, there is plenty of exciting research to be done and numerous challenges still to be met.

Acknowledgements: David Andrew Bohan would like to acknowledge the support of the ANR NGB (ANR-17-CE32-0011) project.

David A. Bohan
Alex Dumbrell

CHAPTER ONE

Linking microbial body size to community co-occurrences and stability at multiple geographical scales in agricultural soils

Pengfa Li[a,b], Alex J. Dumbrell[c], Muhammad Saleem[d], Lu Kuang[a], Ting Li[a], Lu Luan[b], Weitao Li[e], Guilong Li[b], Meng Wu[b], Baozhan Wang[a], Jiandong Jiang[a], Ming Liu[b,*], and Zhongpei Li[b,*]

[a]Department of Microbiology, Key Lab of Microbiology for Agricultural Environment, Ministry of Agriculture, College of Life Sciences, Nanjing Agricultural University, Nanjing, China
[b]State Key Laboratory of Soil and Sustainable Agriculture, Institute of Soil Science, Chinese Academy of Sciences, Nanjing, China
[c]School of Life Sciences, University of Essex, Colchester, Essex, United Kingdom
[d]Department of Biological Sciences, Alabama State University, Montgomery, Alabama, United States
[e]CAS Key Laboratory of Tropical Forest Ecology, Xishuangbanna Tropical Botanical Garden, Chinese Academy of Sciences, Menglun, Mengla, Yunnan, China
*Corresponding authors: e-mail address: mliu@issas.ac.cn; zhpli@issas.ac.cn

Contents

1. Introduction 2
2. Methods 5
   2.1 Sample sites and data collection 5
   2.2 Molecular methods, metabarcoding and bioinformatics 7
   2.3 Determination of microbial body sizes 7
   2.4 Statistical analysis 10
3. Results 11
   3.1 Inferring microbial community structure and body size 11
   3.2 Microbial niche breadth, niche overlap, and migration rate 12
   3.3 Co-occurrence patterns of different microbial groups 14
   3.4 Cohesion of differently sized microorganisms 16
   3.5 Community stability of differently sized microorganisms 16
4. Discussion 19
5. Data accessibility statement 23
Acknowledgements 23
References 23

Advances in Ecological Research, Volume 67
ISSN 0065-2504
https://doi.org/10.1016/bs.aecr.2022.09.002

Abstract

Body size determines individuals' life history and metabolic rates and thus, regulates community-level dynamics. However, whether body size mediates community co-occurrences and stability, especially in complex communities across different microbial trophic levels, remains unknown. Here, we investigate whether body size determines the co-occurrence pattern and stability of microbial communities across local, regional, and continental scales in the paddy soil ecosystems.

Soil samples were collected from rice paddy fields at multiple spatial scales, and soil microbial communities were subsequently sequenced. The microorganisms were then divided into different groups based on taxonomic information at phylum/subphylum level, and the average body size of each microbial group was identified based on propagule size from documented literature. We examined the relationships between microbial body size and various community traits such as potential migration rate, co-occurrence pattern, cohesion, and community stability.

Our results consistently showed that the small-sized microorganisms such as bacteria had significantly higher niche breadth, niche overlap and migration rate at various spatial scales. We found that microbial body size is consistently negatively correlated to negative co-occurrences and community stability. Our results, for the first time, put microbial body size into a broader community ecology framework, and contribute to a greater understanding of how microbial taxa with different body sizes would respond to future changes and perturbations.

1. Introduction

The functions and services of complex natural ecosystems are dependent on a relatively stable microbiome (Cramer and Katz, 2021), which is defined as the degree of variation or rate of turnover in the microbial communities (Tripathi et al., 2018). Understanding the determinants of community stability is highly important to sustain the functions and services delivered by ecological communities (Chen et al., 2021). Ecological stability is influenced by multiple biological features of microbial communities, which nevertheless depends on microbial traits and their cumulative effects on ecosystem functions, from individual- to community- level attributes (Schnabel et al., 2019). Microbial species are characterized by numerous ecological, physiological, and molecular traits and there is an emerging interest in understanding whether they have a role in microbiome functioning (Green et al., 2008; Martiny et al., 2015; McGill et al., 2006). Among these, cell size (hereafter body size) is a key eco-physiological trait that determines life history, metabolism, physiology, and many other aspects of an organism's ecology (Saleem et al., 2013). For example, larger microorganisms (protists or fungi) may have a relatively narrow niche breadth, whereas the niche breadth of smaller microorganisms (bacteria) may be relatively wider (Luan et al., 2020), though empirical evidence is required to confirm this prediction. In addition, a significant relationship between body size and community assembly is a general and widespread pattern in macroecology (plants, animals) (Dumbrell, 2019; Farjalla et al., 2012; Luan et al., 2020). However, whether body size determines community stability at different microbial trophic levels such as bacteria, protists, and fungi, remains unknown.

The ecological pyramids demonstrate the abundance, distribution, and biomass (including individual body-sizes) of different organisms across a hierarchy, and thus, they reveal striking regularities among species and ecological communities (Jacquet et al., 2020). Similar to higher trophic pyramids, smaller microorganisms such as bacteria generally have much higher population density than larger ones such as protists and fungi within microbial trophic levels (Jacquet et al., 2020). While, the dynamic behaviour of microbial groups and systems are determined by intrinsic density-dependent phenomena such as, growth, reproduction, species interactions, and mortality (Trosvik et al., 2010). For example, the density-dependent mortality maintains the diversity and stability of microbial communities (Blazewicz et al., 2020), whereas quorum sensing (QS), a mechanism of microbial species communication also depends on cell density (Abisado et al., 2018). The density-dependent factors are likely to influence microbial community dynamics and stability, though these remain understudied (LaManna et al., 2017). But nevertheless, trophic and non-trophic interactions, space constraints, virulence, and toxin production may also determine the size and stability of microbial communities (Georgiou et al., 2017). Given that smaller organisms have faster growth rates and higher population densities than larger ones (Lindmark et al., 2018), the stronger effects of predation, mortality, antagonisms, competitive exclusion, and other environmental filters (local carrying capacities) on their densities are expected (Faust and Raes, 2012). Meanwhile, the differences in the body size of microbial taxa and their relative abundance may also determine microbial niche overlap. Therefore, we assume that smaller rather than larger-sized microbes would demonstrate more niche overlap and resultantly, they would have higher negative co-occurrences and cohesions (an index indicating the degree of connectivity of microbial species in the community) (Herren and McMahon, 2017) among species in the community.

The influence of body size in ecological interactions that maintain communities, such as antagonism, synergism (competition, facilitation, mutualism), and predation, is apparent (Zaneveld et al., 2017; Zhou and Ning, 2017), though mostly reported in the ecology of macroorgansims.

We do not know yet whether these body-size dependent species-interactions regulate stability in the natural communities of soil microbes (Saleem et al., 2013). As predicted from both theoretical and empirical studies, negative or antagonistic interactions may promote community stability (Coyte et al., 2015; de Vries et al., 2018). In addition to the species co-occurrence and cohesion patterns, community stability may also be influenced by many other factors such as microbial stress tolerance, physiological plasticity, dormancy, stochastic gene expression, dispersal rate, and species diversity, etc. (Shade et al., 2012). These aspects may also be microbial body size-dependent. Given that small-sized microorganisms may have greater physiological plasticity, survival strategies, and dispersal rates (Shade et al., 2012), and thus they may demonstrate a higher community stability, though this assumption has never been tested.

Here, we seek to test the following hypotheses: 1) negative co-occurrences would be more prevalent among smaller- compared with larger-sized microbes across microbial trophic levels; and 2) smaller- rather than larger-sized microbial taxa would demonstrate wider niche breadth, greater niche overlap, and higher community stability (Fig. 1). To test our hypotheses, we studied soil microbial communities in rice-paddy ecosystems at local, regional, and continental scales, while considering the potential spatial-scale dependencies. As the general term "stability" is multifaceted, we define community stability as the degree of structural variations in the microbial communities over space (i.e. variability), according to the "ecological checklist" of stability statement (Grimm and Wissel, 1997; Shade et al., 2012). This definition is more relevant to this research because we only intend to compare the degree of spatial structure changes between different microbial groups, rather than focusing on the response of individual microbial groups to a disturbance. Therefore, we

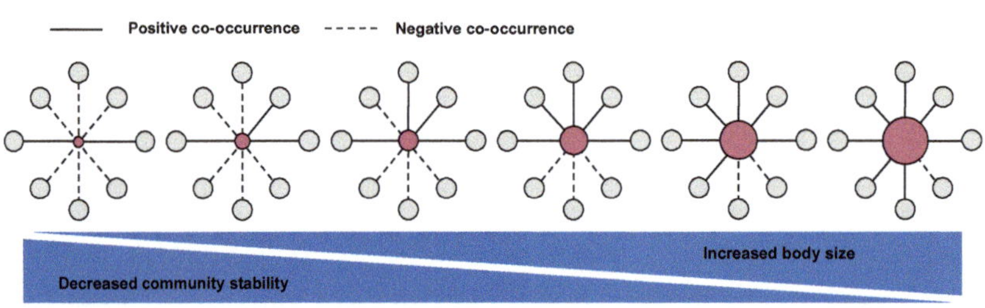

Fig. 1 Hypothetical illustration of microbial co-occurrence patterns and community stability as a function of a gradient in microbial body size. The pink circles represent the size of different microorganisms, and the grey circles represent all other microorganisms. Increasing microbial body size would reduce negative co-occurrences among microbes and promote positive co-occurrences, which ultimately would lead to community instability. Solid and dashed lines represent positive and negative co-occurrences, respectively.

calculated average Bray-Curtis dissimilarity as a proxy for predicting community stability, in the sense that a **higher** average Bray-Curtis dissimilarity should indicate a **lower** community stability over space and vice versa (Zaneveld et al., 2017).

2. Methods

2.1 Sample sites and data collection

At the **local scale**, the soil samples were collected from Yingtan Red Soil Ecological Experiment Station (28°15′30″ N, 116°55′30″ E; Fig. 2), Chinese Academy of Sciences in Yujiang County, Jiangxi Province of China. This site has a typical subtropical monsoon climate with a mean annual temperature of 17.6 °C and an annual precipitation of 1795 mm. The sampling field was a double-cropped rice (*Oryza sativa* L.) cropping system that included early and late season rice. Planting dates were early April to late October, and the rest of the year was fallow. Before the early cultivation season of 2019, five soil cores (5 cm × 10 cm × 18 cm, free from rice roots) were collected using a W-shaped transect from each plot and then pooled to form one composite sample. In total, twenty-seven soil samples were collected. The distance between any two samples is less than 100 m. Samples were stored at −40 °C until further processing.

At the **regional scale**, soil samples were collected near the end of December 2017 from red paddy soils in Yujiang (Jiangxi Province, China - 116°41′ E to 117°09′ E, and 28°04′ N to 28°37′ N; Fig. 2), where >85% of the cultivated land is paddy fields. Sampling sites were chosen to satisfy the following conditions: 1) the whole region needed to be covered; 2) the main parent material of the soils needed to be included; and, 3) field management including cropping system and fertilizer applications should be uniform. Based on these principles, twenty-six sites were selected, with pairwise geographical distances ranging from 1.3 km to 50.7 km (Fig. 2). The soil samples were collected after the harvest and in the absence of water flooding. Within each site, five soil cores (5 cm × 10 cm × 18 cm, free from rice roots) were collected using a W-shaped transect from each plot and then pooled to form one composite sample. All samples were stored at −40 °C before further processing.

At the **continental scale**, the data were collected from a published article, which quantified the soil microbiome of paddy fields (Fig. 2) (Jiao et al., 2020). Briefly, 122 rice soil samples ranging from 18.30°N to 48.35°N and 87.61°E to 99.91°E across Eastern China were collected in July–September 2017.

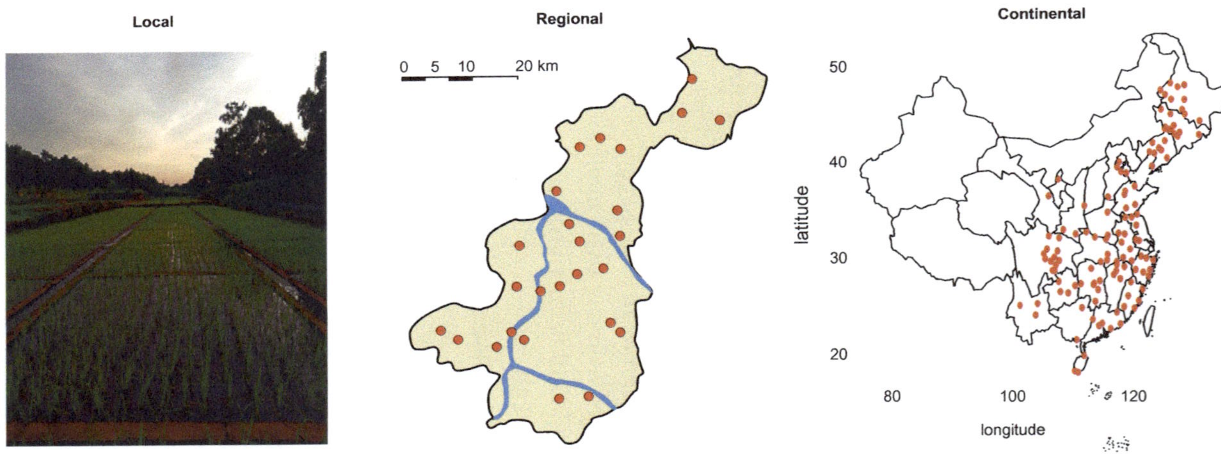

Fig. 2 Sampling sites at local, regional, and continental scale. The red dots represent sampling locations.

2.2 Molecular methods, metabarcoding and bioinformatics

For samples at local and regional scales, the soil DNA was extracted from 0.5 g of soil (fresh weight) using a Fast®DNA SPIN Kit (MP Biomedicals, CA, USA) and then subsequently purified using a PowerClean® DNA Clean-up Kit (MoBio, CA, USA) according to the manufacturers' instructions. The concentration and quality of the extracted DNA were measured using a NanoDrop ND-1000 spectrophotometer (NanoDrop Technologies, DE, USA). Three commonly used primer sets were applied to metabarcode microbial communities, targeting bacterial 16S rRNA genes (Biddle et al., 2008), the fungal ITS (Gardes and Bruns, 1993), and protist 18S rRNA genes (Stoeck et al., 2010), respectively. Detailed information about primers and PCR conditions for each primer set are shown (Table 1). The purified amplicons were pooled in equimolar concentrations, and then sent for paired-end sequencing on an Illumina MiSeq (300 bp × 300 bp) at Majorbio Bio-Pharm Technology Co. Ltd. (Shanghai, China).

For continental-scale data, raw sequences (protists were not sequenced) were downloaded from NCBI Sequence Read Archive (SRA) using the SRA toolkit "*prefetch*", and the sequences were then transformed to *fastq* format for downstream analyses using the SRA toolkit "*fastq-dump*".

Raw sequence data were analysed using the Quantitative Insights into Microbial Ecology (QIIME) pipeline (v1.9.1) (http://qiime.org/) (Caporaso et al., 2010). Paired-end reads were merged using the FLASH (Magoc and Salzberg, 2011). Reads with length of <200 bp or with average quality scores of <25 were removed. The UPARSE software was used for chimera removal, and operational taxonomic units (OTUs) were clustered at 97% sequence similarity and a representative sequence of each OTU was selected and used for taxonomic assignments (Edgar, 2013). The taxonomic identity of the bacteria, fungi, and protist OTUs was determined based on comparisons against the RDP (http://rdp.cme.msu.edu/) (Wang et al., 2007), UNITE (v7) (https://unite.ut.ee/) (Nilsson et al., 2019), and PR2 databases (http://bigd.big.ac.cn/) (Guillou et al., 2013) respectively using the RDP naïve Bayesian classifier. It should be noted that some non-protist OTUs originating from other eukaryotes were removed from the 18S datasets. The OTU data were rarefied for all microbial groups to an even sequencing depth (6677 sequences per sample) based on the sample with the minimum numbers of reads (McKnight et al., 2019).

2.3 Determination of microbial body sizes

The body-size data for species-level taxonomic assignation and/or morphological descriptions does not exist for most microbes. Thus, we distinguished

Table 1 Detailed information about primers and PCR conditions

	Primers	Primer sequence (5′- 3′)	Target gene	Target subfragment	PCR reaction condition
Bacteria	Forward: 515F	GTGCCAGCMGCCGCGGTAA	16S	V4–V5	An initial denaturation at 95 °C for 3 min, followed by 27 cycles of 30 s at 95 °C, annealing for 30 s at 55 °C and elongation for 45 s at 72 °C, the last step being extension at 72 °C for 10 min.
	Reverse: 907R	CCGTCAATTCCTTTGAGTTT			
Fungi	Forward: ITS1F	CTTGGTCATTTAGAGGAAGTAA	ITS	ITS1	An initial denaturation at 95 °C for 3 min, followed by 35 cycles of 30 s at 95 °C, annealing for 30 s at 55 °C and elongation for 45 s at 72 °C, the last step being extension at 72 °C for 10 min
	Reverse: ITS2	GCTGCGTTCTTCATCGATGC			
Protist	Forward: TAReukFWD1F	CCAGCASCYGCGGTAATTCC	18S	V4	An initial denaturation at 95 °C for 5 min, followed by 10 cycles of 30 s at 94 °C, annealing for 45 s at 57 °C and elongation for 60 s at 72 °C, and followed by 25 cycles of 30 s at 94 °C, annealing for 45 s at 45 °C, 47 °C, 48 °C, 49 °C, respectively and elongation for 60 s at 72 °C, the last step being extension at 72 °C for 2 min.
	Reverse: TAReukREV3R	ACTTTCGTTCTTGATYRA			

groups based on their taxonomic affiliation at the phylum or subphylum level, and their body sizes were identified based on propagule size through literature estimates. Recent research calculated the average microbial body sizes based on approximately 576 dominant genera, covering almost all common soil microbial phyla or subphyla (Luan et al., 2020). We also calculated and added average body sizes of some microbial groups based on our own OTU tables. The microbial groups and the average body sizes used in this study are shown in Table 2. We restricted our analysis to the most abundant phyla or subphyla (i.e., relative abundance >1% or very close to 1%), which assumed that the broadly defined functional traits such as body size and trophic status are generally conserved within the microbial taxa at phyla level (Geisen et al., 2015; Luan et al., 2020; Mulder and Elser, 2009; Zinger et al., 2019). Overall, we investigated 25, 32, and 22 microbial groups at local, regional, and continental scales, respectively (Table 2).

Table 2 Average body size of the selected microbial groups. The "+" and "-" represent whether the group was analysed or not at a particular spatial scale, respectively

Microbial group	Average body size (um)	Local scale	Regional scale	Continental scale
Acidobacteria	0.5	+	+	+
Actinobacteria	4	+	+	+
Alphaproteobacteria	0.4	+	+	+
Armatimonadetes	1	+	+	–
Bacteroidetes	0.5	+	+	+
Betaproteobacteria	5	+	+	+
Chlorobi	0.8	+	–	–
Chloroflexi	1.2	+	+	+
Cyanobacteria	4	+	+	–
Deltaproteobacteria	0.5	+	+	+
Gammaproteobacteria	5	+	+	+
Nitrospirae	5	+	+	+
Planctomycetes	1	+	+	+
Elusimicrobia	0.2	–	+	–
Firmicutes	3	–	+	+

Continued

Table 2 Average body size of the selected microbial groups. The "+" and "-" represent whether the group was analysed or not at a particular spatial scale, respectively—cont'd

Microbial group	Average body size (um)	Local scale	Regional scale	Continental scale
Gemmatimonadetes	0.5	–	+	+
Ignavibacteriae	5	–	+	+
Verrucomicrobia	0.5	–	+	+
Dothideomycetes	28.3	+	+	+
Eurotiomycetes	18.3	+	+	+
Leotiomycetes	12.5	+	+	+
Sordariomycetes	21	+	+	+
Basidiomycota	10	+	+	+
Mortierellomycota	14.3	+	+	+
Rozellomycota	1.1	–	+	+
Chytridiomycota	4.85	–	+	+
Glomeromycota	130	–	+	–
Cercozoa	13.8	+	+	–
Chlorophyta	18.1	+	+	–
Ciliophora	71.9	+	+	–
Conosa	40.2	+	+	–
Lobosa	22	+	+	–
Ochrophyta	40.5	+	+	–

2.4 Statistical analysis

We performed network analysis to explore the microbial co-occurrence patterns using the plugin CoNet in Cytoscape 3.5.1 (Shannon et al., 2003). Robust correlations between two OTUs were defined as those with Spearman's correlation coefficients (ρ) >0.6 and with false discovery rate-corrected at P values <0.01. With these, we constructed a correlation network in which each node represented one OTU, and each edge represented a strong and significant correlation between two nodes. The topology parameters of each network were determined in Cytoscape using the NetworkAnalyzer (Shannon et al., 2003).

We calculated the cohesion pattern to investigate community interconnectedness of each microbial group (Herren and McMahon, 2017). Briefly, we first calculated pairwise correlations between all taxa in each microbial group, and then used a null model to account for how the features of microbial datasets might affect correlations, and we subtracted these values. For each taxon, we averaged the positive and negative corrected correlations separately and recorded these values as the positive and negative connectedness values. Cohesion values were obtained by multiplying the relative abundance table by the connectedness values. Finally, we got two metrics of cohesion, corresponding to positive and negative values. Detailed mechanism and diagram generating cohesion metrics are available elsewhere (Herren and McMahon, 2017).

The Bray-Curtis dissimilarity of each microbial group was calculated in QIIME using the "*beta_diversity.py*" script to reflect community stability. Niche breadth and niche overlaps were calculated according to the Levin's niche breadth index and Levin's niche overlap index respectively (Li et al., 2020). A dominance test was used to determine the potential migration rate (m) of each microbial group (Sloan et al., 2007). The m is a parameter for evaluating the probability that a random loss of an individual in a local community would be replaced by dispersal from the metacommunity, and, therefore, it is a measure of dispersal limitation. Higher m values indicate that microbial communities have higher dispersal potential (Jiao et al., 2020). The formula is as follows:

$$Freq_i = 1 - I(1/N|N*m*pi,\ N*m*(1 - pi))$$

where $Freq_i$ is the occurrence frequency of taxon i across communities; N is the number of individuals per community; m is the estimated migration rate; p_i is the average relative abundance of taxon i across communities; and I is the probability density function of beta distribution.

3. Results

3.1 Inferring microbial community structure and body size

We investigated soil microbial communities at local (27 samples), regional (26 samples), and continental scales (122 samples) using the metabarcoding data (Fig. 2). At the local scale, the metabarcoding analysis revealed 6677 sequences of bacterial, fungal, and protist communities per sample after excluding singletons and rarefying to an even depth. These sequences were clustered into bacterial (4641), fungal (2478), and protistan (2044) OTUs, respectively. We sorted twenty-five microbial groups based on their body size, while these groups included 13 bacterial, six fungal groups and six protistan groups at a higher-level taxonomy (Table 2). These groups were used for the downstream analyses. At regional scale, the metabarcoding analysis revealed 7664 sequences of bacterial, fungal, and protist communities per sample after excluding singletons and rarefying to even depth. These sequences were clustered into 12,353 bacterial, 6634 fungal, and 1969 protistan OTUs, respectively. Thirty-two groups including 17 bacterial, 9 fungal, and 6 protistan groups were used for downstream analyses at regional scale (Table 2). At the continental scale, the metabarcoding data yielded 8315 sequences of bacterial and fungal communities per sample after excluding singletons and rarefying to even depth, but did not contain any protistan data. These sequences were clustered into 18,364 bacterial and 6166 fungal OTUs. From this data, we sorted twenty-two groups that included 13 bacterial and nine fungal groups for the downstream analyses.

In total, we studied 33 microbial groups in this study. Our results showed that body sizes of these 33 microbial groups ranged from 0.2 μm to 130 μm in the paddy soils. The bacterial cell size varied between 0.2 μm (*Elusimicrobia*) and 5 μm (*Betaproteobacteria, Gamaproteobacteria, Nitrospirae, and Ignavibacteriae*). The size of fungal spores ranged between 1.1 μm (Rozellomycota) and 130 μm (*Glomeromycota*). The size of protistan cells ranged from 13.8 μm (*Cercozoa*) to 71.9 μm (*Ciliophora*) (Table 2).

3.2 Microbial niche breadth, niche overlap, and migration rate

The bacteria consistently had significantly higher niche breadth, niche overlap, and migration rate than fungi and protists at all three spatial scales, aside from a non-significant difference in niche breadth between bacteria and protists at the regional scale (Fig. 3A-C). The protists had significantly higher niche breadth, niche overlap, and migration rate than fungi at the local scale; but at the regional scale, protists and fungi had similar niche breadth and niche breadth, and fungi had higher migration rate (Fig. 3A-C).

The microbial body size consistently negatively correlated with their niche breadth (Fig. 3D; **local scale**: $r = -0.566$, $P = 0.003$, $n = 25$; **regional scale**: $r = -0.216$, $P = 0.235$, $n = 32$; **continental scale**: $r = -0.628$, $P = 0.003$, $n = 22$) and niche overlap (Fig. 3E; **local scale**: $r = -0.582$, $P = 0.002$, $n = 25$; **regional scale**: $r = -0.573$, $P = 0.001$, $n = 32$; **continental scale**: $r = -0.455$, $P = 0.038$, $n = 22$), thus suggesting

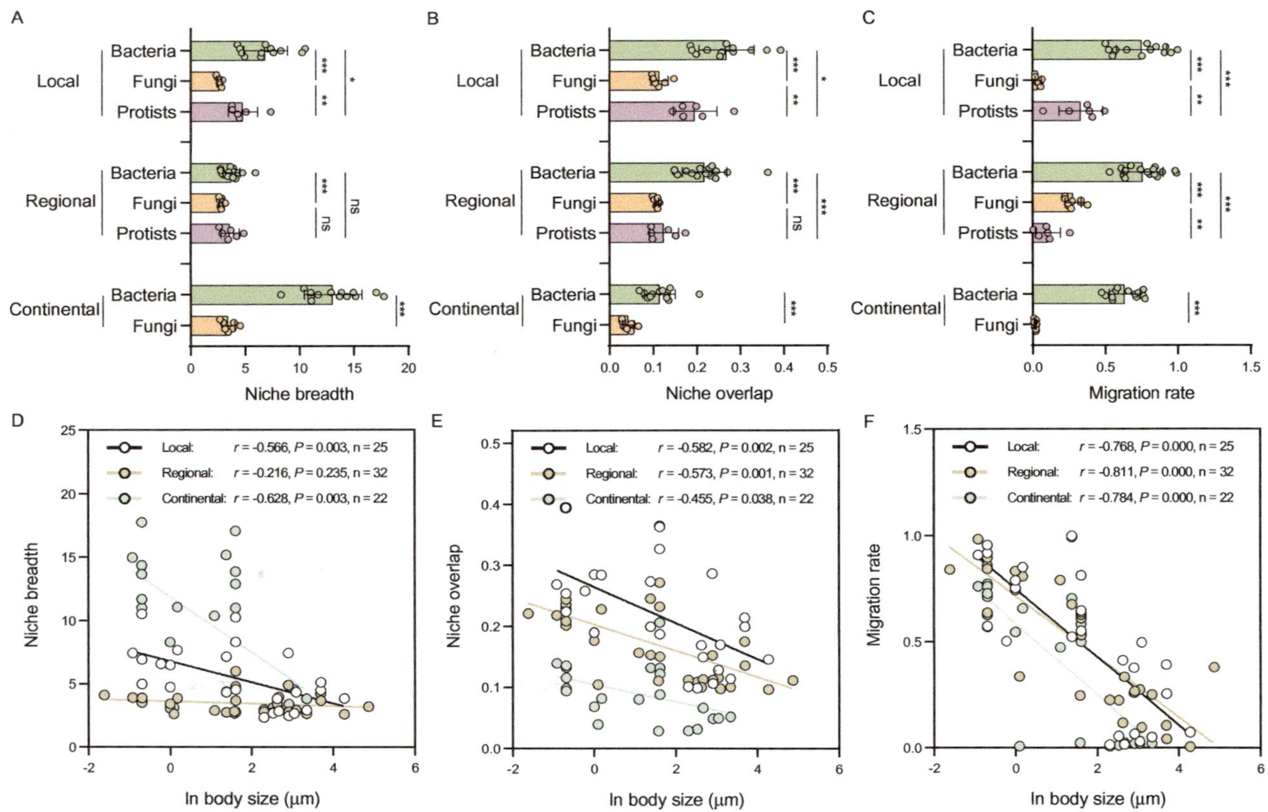

Fig. 3 See figure legend on next page.

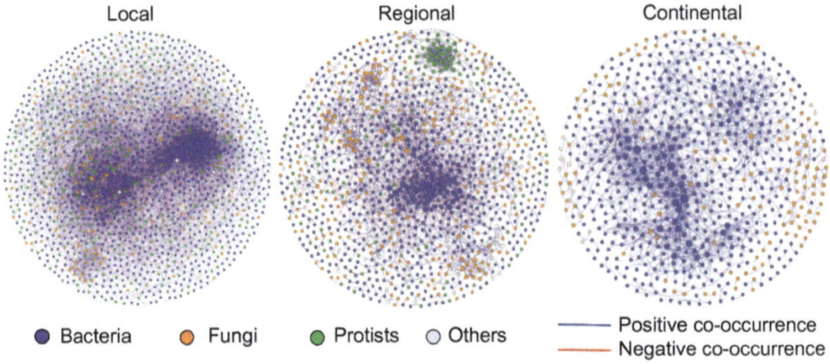

Fig. 4 Co-occurrence networks at local, regional, and continental scale. The networks were constructed under the parameters $\rho > 0.6$ and corrected *P*-values <0.01. Each node represents an OTU, and each line represents a significant and strong co-occurrence. The size of the nodes is plotted based on the total co-occurrences of each node.

that smaller-sized microbes occupy a wider niche breath, but their niches are more overlapped than the observed in larger organisms. Furthermore, our results showed that microbial body size consistently negatively correlated with the migration rate (Fig. 3F; **local scale**: $r = -0.768$, $P < 0.001$, $n = 25$; **regional scale**: $r = -0.811$, $P < 0.001$, n = 32; **continental scale**: $r = -0.784$, $P < 0.001$, n = 22).

3.3 Co-occurrence patterns of different microbial groups

We calculated co-occurrence parameters of microbial groups after constructing the multitrophic co-occurrence network (Fig. 4). Our results indicated that microbial body size was consistently negatively correlated with the average negative co-occurrences and PNC across all three spatial scales (Fig. 5A-C; **local scale**: negative co-occurrences, $r = -0.690$, $P < 0.001$, $n = 25$; PNC, $r = -0.679$, $P < 0.001$, n = 25; **regional scale**: negative

Fig. 3 Relationships between microbial body sizes and niche breadth, niche overlap, and migration rate. The average niche breadth (A), niche overlap (B), and migration rate (C) within each trophic level at various spatial scales are provided in the top row. The Relationships between microbial body sizes and niche breadth (D), niche overlap (E), and migration rate (F) at various spatial scales are provided in the bottom row. Significant differences in the niche breadth, niche overlap, and migration rate between trophic level were determined by the nonparametric Mann-Whitney *U* tests. *, **, and *** represents significant difference under $P < 0.05$, 0.01, and 0.001, respectively. 'ns' represents non-significant difference. The "n" represents the sample size of microbial groups. We applied one-side *F* and two-side *t*-tests, and then calculated *P* values as shown.

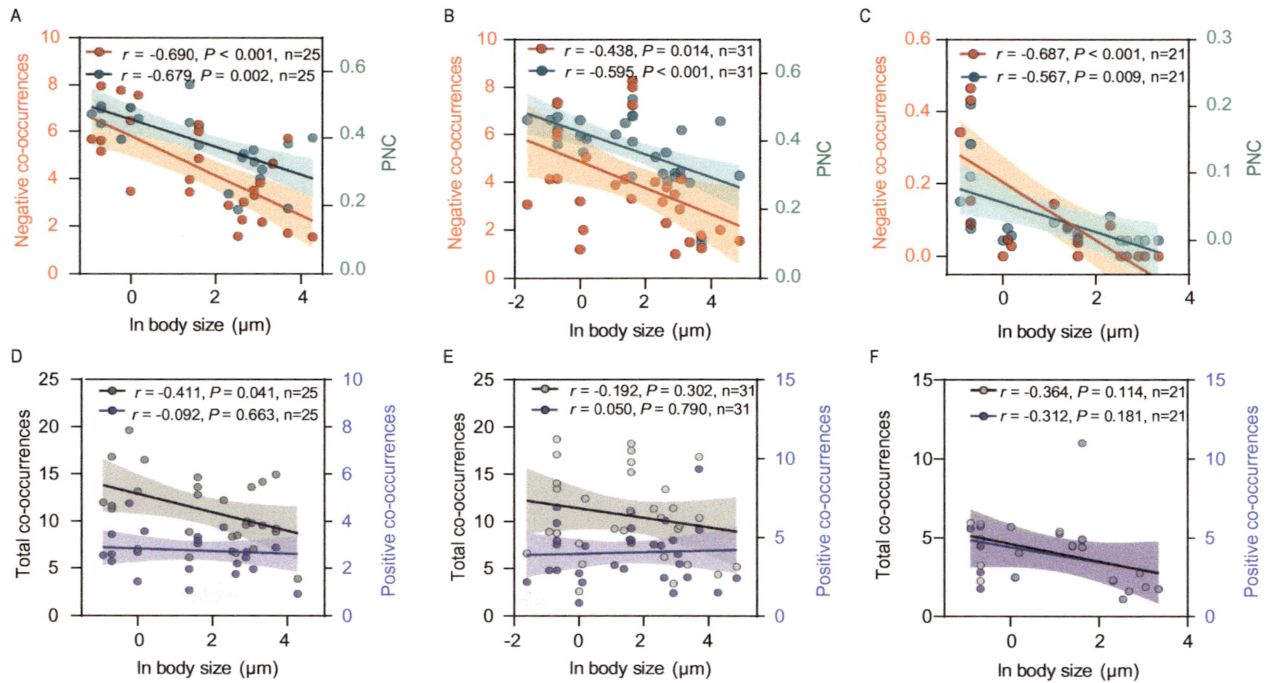

Fig. 5 See figure legend on next page.

co-occurrences, $r = -0.438$, $P = 0.014$, $n = 31$; PNC, $r = -0.595$, $P < 0.001$, n = 31; **continental scale**: negative co-occurrences, $r = -0.687$, $P < 0.001$, $n = 21$; PNC, $r = -0.567$, $P < 0.001$, n = 21). However, the correlations between microbial body size and both total and positive co-occurrences were consistently weak at all three spatial scales investigated (Fig. 5D-F).

3.4 Cohesion of differently sized microorganisms

Microbial body size was consistently positively correlated with the negative cohesion at all spatial scales (Fig. 6A-C; **local scale**: $r = 0.442$, $P = 0.027$, n = 25; **regional scale**: $r = -0.401$, $P = 0.023$, $n = 32$; **continental scale**: $r = -0.518$, $P = 0.014$, $n = 22$), thus suggesting that smaller microorganisms had consistently higher negative cohesion than larger ones. Microbial body size was also significantly negatively correlated with positive cohesion at the continental scale (Fig. 6F; $r = -0.537$, $P = 0.010$, n = 22), but not at the local or regional scales (Fig. 6D-E).

3.5 Community stability of differently sized microorganisms

The Bray-Curtis dissimilarity of each microbial group was also calculated to predict community stability, and the results showed that the microbial body size consistently positively correlated with the Bray-Curtis dissimilarity across all three spatial scales (Fig. 7; **local scale**: $r = 0.542$, $P = 0.005$, n = 25; **regional scale**: $r = 0.483$, $P = 0.005$, n = 32; **continental scale**: $r = 0.493$, $P = 0.023$, n = 22).

In addition, our results showed a significant relationship between negative co-occurrence and Bray-Curtis dissimilarity at local and regional scales, but not at the continental scale (Fig. 7; **local scale**: negative co-occurrences,

Fig. 5 Relationships between microbial body sizes and community co-occurrences. The Relationships between microbial body sizes and negative co-occurrences and PNC at local (A), regional (B), and continental (C) scales are provided in the top row. The red dots and lines indicate the relationships between microbial body sizes and negative co-occurrences, while the dark blue dots and lines describe the relationships between microbial body sizes and the PNC. The Relationships between microbial body sizes and total and positive co-occurrences at local (D), regional (E), and continental (F) scale are provided in the bottom row. The grey dots and lines indicate the relationships between microbial body sizes and total co-occurrences, while the blue dots and lines describe the relationships between microbial body sizes and the positive co-occurrences. The "*n*" represents the sample size of microbial groups. We applied one-side *F* and two-side *t*-tests, and then calculated *P* values as shown.

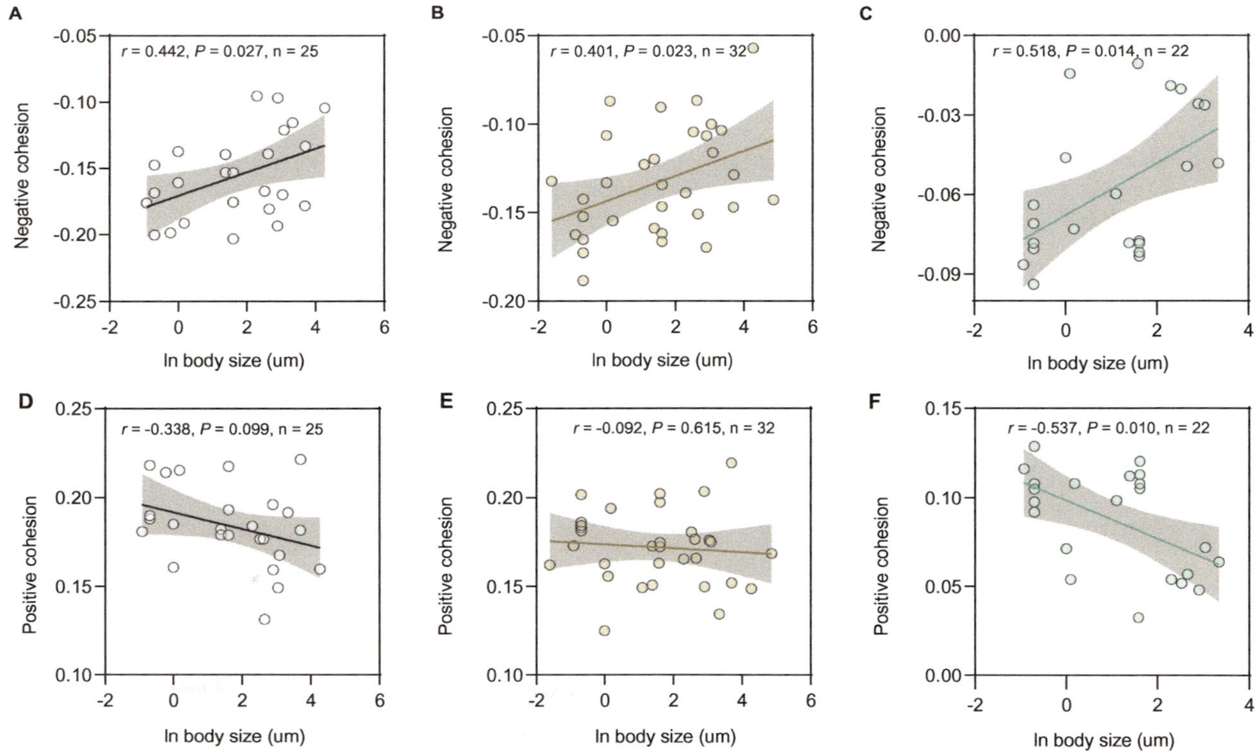

Fig. 6 Relationships between microbial body sizes and community cohesion. The Relationships between microbial body sizes and negative cohesion at local (A), regional (B), and continental (C) scale are provided in the top row. The Relationships between microbial body sizes and positive cohesion at local (D), regional (E), and continental (F) scale are provided in the bottom row. The "n" represents the sample size of microbial groups. We applied one-side *F* and two-side *t*-tests, and then calculated *P* values as shown.

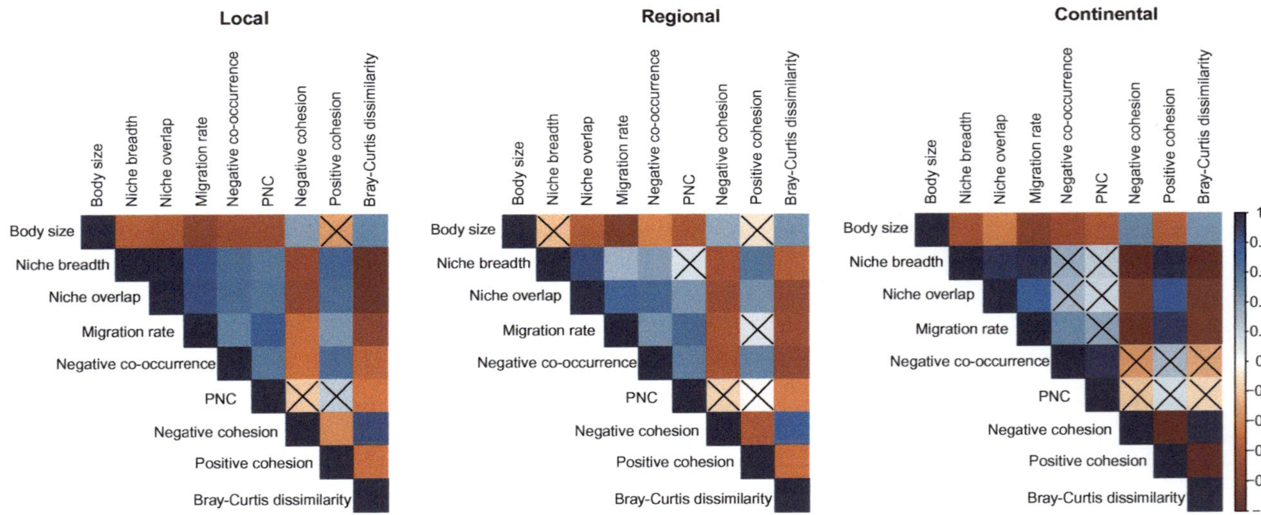

Fig. 7 The relationships of microbial body sizes and average Bray-Curtis dissimilarity at various scales. The relationship of microbial body size with niche breadth, niche overlap, migration rate, negative co-occurrences, community cohesion, and average Bray-Curtis dissimilarity at local, regional, and continental scales. We applied one-side *F* and two-side *t*-tests, and then calculated the *P* values. × represents non-significant relationship under $P < 0.05$.

$r = -0.534$, $P = 0.006$, n=25; PNC, $r = -0.481$, $P = 0.015$, n=25; **regional scale**: negative co-occurrences, $r = -0.722$, $P < 0.001$, n=31; PNC, $r = -0.452$, $P = 0.011$, n=31; **continental scale**: negative co-occurrences, $r = -0.304$, $P = 0.192$, n=21; PNC, $r = -0.158$, $P = 0.507$, n=21).

Furthermore, our results showed a very strong relationship between negative cohesion and Bray-Curtis dissimilarity at all three scales (Fig. 7; **local scale**: $r = 0.835$, $P < 0.001$, n=25; **regional scale**: $r = 0.764$, $P < 0.001$, n=32; **continental scale**: $r = 0.963$, $P < 0.001$, n=22). Similarly, there was also a significant relationship between positive cohesion and Bray-Curtis dissimilarity at all three scales (Fig. 6; **local scale**: $r = -0.507$, $P = 0.010$, n=25; **regional scale**: $r = -0.510$, $P = 0.003$, n=32; **continental scale**: $r = -0.960$, $P < 0.001$, n=22).

The microbial niche breadth, niche overlap, and potential migration rate were consistently positively correlated with the negative co-occurrences, PNC, and positive cohesion (Fig. 7). On the contrary, the niche breadth, niche overlap, and potential migration rate were consistently negatively correlated with the negative cohesion and Bray-Curtis dissimilarity at all three scales (Fig. 7).

4. Discussion

Using high-throughput sequencing and ecological models, this study discerns the role of body size in microbial co-occurrence, community cohesion, and stability patterns at local, regional, and continental scales. We show that body size is the primary determinant of microbial co-occurrence patterns and community cohesion and stability. Our results consistently demonstrated that small-sized microorganisms exhibited higher negative co-occurrences and negative cohesion, and subsequently a higher community stability across spatiotemporal scales. This study puts microbes into a common body-sized based community ecology framework that has been previously applied to the macroorganisms to determine their interactions and roles in food webs (Woodward et al., 2005).

Our results show that microbial body size consistently correlated with the average negative co-occurrences, PNC, and negative cohesion across all three spatial scales (Fig. 5, Fig. 6). These results support our first hypothesis that negative rather than positive co-occurrences and cohesion are more prevalent among small-sized microorganisms. The decreasing negative co-occurrences and negative cohesion with increasing body sizes implies that small-sized microorganisms may experience relatively higher competition, antagonistic interactions, and predation pressure in the community (Saleem et al., 2013). For instance, our analysis indicated a significant niche overlap between small-sized microorganisms (Fig. 3), thus implying that these fast-growing microbes may demonstrate intense intra- and inter- specific competition under resource limited conditions (Allen et al., 2006; Berendsen et al., 2012). Moreover, predation exerts a fundamental selective pressure on small-sized microbes, while trophic regulation of microbial taxa may influence microbial populations and interactions (Allen et al., 2006; Saleem et al., 2013). For instance, the selective removal of certain body-sized microbial taxa may significantly reduce their population densities, while it may release other taxa from predation and competition pressures. Thus, these microbial taxa may expand their niches and population densities (Moksnes et al., 1998; Pennings, 1990). Consequently, such a trade-off could result in the evolution of novel interactions among species that may lead to strong interactions and cohesion among species in the community.

Our results also confirm our second hypothesis that microbial communities dominated by small-sized microorganisms may demonstrate a higher community stability (Fig. 7). The decreasing community stability with increasing microbial body sizes hints that smaller- as opposed to larger- sized taxa may have: (1) a wider range of strategies to adjust to a broad range of environmental conditions; and, (2) higher dispersal potential, propensity, and ecological ranges. Moreover, our results support previous empirical and theoretical predictions that smaller-sized microbes, for instance bacteria, often have a wider niche breadth than larger-sized microorganisms such as protists and fungi (Fig. 3) (Finlay, 2002; Langenheder et al., 2005; Luan et al., 2020), which nevertheless, suggests an increased niche overlap in these taxa (Fig. 3). Generalists are typically small-sized microbes with wider niche breadth, thus implying that they can frequently evolve, adapt or acclimate to changing environmental conditions (Jiao et al., 2020). In contrast, large-sized microbes with a narrower niche breadth and higher positive co-occurrences may respond in tandem to environmental fluctuations that may compromise their community stability (Coyte et al., 2015; de Vries et al., 2018). The environmental fluctuations resulting from locally less favourable conditions, can cause mass effects that may be more prominent on the large-sized taxa with narrow niches and limited dispersal propensity (Shmida and Wilson, 1985). Moreover, dispersal potential is a size-dependent factor that may influence community stability. Although historically microorganisms are categorized as non-dispersal limited or cosmopolitan, recent research has consistently challenged this narrative (Farjalla et al., 2012). In a manner consistent with recent predictions (Luan et al., 2020), we found a significant decrease in the dispersal potential in the large-sized microorganisms such as protists and fungi (Fig. 3), suggesting the expectation that these taxa might be more sensitive to climate and land-use change (Saleem and Moe, 2014). A higher dispersal propensity of smaller-sized microorganisms would potentially homogenize microbial populations, thus creating similar microbial communities across different patches and scales (Fodelianakis et al., 2019). However, we could not directly link a given taxon's dispersal potential to the microbial negative co-occurrence patterns. This is most likely because highly motile organisms that can rapidly disperse into the surrounding environment are less likely to interact with other individuals of the same species, either via competition or cooperation (Hibbing et al., 2010; Reichenbach et al., 2007).

The co-occurrence and connectivity of microbial taxa have been previously hypothesized to influence the community dynamics (Coyte et al., 2015; Nilsson and McCann, 2016), and our study clearly demonstrated that community "negative patterns" especially negative cohesion can well predict the community's stability. Our results are in line with findings from a previous study on phytoplankton communities in the Lake Mendota, which also reported a significant relationship between negative cohesion and community stability (Herren and McMahon, 2017). These results may well indicate that negative interactions (as indicated by either co-occurrences or cohesion indices) between microbial species are arranged non-randomly to interact with each other, thereby stabilizing the community composition. While, the strong negative interactions between microbial species would act as a buffer as opposed to a source of community variations. From the perspective of correlation parameters, the negative cohesion (Fig. 7; r ranges from 0.764 to 0.963) may be more powerful than negative co-occurrences (Fig. 7; r ranges from -0.304 to -0.722) in predicting microbial community stability. This is possibly because the co-occurrence network analysis only considers significant and strong correlations between microbial taxa, whereas the cohesion index considered all pairwise correlations between microbial taxa. Natural ecosystems are always dominated by many weak interactions and a few strong interactions, while these prevalent weak interactions are also ecologically important in affecting community variation and stability (McCann et al., 1998). Therefore, the co-occurrence network analysis overlooks these important weak and intermediate interactions, resulting in a relatively weak correlation between network parameters and community stability.

On the contrary, all possible interactions, either weak, intermediate or strong, are considered in the cohesion matrix (Herren and McMahon, 2017), strengthening its power in predicting community stability.

We acknowledge that correlation between community stability and "negative patterns" might also be related to species diversity (Shade et al., 2012), which has not been taken into consideration in this study. This is because comparing species diversity at phylum or class level using metabarcoding sequencing is not a tractable and representative method as we cannot rarefy all investigated microbial groups into an even sequencing depth. However, not considering diversity does not affect our conclusions regarding microbial co-occurrence and cohesion patterns because these indices were calculated using average parameters of each microbial taxon (co-occurrence), or average pairwise correlations between taxa (cohesion).

Finally, our findings robustly demonstrate that microbial "negative patterns" and community stabilities are tightly linked to their body size, irrespective of the spatial scale. Thus, this study provides the foundations for a mechanistic and predictive role of body size in microbial community ecology. Specifically, our results provide solid evidence that body size is the primary determinant of microbial co-occurrence, cohesion, and community stability in the paddy soil ecosystems across spatiotemporal scales. Meanwhile, we acknowledge that there might be discrepancies resulting from variations in the microbial sizes. Therefore, further studies testing the body-size dependent interactions in microbial communities would be required to generalize the role of body size in the microbial community ecology. Particularly, the manipulated experiments representing microbial communities of differently sized taxa at various trophic levels will further add into the understanding of microbial community assembly, co-occurrence, cohesion, and stability. Our findings also highlight the need for more accurate quantification of body sizes among microbial taxa in situ, and the integration of these data with microbial community parameters to discern the significance of body sizes in microbial community assembly and stability. Despite these various limitations, our study provides a foundation for a mechanistic understanding of microbial communities and trophic groups with respect to body size and community stability. Moreover, our results may contribute to a greater understanding of how microbial taxa with different body sizes would respond to future changes and perturbations. In particular, the results suggest the expectation that fungal and protistan taxa will be more sensitive to perturbations than bacterial communities in the agricultural soils. Overall, we anticipate that body size will determine the stability of bacterial, protist, and fungal communities at all trophic levels.

5. Data accessibility statement

Raw sequence data used in this study are archived in the NCBI Sequence Read Archive (SRA) under the accession numbers SRP260652 (16S rRNA gene sequences at local scale), SRP260858 (ITS sequences at local scale), SRP260878 (18S rRNA sequences at local scale), SRP301389 (16S rRNA gene sequences at regional scale), SRP200912 (ITS sequences at regional scale), SRP301388 (18S rRNA sequences at regional scale), and SRP199784 (16S rRNA gene sequences and ITS sequences at continental scale).

Acknowledgements

We thank the authors whose data were used in this research. This study was supported by the National Natural Science Foundation of China (42207349, 42177294), Natural Science Foundation of Jiangsu Province (BK20221005), China Postdoctoral Science Foundation (2022M711653), and Jiangsu Funding Program for Excellent Postdoctoral Talent (2022ZB331).

References

Abisado, R.G., Benomar, S., Klaus, J.R., Dandekar, A.A., Chandler, J.R., 2018. Bacterial quorum sensing and microbial community interactions. mBio 9, e01749–18.

Allen, C.R., Garmestani, A.S., Havlicek, T.D., Marquet, P.A., Peterson, G.D., Restrepo, C., Stow, C.A., Weeks, B.E., 2006. Patterns in body mass distributions: sifting among alternative hypotheses. Ecol. Lett. 9, 630–643.

Berendsen, R.L., Pieterse, C.M.J., Bakker, P.A.H.M., 2012. The rhizosphere microbiome and plant health. Trends Plant Sci. 17, 478–486.

Biddle, J.F., Fitz-Gibbon, S., Schuster, S.C., Brenchley, J.E., House, C.H., 2008. Metagenomic signatures of the Peru margin subseafloor biosphere show a genetically distinct environment. Proc. Natl. Acad. Sci. U. S. A. 105, 10583–10588.

Blazewicz, S.J., Hungate, B.A., Koch, B.J., Nuccio, E.E., Morrissey, E., Brodie, E.L., Schwartz, E., Pett-Ridge, J., Firestone, M.K., 2020. Taxon-specific microbial growth and mortality patterns reveal distinct temporal population responses to rewetting in a California grassland soil. ISME J. 14, 1520–1532.

Caporaso, J.G., Kuczynski, J., Stombaugh, J., Bittinger, K., Bushman, F.D., Costello, E.K., Fierer, N., Pena, A.G., Goodrich, J.K., Gordon, J.I., Huttley, G.A., Kelley, S.T., Knights, D., Koenig, J.E., Ley, R.E., Lozupone, C.A., McDonald, D., Muegge, B.D., Pirrung, M., Reeder, J., Sevinsky, J.R., Tumbaugh, P.J., Walters, W.A., Widmann, J., Yatsunenko, T., Zaneveld, J., Knight, R., 2010. QIIME allows analysis of high-throughput community sequencing data. Nat. Methods 7, 335–336.

Chen, L.T., Jiang, L., Jing, X., Wang, J.L., Shi, Y., Chu, H.Y., He, J.S., 2021. Above- and belowground biodiversity jointly drive ecosystem stability in natural alpine grasslands on the Tibetan plateau. Glob. Ecol. Biogeogr. 30, 1418–1429.

Coyte, K.Z., Schluter, J., Foster, K.R., 2015. The ecology of the microbiome: networks, competition, and stability. Science 350, 663–666.

Cramer, A.N., Katz, S.L., 2021. Primary production and habitat stability organize marine communities. Glob. Ecol. Biogeogr. 30, 11–24.

de Vries, F.T., Griffiths, R.I., Bailey, M., Craig, H., Girlanda, M., Gweon, H.S., Hallin, S., Kaisermann, A., Keith, A.M., Kretzschmar, M., Lemanceau, P., Lumini, E., Mason, K.E., Oliver, A., Ostle, N., Prosser, J.I., Thion, C., Thomson, B., Bardgett, R.D., 2018. Soil bacterial networks are less stable under drought than fungal networks. Nat. Commun. 9, 3033.

Dumbrell, A.J., 2019. Size matters in regulating the biodiversity of tropical forest soils. Mol. Ecol. 28, 525–527.

Edgar, R.C., 2013. UPARSE: highly accurate OTU sequences from microbial amplicon reads. Nat. Methods 10, 996.

Farjalla, V.F., Srivastava, D.S., Marino, N.A.C., Azevedo, F.D., Dib, V., Lopes, P.M., Rosado, A.S., Bozelli, R.L., Esteves, F.A., 2012. Ecological determinism increases with organism size. Ecology 93, 1752–1759.

Faust, K., Raes, J., 2012. Microbial interactions: from networks to models. Nat. Rev. Microbiol. 10, 538–550.

Finlay, B.J., 2002. Global dispersal of free-living microbial eukaryote species. Science 296, 1061–1063.

Fodelianakis, S., Lorz, A., Valenzuela-Cuevas, A., Barozzi, A., Booth, J.M., Daffonchio, D., 2019. Dispersal homogenizes communities via immigration even at low rates in a simplified synthetic bacterial metacommunity. Nat. Commun. 10, 1314.

Gardes, M., Bruns, T.D., 1993. ITS primers with enhanced specificity for Basidiomycetes - application to the identification of mycorrhizae and rusts. Mol. Ecol. 2, 113–118.

Geisen, S., Rosengarten, J., Koller, R., Mulder, C., Urich, T., Bonkowski, M., 2015. Pack hunting by a common soil amoeba on nematodes. Environ. Microbiol. 17, 4538–4546.

Georgiou, K., Abramoff, R.Z., Harte, J., Riley, W.J., Torn, M.S., 2017. Microbial community-level regulation explains soil carbon responses to long-term litter manipulations. Nat. Commun. 8, 1223.

Green, J.L., Bohannan, B.J., Whitaker, R.J., 2008. Microbial biogeography: from taxonomy to traits. Science 320, 1039–1043.

Grimm, V., Wissel, C., 1997. Babel, or the ecological stability discussions: an inventory and analysis of terminology and a guide for avoiding confusion. Oecologia 109, 323–334.

Guillou, L., Bachar, D., Audic, S., Bass, D., Berney, C., Bittner, L., Boutte, C., Burgaud, G., de Vargas, C., Decelle, J., del Campo, J., Dolan, J.R., Dunthorn, M., Edvardsen, B., Holzmann, M., Kooistra, W.H.C.F., Lara, E., Le Bescot, N., Logares, R., Mahe, F., Massana, R., Montresor, M., Morard, R., Not, F., Pawlowski, J., Probert, I., Sauvadet, A.L., Siano, R., Stoeck, T., Vaulot, D., Zimmermann, P., Christen, R., 2013. The protist ribosomal reference database (PR2): a catalog of unicellular eukaryote small sub-unit rRNA sequences with curated taxonomy. Nucleic Acids Res. 41, D597–D604.

Herren, C.M., McMahon, K.D., 2017. Cohesion: a method for quantifying the connectivity of microbial communities. ISME J. 11, 2426–2438.

Hibbing, M.E., Fuqua, C., Parsek, M.R., Peterson, S.B., 2010. Bacterial competition: surviving and thriving in the microbial jungle. Nat. Rev. Microbiol. 8, 15–25.

Jacquet, C., Gounand, I., Altermatt, F., 2020. How pulse disturbances shape size-abundance pyramids. Ecol. Lett. 23, 1014–1023.

Jiao, S., Yang, Y.F., Xu, Y.Q., Zhang, J., Lu, Y.H., 2020. Balance between community assembly processes mediates species coexistence in agricultural soil microbiomes across eastern China. ISME J. 14, 202–216.

LaManna, J.A., Belote, R.T., Burkle, L.A., Catano, C.P., Myers, J.A., 2017. Negative density dependence mediates biodiversity-productivity relationships across scales. Nat. Ecol. Evol. 1, 1107–1115.

Langenheder, S., Lindstrom, E.S., Tranvik, L.J., 2005. Weak coupling between community composition and functioning of aquatic bacteria. Limnol. Oceanogr. 50, 957–967.

Li, P.F., Li, W.T., Dumbrell, A.J., Liu, M., Li, G.L., Wu, M., Jiang, C.Y., Li, Z.P., 2020. Spatial variation in soil fungal communities across paddy fields in subtropical China. mSystems 5, e00704–19.

Lindmark, M., Huss, M., Ohlberger, J., Gardmark, A., 2018. Temperature-dependent body size effects determine population responses to climate warming. Ecol. Lett. 21, 181–189.

Luan, L., Jiang, Y.J., Cheng, M.H., Dini-Andreote, F., Sui, Y.Y., Xu, Q.S., Geisen, S., Sun, B., 2020. Organism body size structures the soil microbial and nematode community assembly at a continental and global scale. Nat. Commun. 11, 6406.

Magoc, T., Salzberg, S.L., 2011. FLASH: fast length adjustment of short reads to improve genome assemblies. Bioinformatics 27, 2957–2963.

Martiny, J.B., Jones, S.E., Lennon, J.T., Martiny, A.C., 2015. Microbiomes in light of traits: a phylogenetic perspective. Science 350, aac9323.

McCann, K., Hastings, A., Huxel, G.R., 1998. Weak trophic interactions and the balance of nature. Nature 395, 794–798.

McGill, B.J., Enquist, B.J., Weiher, E., Westoby, M., 2006. Rebuilding community ecology from functional traits. Trends Ecol. Evol. 21, 178–185.

McKnight, D.T., Huerlimann, R., Bower, D.S., Schwarzkopf, L., Alford, R.A., Zenger, K.R., 2019. Methods for normalizing microbiome data: an ecological perspective. Methods Ecol. Evol. 10, 389–400.

Moksnes, P.O., Pihl, L., van Montfrans, J., 1998. Predation on postlarvae and juveniles of the shore crab Carcinus maenas: importance of shelter, size and cannibalism. Mar. Ecol. Prog. Ser. 166, 211–225.

Mulder, C., Elser, J.J., 2009. Soil acidity, ecological stoichiometry and allometric scaling in grassland food webs. Glob. Chang. Biol. 15, 2730–2738.

Nilsson, R.H., Larsson, K.H., Taylor, A.F.S., Bengtsson-Palme, J., Jeppesen, T.S., Schigel, D., Kennedy, P., Picard, K., Glockner, F.O., Tedersoo, L., Saar, I., Koljalg, U., Abarenkov, K., 2019. The UNITE database for molecular identification of fungi: handling dark taxa and parallel taxonomic classifications. Nucleic Acids Res. 47, D259–D264.

Nilsson, K.A., McCann, K.S., 2016. Interaction strength revisited-clarifying the role of energy flux for food web stability. Theor. Ecol. 9, 59–71.

Pennings, S.C., 1990. Predator-prey interactions in opisthobranch gastropods - effects of prey body size and habitat complexity. Mar. Ecol. Prog. Ser. 62, 95–101.

Reichenbach, T., Mobilia, M., Frey, E., 2007. Mobility promotes and jeopardizes biodiversity in rock-paper-scissors games. Nature 448, 1046–1049.

Saleem, M., Fetzer, I., Harms, H., Chatzinotas, A., 2013. Diversity of protists and bacteria determines predation performance and stability. ISME J. 7, 1912–1921.

Saleem, M., Moe, L.A., 2014. Multitrophic microbial interactions for eco- and agro-biotechnological processes: theory and practice. Trends Biotechnol. 32, 529–537.

Schnabel, F., Schwarz, J.A., Danescu, A., Fichtner, A., Nock, C.A., Bauhus, J., Potvin, C., 2019. Drivers of productivity and its temporal stability in a tropical tree diversity experiment. Glob. Chang. Biol. 25, 4257–4272.

Shade, A., Peter, H., Allison, S.D., Baho, D.L., Berga, M., Burgmann, H., Huber, D.H., Langenheder, S., Lennon, J.T., Martiny, J.B., Matulich, K.L., Schmidt, T.M., Handelsman, J., 2012. Fundamentals of microbial community resistance and resilience. Front. Microbiol. 3, 417.

Shannon, P., Markiel, A., Ozier, O., Baliga, N.S., Wang, J.T., Ramage, D., Amin, N., Schwikowski, B., Ideker, T., 2003. Cytoscape: a software environment for integrated models of biomolecular interaction networks. Genome Res. 13, 2498–2504.

Shmida, A., Wilson, M.V., 1985. Biological determinants of species-diversity. J. Biogeogr. 12, 1–20.

Sloan, W.T., Woodcock, S., Lunn, M., Head, I.M., Curtis, T.P., 2007. Modeling taxa-abundance distributions in microbial communities using environmental sequence data. Microb. Ecol. 53, 443–455.

Stoeck, T., Bass, D., Nebel, M., Christen, R., Jones, M.D.M., Breiner, H.W., Richards, T.A., 2010. Multiple marker parallel tag environmental DNA sequencing reveals a highly complex eukaryotic community in marine anoxic water. Mol. Ecol. 19, 21–31.

Tripathi, B.M., Stegen, J.C., Kim, M., Dong, K., Adams, J.M., Lee, Y.K., 2018. Soil pH mediates the balance between stochastic and deterministic assembly of bacteria. ISME J. 12, 1072–1083.

Trosvik, P., Rudi, K., Straetkvern, K.O., Jakobsen, K.S., Naes, T., Stenseth, N.C., 2010. Web of ecological interactions in an experimental gut microbiota. Environ. Microbiol. 12, 2677–2687.

Wang, Q., Garrity, G.M., Tiedje, J.M., Cole, J.R., 2007. Naive Bayesian classifier for rapid assignment of rRNA sequences into the new bacterial taxonomy. Appl. Environ. Microbiol. 73, 5261–5267.

Woodward, G., Ebenman, B., Emmerson, M., Montoya, J.M., Olesen, J.M., Valido, A., Warren, P.H., 2005. Body size in ecological networks. Trends Ecol. Evol. 20, 402–409.

Zaneveld, J.R., McMinds, R., Thurber, R.V., 2017. Stress and stability: applying the Anna Karenina principle to animal microbiomes. Nat. Microbiol. 2, 17121.

Zhou, J.Z., Ning, D.L., 2017. Stochastic community assembly: does it matter in microbial ecology? Microbiol. Mol. Biol. Rev. 81, e00002–e00017.

Zinger, L., Taberlet, P., Schimann, H., Bonin, A., Boyer, F., De Barba, M., Gaucher, P., Gielly, L., Giguet-Covex, C., Iribar, A., Rejou-Mechain, M., Raye, G., Rioux, D., Schilling, V., Tymen, B., Viers, J., Zouiten, C., Thuiller, W., Coissac, E., Chave, J., 2019. Body size determines soil community assembly in a tropical forest. Mol. Ecol. 28, 528–543.

CHAPTER TWO

The functional microbiome of grapevine throughout plant evolutionary history and lifetime

Paola Fournier[a,b,†], Lucile Pellan[b,†], Didac Barroso-Bergadà[c], David A. Bohan[c], Thierry Candresse[d], François Delmotte[b], Marie-Cécile Dufour[b], Virginie Lauvergeat[e], Claire Le Marrec[f], Armelle Marais[d], Guilherme Martins[f], Isabelle Masneuf-Pomarède[f], Patrice Rey[g], David Sherman[h], Patrice This[i], Clémence Frioux[h], Simon Labarthe[a,h], and Corinne Vacher[a,*]

[a]INRAE, Univ. Bordeaux, BIOGECO, Pessac, France
[b]INRAE, Bordeaux Sciences Agro, ISVV, SAVE, Villenave-d'Ornon, France
[c]INRAE, Université Bourgogne Franche-Comté, Agroécologie, Dijon, France
[d]Univ. Bordeaux, INRAE, UMR BFP, Villenave d'Ornon, France
[e]EGFV, Univ. Bordeaux, Bordeaux Sciences Agro, INRAE, ISVV, Villenave d'Ornon, France
[f]Univ. Bordeaux, UMR oenologie, INRAE, Bx INP, Bordeaux Sciences Agro, ISVV, Villenave d'Ornon, France
[g]Université de Pau et des Pays de l'Adour, E2S UPPA, CNRS, Institut des Sciences Analytiques et de Physicochimie pour l'Environnement et les Matériaux—UMR 5254, IBEAS, Pau, France
[h]Inria, Univ. Bordeaux, INRAE, Talence, France
[i]UMR AGAP Institut, Univ Montpellier, CIRAD, INRAE, Institut Agro, Montpellier, France
*Corresponding author: e-mail address: corinne.vacher@inrae.fr

Contents

1. Introduction 28
2. The grapevine functional microbiome throughout evolutionary history 29
   2.1 Microbiome evolution during grapevine domestication and breeding 30
   2.2 Microbial interactions in the centre of origin of major grapevine pathogens 34
   2.3 Microbial dispersal across geographic regions and *Vitis* species 37
3. The grapevine functional microbiome throughout plant lifetime 39
   3.1 Initial microbiome and virome at the graft stage 39
   3.2 Recruitment of the root microbiome from the soil reservoir 42
   3.3 Seasonal assembly of the leaf microbiome in interaction with leaf pathogens 47
   3.4 Dynamics of the berry microbiome to ripening and winemaking 51
   3.5 Wood microbiome dysbiosis during grapevine aging and decline 54
4. Let's make the grapevine microbiome more functional 58

† Co-first authors.

Advances in Ecological Research, Volume 67
ISSN 0065-2504
https://doi.org/10.1016/bs.aecr.2022.09.001

4.1 Field sampling designs and statistical approaches to identify beneficial microbial taxa 58
4.2 Molecular tools to uncover the functional potential of the microbiome 61
4.3 Computational approaches to understand the grapevine holobiont as a functional and dynamic network 64
4.4 Culture-dependent approaches to validate microbial interactions and functions 70
5. Conclusion and perspectives 71
Acknowledgements 74
References 74

Abstract

European grapevine is a complex holobiont composed of two plant genomes, that of the scion (*Vitis vinifera* L.) and the rootstock (*Vitis* spp.), and a multitude of microbial genomes that collectively form the microbiome. The grapevine microbiome has been extensively described over the last decade, primarily using metabarcoding approaches. Unfortunately, metabarcoding data alone provide little information on microbial functions and outcomes of plant-microbe interactions. Here we review knowledge about the microorganisms that have a demonstrated influence, positive or negative, on the performance of the grapevine holobiont. Our review encompasses bacteria, filamentous fungi, yeasts, oomycetes and viruses. It covers aboveground and belowground microorganisms, including arbuscular mycorrhizal and ectomycorrhizal fungi. We focus on taxa and functions that protect the plant against pathogens and pests, promote growth, increase tolerance to abiotic stresses and highlight those involved in disease and decline. As the outcomes of plant-microbe interactions are labile, we examine the dynamics and functions of grapevine-microbiome interactions over both the plant lifetime and the plant evolutionary history, beginning with plant domestication. Based on the knowledge and gaps we identify, we suggest field sampling designs, culture-based experiments, molecular tools and theoretical analysis methods, including shotgun metagenomics and network models, that could be used in future research to uncover and leverage the full functional potential of the grapevine microbiome.

1. Introduction

European cultivated grapevine is a complex and dynamic system of plant-microbe interactions that has been shaped by humankind to produce grapes and wine. Each plant individual is typically composed of two plant genomes, that of the rootstock (*Vitis* spp.) and the scion (*Vitis vinifera* L.), onto which are linked a multitude of microbial and viral genomes. The set of microbial genomes forms the microbiome (Berg et al., 2020; Compant et al., 2019; Saikkonen et al., 2020), while the set of viral genomes forms the virome (Roossinck, 2010). The whole, including the plant genomes, forms the hologenome (Bettenfeld et al., 2021; Theis et al., 2016; Zilber-Rosenberg and Rosenberg, 2008). This concept of each grapevine individual as being a complex and dynamic system of plant-microbe interactions (i.e., an holobiont; Bettenfeld et al., 2021; Vandenkoornhuyse et al., 2015) has not been necessary, for most of human history, to grow vines and produce wine. It could provide solutions in the future, however, to maintain grapevine cultivation in a context of climate change (Gambetta et al., 2020) and pesticide-free agriculture (Jacquet et al., 2022; Pertot et al., 2017).

The objective of this chapter is to present the state of art about the dynamics of grapevine-microbiome interactions, from the evolutionary time scale to the seasonal scale, and to propose some research avenues to increase knowledge on the functions of the grapevine microbiome. The chapter first gives an overview of the evolution of the grapevine microbiome, starting with grapevine domestication (Section 2), and then describes the microbiome dynamics during the lifetime of each plant, from the grafting stage in the nursery through to decline in the vineyard (Section 3). In both sections, we specifically highlight those microorganisms and viruses that have a demonstrated influence, positive or negative, on the performance of the grapevine holobiont (health, growth and berry quality). Based on this state of art, we suggest experimental and theoretical approaches that could lead to a better understanding of the functions of the grapevine microbiome (Section 4). We propose some experiments to identify the microbial taxa and functions that play a key role in grapevine performance under drought conditions and in agro-ecological settings. We describe how to decipher the microbial interaction networks to which these key taxa belong, to understand how they maintain in the system and regulate grapevine performance. We also provide recommendations about the biogeographic regions that could be explored in the future to isolate beneficial microbial taxa, which could then be inoculated to drive the system.

2. The grapevine functional microbiome throughout evolutionary history

This section reviews the knowledge and gaps regarding microbiome evolution during grapevine domestication and breeding, as well as the microbiome dynamics triggered by microbial dispersal across geographic regions and between members of the *Vitis* genus. Based on this state of

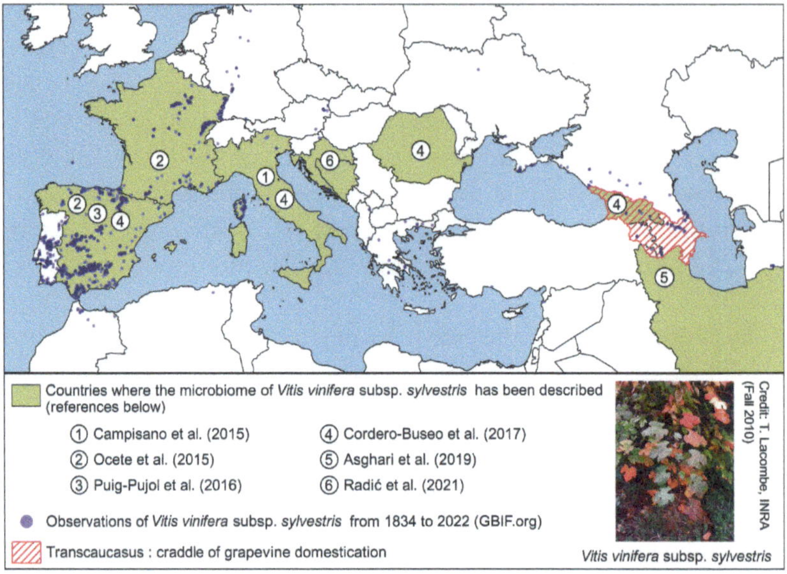

Fig. 1 Area of domestication of European cultivated grapevine, *Vitis vinifera* subsp. *vinifera* (area hatched in pink), and countries where the microbiome of its wild ancestor, *Vitis vinifera* subsp. *sylvestris*, has been described (in green). Campisano et al. (2015) isolated bacterial endophytes from wild grapevine shoots (1). Ocete et al. (2015) explored taxonomic diversity of arbuscular mycorrhizal fungi in wild grapevine rhizosphere (2). Puig-Pujol et al. (2016) isolated yeast strains at the end of spontaneous fermentations of wild grapes (3). Cordero-Bueso et al. (2017) studied yeasts associated with wild grape berries (4). Asghari et al. (2019) analysed bacterial endophytes in healthy roots, stems, leaves and fruits of wild grapevines (5). Radić et al. (2021) characterised fungal communities associated with wild grapevine roots with a focus on mycorrhizae (6). For future research, we recommend exploring the centre of origin of cultivated grapevines (in pink), studying both the microbiome of wild grapevine and that of ancestral grapevine cultivars. Map created using the Free and Open Source QGIS.

art we hypothesise which geographic regions and plant genetic material are most likely to be associated with beneficial microbial taxa and functions (Figs. 1 and 2).

2.1 Microbiome evolution during grapevine domestication and breeding

The Transcaucasus (i.e., the geographic region between Black and Caucasian Sea, which today corresponds to Armenia, Georgia, and Azerbaijan) is the cradle of viticulture (Fig. 1). It is the most probable region

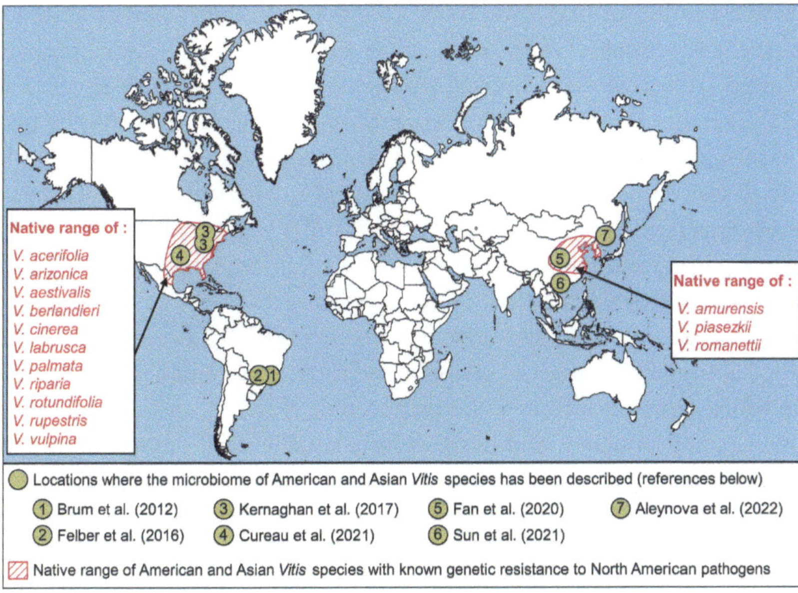

Fig. 2 Geographic locations where the microbiome of American and Asian *Vitis* species, partially or totally resistant to pathogens of North American origin (*Plasmopara viticola*, *Erysiphe necator* and *Phyllosticta ampelicida*), has been described (green circles), as well as the native range of these resistant *Vitis* species according to Wan et al. (2013) (area hatched in pink). Brum et al. (2012) isolated foliar fungal endophytes from *V. labrusca* cv. Niagara Rosada collected in Brazil (1). Felber et al. (2016) isolated foliar endophytic fungi from grapevine cultivars Bordô and Concord (*V. labrusca*) in Brazil (2). Kernaghan et al. (2017) analysed foliar fungal endophytes of a cultivated hybrid grape variety (L'Acadie blanc) and one of its native progenitors (*V. riparia*) in eastern Canada (3). Cureau et al. (2021) explored the taxonomic diversity of berry fungal communities in *V. vinifera*, *V. rotundifolia*, *V. aestivalis*, and in hybrid *Vitis* grape varieties grown in different vineyards in Arkansas (4). Fan et al. (2020) compared endophytic fungi of *V. vinifera* cv. Red Globe (cultivated grape) with those of *V. amurensis* cv. Shuangyou (wild grape) grown in a nursery in China (5). Sun et al. (2021) studied the berry microbiome of six muscadine (*Muscadinia rotundifolia* Michx.) cultivars grown in Guangxi, China (6). Aleynova et al. (2022) compared endophytic bacterial communities of a wild grape population of *V. amurensis* in Russia with those of domesticated *V. vinifera* cultivars from Germany and California (USA) (7). Studies that focused on microorganisms isolated during or at the end of spontaneous grape fermentations are not reported on this map. For future research to identify microbial biocontrol candidates against mildews and black rot, we suggest further exploration of the microbiome of these resistant *Vitis* species in their native range (area hatched in pink), where they have co-evolved with pathogens or with related pathogens. Map created using the Free and Open Source QGIS.

of domestication of grapevine 6.000 to 8.000 years ago (This et al., 2006; Zohary and Hopf, 1993), even if secondary domestication probably occurred later in western Europe (Arroyo-García et al., 2006; Sivan et al., 2021; Terral et al., 2010). *Vitis vinifera* subsp. *vinifera*, is nowadays the main *Vitis* species used for the production of wine, table grapes and raisins worldwide. It was domesticated from a wild ancestor, *Vitis vinifera* subsp. *sylvestris* (hereafter referred to as sylvestris), which naturally occurs in Mediterranean Europe (Fig. 1) and southwestern Asia (Di Vecchi-Staraz et al., 2009). In the wild, sylvestris is a liana that climbs trees, and has very small, black and acidic berries produced by dioecious flowers. During domestication, the grape has undergone marked changes, the most important being the evolution of sex (from dioecious to hermaphroditic flowers), the increase in size of berries and bunches, and changes in berry chemical composition (This et al., 2006).

From the Transcaucasus, cultivated grape spread to Europe, first to the Mediterranean region following the main civilizations (Greeks, Romans, Etruscans, Egyptians), then to northern Europe following the arrival of Christianity. At the beginning, cultivars experienced introgressions from local wild grapes (Myles et al., 2011). Genetic flows in the other direction, from cultivated grapes to wild populations, also occurred (Di Vecchi-Staraz et al., 2009). From the medieval period onwards, cultivars evolved by crossings within *Vitis vinifera* subsp. *vinifera*, most likely in a fortuitous way at the beginning (Bowers et al., 1999), then in a more controlled manner, particularly after the emergence of mildews and phylloxera in Europe (Section 2.2). Nowadays *Vitis vinifera* genetic diversity is huge, with 6000 to 7000 cultivars in the world. However, the total number of cultivars commonly used is much lower, given that half the worldwide plantings were of just 16 varieties in 2016 (Anderson and Aryal, 2013).

In addition to genetic and phenotypic changes to the plant, grapevine domestication was accompanied by an evolution of fermentative microorganisms, particularly *Saccharomyces cerevisiae*. *S. cerevisiae* is the main agent of wine fermentation and one of the best models for understanding the eukaryotic cell (Botstein and Fink, 2011). It is naturally part of the berry microbiome (Section 3.4) and colonises many other ecological niches (Schacherer et al., 2009). *S. cerevisiae* strains naturally occurring in cultivated grapevines are often referred to as vineyard yeasts or domesticated yeasts. The comparison of vineyard and non-vineyard *S. cerevisiae* strains revealed that the oldest lineages and greatest genetic variability are found in the non-vineyard strains (Fay and Benavides, 2005). This is consistent with the hypothesis that *S. cerevisiae* originated in natural environments and was subsequently domesticated by humans. Molecular analyses revealed that domesticated *S. cerevisiae* strains have a single origin. The initial domestication event probably occurred in Mesopotamia, concomitantly with the grapevine domestication event (Legras et al., 2007). Domestication was followed by a marked phenotypic divergence between vineyard and non-vineyard yeasts. For instance, vineyard yeasts show greater resistance to copper sulphate (Fay et al., 2004; Warringer et al., 2011), a chemical compound used as a fungicide against downy mildew since the 1880s, strongly suggesting that vineyard yeast populations have evolved in response to agricultural practices.

How has the microbiome of European cultivated grapevine, *Vitis vinifera* subsp. *vinifera*, evolved during grape domestication? Little is known for many species apart from *S. cerevisiae* (reviewed above). Several studies have suggested that grapevine domestication caused a decrease in microbiome diversity and a change in microbiome functions. Wild sylvestris grapes are inhabited by a greater diversity of endophytic bacteria, epiphytic yeasts and arbuscular mycorrhizae than domesticated grapes (Campisano et al., 2015; Cordero-Bueso et al., 2017; Ocete et al., 2015). Moreover, microbial communities of wild and domesticated grapes differ in their potential for pathogen biocontrol and plant growth promotion (Campisano et al., 2015; Cordero-Bueso et al., 2017). Finally, microorganisms generally associated with humans may have been integrated into the grapevine microbiome during domestication (Campisano et al., 2014; Yousaf et al., 2014). After the initial domestication event, the grapevine microbiome kept evolving as new varieties were created. Molecular analysis of parental relationships among grapevine varieties identified Heunisch weiss, Pinot noir N and Riesling B as the oldest and least evolved varieties, and also the main progenitors of other varieties (Lacombe et al., 2013). Microbiome taxonomic composition of Pinot and Riesling grape varieties has been compared to that of more recent varieties (Bao et al., 2022; Zhang et al., 2020). The results suggested that the compositions of rhizosphere microbial communities of Pinot and Riesling varieties were more similar to one another than those of more recent grape varieties (Bao et al., 2022). Moreover, the bacterial genera *Pseudomonas* and *Rhizobium*, which include plant growth-promoting strains (Section 3.2), were significantly enriched in the rhizosphere of Pinot noir (Bao et al., 2022). Some bacterial genera (*Bacillus*, *Turicibacter* and *Romboutsia*) were also enriched in the leaf microbiome of Pinot noir (Zhang et al., 2020).

Several studies have hypothesised that wild populations of *V. vinifera* subsp. *sylvestris* constitute a reservoir of useful microbial strains, because wild plants often harbour beneficial endophytes that are absent, or less abundant, in domesticated plants (Ofek-Lalzar et al., 2016; Sun et al., 2020). They explored the microbiome of sylvestris and obtained promising results. For instance, the analysis of the root microbiome showed that sylvestris roots were colonised by ectomycorrhizal fungi and that fungal pathogens were completely absent from roots colonised by these fungi (Radić et al., 2021). The analysis of the berry microbiome of sylvestris revealed a high diversity of yeasts, with some of them having promising prospects for use in oenology (Cordero-Bueso et al., 2017, 2022; Puig-Pujol et al., 2016). Yeast strains belonging to four species (*Meyerozyma guilliermondii, Hanseniaspora uvarum, H. clermontiae, and Pichia kluyveri*), all isolated from sylvestris, reduced growth of moulds caused by *Botrytis cinerea, Aspergillus carbonarius*, and *Penicillium expansum* (Cordero-Bueso et al., 2017). Finally, endophytic bacterial strains with biocontrol properties against several pathogens were isolated from sylvestris (Campisano et al., 2015), including strains belonging to *Pantoea* spp. and *Pseudomonas* spp. that showed antagonistic activity against crown gall agents (*Agrobacterium tumefaciens* and *Allorhizobium vitis*) (Asghari et al., 2019).

To identify and isolate beneficial microorganisms that may have been lost during the domestication process, future research should continue to explore the microbiome of *Vitis vinifera* subsp. *sylvestris* and that of ancestral grapevine varieties. We would recommend focusing on sylvestris populations in the centre of origin of cultivated grapevines (i.e., the Transcaucasus region; Fig. 1), where beneficial microbes may have co-evolved with wild progenitors, before being lost in those plants cultivated in the expanded range of grapevine. Such analysis could allow the isolation of plant growth-promoting bacteria (Gutierrez and Grillo, 2022), that might confer tolerance to abiotic stresses and grapevine pathogens of Eurasian origin (such as *Botrytis cinerea*). The case of pathogens introduced from North America is different and is discussed herafter (Section 2.2).

2.2 Microbial interactions in the centre of origin of major grapevine pathogens

In the middle of the 19th century, several fungal pathogen and insect pest species of North American origin crossed the Atlantic and reached Europe, causing the destruction of a large proportion of vineyards (Gessler et al., 2011). These American species included grape phylloxera (introduced in 1863), and the agents of powdery mildew (1848), downy mildew

(1878) and black-rot (1885). Phylloxera is caused by the insect *Daktulosphaira vitifoliae*, powdery mildew by the fungi *Erysiphe necator*, downy mildew by the oomycete *Plasmopara viticola*, and black-rot by the fungi *Phyllosticta ampelicida* (asexual stage of *Guignardia bidwellii*). Scientists who established collections of wild American *Vitis* species in botanical gardens in the early 19th century probably transported these pathogen species to Europe. From there, the diseases spread to other European regions and are now reported in most wine-producing regions of the globe (Bois et al., 2017; Fontaine et al., 2021; Pirrello et al., 2019).

European viticulture was saved by the introduction of native American *Vitis* species (such as *V. riparia*, *V. rupestris* and *V. berlandieri*; Fig. 2) that were resistant to grape phylloxera. Moreover, sulphur, copper and later on, phytosanitary treatments, have been and are still used extensively to protect vines from downy and powdery mildews. The American *Vitis* species were first used to produce rootstocks resistant to phylloxera. Most European cultivated vines, planted since the end of the 19th century, are therefore grafted plants composed of a rootstock tolerant to phylloxera and a scion used to produce berries (Section 3.1). The rootstock and the scion are genetically different and may even belong to different *Vitis* species, which makes European cultivated grapevine an interesting holobiont (Biget, 2021). *Vitis* species of American and Asian origin (Fig. 2) have also been used to produce interspecific hybrids cultivated on their own roots (This et al., 2006). More recently, they have been used as sources of disease-resistance genes in breeding programmes (Merdinoglu et al., 2018) since they are not only resistant to phylloxera but also to other pests and diseases (Cadle-Davidson et al., 2011; Merdinoglu et al., 2018; Staudt and Kassemeyer, 1995; Villano and Aversano, 2020). However, these new disease-resistant varieties are already threatened by pathogen strains able to overcome resistance genes (for downy mildew: Peressotti et al., 2010, for powdery mildew: Feechan et al., 2015).

As with other plant species, the grapevine microbiome offers tremendous genetic variability that has barely been exploited in breeding programmes to date (Gopal and Gupta, 2016). What if American and Asian *Vitis* species still had something to offer to European vineyards? Although this hypothesis has yet to be tested explicitly, the higher resistance of some American *Vitis* accessions to pathogens of North American origin (Fig. 2) could be due to an increased ability of the plant to select a protective microbiome (Gopal and Gupta, 2016; Lyu et al., 2021), in addition to the acquisition of resistance genes in the plant genome. This higher disease-resistance may be the result of a longer coevolution with the pathogens (Jürges et al., 2009). The resistance of some Asian *Vitis* species (Fig. 2) to mildews is more difficult to explain. The resistance of *V. amurensis* to downy mildew, *Plasmopara viticola,* might be due to the presence in Asia of a related pathogen species, *P. amurensis,* which has similar infection mechanisms (Jürges et al., 2009). Recent genetic analyses reopened the question of a possible origin of powdery mildew from Asia (Gur et al., 2021), and might account for the resistance of Asian *Vitis* species to powdery mildew.

Several studies have now characterised the microbiome of American and Asian *Vitis* species. They have found a high richness of endophytes (Fan et al., 2020; Kernaghan et al., 2017) and isolated strains with biocontrol properties against pathogens of North-American origin, as well as other pathogens. For instance, *V. amurensis* cv. Shuangyou in China hosted *Alternaria* strains effective against downy mildew (Musetti et al., 2006). Fungal endophytes isolated from leaves of *V. riparia* in eastern Canada showed high levels of inhibition of grey mould (caused by *Botrytis cinerea*) and black-foot disease (*Cylindrocarpon destructans*). Fungal strains inhibiting grey mould belonged to *Ramularia pratensis, Phoma aliena* and *Fusarium acuminatum*, while strains inhibiting black-foot belonged to *Ramularia* spp., *Phoma* spp., and *Biscogniauxia mediterranea* (Kernaghan et al., 2017). Similarly, endophytes isolated from *V. labrusca* leaves in Brazil showed in vitro antagonistic activity against several grapevine pathogens (Brum et al., 2012; Felber et al., 2016). Isolates of *Flavodon flavus* and *Colletotrichum gloeosporioides* strongly inhibited the root pathogen *Fusarium oxysporum f.* sp. *herbemontis* (Brum et al., 2012). Isolates belonging to *Phoma* spp. inhibited the growth of the causal agent of leaf blight (*Alternaria* spp.), isolates of *Fusarium culmorum* regulated anthracnose (*Sphaceloma* spp.) and *Sordariomycetes* isolates controlled ripe rot of grapes (*Glomerella* spp.) (Felber et al., 2016). The fungi *Acaromyces ingoldii*, previously reported for its plant protection functions, was found to be the most abundant taxon in the berry microbiome of muscadine grape (*Muscadinia rotundifolia* Michx.) in China (Sun et al., 2021). In addition, the bacterial genus *Rahnella* was detected in the microbiome of *M. rotundifolia* and might play a role in the biological control of grapevine crown gall, caused by *Agrobacterium vitis* (Chen et al., 2007).

To move towards pesticide-free viticulture, research on the microbial biocontrol of diseases of North American origin (powdery mildew, downy mildew and black-rot) should be fostered. To identify candidate biocontrol agents, we would recommend analysing the microbiome of American *Vitis* species in their native range (Fig. 2), where they have co-evolved with the pathogens and the rest of the microbiome. It could also be relevant to analyse the microbiome of resistant Asian *Vitis* species (Fig. 2) in areas where they have co-evolved with pathogens related to pathogens of North-American origin. The level of disease-resistance should first be evaluated and the microbiome of all plant compartments colonised by pathogens should be characterised. To highlight microbial taxa that play a role in disease resistance (Section 4.1), the microbiomes of resistant accessions could then be compared to that of sensitive ones in the same geographic region. The inference of microbial interaction networks in resistant *Vitis* species could also generate hypotheses about the microbial taxa or consortia conferring disease resistance (Section 4.3).

2.3 Microbial dispersal across geographic regions and *Vitis* species

As described above, cultivated grapevine is a complex holobiont that incorporates plant and microbial genomes of European, American and Asian origin (Sections 2.1 and 2.2). To understand how this complex system has assembled and evolved since grapevine domestication (Section 2.1), it is necessary to understand the origin and migratory routes of grapevine-associated microbes. To date, research has focused on the migratory flows of the oenological yeast *Saccharomyces cerevisiae* as well as those of major pathogens of grapevine (*Plasmopara viticola*, in particular).

Molecular evidence for the historical presence of *Saccharomyces cerevisiae* in wine fermentation was obtained from pottery jars hidden in the tomb of King Scorpio I, one of the earliest kings of Egypt, dating back to 3150 BC (Cavalieri et al., 2003). At that time, fermentations were only spontaneous (Marsit and Dequin, 2015), suggesting that *S. cerevisiae* was already an integral part of the berry microbiome. Legras et al. (2007) analysed migration routes of *S. cerevisiae* and showed that they correspond to the known migration routes of cultivated grapevine (Grassi and De Lorenzis, 2021). From Lebanon, yeast migration may have taken place across the Mediterranean region towards Italy, Spain and France, through three main routes: (1) the Rhône Valley-Burgundy-Alsace and Nantes route that goes along the Rhône and Loire rivers; (2) the Italy-Cognac route; and, (3) the Danube Valley route. This migration of vineyard yeast strains is accounted for, at least in part, by the trade of grape varieties. For instance, the Muscadet grape variety was imported from Burgundy to Nantes during the 15th century, while Ugni blanc, the dominant grape variety in the

French Cognac region, originated in Italy. Genetically-distant Austrian yeast strains may have been domesticated from local wild sylvestris populations (Section 2.1).

Plasmopara viticola is a complex of five species, each with a unique degree of host specialisation within the American Vitaceae (Fontaine et al., 2021; Rouxel et al., 2013, 2014). Fontaine et al. (2013, 2021) found that all invasive grapevine downy mildew populations worldwide belonged to the small clade *P. viticola* f. sp. *aestivalis,* which parasitizes *Vitis aestivalis* in Northeast America (Fig. 1). *P. viticola* was first introduced into Western Europe in the middle of the 19th century (Section 2.2), and spread to Central and Eastern Europe. Fontaine et al. (2013) and Gobbin et al. (2006) revealed, using population genetic methods, a low genetic diversity and weak population structure for *P. viticola* across European vineyards, suggesting that the strains initially introduced into Europe came from a single source population of North American origin. European populations of downy mildew then served as a source for secondary introductions into other viticultural regions of the world, including China, South Africa, and Australia. A third introduction event occurred later, that spread the disease from Australia to Argentina.

The migration routes of *Erysiphe necator* and *Phyllosticta ampelicida* have been less studied than those of *Plasmopara viticola.* According to Brewer and Milgroom (2010), populations of *E. necator* in Europe are derived from two separate introductions from the Eastern United States. These initial introductions into Europe were followed by secondary introductions from Europe into the Western United States and Australia. Recently, Gur et al. (2021) proposed an additional non-American (possibly Asian) origin of *E. necator.* There is even less knowledge of the invasion pathways of *P. ampelicida* (Narduzzi-Wicht et al., 2014). Rinaldi et al. (2017) suggested that the pathogen was introduced into Northeastern Italy from Eastern Europe and that two introductions occurred in Portugal, one from France or Italy, and another from America. Interestingly, these American pathogens not only colonised the European cultivated grapevine *Vitis vinifera* subsp. *vinifera,* but were also incorporated into the microbiome of its wild ancestor *Vitis vinifera* subsp. *sylvestris* (Section 2.1). As a result, the distribution of wild grapevines has been greatly reduced over the last 150 years and it is now considered an endangered plant species in Europe (Arnold et al., 2005).

Future research should continue unravelling the migratory routes of grapevine pathogens and extend these efforts to other member of the functional microbiome (Section 3). This is crucial to identify those geographic locations where grapevine-pathogen-microbiome co-evolution has occurred, as well as the *Vitis* species involved in this coevolutionary process (Fig. 2). This knowledge is necessary to design microbiome studies informed by the biogeography and evolutionary history of grapevine. Microbiome dynamics on a smaller temporal scale should also be considered and is discussed below (Section 3).

3. The grapevine functional microbiome throughout plant lifetime

This section describes how the microbiome assembles, functions and changes over the course of grapevine life. It highlights the plant-microbe interactions that have a proven, direct and significant influence on the grapevine performance (growth, health and berries quality) at every life stage (Fig. 3). The section deals with both the microbiome (bacteria, filamentous fungi, yeasts and oomycetes) and the virome, including phytoviruses (plant-infecting viruses), mycoviruses (fungi-infecting viruses) and phages (bacteria-infecting viruses). It covers both aboveground and belowground microorganisms, including arbuscular mycorrhizal and ectomycorrhizal fungi.

3.1 Initial microbiome and virome at the graft stage

The life of cultivated grapevine generally begins in a nursery with a graft (Fig. 3). The two components of the graft, the scion and the rootstock, are produced in nursery "mother" plots. Mother plants produce stems that are pruned during the dormant stage in winter to make cuttings that are then stored in a cold room. The rootstock and scion cuttings and the grafted plants then undergo different treatments depending on the country and the nursery (Gramaje and Armengol, 2011). Most often, immediately after grafting, the scion and the graft zone are dipped in a wax solution, which may contain plant growth regulators (mainly auxins) and fungicides. The grafted plants are then incubated for 2–3 weeks under controlled temperature (28 °C) and high humidity (90%), in order to promote callusing. A second waxing step is then performed, before planting in pots in the greenhouse, or in nursery fields. Potted plants can be shipped directly to winegrowers and planted in the field next autumn, while plants kept in nursery fields are usually shipped as 1-year-old, bare-root plants to be planted in spring.

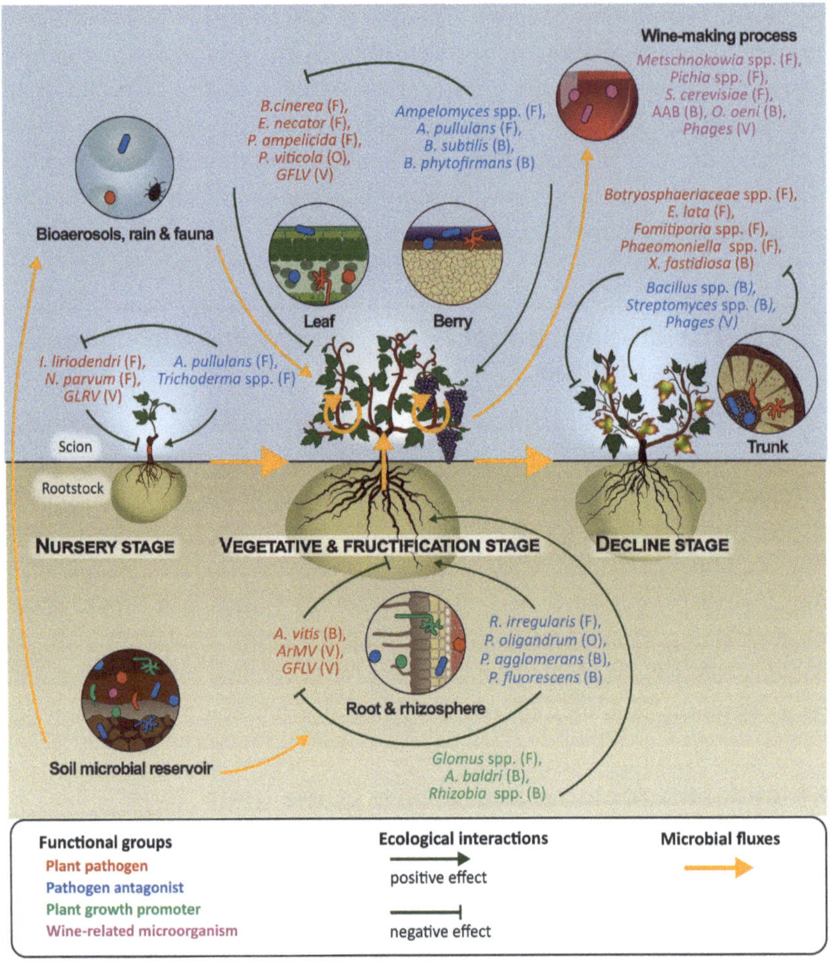

Fig. 3 Microbial functional groups that colonise cultivated grapevine from the nursery to plant decline, as well as ecological interactions (black arrows) and microbial fluxes (yellow arrows) that drive the microbiome dynamics during plant lifetime. Plant-microbe-microbe ecological interactions form complex networks (Section 4) that could not be represented here for clarity. Only a few ecological interactions are represented. For the same reason, we could not represent all the fungal (F), bacterial (B), oomycete (O) species and viruses (V) that significantly influence grapevine performance and wine-making process. We only represented a few well-studied species or species groups. AAB, Acetic Acid Bacteria; GFLV: Grapevine Fanleaf Virus; GLRV, Grapevine Leafroll associated Virus; Other species names: *Agrococcus baldri, Allorhizobium vitis, Aureobasidium pullulans, Bacillus subtilis, Botrytis cinerea, Burkholderia phytofirmans, Candidatus phytoplasma, Erysiphe necator, Eutypa lata, Ilyonectria liriodendri, Neofusicoccum parvum, Oenococcus oeni, Pantoea agglomerans, Phyllosticta ampelicida, Plasmopara viticola, Pseudomonas fluorescens, Pythium oligandrum, Rhizophagus irregularis, Saccharomyces cerevisiae, Xylella fastidiosa.*

The many steps of the grafting process are critical for assuring the quality of the plants produced. First, the choice of the two plant genomes is crucial. Not accounting losses due to incompatibility between genotypes (Tedesco et al., 2022), about 50% of the plants produced are unmarketable because of the low quality of the graft (Carrere et al., 2022). Second, it is fundamental for nurseries to produce plants that do not carry diseases (Waite et al., 2015), in particular phytoviruses. Viruses trigger both graft incompatibility and decline of young vines. For instance, the transmission of grapevine leafroll-associated virus 2 (GLRaV-2; Fig. 3) from scion to rootstock results in graft incompatibility (Rowhani et al., 2017). Rootstock susceptibility to infection depends on the genotype, with genotypes 1616 Couderc, Kober 5BB, 1103 Paulsen being the most susceptible rootstocks (Alkowni et al., 2011; Rowhani et al., 2017; Uyemoto et al., 2000, 2001). Reverse incompatibility, where the rootstock is the source of a latent virus, has not been observed in grapevine yet. The decline of young vines triggered by joint infection with GLRaV-2 and grapevine virus B (GVB) infection, or GLRaV-1 and grapevine virus A (GVA), was also observed (Rowhani et al., 2017). The mechanisms underlying such declines have not been clearly determined yet, but they might involve the production of viral small interfering RNAs (vsiRNAs) by the plant cell machinery, which in turn would affect plant gene expression (Smith et al., 2011). Management of these phytoviruses relies primarily on preventive measures (such as the use of clean propagation and planting materials) and robust diagnostics (Maliogka et al., 2015; Martelli, 1999). In addition, hot water treatment (HWT) of the grafted plant (Eichmeier et al., 2018) is sometimes performed to reduce infection by phytoplasmas (flavescence dorée, bois noir, grapevine yellows) and by fungi responsible for grapevine trunk diseases (Section 3.5). *Trichoderma* spp., which are soil microbes with an antagonistic activity against some pathogenic fungi, are sometimes inoculated by nurseries (Eichmeier et al., 2018; Jaarsveld et al., 2021). Finally, mycorrhiza (*Rhizophagus irregularis,* in particular) can also be inoculated at the nursery stage (Fig. 3), to increase grapevine resistance to biotic and abiotic stresses by promoting the structure and development of the root system (Bettenfeld et al., 2021).

All these manipulations and treatments of scions and rootstocks in the nursery can influence the initial grapevine microbiome and its future trajectory of assembly. Gramaje et al. (2022) analysed the fungal communities associated with rootstocks harvested in two mother plots located 800 m appart. No difference in fungal community composition between these two locations was detected, but the fungal communities varied along the grafting process (after cutting from mother plants, after cold storage and hydration, after grafting and callusing, and in dormant grafted plants). The core mycobiome was composed of the genera *Cadophora, Cladosporium, Penicillium* and *Alternaria* in both rootstocks. Numerous genera associated with grapevine trunk diseases (*Cadophora, Dactylonectria, Diaporthe, Diplodia, Ilyonectria, Neofusicoccum, Phaeoacremonium and Phaeomoniella*; Section 3.5) and potential biocontrol activities (*Aureobasidium pullulans*) were identified (Fig. 3). The pathogenic genus *Neofusicoccum* was persistent throughout the propagation process (Fig. 3), confirming the role of rootstock as a primary source of many pathogens. Lade et al. (2022) compared the contribution of nursery, scion variety and rootstock variety to fungal community composition in the graft zone and found that the nurseries (i.e., plant production practices) played a major role, followed by rootstocks and scion varieties.

To better understand and manage the assembly trajectory of the grapevine microbiome, special attention should be paid to the production practices of grafted plants in nurseries. We recommend using the framework of microbial community coalescence (Rillig et al., 2015) to analyse the influence of grafting on the microbiome. The microbiome of the grafted plant would then be considered as a joining of previously separate communities, forming a new entity that is not easily separable and with properties distinct from the progenitor communities (Rillig et al., 2015). At the graft stage, major physiological changes occur (Cookson et al., 2013; Prodhomme et al., 2019), including vascular connection between rootstock and scion that influences sap flow and associated fluxes of microorganisms (Deyett and Rolshausen, 2020). These physiological processes have barely been studied although they could have a significant impact on the initial microbiome of grafted plants and, through priority effects (Zhou and Ning, 2017), on the recruitment of the microbiome after planting (Section 3.2).

3.2 Recruitment of the root microbiome from the soil reservoir

Winegrowers usually receive grafted vines (potted or bare-rooted) from nurseries (Section 3.1) for planting in their vineyards. The young plants then begin to recruit microorganisms from the soil (Fig. 3), which is the major reservoir of microorganisms for the plant microbiome (Griggs et al., 2021; Liu et al., 2019; Swift et al., 2021). Recruitment is based on the active release by the roots of a wide range of carbon-containing compounds known as rhizodeposits (Chaparro et al., 2014; del Martínez-Diz et al., 2019). The rhizodeposits represent nearly 40% of the photosynthates produced by a plant (Bais et al., 2006). These compounds serve as a substrate for the development of microorganisms and make the rhizosphere (i.e., the transition zone between soil and plant roots) much more attractive for microbes than surrounding soil. The nature and quantity of rhizodeposits depend on the plant genotype and vary across seasons and throughout plant life, allowing the plant to actively shape the composition of rhizosphere microbial communities (Berlanas et al., 2019). Beneficial microorganisms may for instance be recruited under stressful conditions, a mechanism known as the "cry for help" reaction (Rizaludin et al., 2021). Some rhizospheric microorganisms develop physical interactions with the root (case of mycorrhizae, for instance) and some of them can even enter the root endosphere. Once in the root endosphere, some microorganisms colonise the vascular tissues and disseminate in the aerial parts of the plant (Fig. 3), sometimes reaching leaves and berries (Darriaut et al., 2022; Liu et al., 2019). Not all microorganisms can colonise the plant internal tissues from surrounding soil, which creates a decreasing gradient of microbial diversity from the outside of the root towards the inside (del Martínez-Diz et al., 2019; Marasco et al., 2018; Zarraonaindia et al., 2015).

Many studies have examined the taxonomic composition of the grapevine rhizosphere and root microbiome. The same major taxonomic groups are generally represented. Pseudomonadota, Actinomycetota and Acidobacteriota (formerly named Proteobacteria, Actinobacteria and Acidobacteria; Oren and Garrity, 2021) are the dominant bacterial phyla, followed by Bacillota and Bacteroidota (i.e., Firmicutes and Bacteroidetes) (Bettenfeld et al., 2021; Dries et al., 2021; Novello et al., 2017; Zarraonaindia et al., 2015). Ascomycota, followed by Basidiomycota, are the dominant fungal phyla (Bettenfeld et al., 2021; Liu and Howell, 2021). At a finer taxonomic scale, the microbiome composition varies with soil physical and chemical features (Bettenfeld et al., 2021; Signorini et al., 2021), geography and climate (Hernandez and Menéndez, 2019; Liu et al., 2019), plant age and rootstock genotype (Berlanas et al., 2019; Ji et al., 2019), and vineyard management practices (Likar et al., 2017) including applications of plant protection products and fertilizers (Canfora et al., 2018; Zaller et al., 2018) and the presence of vegetation cover (Vukicevich et al., 2018).

Root-microbe interactions can foster vine growth and productivity through various mechanisms. Plant Growth-Promoting (PGP) microbes (Fig. 3) can solubilise and make available nutrients to the roots (Darriaut et al., 2022). The mineral elements needed by the plant (phosphorus, nitrogen and phosphate) are naturally present in the soil but cannot be efficiently or directly assimilated by plants because they are complexed with other molecules. Other PGP bacteria produce enzymes (phosphatases) or organic acids (e.g., hydrogen cyanide; Rijavec and Lapanje, 2016) that solubilise inorganic phosphate, while others produce ammonium (NH4+) or convert nitrites (NO2−) to nitrates (NO3−), that are easily absorbed by the plant. Bacteria, including *Bacillus herbersteinensis, Bacillus licheniformis, Micrococcus* spp., *Pseudomonas* spp., and *Pantoea agglomerans*, can combine these mechanisms (Asghari et al., 2020; Baldan et al., 2015). In the grapevine rhizosphere, the bacterial genera *Pseudomonas, Streptomyces* and *Rhizobia* are those most involved in phosphorus and nitrogen cycles (Bona et al., 2018). Some fungal genera are also involved in phosphate solubilisation, such as the genus *Mortierella* (Carbone et al., 2021; Liu and Howell, 2021). Rhizosphere PGP bacteria (Fig. 3) can also produce hormones, such as auxin, which directly stimulates plant growth (Morales-Cedeño et al., 2021). For instance, *Agrococcus baldri* and *Paenibacillus* spp. produce hormones that, even at low concentration, cause root elongation, branching and increase root diameter and density (Baldan et al., 2015). Bacterial species, such as *Bacillus amyloliquefaciens*, can produce microbial siderophores, which bind with iron, and contribute to plant iron nutrition (Mardanov et al., 2019). Other endophytic bacteria associated with root tissues (*Pseudomonas* spp., in particular), produce enzymes (ACC-deaminases) that cleave ACC (1-aminocyclopropane-1-carboxylic acid), which is the immediate precursor of ethylene in plants. This reduces the production of ethylene and limits the damage caused by biotic and abiotic stresses (Marasco et al., 2018; Saraf et al., 2010).

Arbuscular mycorrhizal fungi (AMF) also play an important role in water and nutrient absorption. Belonging mainly to the *Glomeromycota* phylum, these endomycorrhizal fungi establish symbioses with vine roots (Balestrini et al., 2010). Hyphae increase the exchange surface between the vine roots and the soil. They also activate phosphorus and nitrogen transporters in root cortical cells (Trouvelot et al., 2015). AMF are the most abundant mycorrhizae associated with grapevine roots and include species belonging to the genera *Acaulospora, Diversispora, Funneliformis, Glomus* and *Rhizophagus* (Darriaut et al., 2022; Liu and Howell, 2021). They are actively recruited by vine roots during water stress (Carbone et al., 2021).

Grapevine roots are not only colonised by beneficial microorganisms such as PGP and AMF, but also by bacterial, fungal and viral pathogens (Fig. 3). The bacterial pathogen *Allorhizobium vitis,* the causal agent of crown gall, is frequently isolated from diseased plants in the vineyard (Habbadi et al., 2020). The bacteria harbours tumour-inducing plasmids (pTi), part of which integrate into the plant genome (Transfert-DNA, T-DNA) to induce the development of root or crown galls and the synthesis of opines, which the bacteria utilises as nutrients. The spread of pTI plasmids can be facilitated by bacteria belonging to the *Novosphingobium* genus, which are therefore pathogen stimulants (Gan et al., 2019). Black-foot disease, caused by *Ilyonectria* spp. (*I. liriodendri*, in particular), is responsible for root necrotic lesions, with a reduction in root biomass, besides other symptoms that include reduction of internodes, loss of vigour, patchy foliage and smaller leaves (Bleach et al., 2021). The fungal genus *Ilyonectria* is found in many metabarcoding analyses of vineyard soil because it can persist for several years after the removal of infected vines (de Brito et al., 2019; Deyett and Rolshausen, 2020; Liu and Howell, 2021; Rivas et al., 2022). Several fungal genera associated with grapevine trunk diseases (*Botryosphaeria* and *Diplodia*; Section 3.5), Petri disease (*Cadophora, Phaeoacremonium* and *Phaeomoniella*) and Diaporthe dieback (*Diaporthe*) have been detected in asymptomatic root tissues (del Martínez-Diz et al., 2019; Deyett and Rolshausen, 2020; Liu and Howell, 2021; Rivas et al., 2022). *Candidatus phytoplasma* spp., the causal agent of bois noir disease, was also detected in healthy root tissues (Marasco et al., 2018), indicating that the root endosphere may be an important reservoir for latent pathogens (del Martínez-Diz et al., 2019; Liu and Howell, 2021). Finally, grapevine roots can be infected by phytoviruses, some of which are transmitted by soil nematodes. For instance, grapevine fanleaf virus (GFLV) and arabis mosaic virus (ArMV), the main causal agents of grapevine fanleaf disease, are transmitted by the nematode species *Xiphinema index* and *Xiphinema italiae* in the case of GFLV, and *Xiphinema diversicaudatum* in the case of ArMV (Oliver and Fuchs, 2011). The nematodes feed on the tips of growing roots by pricking them with a stylet. By successively feeding on neighbouring vines, they disseminate the virus retained in their oesophageal tracts (Maliogka et al., 2015), producing the often patchy distribution of virus-infected plants.

The root microbiome also harbours microorganisms that directly inhibit pathogen growth, through antagonistic microbial interactions, or indirectly limit disease severity or frequency by stimulating the plant immune system or by acting on vector populations (Fig. 3). Microbial interactions responsible for pathogen biocontrol include antibiosis, hyperparasitism, spatial and nutritional competition, interference with pathogen signalling, and disease symptom reduction by degradation of pathogen toxins or virulence factors (Compant et al., 2013). For instance, the biocontrol activity of *Bacillus amyloliquefasciens* mainly relies on antibiosis. This bacteria synthesises a wide range of antimicrobial compounds (surfactin, plantazolicin, macrolactin, bacillaene, fengycin, difficidin, bacillibactin, and bacilysin) that regulate the growth of several grapevine pathogen species, including *Botrytis cinerea, Phaeoacremonium aleophilum, Phaeomoniella chlamydospora* (Mardanov et al., 2019). Root microbes also stimulate plant defences, by inducing oxidative bursts, activating signalling pathways (salicylic acid, ethylene/jasmonic acid), and increasing the synthesis of pathogenesis-related (PR) proteins, lytic enzymes, polyamines and phytoalexins (Héloir et al., 2019). This stimulation increases the capacity of plants to react to ongoing infections, or to future ones by priming the immune system. Bacteria of the *Burkholderia* genus increase the production of PR proteins when inoculated on the roots of grapevine infected with *B. cinerea* (Esmaeel et al., 2020), and the strain *Burkholderia phytofirmans* PsJN can migrate to the aerial parts of grapevine and form biofilms that directly limit the spread of *B. cinerea* (Miotto-Vilanova et al., 2016). Many other root microorganisms have a proven antagonistic effect on grapevine pathogens but their mode of action is not always known. Other bacteria with a biocontrol activity include *Streptomyces* spp., that reduce young grapevine decline and black-foot disease (González-García et al., 2019; Pilar Martínez-Diz et al., 2021), and *Bacillus subtilis*, which directly limits the growth of fungal pathogens related with grapevine trunk diseases (Berlanas et al., 2019). The latter also stimulates plant defences, similar to *Pseudomonas fluorescens, Pantoea agglomerans, Acinetobacter lwoffii* (Aziz et al., 2016, 2020; Magnin-Robert et al., 2007). Root fungi, in particular AMF, also play a role in disease biocontrol. AMF of the *Glomus* genus and *Rhizophagus irregularis* reduce the severity of black-foot disease (Berlanas et al., 2019; Darriaut et al., 2022). AMF may also protect grapevines from GFLV by inhibiting the proliferation and penetration of its nematode vector (*X. index*) in roots (Hao et al., 2018). In addition, the fungal species *Trichoderma atroviride* and *Aureobasidium pullulans* are antagonists of the fungal pathogens associated with Petri disease (*Phaeomoniella chlamydospora and Phaeoacremonium minimum*), and of *Diplodia seriata* which is associated with Botryosphaeria dieback (del Martínez-Diz et al., 2019). *T. atroviride* also fosters plant defences (Stempien et al., 2020). Finally, the oomycete *Pythium oligandrum* has mostly indirect negative effects on fungal pathogens associated with Petri disease (Yacoub et al., 2016).

This state of art demonstrates that grapevine roots constitute a rich microbial compartment that is quite well understood from a functional point of view. Many microorganisms that have a positive (such as PGP bacteria, AMF and biocontrol agents) or negative (such as *Allorhizobium vitis* or phytoviruses) influence on grapevine have been identified and their mode of action has been described. Shotgun metagenomics and metabolomics associated with metabolic network modelling (Section 4.3) could further advance this understanding by allowing us to build predictive models of the nutritional interactions between grapevine and microbes in the rhizosphere (Jacoby et al., 2017). These models could inform the design of microbial consortia to inoculate to boost plant growth and defence. To optimise soil management, future research should focus on belowground-aboveground microbial interactions in vineyards. They could first assess the effects of soil management on soil microbiome (Bettenfeld et al., 2021; Signorini et al., 2021), and then the cascading effects of the soil microbiome on the root, trunk, leaf and berry microbiomes (Sections 3.3 and 3.4) and their consequences on grapevine health and wine quality.

3.3 Seasonal assembly of the leaf microbiome in interaction with leaf pathogens

The grapevine leaf microbiome is dynamic: it assembles in spring, changes during the vegetative season and then disassembles during leaf senescence. In spring, the buds burst and let the first shoots appear. As vines are lianas, the shoots hang on to the trellis with their tendrils and grow by producing new leaves, at a rate of about one every 3 days in the Bordeaux area, France (Calonnec et al., 2008). At a given time, each vine thus displays a mosaic of leaves of different ages. The leaves at the distal end of shoots are the youngest and the leaves near the stock are the oldest (see Fan et al., 2020). As the shoot grows, the young leaves expand, become mature, and are colonised by epiphytic microorganisms (the phyllosphere microbiome *sensu stricto*; Behrens and Fischer, 2022; Vacher et al., 2016a) and endophytic microorganisms (Pacifico et al., 2019) that collectively form the leaf microbiome. The latter is largely dominated by fungi belonging to the Ascomycota and by bacteria belonging to the Actinomycetota, followed by Pseudomonadota and Bacillota (Bettenfeld et al., 2021; Fort et al., 2016; Oren and Garrity, 2021; Singh et al., 2018). The fungal species *Aureobasidium pullulans* was dominant at the end of growing season in several independent metabarcoding studies, representing more than half of the fungal community in old leaves (Grube et al., 2011; Knapp et al., 2021).

Leaf microbiome dynamics depends on many factors, and first and foremost on the specific growth pattern of grapevine (Calonnec et al., 2008, 2018). Indeed, the date at which a leaf emerges determines its morphology and physiology (Chitwood et al., 2016; Majer and Hideg, 2012) and also the microbial fluxes and environmental perturbations to which it will subsequently be exposed. Leaves that emerge first are colonised by microorganisms that overwintered in buds or on woody parts (case of primary infections by the fungal pathogen *Erysiphe necator*, Gadoury et al., 2012), or in senescent leaf debris that remained on the soil (case of primary infections by the oomycete pathogen *Plasmopara viticola*, Burruano, 2000). They are also colonised by microorganisms circulating in the xylem or in the phloem (Fig. 3). These circulating microorganisms may be beneficial, as in the case of the bacterium *Burkholderia phytofirmans* (Compant et al., 2008) or they may be pathogens, like the bacterial pathogen *Xylella fastidiosa* (Deyett and Rolshausen, 2019) and viruses and phytoplasmas (Laimer et al., 2009; Zherdev et al., 2018). Wind and rain disseminate microorganisms from multiple sources (vines, soil, adjacent semi-natural and natural habitats) (Fig. 3) and may also contribute to the initial colonisation of the phyllosphere (Abdelfattah et al., 2019; Griggs et al., 2021). By contrast, those leaves that emerge later in the season are in direct contact with other grapevine leaves that already have a more developed microbiome. Leaf-to-leaf transmission of the microbiome (Fig. 3), which may transiently involve the atmosphere or other plant organs, is the dominant microbial colonisation pathway in grapevine leaves (Abdelfattah et al., 2019). This transmission is responsible for secondary infection cycles in leaf pathogens. Late-emerging leaves also face greater activity from fauna (arthropods, birds) that are underestimated microbial vectors (Griggs et al., 2021; Stefanini and Cavalieri, 2018). Finally, it should be noted that leaves of different ages do not provide the same habitat for microorganisms. Leaf age determines the structure of the cuticle and epicuticular waxes, the amount of available nutrients, the concentration of secondary metabolites, all of which regulate the growth of microbial populations (Griggs et al., 2021; Shakir et al., 2021). Leaf age also determines immune status as microbial colonisation can stimulate defences and prime the leaf to be more resistant to subsequent infection (Chaudhry et al., 2021; Shakir et al., 2021; Section 3.2). The effect of leaf age on disease development is well undertsood and is termed ontogenic resistance. It has been demonstrated for several grapevine leaf pathogens (Calonnec et al., 2018; Qiu et al., 2015; Steimetz et al., 2012). All of these factors lead to intra-canopy variation in the composition of the grapevine leaf microbiome (Calonnec et al., 2008; Fan et al., 2020) as well as strong seasonal dynamics. Seasonal dynamics, however, are less pronounced in bacterial communities than in fungal communities, whose diversity decreases markedly over the growing season (Fort et al., 2016; Gobbi et al., 2020; Liu and Howell, 2021; Pinto et al., 2014; Singh et al., 2018).

In addition to seasonal intra-canopy dynamics, the grapevine leaf microbiome exhibits high spatial variability. Its composition changes significantly from one geographical region to another (Oliveira et al., 2018; Singh et al., 2018) and from one vineyard plot to another (Barroso-Bergadà et al., 2021, but see Knapp et al., 2021). These spatial variations are explained by vine-related (scion variety, cultural practices) and environmental factors (climate and microclimate, landscape structure) (Bettenfeld et al., 2021). Several studies have attempted to establish a hierarchy between these factors. For example, variety has a significant effect on leaf microbiome composition (Singh et al., 2018, 2019; Zhang et al., 2020), but its effect is much smaller than that of geographic and climatic factors (Singh et al., 2018). The effect of phytosanitary treatments appears to be less important than seasonal dynamics (Gobbi et al., 2020), in agreement with results demonstrating high resilience of the phyllosphere microbiome to phytosanitary treatments (Perazzolli et al., 2014). The cropping system as a whole, however, has a substantial influence on the diversity and composition of leaf microbiome, with significant differences found between organic and conventional farming (Barroso-Bergadà et al., 2021; Castañeda et al., 2018; Kernaghan et al., 2017; Miura et al., 2019). Finally, several studies investigated whether forest patches or orchards adjacent to grapevine plots structure the grapevine leaf microbiome, with varying results (Castañeda et al., 2018; Castañeda and Barbosa, 2017; Chandra et al., 2020; Fort et al., 2016; Miura et al., 2019).

Among the multitude of microorganisms that colonise grapevine leaves, some affect plant health and growth. Pathogens and their mode of action have been the most studied (Armijo et al., 2016). Powdery and downy mildews (respectively caused by the fungal species *Erysiphe necator* and the oomycete species *Plasmopara viticola*; Fig. 3) are the most important diseases of grapevine (Bois et al., 2017). Black-rot (caused by the fungi *Phyllosticta ampelicida*, Fig. 3) is a foliar disease that is becoming increasingly important in Europe, concomitant with efforts to reduce fungicide use (Molitor and Beyer, 2014). All three pathogens are introduced species of

North-American origin (Sections 2.2 and 2.3). When these pathogen species colonise the grapevine leaf, they encounter the resident microbiome and virome, which can act as a barrier to their growth through direct ecological interactions (Chaudhry et al., 2021; Hacquard et al., 2017), such as exploitative competition, interference competition (antibiosis), and hyperparasitism. Once this first barrier is overcome, pathogens are confronted by the plant's immune system, which is itself stimulated by the microbiome, especially the root microbiome (Section 3.2).

Natural antagonists of downy mildew (Fig. 3) have been identified using culture-dependent methods. Several bacterial and fungal species, isolated from grapevine leaves or previously detected in the leaf microbiome, have a negative effect on *Plasmopara viticola* under experimental conditions. For fungi these include *Acremonium byssoides, Alternaria alternata, Beauveria bassiana, Epicoccum nigrum* (Burruano et al., 2008, 2016; Kortekamp, 1997; Musetti et al., 2006; Rondot and Reineke, 2019; Taguiam et al., 2021; Zanzotto and Morroni, 2016), and several *Fusarium* species (Bakshi et al., 2001; Falk, 1996). The genus *Bacillus*, including *B. subtilis* and *B. pumilus*, predominates among *P. viticola* antagonistic bacteria (Bruisson et al., 2019; Dagostin et al., 2011; Zanzotto and Morroni, 2016; Zhang et al., 2017). Moreover, root colonisation by the arbuscular mycorrhizal fungus *Rhizophagus irregularis* (Sections 3.1 and 3.2) increases grapevine resistance to *P. viticola* in leaves (Cruz-Silva et al., 2021). Finally, the virome of *P. viticola* has been analysed using metatranscriptomics in search for hypovirulence-inducing mycoviruses (Chiapello et al., 2020). However, despite the description of many new mycoviruses (283 new viruses associated with lesions caused by *P. viticola*), the limited availability of phenotypic data on fungal hosts has so far prevented the association of particular viruses with variations in fungal virulence.

Pathogen-microbiome-virome interactions are not well documented for powdery mildew (Panstruga and Kuhn, 2019), but some antagonistic interactions (Fig. 3) have been demonstrated experimentally. Several natural hyperparasites of *Erysiphe necator* have been identified among the *Ampelomyces, Lecanicillium,* and *Acremonium* genera (Falk, 1995; Ghule et al., 2019; Kiss, 2003). As with downy mildew, powdery mildew growth is reduced by *B. subtilis* and *B. pumilus* (Compant et al., 2013; Crisp et al., 2006). Network inference (Section 4.3) has suggested an antagonistic interaction between powdery mildew and the yeast *Buckleyzyma aurantiaca* (Pauvert et al., 2020), which remains to be experimentally validated with culture-dependant approaches (Section 4.4). Moreover, as for *Plasmopara viticola*, the virome of *E. necator* has been studied in search of hypovirulence-inducing mycoviruses (Pandey et al., 2018). Mycoviruses have been detected but an effect on fungal virulence and disease development has yet to be been demonstrated.

Future studies should reinforce the searching for members of the resident microbiome and virome that naturally regulate powdery and downy mildews, and identify the biodiversity management practices that favour these natural antagonists. Such results are urgently needed to move towards pesticide-free viticulture (Jacquet et al., 2022), as mildews are currently controlled by chemical pesticides whose use should be drastically reduced to preserve environmental and human health. Based on the co-evolutionary history of *Vitis* spp. and mildew agents (Section 2.2), we recommended exploring the microbiome and the virome of mildew-resistant *Vitis* species in their native range (Fig. 2) to identify biocontrol candidates. In addition, we would recommend comparing the leaf microbiome of European cultivated grapevines that are rarely *vs.* frequenly infected with mildews, according to epidemiological records (Section 4.1). The upper soil and bark microbiomes could also be analysed as they constitute the overwintering sites of the pathogens. Identifying the microbial taxa that are enriched in mildew-resistant *Vitis* species and in rarely-infected plots, and deciphering the microbial interaction networks (Section 4.3), could provide interesting insights into the natural regulation of mildews.

3.4 Dynamics of the berry microbiome to ripening and winemaking

Like the leaf microbiome (Section 3.3), the berry microbiome varies significantly in abundance and composition throughout the season. Population sizes of both fungal and bacterial communities on the berry surface increase with ripening and reach a maximum when the berries become overripe. The community composition varies significantly during the season, with a decrease in *Basidiomycota (Aureobasidium, Cryptococcus, and Rhodotorula* spp.) and an increase in *Ascomycota* (*Hanseniaspora, Metschnokowia, Pichia* spp.) (Barata et al., 2012). As the season progresses, Gram-negative bacteria (mostly *Pseudomonas spp.*) decrease in abundance whereas Gram-positive communities (mostly *Micrococcus* spp.) increase (Martins et al., 2012). The morphology and chemical composition of berries are major drivers of these seasonal variations in microbiome composition (Liu and Howell, 2021). Indeed, the water activity on grapevine berry skin and the composition of exudates change along the grapevine berry ripening process, with an increase in sugar content, a decrease in pH and acidity (Griggs et al., 2021; Martins et al., 2014, 2020). Other drivers include agricultural practices during the fructification period (phytosanitary treatments, pruning), as well as the climatic conditions from fruit set to harvest (Belessi et al., 2022; Bokulich et al., 2014; Cordero-Bueso et al., 2011a,b; Ding et al., 2021; Martins et al., 2014, 2020; Pinto et al., 2014). In addition to these seasonal variations, the berry microbiome varies markedly in space. Over the last decade, many studies conducted at different spatial scales (world, countries, regions and vineyards) have provided evidence for a biogeographic differentiation of the berry microbiome (Bokulich et al., 2014; Gobbi et al., 2022; Jara et al., 2016; Li et al., 2022; Miura et al., 2017; Vitulo et al., 2019). Based on these findings, these authors have developed the concept of "*microbial* terroir" that proposes a causal relationship between grapevine microbial biogeography and regional wine characteristics.

At the ripening stage, the grapevine berries harbour fermentative microorganisms that will subsequently contribute to the winemaking process and to wine quality (Fig. 3). These microorganisms include *Saccharomyces cerevisiae* (Sections 2.1 and 2.3), non-*Saccharomyces* yeasts and Lactic Acid Bacteria (LAB). *S. cerevisiae* is the species most responsible for alcoholic fermentation. Although its populations reach their highest abundance at berry maturity (10^4–10^6 cfu/g), the yeast remains rare at the grapevine berry surface, thus raising questions about its origin (Börlin et al., 2020; Mortimer and Polsinelli, 1999). The abundance and diversity of fermentative yeasts on the grapevine berry surface is impacted by fungicide treatments (Cordero-Bueso et al., 2014) and infection by *Botrytis cinerea*, which modifies sugar availability (Fleet, 2003; Nisiotou and Nychas, 2007). Low numbers ($<10^3$ cfu/g) of cultivable LAB are also present on sound fruit and end up in must during the early stages of wine processing. Within this group of bacteria, *Oenococcus oeni* (Fig. 3) is responsible for a crucial process called malo-lactic fermentation (MLF), where L-malic acid in wine is converted to softer L-lactic acid. The reduction of acidity, which spontaneously occurs after alcoholic fermentation, is beneficial to the quality of wines made in cool winegrowing regions. Yet, MLF only starts when the *O. oeni* population reaches 10^6 cfu/mL. Understanding the factors reducing the bacterial biomass is crucial, in order to avoid slow or blocked MLF. Phages (Fig. 3) are one of these factors, as they may modulate *O. oeni* population size, and a number of temperate and virulent phages infecting *O. oeni* have been recently isolated from crushed grapes and later winemaking stages (Claisse et al., 2021).

The berry microbiome may also contain microorganisms that alter wine quality, such as Acetic Acid Bacteria (AAB) or fungi that produce off-flavours or mycotoxins. AAB (*Acetobacter* spp., *Gluconobacter* spp.) (Fig. 3) are wine spoilage bacteria that are naturally present in the microbiome of grapevine berries (Fleet, 1993). They cause pre- and post-harvest yield losses and render berries unsuitable for wine production. A relevant example is sour rot of grape (Lleixa et al., 2018), which is a polymicrobial disease requiring the presence of *Drosophila* flies (disease vector), wounded fruits, various yeasts that convert grape sugars to ethanol, and AAB that subsequently oxidise the ethanol to acetic acid. AAB co-exist with phages on ripe berries and a member of the *Tectiviridae* family has been recently characterised (Philippe et al., 2018). The control of AAB is challenging as these bacteria show great capacities to persist during wine making. Phages and their derived enzymes called endolysins may offer new solutions to their control.

Fungi naturally present on ripened grapevine berries can also produce off-flavours, in particular earthy or mouldy odours in grape juice and wines. This is the case of *Botrytis cinerea* (Fig. 3), the necrotrophic fungal pathogen responsible for grey mould (Armijo et al., 2016). *B. cinerea* produces cellulases, pectinases, proteases and a *p*-phenol oxygen oxidoreductase able to oxidise phenolics compounds, which is at the origin of severe alteration of wine colour. *B. cinerea* also causes an enhancement of phenylpropanoid metabolism, terpenes and fatty acid aroma precursors at the grapevine berry level. It can be associated with *Penicillium expansum*, then leading to the production of geosmin, an earthy off-aroma (La Guerche et al., 2005, 2007). However, *B. cinerea* can also colonise berries under its "noble rot" form which positively develops the berry aroma potential (Thibon et al., 2010, 2011). Natural antagonists of *B. cinerea* have been searched for within the leaf and berry microbiome to develop microbial biocontrol. They include *Aureobasidium pullulans*, which acts against *B. cinerea* through nutritive and spatial competition and antibiosis (Carmichael et al., 2019; Martini et al., 2009, p. 200; Rathnayake et al., 2018). They also include some bacteria (*Bacillus* spp., Actinomycetes) and yeasts (*Metchnikovia* spp., *Pichia* spp.) (Loqman et al., 2009; Raspor et al., 2010; Santos and Marquina, 2004). Mycoviruses infecting *B. cinerea* (Pearson and Bailey, 2013; Ruiz-Padilla et al., 2021) have been searched for to foster biocontrol. Although a rich virome was discovered, only a few mycoviruses were associated with a reduced virulence of the fungal pathogen (i.e., an hypovirulence phenotype) (Khalifa and MacDiarmid, 2021; Wu et al., 2010; Yu et al., 2015). These mycoviruses belonged to several families, including *Narnaviridae* (single-stranded RNA virus) and *Genomoviridae* (single-stranded DNA virus). An unclassified double-stranded RNA virus was also found. Finally, fungal pathogens producing mycotoxins (*Aspergillus carbonarius*, *Aspergillus niger*) can also develop in mature berries. Mycotoxin contamination can be limited by specific practices (Hocking et al., 2007; Perera et al., 2021), including biocontrol by *A. pullulans* (Bozoudi and Tsaltas, 2018).

Grapevine berries constitute a pivotal microbial compartment because the dynamics of their microbiome depends on the agricultural practices and the climatic conditions in the vineyard, and determines the quality of the end product. Future research should therefore continue investigating the seasonal dynamics of microbial interactions and functions on the grapevine berry surface (Section 4.3), to elucidate their impact on berry quality and wine properties. Significant advances using metabolic networks (Section 4.3) were made made for *Oenococcus oeni* and should be extended the the microbial community level, to understand the interactions between *O. oeni* and other microorganisms (like that *Saccharomyces cerevisiae*, LAB and AAB) that belong to the berry microbiome and are involved in winemaking (Mendoza et al., 2017). Conversely, the impact of the winemaking process on the berry microbiome dynamics could be assessed. Indeed, during winemaking, fermentative microorganisms reach very high population levels in cellars and can be disseminated in the vineyard through CO_2-extraction systems and insect vectors.

3.5 Wood microbiome dysbiosis during grapevine aging and decline

Cultivated grapevines are perennial plants that age and are eventually removed when they are no longer productive. Nowadays, a major cause of grapevine decline (Fig. 3) are Grapevine Trunk Diseases (GTDs), a group of diseases which re-emerged in the late 1990s and cause wood degradation and necrosis within the inner structures of mature grapevines (Bertsch et al., 2013; Bruez et al., 2013; Mugnai et al., 1999). GTDs include *Botryosphaeria* dieback, *Diaporthe* dieback, *Eutypa* dieback, *Phomopsis* dieback, Petri disease, black-foot disease and Esca complex diseases. Pathogenic fungi associated with GTDs belong to several genera, including *Phaeomoniella, Phaeoacremonium, Fomitiporia, Eutypa*, and to the *Botryosphaeriaceae* family (Cobos et al., 2022; Fig. 3). However, these fungi have been isolated from both healthy and necrotic wood tissues of both GTD-symptomatic and asymptomatic grapevines. Their role in GTDs has therefore been questioned

(Hofstetter et al., 2012) and other microorganisms potentially involved in GTDs have been isolated using culture-dependent methods (Elena et al., 2018; Kraus et al., 2019). Cobos et al. (2022) reported that in 2018, 133 fungal species belonging to 34 genera (mainly ascomycetes and basidiomycetes) had been associated with GTDs worldwide, this number increasing yearly. Bacteria inhabiting grapevine wood have also been isolated, identified and their functions studied. For instance, Bruez et al. (2015) identified 26 bacterial genera inhabiting the various wood tissues, the most abundant being *Bacillus* (34% of the bacterial strains), then *Pantoea* (12%), *Paenibacillus* (9%), and *Enterobacter* (6%). Bacterial communities differed between anatomical parts (i.e., trunk or cordon) and tissue types (i.e., necrotic or non-necrotic). Functions of bacteria colonising necrotic and non-necrotic tissues differed, as they metabolised differently the carbon substrates. Over the last decade, next-generation sequencing (Del Frari et al., 2019; Eichmeier et al., 2018; Gramaje et al., 2022; Lade et al., 2022) and metatranscriptomic analyses (Nerva et al., 2022; Paolinelli et al., 2022) have provided a deeper resolution in the grapevine wood fungal and bacterial microbiomes. Depending on the grapevine organs, years, seasons and wood health status (i.e., healthy or necrotic), variations in the wood microbiome composition were observed, for both fungi and bacteria (Bruez et al., 2014, 2020).

Wood necrosis is often observed in mature grapevines (15–25-year-old), the difference between healthy and GTDs-plants being the quantity of wood necrosis. White-rot necrotic wood, in particular, is more abundant in diseased plants (Maher et al., 2012; Ouadi et al., 2019, 2021). This observation raised a key question: why does a substantial percentage of wood become necrotic in mature plants? This question has recently been investigated in the case of Esca, the most frequent GTD. Microbiome changes occurring at the onset of Esca were investigated using metabarcoding approaches (Bruez et al., 2020). The results showed a decrease in the diversity of fungal communities in the white-rot of diseased trunks, linked with the tissue colonisation by the fungal pathogen *Fomitiporia mediterranea*. Because fungi and bacteria coexist in the wood, the authors hypothesised that fungal-bacteria associations were involved in the wood degradation. In agreement with this hypothesis, associations between GTD-fungal pathogens (*F. mediterranea* and *Phaeomoniella chlamydospora*) and bacterial species (*Sphingomonas* spp. and *Mycobacterium* spp.) were detected in the cordons of young grapevines expressing Esca-foliar symptoms. More recently, Haidar et al. (2021) showed that *F. mediterranea* and a novel species of bacteria isolated from grapevine wood (Haidar et al., submitted) acted synergistically to enhance wood structure degradations associated with Esca.

Plant-microbe and microbe-microbe interactions responsible for other GTDs (*Botryosphaeria* dieback, *Diaporthe* dieback) have also been studied. Several fungal pathogens of the genera *Botryosphaeria*, *Diplodia*, *Lasiodiplodia* and *Neofusicoccum* have been identified as causal agents of *Botryosphaeria* dieback. This complex of xylem-inhabiting fungi can remain latent in the roots before producing phytotoxic compounds and degrading enzymes (xylanases, cellulases, β-glucanases) in stems and symptomatic leaves (del Martínez-Diz et al., 2019; Gramaje et al., 2018; Marais et al., 2021). Fungi of the genus *Diaporthe* are responsible for *Diaporthe* dieback, a disease that causes, among others, cane bleaching, swelling arms, trunk cankers, shoots breaking off at the base, loss of vigour, and reduced bunch size (del Martínez-Diz et al., 2019; Guarnaccia et al., 2018). These symptoms can allow the implantation of opportunistic microorganisms such as *Didymella negriana*, which is capable of producing mycotoxins (Liu and Howell, 2021; Stranska et al., 2022).

Wood bacteria, fungi and mycoviruses with potential biocontrol activities against fungal pathogens associated GTDs have been investigated (Haidar et al., 2021; Marais et al., 2021; Mondello et al., 2018; Rezgui et al., 2016). Bekris et al. (2021) analysed the wood microbiome of asymptomatic and symptomatic plants in three Greek grapevine varieties, using differential abundance analysis (Section 4.1) and network inference methods (Section 4.3). Their results showed that *Bacillus* and *Streptomyces* genera dominated the bacterial microbiome of asymptomatic plants and had a negative co-occurrence pattern with GTD-pathogens belonging to the *Phaeomoniella*, *Phaeoacremonium* and *Seimatosporium* genera. The authors concluded that these bacterial taxa may play a role in the suppression of GTDs. The suppressive role of the wood microbiome was also suggested by Bruez et al. (2016), as they noticed that pathogenic and non-pathogenic fungi reach an equilibrium in the functional tissues of old grapevines. Mycoviruses may also play a role in the suppression of GTDs. For instance, mycoviruses were detected in *Neofusicoccum parvum* isolates from grapevines without symptoms of Esca (Nerva et al., 2019) and in hypovirulent isolates (Marais et al., 2021), suggesting that mycoviruses may reduce the aggressiveness of this *Botryosphaeriaceae* species. It is noteworthy that viruses could also be helpful to fight pathogenic bacteria that colonise xylem vessels, like *Xylella fastidiosa* (Fig. 3), the causal agent of Pierce's Disease, and

Xylophilus ampelinus, the causal agent of bacterial blight. Biocontrol protocols using virulent phages have been successfully implemented *in planta* in Texas, USA, to control *Xylella fastidiosa* (Das et al., 2015). This approach is promising, especially since the pathogen is vectored by an insect (the glassy-winged sharpshooter, *Homalodisca vitripennis*) (Bhowmick et al., 2016) which has the capacity to uptake *X. fastidiosa* phages (Clavijo-Coppens et al., 2021).

Deciphering plant-microbe-virus interactions occurring in the grapevine wood is crucial for future monitoring and management of GTDs. Specically, tipping points in the wood microbiome dynamics related with the onset of disease symptoms could inform disease-monitoring programmes (Section 4.3). For future research, we would recommend analysing the taxonomy and functions of the wood microbiome, as well as the changes in microbial interactions networks (Section 4.3), just before the appearance of the first Esca-foliar symptoms to answer the following questions: (i) is the onset of the disease marked by an increase in the heterogeneity of the microbiome between grapevines, as stated in the "Anna Karenina principle" (Zaneveld et al., 2017)? (ii) is the onset of the disease marked by a collapse in microbial interaction networks? (iii) is there a change in the functional composition of the microbiome, marked by an increase in the abundance of pathogenic species?

Answering these questions about the wood microbiome and those raised earlier for the other microbial compartments (roots, leaves, and berries) will require the study of the complex microbial communities associated with cultivated grapevines, to identify the environmental drivers of plant-microbe-virus interactions. It will also require the analysis of simpler, synthetic microbial consortia in controlled conditions, to get mechanistic insights into microbial interactions. Both approaches will be linked. Studying natural communities will require dedicated field sampling designs to identify key taxons and functions associated with grapevine performance. These key taxa and functions will then help design culture-based experiments coupled with modelling approaches to leverage omics data and decipher microbial interactions at the genome scale. These models will be used to optimise the microbial consortia (Busby et al., 2017; Toju et al., 2018) that could ultimately be inoculated to foster grapevine performance under conditions of climate change and reduced pesticide use. Some of the tools to address these challenges and make the grapevine microbiome more functional are described in the next section (Section 4).

4. Let's make the grapevine microbiome more functional

This section proposes experimental and computational approaches that could help uncover the functioning of the grapevine microbiome, with the goal of integrating it into a more sustainable viticulture. We propose field sampling designs and statistical approaches that exploit what we know of the evolution and dynamics of the grapevine microbiome (Sections 2 and 3) and aim to identify microbial taxa or consortia that may play a role in vine performance and vineyard ecosystem health. We then review the molecular tools and computational approaches that work together to analyse the grapevine holobiont as a dynamic, interactive, and functional system. We emphasise throughout the critical need for culture-dependent approaches to validate the effects of specific microbial taxa and functions on vine performance and vineyard ecosystem functioning. Each subsection sets out research hypotheses or questions about microbiome functioning, describes methodological frameworks that exist or could be developed to address these questions, and finally suggests how to apply them to the grapevine system.

4.1 Field sampling designs and statistical approaches to identify beneficial microbial taxa

Grapevine hosts many hundreds of microbial taxa (Knapp et al., 2021; Morgan et al., 2017; Zarraonaindia et al., 2015 and Section 3) of which only a few dozen have known roles and modes of action (Fig. 3). Here we propose field sampling designs and statistical approaches that could be applied to metabarcoding datasets to elucidate the links between the presence or abundance of particular microbial taxa, and functional traits measured at the plant- or ecosystem-scale. In the current context of climate change and need to limit pesticide use (Jacquet et al., 2022), we ask which field sampling designs and statistical approaches are suitable for identifying microorganisms enhancing plant tolerance to drought or providing a barrier effect against microbial pathogens?

Question #1: How to identify microbial taxa enhancing plant tolerance to drought or providing a barrier effect against microbial pathogens from metabarcoding data?

To address this question, the composition of grapevine microbiome samples can be analysed using metabarcoding approaches and then statistically related to measurements of grapevine physiology or vineyard ecosystem functioning. For example, the microbiome can be related to ecophysiological traits that indicate grapevine tolerance to drought (Gambetta et al., 2020; Vacher et al., 2022), or to the frequency and intensity of diseases at the plot level in epidemiological surveys (Chen et al., 2019), or to more integrative agricultural variables such as yield potential. The grapevine microbiome can then be considered as an explanatory factor of the variation in the functional trait of interest (e.g., drought tolerance, disease resistance, yield). This variation can be analysed at different temporal and spatial scales (Fig. 4): within

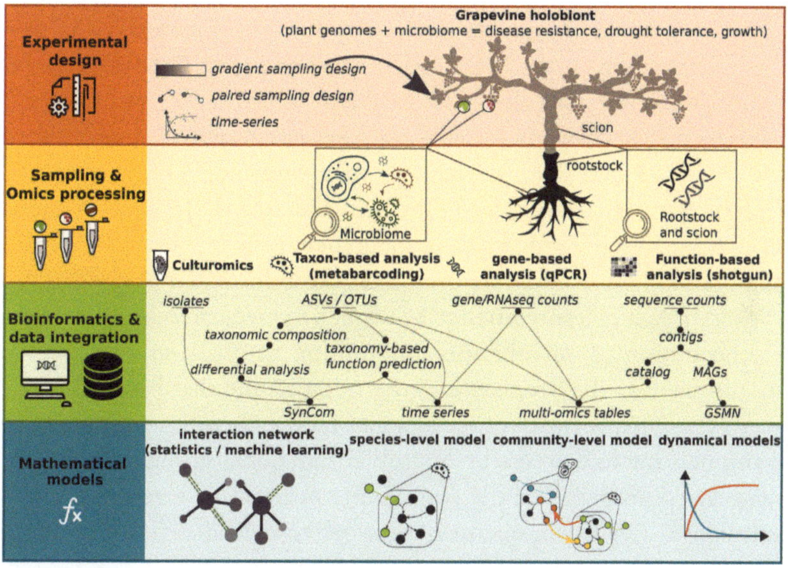

Fig. 4 Workflow, from field experiments to mathematical models of omics data, to decipher the complex plant-microbe-microbe interactions forming the grapevine holobiont and to identify beneficial microbial taxa. The experimental design sets up the sampling plots, hosts, compartments and time-points, according to spatial and temporal variations in a functional trait of interest (e.g., grapevine drought tolerance, disease resistance, yield). Grapevine samples (roots, trunk, leaves or berries) are then collected, cultivated, amplified and sequenced, and omics data are processed. Omics approaches can be culture-based (culturomics), taxon-based (metabarcoding), gene-based (microfluidic qPCR chips, for instance) or function-based (shotgun metagenomics). Bioinformatic analysis and data integration pipelines result in multi-omics tables, time series, SynComs or GSMN construction. This data can be used as input data for interaction network inference, species or community-level models and dynamical systems representing microbial populations' dynamics under various environmental conditions. ASV, Amplicon Sequence Variant; GSNM, Genome-Scale Metabolic Network; MAG, Metagenome-Assembled Genomes; OTU, Operational Taxonomic Unit; SynCom, Synthetic Community.

the lifetime of a cultivated vine individual (Fig. 3), between vines planted in the same plot (e.g., Darriaut et al., 2021), between vineyard plots, between cultivated vines and their wild ancestors (Section 2.1), or between cultivated vines and disease-resistant *Vitis* species (Section 2.2).

The field sampling design should determine the statistical analyses to be used. One sampling approach is to examine grapevine individuals or vineyard plots along a gradient of a previously measured functional trait of interest. Another would sample pairs of individuals or pairs of plots, of contrasted functional traits (e.g., drought-tolerant vs. drought-sensitive, disease-resistant vs. disease-susceptible, high- vs. low-yielding), other effects being otherwise controlled (same rootstock and scion cultivars, similar plant age, soil type and farming system). These two sampling strategies (Fig. 4) rely on spatial variation in the functional trait of interest. The first maximises the variation in the trait along the gradient, while the second reduces the environmental heterogeneity within a pair. Longitudinal, time-series surveys (Fig. 4) would complete this picture, and highlight joint temporal variations between the microbiome and the functional trait of interest. This combination is highly relevant for grapevine given the strong dynamics in the microbiome at multiple timescales (Sections 2 and 3).

Several statistical tools can then be applied to metabarcoding data to assess the contribution of microbiome composition to variation in a functional trait, and identify the properties of the microbiome that have explanatory power. These explanatory properties can be the occurrence or abundance of particular microbial taxa or microbial functional groups (Section 4.2), or aggregated properties at the community or network levels such as community alpha-diversity or microbial network connectance (Barroso-Bergadà et al., 2021). The compositional dissimilarity matrix between samples (community beta-diversity) can be converted into explanatory vectors using the method of PCNM eigenvectors, developed for spatial distance matrices (Borcard and Legendre, 2002). The statistical model will have the functional trait of interest as the dependent variable and the microbiome properties and environmental factors as explanatory variables (e.g., Asad et al., 2022 for yield; Cambon et al., 2022 for drought tolerance; Pérez-Valera et al., 2020 for an example of paired samples). When the microbiome contributes significantly to the functional trait of interest, identifying which microbial taxa are the main players can be done using the TITAN2 method (Baker et al., 2015) or differential abundance analysis methods (Nearing et al., 2022; Weiss et al., 2017). Machine learning or deep learning methods can also be applied (Xu et al., 2022); for example, a Random Forest algorithm has been used to identify combinations of microbial taxa indicative of yield (Yergeau et al., 2020). For grapevine, statistical identification of subsets of microbial taxa potentially involved in disease resistance or drought tolerance would be relevant to set up experimental, culture-dependant approaches (Sections 4.4). Finally, to test specific hypotheses on the direct and indirect effects of the explanatory factors (including microbiome) on the functional trait of interest, structural equation models (SEM) have been developed (Jassey et al., 2018). For grapevine, a descriptive SEM would be highly relevant for assessing the direct and indirect effects (through the microbiome) of variety and environmental factors (drought, agricultural practices) on disease severity and frequency.

4.2 Molecular tools to uncover the functional potential of the microbiome

Microbiome samples collected from contrasted environments following the field sampling designs described above (Section 4.1) can be further analysed using culture-independent molecular tools to uncover the microbial genomes present in the sample, analyse the microbial functions they carry, estimate precisely variation in their abundance and link this variation with plant gene expression patterns. Here, we review the molecular tools that could be used to address the following questions: How to characterise the functional potential of the grapevine microbiome with shotgun metagenomics; and, how to decipher the functional coupling between grapevine and its microbiome with quantitative approaches?

Question #2: How to characterise the functional potential of the grapevine microbiome with shotgun metagenomics?

Metataxonomics using metabarcoding of amplified marker DNA sequences has been widely used to inventory the microbial taxa present in a sample (Marchesi and Ravel, 2015), including in grapevine (Sections 2 and 3). Bioinformatic tools exploit this data to generate abundance tables of microbial Operational Taxonomic Units (OTUs) or Amplicon Sequence Variants (ASVs) in samples (Callahan et al., 2016, 2017; Estaki et al., 2020; Hildebrand et al., 2014; Pauvert et al., 2019; Schloss et al., 2009). Metabarcoding has strong limitations, however, including primer amplification bias, copy number variation of the target sequence across taxa, and limits to taxonomic resolution, which may prevent the accurate characterisation of taxa composition and abundance. Although primarily dedicated to taxonomic characterisation, metabarcoding data can nevertheless be used to get insight into the functional potential of a microbiome with taxonomy-based function prediction methods (Belcour et al., 2022; Douglas et al., 2020; Louca et al., 2016; Nguyen et al., 2016; Wemheuer et al., 2020; Zanne et al., 2020). These methods build upon the association of taxonomic clades to known functional guilds, or to metabolic functions or pathways obtained from public databases of annotated genomes (Section 4.3).

In contrast to metabarcoding approaches, shotgun metagenomics (Fig. 4) does not sequence DNA markers but the total DNA material of a sample. Significant data processing is required (Bickhart et al., 2022). Reference-based profiling (Section 4.3) compares these reads to reference genomes. In genome-resolved metagenomics, quality-filtered reads are assembled into contigs (Li et al., 2015; Nurk et al., 2017), which can be parsed into genes that are binned into metagenome-assembled genomes (MAGs) (Borderes et al., 2021; Kang et al., 2019; Wu et al., 2016; Yue et al., 2020). Genomes can be resolved for uncultured bacterial and eukaryotic members of the microbiome, and the precise function carried by a species in the sample can be investigated. The limitations of MAG reconstruction come about as a consequence of the diversity of species found in the same sample (Lemos et al., 2021; Nayfach et al., 2020). Improvements to long-read sequencing such as PacBio or Nanopore technologies lead to better metagenome assembly. They can be combined synergistically with low error rate short reads to accurately retrieve repetitive genome regions (Chen et al., 2021; Damme et al., 2021). Long reads using the most recent technologies can however suffice alone for reconstructing high-quality MAGs (Sereika et al., 2022). The principal bias in shotgun metagenomics is the taxonomic and functional incompleteness in reference databases that can add noise to the taxonomic assignment and the functional annotation for underrepresented species and functions. In addition to shotgun metagenomics, metabolomics and metatranscriptomics (Paolinelli et al., 2022) approaches identify metabolites or expressed genes, respectively, in the host and the microbiome.

While genome-resolved metagenomics applications to soil and plants are becoming more frequent (Wu et al., 2022a,b), few studies have yet applied it to grapevine. Metagenomics has been used to analyse the microbial composition of grapevine berries and to reconstruct 13 near-complete draft genomes (Salvetti et al., 2016). Azevedo-Silva et al. (2021) discussed the feasibility of applying shotgun metagenomics to grapevine leaf, rhizosphere and soil characterisation and proposed a protocol. For grapevine, we recommend using metataxonomics before shotgun metagenomics to first target the environmental samples worthy of deeper sequencing (Fig. 4). We argue that, while metagenomics has some biases, it has fewer than amplicon sequencing and the substantial amount of information it brings makes it worthy of consideration for deeper analyses of the grapevine microbiome.

Question #3: How to decipher the functional coupling between grapevine and its microbiome with quantitative approaches?

Metabarcoding and metagenomics data are by nature compositional, meaning that we measure the relative abundance of microbial taxa or functions within a sample rather than their absolute abundance. To produce absolute counts requires complementary, quantitative methodologies (Props et al., 2017). Real-time quantitative PCR (RT-qPCR) has developed into a powerful and accurate technique for measuring absolute abundance by assessing quantitative changes in taxonomic or phylogenetic marker genes across spatial and temporal scales (Heid et al., 1996), supplanting the classical colony counting method, which is hindered by the low proportion of cultivable microorganisms (Lou et al., 2018). Development of automated, high-throughput qPCR platforms (Fig. 4) have increased, 10-fold, the number of qPCR reactions that can be simultaneously performed on a single instrument. Microfluidic qPCR chips with the Biomark® HD system, for example, can perform 9216 qPCR reactions (96 samples amplified with 96 primer sets) in only 4h.

Different microfluidic qPCR chips have been used in pioneer studies of gene expression in grapevines. The NeoVigen chip (Dufour et al., 2016), developed to quantify defence gene expressions in grapevine, quantifies plant genes involved in pathogenesis-related protein synthesis, secondary metabolites synthesis, the indole pathway, the oxidoreduction system, the salicylic acid, ethylene, and jasmonic acid pathways, and cell wall reinforcement. The BioStim chip (Bodin et al., 2020) has extended the gene set by including, among others, genes involved in primary pathways and phytohormones pathways (auxin, cytokinin, gibberellin, abscisic acid). These chips have been used to decipher the mechanisms underlying the effects of biostimulants on grapevine resistance to multiple stresses. The workflow can be adapted to different types of microbial marker genes, whether taxonomic (Dreier et al., 2021; Kleyer et al., 2017) or functional (Crane et al., 2018). The technology could therefore provide simultaneous quantitative snapshots of the microbiome taxonomic and functional composition and of the host response (Noman et al., 2021; Zolti et al., 2020).

To decipher the functional coupling between the grapevine host and its microbiome, we recommend parallel development of microfluidic chips targeting plant genes involved in plant growth or stress resistance, and chips targeting specific microbial taxa and functions. Considerable progress has been made for the plants (Bodin et al., 2020; Dufour et al., 2016), but a chip has yet to be created for microbiomes. Several challenges must be overcome to choose the most relevant microbial targets, design primer sets, and validate the chip. Target selection must deal with the multitude of microbial taxa colonising the plant and the multitude of functions they carry (Section 3). Chip validation must use reference species and synthetic microbial communities (Section 4.4) before applying the chip to environmental samples. Once developed, this tool could link microorganisms to plant health biomarkers, for instance to identify beneficial (phyto-probiotic) microbial agents that foster plant health or stimulate plant immunity, or potential antagonists of vine pathogens that are candidates for next-generation biocontrol products.

4.3 Computational approaches to understand the grapevine holobiont as a functional and dynamic network

Computational methods exploit the wealth of -omics data produced by high-throughput methods (Sections 4.2) to infer networks of ecological interactions between microbial taxa, unravel the trophic and metabolic relationships that underlie them, and analyse how they drive microbiome dynamics at different time scales (Sections 2 and 3). The various types of–omics data (metataxonomics, shotgun metagenomics, metranscriptomics, metaproteomics, metabolomics) are complementary and can be used to reconstruct ecological networks, metabolic networks, protein-protein interaction networks and gene regulatory networks (Liu et al., 2021). Here we describe the tools to specifically answer the following questions: How to characterise ecological interactions between grapevine-associated microorganisms using metabarcoding data? How to decipher metabolic interactions within the grapevine microbiome using shotgun metagenomics? How to infer the microbial interactions that shape microbiome dynamics from time-series data?

Question #4: How to identify and characterise ecological interactions between grapevine-associated microorganisms using metabarcoding data?

Microbial ecologists have long used metabarcoding to learn ecological interactions between microbial taxa. They assume that since past or ongoing interactions between microbial taxa may impact their frequencies, patterns in relative shifts in frequency may provide information about the structure of the ecological interaction network (Cobo-Díaz et al., 2022; Jakuschkin et al., 2016; Kerdraon et al., 2019; Poudel et al., 2016; Vacher et al., 2016b). However, this process is not straightforward: metabarcoding data can be biased by amplification and sequencing artefacts, and are always compositional (Gloor et al., 2017; Section 4.2). One of the first tools for inferring interaction networks from relative sequence counts was SparCC (Friedman and Alm, 2012), which avoided the compositional effect by using log-transformation components to obtain correlation. Other tools like SPIEC-EASI (Kurtz et al., 2015) or CCLasso (Fang et al., 2015) refined correlation-based measures. PLN (Chiquet et al., 2019) enhances inference accuracy by exploiting sample covariance, while MPLasso (Lo and Marculescu, 2017) introduces the use of prior microbiological knowledge through data-mining of external sources of information. HMSC (Ovaskainen et al., 2017) also integrates covariates on samples and taxa and was recently used to infer microbial networks (Abrego et al., 2020; Fort et al., 2021). These tools are extremely robust to noise in experimental data and can be run rapidly on even quite large datasets. Most of them are available in R, which facilitates their use (Dohlman and Shen, 2019; Tikhonov et al., 2020). However, these tools define no clear link between the positive and negative interactions they find, and the ecological mechanisms that may generate them (Röttjers and Faust, 2018).

Explainable machine learning can relate ecological mechanisms to inferred interaction networks (Tamaddoni-Nezhad et al., 2021). Different interaction types, such as competition and mutualism, lead to different changes in species frequencies and patterns (Fig. 4), which can be learned (Derocles et al., 2018; Tshikantwa et al., 2018). Explainable machine learning transcribes these criteria into logical rules that directly infer ecological interactions, with little to no human intervention in the interpretation step (Barroso-Bergada et al., 2022). Hypotheses of ecological interactions between taxa pairs are generated through a process that simultaneously detects interactions and classifies them as competition, mutualism, amensalism or commensalism. This direct approach is particularly valuable for reconstructing microbial networks in previously unstudied ecosystems, where human knowledge for interpretation may effectively be lacking.

For grapevine, this approach was recently applied to microbial communities sampled from symptomatic and asymptomatic leaves of European cultivated grapevine during downy mildew epidemics. After a thorough assessment of the method using simulated datasets (Weiss et al., 2016), preliminary interaction networks were computed and showed potential antagonists of the causal agent *Plasmopara viticola*—some known, some new—constituting options to control mildew development. We expect that this approach will identify sub-networks of interest, centred on pathogen species and their direct antagonists or facilitators, whose subsequent modelling could lead to better understanding of pathogen-microbiome interaction dynamics.

Question #5: How to decipher metabolic interactions within the grapevine microbiome using shotgun metagenomics?

The microbial functions present in a given sample can be assessed from shotgun metagenomic surveys or from metataxonomic data after taxonomy-based function prediction (Section 4.2). An advantage of the former is functional profiling of communities, using assembly-independent reference-based approaches, or de-novo reference-free assembled sequences (Quince et al., 2017). Both approaches delineate microbial species and strains, needed for reconstructing microbial metabolic networks. These networks are then used in mathematical models (kinetic, constraint-based and discrete models) to predict metabolic reactions at the species- or community-level.

Functional profiling of microbial communities using assembly-independent reference-based approaches directly compares reads to reference databases (Beghini et al., 2021) to identify genes and thereby functions. Functions or pathways associated with samples are then compared or used in statistical learning in order to reveal higher level functional patterns between sample groups (Raguideau et al., 2016). Unfortunately, reads cannot be linked to their genome of origin, impeding the interpretation of these patterns in terms of metabolite-mediated trophic interactions between microbial taxa. Moreover, plant and grapevine-associated reference catalogues are less available than, for instance, the gut microbiome. Hence, genome-resolved metagenomics, below, seems to be a better option for grapevine.

Reference-free de novo assembly allows functional annotation of genes (Cantalapiedra et al., 2021; Seemann, 2014), functional characterisation of the corresponding organisms, and genome-scale metabolic network (GSMN) reconstructions (Fig. 4). GSMN consist of all biochemical and transport reactions predicted from an individual microbial genome (Gu et al., 2019) and can be reconstructed by different tools (Arkin et al., 2018; Karp et al., 2019; Seaver et al., 2021) or even directly connected to MAG reconstruction (Zorrilla et al., 2021). In most cases,

GSMN reconstruction involves manual curation to ensure the quality of the predictions they provide (Karp et al., 2018). GSMN can be associated with mathematical models to predict the metabolism of a species or a community of organisms in an environment. The main difficulty is to achieve high-quality GSMNs, since (i) MAGs are frequently incomplete, even worse so when under-studied microbiota are under-represented in the gene annotation databases; and (ii) GSMN reconstruction from the MAG can be imperfect. Manually curated GSMNs are best but the number of genomes in a microbiota and limited knowledge about the microorganisms argues in favour of automatisation (Belcour et al., 2020). Fortunately, the multiplication of shotgun sequencing studies of the grapevine microbiome will increase the size and quality of grapevine-related annotated gene catalogues and favour better MAGs-based GSMN automatic reconstruction

Kinetic modelling uses differential equations models of the kinetics of biochemical reactions throughout the metabolic network of an organism (Ross and McMeekin, 1994), permitting microbial phenotype prediction that accounts for enzyme and metabolite concentration dynamics. While it has been applied to model organisms such as *Escherichia coli*, it is not widely applied due to the need to obtain biochemical measurements to parameterize the models (Khodayari et al., 2014). An alternative are stoichiometric models, also known as constraint-based models. They translate a GSMN into a linear system of constraints for an optimisation problem used to predict metabolic fluxes across the GSMN under simplificatory assumptions such as steady state in the system, i.e., the non-accumulation of internal metabolites and their constant concentration over time. Flux Balance Analysis (FBA) or Flux Variability Analysis (FVA) compute the activity rates of metabolic reactions while optimising an objective function, typically the maximisation of the microbial population growth (Orth et al., 2010). FBA models can be coupled with differential equations through dynamic FBA (dFBA) (Mahadevan et al., 2002). These methods have been widely used and extended to the study of microbial communities (Biggs et al., 2015; Frioux et al., 2020) and to study the metabolism of oenological yeasts (Eder et al., 2020; Nidelet et al., 2016).

Discrete models are an alternative to constraint-based models to capture the metabolic potential of a GSMN, in a nutritional environment ideally characterised by metabolomics. Network expansion (Ebenhöh et al., 2004) computes the set of reachable metabolites from a given set of metabolic seeds and has been used as a building block for screening the metabolism of microbial communities (Belcour et al., 2020). Complexity is reduced by selecting minimal communities that satisfy a metabolic objective, which may take the host metabolism into account (Frioux et al., 2018). Hypotheses arising from these minimal communities can be tested using synthetic communities (SynComs; Marín et al., 2021) selected to provide a targeted function. These approaches can be coupled with dFBA: discrete models are used for systematic screening and selection of SynComs, then dFBA is used on the selected SynComs to decipher the dynamics of the community at a molecular level.

GSMN were used to study soil and rhizosphere microbiome (Jacoby et al., 2017), and to model winemaking process (Mendoza et al., 2017). However, to our knowledge, GSMNs of grapevine phyllosphere or berry microbiomes remain an open challenge, of great potential for improving functional characterisation of microbiomes. For grapevine, kinetic modelling would permit, for example, integration of multi-omics data to decipher and validate the dynamical interactions between a microbial pathogen and putative antagonists. Minimal communities identified by discrete modelling, for example, to compete against microbial grapevine pathogens, or to promote oenological yeast growth by maintaining a tailored nutritional niche, can be tested with SynComs (Section 4.4).

Question #6: How to infer microbial interactions shaping microbiome dynamics from time-series?

Four fundamental, intrinsically dynamical ecological processes drive microbial communities: evolutionary diversification, selection, dispersal, and ecological drift (Nemergut et al., 2013; Vellend, 2010). The contribution of these deterministic and stochastic processes, complementary to historical contingency and priority effects, can now be evaluated cost-effectively using -omics time-series data (Ning et al., 2019; Vass et al., 2020; Zhou and Ning, 2017). Longitudinal time-series data enable the study of the dynamical interactions that determine selection in the microbial ecosystem.

One increasingly common approach is the use of Generalised Lotka-Volterra models (GLV) to decipher the dynamic interactions shaping community dynamics (Faust et al., 2018). GLV models decompose microbial growth into an intrinsic growth rate that models population fitness and interspecies interactions modelling ecological processes such as competition or parasitism (Fig. 4). These interactions are fitted to reproduce temporal data, which permits an ecological interpretation of the community dynamics. Despite simplifying assumptions such as constant environmental constraints, GLV models are suitable for studying microbiome stability, microbial population persistence and microbial community robustness (Gonze et al., 2018). They model dynamics that correlation-based methods cannot (Carr et al., 2019). They have been used to study the dynamics of the gut microbiome during primary succession (Marino et al., 2014) and pathogenic infection (Stein et al., 2013), and to identify the assembly rules that determine the long-term composition of synthetic communities (Friedman et al., 2017). Computational tools for inferring GLV interactions have been proposed by Bucci et al. (2016) and Kuntal et al. (2019).

GLV models need specific experimental design to guarantee accuracy of microbial network inference (Remien et al., 2021). GLV models describe the evolution of absolute population, so model parametrization requires time-series of absolute counts (e.g., qPCR or flux cytometry, Section 4.2). Abundance proxies can be used, such as compositional metabarcoding data renormalized in each sample by the dosage of total DNA (Stein et al., 2013). The need for absolute population counts has been relaxed with an alternative GLV formulation for compositional data (Joseph et al., 2020). GLV models are better suited to analyse marked changes of the microbial community, for example, after strong perturbations such as pesticide treatment or colonisation by exogenous microbes such as pathogens (Remien et al., 2021), although Xiao et al. (2017) does address the steady-state microbiome.

GLV models have intrinsic limitations. First, growth and interspecies interactions are treated phenomenologically without regard to their underlying mechanisms, contrary to other models that supplement pairwise ecological interactions with metabolite-mediated interactions (Momeni et al., 2017). Second, microbial populations are treated as well-mixed without regard to spatial structures caused by nutritional niches (Prosser et al., 2007), which can impair GLV estimates (Armitage and Jones, 2019). Other population dynamics modelling frameworks avoid these drawbacks: GSMN, above, are particularly suitable for describing the metabolism of species subject to trophic interactions such as competition for resources, or cross-feeding in small consortia (Widder et al., 2016); spatially explicit microbial population dynamics models can take into account spatial heterogeneities (Labarthe et al., 2019). These models require, for parameterization, experimental measurements of reduced synthetic communities in controlled environments.

For grapevine, GLV models have not yet been used, but have been to model microbial dynamics in other plant species (Becerra-Lucio et al., 2021). GLV models could be coupled with machine-learning approaches to select pathogen-associated sub-networks before studying their dynamics. They could also be used to study the dynamics of minimal communities identified by discrete modelling.

4.4 Culture-dependent approaches to validate microbial interactions and functions

All statistical and computational approaches described above generate hypotheses about the microbiome and its dynamics that must be experimentally validated. Therefore, we ask, how to isolate and culture microorganisms from a microbiome to study their functions and their interactions?

Question #7: How to isolate and culture microorganisms from the grapevine microbiome to study their functions and their interactions?

Culturomics, a high-throughput combination of multiple culture conditions and rapid identification of microorganisms (Fig. 4), has revitalised culture-dependent methods. The landmark studies by Lagier et al. (2015, 2016, 2018) combined high-throughput cultivation with cost-effective rapid identification by Matrix-Assisted Laser Desorption/Ionisation Time Of Flight Mass Spectrometry (MALDI-TOF MS), and increased the number of microorganisms identified in the human gut microbiome by more than 50%. Large-scale isolation strategies have since been developed, including new culture media and advanced cultivation technologies. Specific protocols consisting of successive dilutions of a microbial population to the point of no growth, and technologies such as microdroplet and microfluidics, have been proposed to enhance isolation efficiency (Button et al., 1993; Kaminski et al., 2016; Zhang et al., 2021). For the plant microbiome, plant-based media that isolate microorganisms in environmental and nutritional conditions close to their natural habitats have been successfully developed to improve cultivability (Mourad et al., 2018; Sarhan et al., 2018, 2019).

The success of culturomics in improving the catalogue of isolated human bacteria and plant endophytic bacterial communities (Bilen et al., 2018; Riva et al., 2022) holds promise for the grapevine microbiome, but the task ahead is huge. Fewer than 3000 plant-associated bacterial genomes are available, out of 70,000 known bacterial genomes (Joint Genome Institute Integrated Microbial Genomes; https://img.jgi.doe.gov/; Trivedi et al., 2021). High-quality sequencing, assembly, and annotation of pure microbial strain genomes is needed to improve gene and protein databases and knowledge of the genome-based mechanisms underlying plant-microbe interactions (Levy et al., 2018; Sarhan et al., 2019; Section 4.3). Genome-resolved metagenomics produces uncurated draft genomes whose gaps, assembly errors and contamination from other genomes limit their usefulness (Chen et al., 2020); obtaining reference bacterial genome sequences (Trivedi et al., 2021; Section 4.3) will improve quality. Microbial culture collections make it possible to screen for microbial phenotypic traits that foster plant growth (growth promoting bacteria) and plant health (biocontrol agents) and they are imperative for validation of candidate genes. Cultivation and co-cultivation strategies are needed for experimental validation of predictive models of plant-microbiome and pathogen-microbiome interactions, and to provide deeper insight into the functional potential of beneficial plant-associated microbes (Nai and Meyer, 2018; Trivedi et al., 2020).

5. Conclusion and perspectives

The grapevine microbiome has developed over thousands of years, integrating taxa from Asia, Europe and America, under the pressures of human selection and pathogen invasions. There are many gaps in this history, but the numerous studies of the grapevine microbiome show that it contains beneficial microorganisms (plant growth-promoting bacteria, mycorrhizae, biocontrol agents, microorganisms of oenological interest) whose mode of action is sometimes well documented. These beneficial microorganisms belong to a dynamic microbiome that significantly varies throughout the plant lifetime and throughout the year. The initial microbiome of grafted plants is determined by the microbiomes of mother plants that produce rootstocks and scions. Those initial microbiomes may contain phytoviruses and fungal pathogens, but also beneficial microorganisms increasing tolerance to biotic and abiotic stressors. Some of the beneficial microorganisms (for instance, the arbuscular mycorrhizal fungi *Rhizophagus irregularis*) can be inoculated at the nursery stage to strenghten the microbiome. Once planted, roots actively release exudates that attract the rhizosphere microbiome from the soil reservoir. Beneficial microorganisms recruited by grapevine roots include plant-growth promoting bacteria (e.g., *Pseudomonas* spp., *Rhizobia* spp.) that make mineral nutrients (phosphorus, nitrogen and phosphate) available to the plant, arbuscular mycorrhizal fungi (e.g., *Glomus* spp.) that help the plant to cope with water stress, and microorganisms whose biocontrol activity (e.g., *Bacillus* spp., *Pythium oligandrum*) relies on various types of microbial interactions including antibiosis, hyperparasitism, spatial and nutritional competition and interference with pathogen signalling. Some beneficial microbes may enter the root and disseminate in the aerial parts, such as the bacterial endophyte *Burkholderia phytofirmans*. The circulating microorganisms, that may also be pathogens like the bacteria *Xylella fastidiosa,* can reach leaves and integrate their highly dynamic microbiome, driven by seasonal variations. Throughout the vegetative season, leaves are colonised by major grapevine pathogens, such as the causal agents of mildews (*Plasmopara viticola* and *Erysiphe necator*) which are both introduced pathogens of North-American origin. Leaves also harbour microorganisms with biocontrol activity (e.g., *Bacillus subtilis, Aureobasidium pullulans, Ampelomyces* spp.). In mature grapevines, wood degradation associated with complex sets of microorganisms (e.g., *Phaeomoniella* spp., *Fomitiporia* spp., *Eutypa lata, Botryosphaeriaceae* spp.) may occur and lead to plant death. The search for biocontrol agents among mycoviruses, bacteria and fungi present in the wood is an active field of research requiring the inference of microbial interaction networks. Finally, the berry microbiome varies markedly during the development of the berry and contains, at the ripening stage, some fermentative yeasts and bacteria (e.g., *Saccharomyces cerevisiae, Oenococcus oeni)* positively involved in winemaking, but also undesirable microorganisms such as fungi producing off-flavours (e.g., the fungal pathogen *Botrytis cinerea*) and acetic acid bacteria leading to wine defaults.

Omics and associated computational approaches can further improve our understanding of the functioning of the grapevine microbiome, to ultimately design beneficial microbial consortia that could be inoculated to foster grapevine performance (health, growth and berry quality) under conditions of climate change and reduced pesticide use. To reach this objective, we propose to identify beneficial microbial taxa by performing field sampling campaigns informed by grapevine evolutionary history and phenotypic measurements (disease resistance, drought tolerance). Statistical approaches could then be applied to the metabarcoding datasets to identify subsets of microbial taxa whose presence or abundance is significantly related to grapevine phenotypic variations. Shotgun metagenomics could then be used to uncover microbial genomes and functions in selected samples, while microfluidic qPCR chips could be used to accurately estimate the abundance of selected microbial taxa. These data would then feed mathematical models to decipher metabolic and ecological interactions in small microbial consortia and analyse how these interactions drive microbial community dynamics and variations in plant phenotypic traits. The high-throughput cultivation of environemental samples in parallel with sequencing will be crucial to isolate microbial strains of interest and experimentally validate hypotheses generated by mathematical models.

Based on current knowledge and available methods, we propose the following research avenues to make the grapevine microbiome more functional:

A. *Design microbiome studies informed by evolutionary history*
- Analyse the microbiome of wild grapevine and ancestral grapevine varieties to discover beneficial microorganisms that may have been lost during domestication and breeding
- Analyse the microbiome of American *Vitis* species to discover potential antagonists of North-American pathogens

B. *Consider grapevine as a dynamic system of interactions*
- Investigate how rootstock-scion interactions in the graft zone shape the microbiome of the whole plant
- Decipher the microbial interactions that regulate the dynamics of pathogen populations
- Assess the cascading effects of the soil microbiome on the root, trunk, leaf and berry microbiome, and their consequences on plant health
- Analyse the microbial fluxes that govern the seasonal dynamics of the berry microbiome, and their consequences on wine properties
- Identify and monitor the microbiome changes that precede the development of trunk diseases
- Identify the viruses (mycoviruses and phages, in particular) that shape the dynamics of the grapevine microbiome

C. *Combine metagenomics, culturomics and computational approaches*
- Use field experiments and metabarcoding data to highlight small consortia of microbial taxa related to variations in grapevine phenotypic traits
- Create and maintain microbial culture collections to perform controlled experiments with the selected microbial consortia
- Develop shotgun metagenomics and metabolic network modelling to understand the functionning of those microbial consortia
- Monitor selected microbial populations with quantitative approaches to model and predict their dynamics in response to environmental changes and management scenarios.

These research avenues are intertwined and not independent of each other: they require the creation of multidisciplinary research consortia with complementary expertise to define the scientific questions, experimental designs and analysis pipelines.

Acknowledgements

We are very grateful to Maureen Pellan for the design of Fig. 3. The authors acknowledge the support of the French National Research Agency (ANR) under the grants 20-PCPA-0010 (PPR VITAE) and ANR-17-CE32-0011 (NGB), and of FranceAgrimer and CNIV under the grant n°2018-52537 (Vitirhizobiome). LP received funding from *Fondation Bordeaux Université*.

References

Abdelfattah, A., Sanzani, S.M., Wisniewski, M., Berg, G., Cacciola, S.O., Schena, L., 2019. Revealing cues for fungal interplay in the plant–air interface in vineyards. Front. Plant Sci. 10, 922.

Abrego, N., Roslin, T., Huotari, T., Tack, A.J.M., Lindahl, B.D., Tikhonov, G., Somervuo, P., Schmidt, N.M., Ovaskainen, O., 2020. Accounting for environmental variation in co-occurrence modelling reveals the importance of positive interactions in root-associated fungal communities. Mol. Ecol. 29, 2736–2746.

Aleynova, O.A., Nityagovsky, N.N., Dubrovina, A.S., Kiselev, K.V., 2022. The biodiversity of grapevine bacterial endophytes of *Vitis amurensis* Rupr. Plan. Theory 11, 1128.

Alkowni, R., Zhang, Y.-P., Rowhani, A., Uyemoto, J.K., Minafra, A., 2011. Biological, molecular, and serological studies of a novel strain of grapevine leafroll-associated virus 2. Virus Genes 43, 102–110.

Anderson, K., Aryal, N.R., 2013. Which Winegrape Varieties are Grown Where? A Global Empirical Picture. University of Adelaide Press.

Arkin, A.P., Cottingham, R.W., Henry, C.S., Harris, N.L., Stevens, R.L., Maslov, S., Dehal, P., Ware, D., Perez, F., Canon, S., Sneddon, M.W., Henderson, M.L., Riehl, W.J., Murphy-Olson, D., Chan, S.Y., Kamimura, R.T., Kumari, S., Drake, M.M., Brettin, T.S., Glass, E.M., Chivian, D., Gunter, D., Weston, D.J., Allen, B.H., Baumohl, J., Best, A.A., Bowen, B., Brenner, S.E., Bun, C.C., Chandonia, J.-M., Chia, J.-M., Colasanti, R., Conrad, N., Davis, J.J., Davison, B.H., DeJongh, M., Devoid, S., Dietrich, E., Dubchak, I., Edirisinghe, J.N., Fang, G., Faria, J.P., Frybarger, P.M., Gerlach, W., Gerstein, M., Greiner, A., Gurtowski, J., Haun, H.L., He, F., Jain, R., Joachimiak, M.P., Keegan, K.P., Kondo, S., Kumar, V., Land, M.L., Meyer, F., Mills, M., Novichkov, P.S., Oh, T., Olsen, G.J., Olson, R., Parrello, B., Pasternak, S., Pearson, E., Poon, S.S., Price, G.A., Ramakrishnan, S., Ranjan, P., Ronald, P.C., Schatz, M.C., Seaver, S.M.D., Shukla, M., Sutormin, R.A., Syed, M.H., Thomason, J., Tintle, N.L., Wang, D., Xia, F., Yoo, H., Yoo, S., Yu, D., 2018. KBase: The United States department of energy systems biology knowledgebase. Nat. Biotechnol. 36, 566–569.

Armijo, G., Schlechter, R., Agurto, M., Muñoz, D., Nuñez, C., Arce-Johnson, P., 2016. Grapevine pathogenic microorganisms: understanding infection strategies and host response scenarios. Front. Plant Sci. 7, 382.

Armitage, D.W., Jones, S.E., 2019. How sample heterogeneity can obscure the signal of microbial interactions. ISME J. 13, 2639–2646.

Arnold, C., Schnitzler, A., Douard, A., Peter, R., Gillet, F., 2005. Is there a future for wild grapevine (*Vitis vinifera* subsp. *silvestris*) in the Rhine Valley? Biodivers. Conserv. 14, 1507–1523.

Arroyo-García, R., Ruiz-García, L., Bolling, L., Ocete, R., López, M.A., Arnold, C., Ergul, A., SöylemezoLu, G., Uzun, H.I., Cabello, F., Ibáñez, J., Aradhya, M.K., Atanassov, A., Atanassov, I., Balint, S., Cenis, J.L., Costantini, L., Gorislavets, S., Grando, M.S., Klein, B.Y., McGovern, P.E., Merdinoglu, D., Pejic, I., Pelsy, F., Primikirios, N., Risovannaya, V., Roubelakis-Angelakis, K.A., Snoussi, H., Sotiri, P., Tamhankar, S., This, P., Troshin, L., Malpica, J.M., Lefort, F., Martinez-Zapater, J.M., 2006. Multiple origins of cultivated grapevine (*Vitis vinifera* L. ssp. *sativa*) based on chloroplast DNA polymorphisms. Mol. Ecol. 15, 3707–3714.

Asad, N.I., Tremblay, J., Dozois, J., Mukula, E., L'Espérance, E., Constant, P., Yergeau, E., 2022. Predictive microbial-based modelling of wheat yields and grain baking quality across a 500 km transect in Québec. FEMS Microbiol. Ecol. 97, fiab160.

Asghari, S., Harighi, B., Mozafari, A.A., Esmaeel, Q., Ait Barka, E., 2019. Screening of endophytic bacteria isolated from domesticated and wild growing grapevines as potential biological control agents against crown gall disease. BioControl 64, 723–735.

Asghari, S., Harighi, B., Ashengroph, M., Clement, C., Aziz, A., Esmaeel, Q., Ait Barka, E., 2020. Induction of systemic resistance to *Agrobacterium tumefaciens* by endophytic bacteria in grapevine. Plant Pathol. 69, 827–837.

Azevedo-Silva, D., Rasmussen, J.A., Carneiro, M., Gilbert, M.T.P., Azevedo, H., 2021. Feasibility of applying shotgun metagenomic analyses to grapevine leaf, rhizosphere and soil microbiome characterisation. Aust. J. Grape Wine Res. 27, 519–526.

Aziz, A., Verhagen, B., Magnin-Robert, M., Couderchet, M., Clément, C., Jeandet, P., Trotel-Aziz, P., 2016. Effectiveness of beneficial bacteria to promote systemic resistance of grapevine to gray mold as related to phytoalexin production in vineyards. Plant and Soil 405, 141–153.

Aziz, A., Perazzolli, M., Gramaje, D., Zyprian, E.M., Cantu, D., 2020. Editorial: recent advances on grapevine-microbe interactions: from signal perception to resistance response. Front. Plant Sci. 11, 1164.

Bais, H.P., Weir, T.L., Perry, L.G., Gilroy, S., Vivanco, J.M., 2006. The role of root exudates in rhizosphere interactions with plants and other organisms. Annu. Rev. Plant Biol. 57, 233–266.

Baker, M.E., King, R.S., Kahle, D., 2015. TITAN2: Threshold Indicator Taxa Analysis. R Package Version 2.1.

Bakshi, S., Sztejnberg, A., Yarden, O., 2001. Isolation and characterization of a cold-tolerant strain of *Fusarium proliferatum*, a biocontrol agent of grape downy mildew. Phytopathology 91, 1062–1068.

Baldan, E., Nigris, S., Romualdi, C., D'Alessandro, S., Clocchiatti, A., Zottini, M., Stevanato, P., Squartini, A., Baldan, B., 2015. Beneficial bacteria isolated from grapevine inner tissues shape *Arabidopsis thaliana* roots. Plos one 10, e0140252.

Balestrini, R., Magurno, F., Walker, C., Lumini, E., Bianciotto, V., 2010. Cohorts of arbuscular mycorrhizal fungi (AMF) in *Vitis vinifera*, a typical Mediterranean fruit crop: cohorts of AM fungi in vineyards. Environ. Microbiol. Rep. 2, 594–604.

Bao, L., Sun, B., Wei, Y., Xu, N., Zhang, S., Gu, L., Bai, Z., 2022. Grape cultivar features differentiate the grape rhizosphere microbiota. Plants 11, 1111.

Barata, A., Malfeito-Ferreira, M., Loureiro, V., 2012. The microbial ecology of wine grape berries. Int. J. Food Microbiol. 153, 243–259.

Barroso-Bergadà, D., Pauvert, C., Vallance, J., Delière, L., Bohan, D.A., Buée, M., Vacher, C., 2021. Microbial networks inferred from environmental DNA data for biomonitoring ecosystem change: Strengths and pitfalls. Mol. Ecol. Resour. 21, 762–780.

Barroso-Bergada, D., Tamaddoni-Nezhad, A., Muggleton, S.H., Vacher, C., Galic, N., Bohan, D.A., 2022. Machine learning of microbial interactions using abductive ILP and hypothesis frequency/compression estimation. In: Katzouris, N., Artikis, A. (Eds.), Inductive Logic Programming. Lecture Notes in Computer Science, Springer International Publishing, Cham, pp. 26–40.

Becerra-Lucio, A.A., Labrín-Sotomayor, N.Y., Becerra-Lucio, P.A., Trujillo-Elisea, F.I., Chávez-Bárcenas, A.T., Machkour-M'Rabet, S., Peña-Ramírez, Y.J., 2021. Diversity and interactomics of bacterial communities associated with dominant trees during tropical forest recovery. Curr. Microbiol. 78, 3417–3429.

Beghini, F., McIver, L.J., Blanco-Míguez, A., Dubois, L., Asnicar, F., Maharjan, S., Mailyan, A., Manghi, P., Scholz, M., Thomas, A.M., Valles-Colomer, M., Weingart, G., Zhang, Y., Zolfo, M., Huttenhower, C., Franzosa, E.A., Segata, N., 2021. Integrating taxonomic, functional, and strain-level profiling of diverse microbial communities with bioBakery 3. Elife 10, e65088.

Behrens, F.H., Fischer, M., 2022. Evaluation of different phyllosphere sample types for parallel metabarcoding of fungi and oomycetes in *Vitis vinifera*. Phytobiomes J. 6, 207–213 (PBIOMES-11-21-0072-SC).

Bekris, F., Vasileiadis, S., Papadopoulou, E., Samaras, A., Testempasis, S., Gkizi, D., Tavlaki, G., Tzima, A., Paplomatas, E., Markakis, E., Karaoglanidis, G., Papadopoulou, K.K., Karpouzas, D.G., 2021. Grapevine wood microbiome analysis identifies key fungal pathogens and potential interactions with the bacterial community implicated in grapevine trunk disease appearance. Environ. Microbiome 16, 23.

Belcour, A., Frioux, C., Aite, M., Bretaudeau, A., Hildebrand, F., Siegel, A., 2020. Metage2Metabo, microbiota-scale metabolic complementarity for the identification of key species. Elife 9, e61968.

Belcour, A., Ruiz, B., Frioux, C., Blanquart, S., Siegel, A., 2022. EsMeCaTa: Estimating metabolic capabilities from taxonomic affiliations. bioRxiv. 2022.03.16.484574.

Belessi, C., Chalvantzi, I., Marmaras, I., Nisiotou, A., 2022. The effect of vine variety and vintage on wine yeast community structure of grapes and ferments. J. Appl. Microbiol. 132, 3672–3684.

Berg, G., Rybakova, D., Fischer, D., Cernava, T., Champomier Vergès, M.-C., Charles, T., Chen, X., Cocolin, L., Eversole, K., Corral, G.H., Kazou, M., Kinkel, L., Lange, L., Lima, N., Loy, A., Macklin, J.A., Maguin, E., Mauchline, T., McClure, R., Mitter, B., Ryan, M., Sarand, I., Smidt, H., Schelkle, B., Roume, H., Kiran, G.S., Selvin, J., de Souza, R.S.C., van Overbeek, L., Singh, B.K., Wagner, M., Walsh, A., Sessitsch, A., Schloter, M., 2020. Microbiome definition re-visited: old concepts and new challenges. Microbiome 8, 103.

Berlanas, C., Berbegal, M., Elena, G., Laidani, M., Cibriain, J.F., Sagües, A., Gramaje, D., 2019. The fungal and bacterial rhizosphere microbiome associated with grapevine rootstock genotypes in mature and young vineyards. Front. Microbiol. 10, 1142.

Bertsch, C., Ramírez-Suero, M., Magnin-Robert, M., Larignon, P., Chong, J., Abou-Mansour, E., Spagnolo, A., Clément, C., Fontaine, F., 2013. Grapevine trunk diseases: complex and still poorly understood: *Grapevine trunk diseases*. Plant Pathol. 62, 243–265.

Bettenfeld, P., Cadenai Canals, J., Jacquens, L., Fernandez, O., Fontaine, F., van Schaik, E., Courty, P.-E., Trouvelot, S., 2021. The microbiota of the grapevine holobiont: a key component of plant health. J. Adv. Res. 40, 1–15. S2090123221002617.

Bhowmick, T.S., Das, M., Heinz, K.M., Krauter, P.C., Gonzalez, C.F., 2016. Transmission of phage by glassy-winged sharpshooters, a vector of *Xylella fastidiosa*. Bacteriophage 6, e1218411.

Bickhart, D.M., Kolmogorov, M., Tseng, E., Portik, D.M., Korobeynikov, A., Tolstoganov, I., Uritskiy, G., Liachko, I., Sullivan, S.T., Shin, S.B., Zorea, A., Andreu, V.P., Panke-Buisse, K., Medema, M.H., Mizrahi, I., Pevzner, P.A., Smith, T.P.L., 2022. Generating lineage-resolved, complete metagenome-assembled genomes from complex microbial communities. Nat. Biotechnol. 40, 1–9.

Biget, M., 2021. La vigne comme holobionte: validité du concept, dynamique temporelle, déterminants environnementaux et application à la définition du terroir microbiologique. Doctoral dissertation. University of Rennes 1, 74–104.

Biggs, M.B., Medlock, G.L., Kolling, G.L., Papin, J.A., 2015. Metabolic network modeling of microbial communities. Wiley Interdiscip. Rev. Syst. Biol. Med. 7, 317–334.

Bilen, M., Dufour, J.-C., Lagier, J.-C., Cadoret, F., Daoud, Z., Dubourg, G., Raoult, D., 2018. The contribution of culturomics to the repertoire of isolated human bacterial and archaeal species. Microbiome 6, 94.

Bleach, C., Ridgway, H.J., Jaspers, M.V., Jones, E.E., 2021. The relative susceptibility of grapevine rootstocks to black foot disease is dependent on inoculum pressure. N. Z. Plant Prot. 74, 62–69.

Bodin, E., Bellée, A., Dufour, M.-C., André, O., Corio-Costet, M.-F., 2020. Grapevine stimulation: a multidisciplinary approach to investigate the effects of biostimulants and a plant defense stimulator. J. Agric. Food Chem. 68, 15085–15096.

Bois, B., Zito, S., Calonnec, A., 2017. Climate vs grapevine pests and diseases worldwide: the first results of a global survey. OENO One 51, 133–139.

Bokulich, N.A., Thorngate, J.H., Richardson, P.M., Mills, D.A., 2014. Microbial biogeography of wine grapes is conditioned by cultivar, vintage, and climate. Proc. Natl. Acad. Sci. U. S. A. 111, 139–148.

Bona, E., Massa, N., Novello, G., Boatti, L., Cesaro, P., Todeschini, V., Magnelli, V., Manfredi, M., Marengo, E., Mignone, F., Berta, G., Lingua, G., Gamalero, E., 2018. Metaproteomic characterization of *Vitis vinifera* rhizosphere. FEMS Microbiol. Ecol. 95, 1–15.

Borcard, D., Legendre, P., 2002. All-scale spatial analysis of ecological data by means of principal coordinates of neighbour matrices. Ecol. Model. 153, 51–68.

Borderes, M., Gasc, C., Prestat, E., Galvão Ferrarini, M., Vinga, S., Boucinha, L., Sagot, M.-F., 2021. A comprehensive evaluation of binning methods to recover human gut microbial species from a non-redundant reference gene catalog. NAR Genom. Bioinform. 3, lqab009.

Börlin, M., Claisse, O., Albertin, W., Salin, F., Legras, J.L., Masneuf-Pomarede, I., 2020. Quantifying the effect of human practices on *S. cerevisiae* vineyard metapopulation diversity. Sci. Rep. 10 (1), 16214.

Botstein, D., Fink, G.R., 2011. yeast: an experimental organism for 21st century biology. Genetics 189, 695–704.

Bowers, J., Boursiquot, J.-M., This, P., Chu, K., Johansson, H., Meredith, C., 1999. Historical genetics: the parentage of chardonnay, gamay, and other wine grapes of Northeastern France. Science 285, 1562–1565.

Bozoudi, D., Tsaltas, D., 2018. The multiple and versatile roles of *Aureobasidium pullulans* in the vitivinicultural sector. Fermentation 4, 85.

Brewer, M.T., Milgroom, M.G., 2010. Phylogeography and population structure of the grape powdery mildew fungus, *Erysiphe necator*, from diverse *Vitis* species. BMC Evol. Biol. 10, 268.

Bruez, E., Lecomte, P., Grosman, J., Doublet, B., Bertsch, C., Ugaglia, A., Teissedre, P.-L., Costa, J.-P.D., Guerin-Dubrana, L., Rey, P., 2013. Overview of grapevine trunk diseases in France in the 2000s. Phytopathol. Mediterr. 52, 262–275.

Bruez, E., Vallance, J., Gerbore, J., Lecomte, P., Da Costa, J.-P., Guerin-Dubrana, L., Rey, P., 2014. Analyses of the temporal dynamics of fungal communities colonizing the healthy wood tissues of esca leaf-symptomatic and asymptomatic vines. PLoS One 9, e95928.

Bruez, E., Haidar, R., Alou, M.T., Vallance, J., Bertsch, C., Mazet, F., Fermaud, M., Deschamps, A., Guerin-Dubrana, L., Compant, S., Rey, P., 2015. Bacteria in a wood fungal disease: characterization of bacterial communities in wood tissues of esca-foliar symptomatic and asymptomatic grapevines. Front. Microbiol. 6, 1137.

Bruez, E., Baumgartner, K., Bastien, S., Travadon, R., Guérin-Dubrana, L., Rey, P., 2016. Various fungal communities colonise the functional wood tissues of old grapevines externally free from grapevine trunk disease symptoms: fungal microflora of GTD-free old vines. Aust. J. Grape Wine Res. 22, 288–295.

Bruez, E., Vallance, J., Gautier, A., Laval, V., Compant, S., Maurer, W., Sessitsch, A., Lebrun, M., Rey, P., 2020. Major changes in grapevine wood microbiota are associated with the onset of esca, a devastating trunk disease. Environ. Microbiol. 22, 5189–5206.

Bruisson, S., Zufferey, M., L'Haridon, F., Trutmann, E., Anand, A., Dutartre, A., De Vrieze, M., Weisskopf, L., 2019. Endophytes and epiphytes from the grapevine leaf microbiome as potential biocontrol agents against phytopathogens. Front. Microbiol. 10, 2726.

Brum, M.C.P., Araújo, W.L., Maki, C.S., Azevedo, J.L., 2012. Endophytic fungi from *Vitis labrusca* L. ('Niagara Rosada') and its potential for the biological control of *Fusarium oxysporum*. Genet. Mol. Res. 11, 4187–4197.

Bucci, V., Tzen, B., Li, N., Simmons, M., Tanoue, T., Bogart, E., Deng, L., Yeliseyev, V., Delaney, M.L., Liu, Q., Olle, B., Stein, R.R., Honda, K., Bry, L., Gerber, G.K., 2016. MDSINE: microbial dynamical systems INference engine for microbiome time-series analyses. Genome Biol. 17, 121.

Burruano, S., 2000. The life-cycle of *Plasmopara viticola*, cause of downy mildew of vine. Mycologist 14, 179–182.

Burruano, S., Alfonzo, A., Lo Piccolo, S., Conigliaro, G., Mondello, V., Torta, L., Moretti, M., Assante, G., 2008. Interaction between *Acremonium byssoides* and *Plasmopara viticola* in *Vitis vinifera*. Phytopathol. Mediterr. 47, 122–131.

Burruano, S., Mondello, V., Conigliaro, G., 2016. Endophytic fungi in asymptomatic *Vitis vinifera* L. and their effects on *Plasmopara viticola*. In: Compant, S., Mathieu, F. (Eds.), Biocontrol of Major Grapevine Diseases: Leading Research. CABI, Wallingford, pp. 98–112.

Busby, P.E., Soman, C., Wagner, M.R., Friesen, M.L., Kremer, J., Bennett, A., Morsy, M., Eisen, J.A., Leach, J.E., Dangl, J.L., 2017. Research priorities for harnessing plant microbiomes in sustainable agriculture. PLoS Biol. 15, e2001793.

Button, D.K., Schut, F., Quang, P., Martin, R., Robertson, B.R., 1993. Viability and isolation of marine bacteria by dilution culture: theory, procedures, and initial results. Appl. Environ. Microbiol. 59, 881–891.

Cadle-Davidson, L., Chicoine, D.R., Consolie, N.H., 2011. Variation within and among *Vitis* spp. for foliar resistance to the powdery mildew pathogen *Erysiphe necator*. Plant Dis. 95, 202–211.

Callahan, B.J., McMurdie, P.J., Rosen, M.J., Han, A.W., Johnson, A.J.A., Holmes, S.P., 2016. DADA2: high-resolution sample inference from Illumina amplicon data. Nat. Methods 13, 581–583.

Callahan, B.J., McMurdie, P.J., Holmes, S.P., 2017. Exact sequence variants should replace operational taxonomic units in marker-gene data analysis. ISME J. 11, 2639–2643.

Calonnec, A., Cartolaro, P., Naulin, J.-M., Bailey, D., Langlais, M., 2008. A host-pathogen simulation model: powdery mildew of grapevine. Plant Pathol. 57, 493–508.

Calonnec, A., Jolivet, J., Vivin, P., Schnee, S., 2018. Pathogenicity traits correlate with the susceptible *Vitis vinifera* leaf physiology transition in the biotroph fungus *Erysiphe necator*: an adaptation to plant ontogenic resistance. Front. Plant Sci. 9, 1808.

Cambon, M.C., Cartry, D., Chancerel, E., Ziegler, C., Levionnois, S., Coste, S., Stahl, C., Delzon, S., Buée, M., Burban, B., Cazal, J., Fort, T., Goret, J.-Y., Heuret, P., Léger, P., Louisanna, E., Ritter, Y., Bonal, D., Roy, M., Schimann, H., Vacher, C., 2022. Drought tolerance traits in neotropical trees correlate with the composition of phyllosphere fungal communities. Phytobiomes J. (PBIOMES-04-22-0023-R).

Campisano, A., Antonielli, L., Pancher, M., Yousaf, S., Pindo, M., Pertot, I., 2014. Bacterial endophytic communities in the grapevine depend on pest management. PLoS One 9, e112763.

Campisano, A., Pancher, M., Puopolo, G., Puddu, A., Lòpez-Fernàndez, S., Biagini, B., Yousaf, S., Pertot, I., 2015. Diversity in endophyte populations reveals functional and taxonomic diversity between wild and domesticated grapevines. Am. J. Enol. Vitic. 66, 12–21.

Canfora, L., Vendramin, E., Felici, B., Tarricone, L., Florio, A., Benedetti, A., 2018. Vineyard microbiome variations during different fertilisation practices revealed by 16s rRNA gene sequencing. Appl. Soil Ecol. 125, 71–80.

Cantalapiedra, C.P., Hernández-Plaza, A., Letunic, I., Bork, P., Huerta-Cepas, J., 2021. eggNOG-mapper v2: functional annotation, orthology assignments, and domain prediction at the metagenomic scale. Mol. Biol. Evol. 38, msab293.

Carbone, M.J., Alaniz, S., Mondino, P., Gelabert, M., Eichmeier, A., Tekielska, D., Bujanda, R., Gramaje, D., 2021. Drought influences fungal community dynamics in the grapevine rhizosphere and root microbiome. J. Fungi 7, 686.

Carmichael, P.C., Siyoum, N., Chidamba, L., Korsten, L., 2019. Exploring the microbial communities associated with *Botrytis cinerea* during berry development in table grape with emphasis on potential biocontrol yeasts. Eur. J. Plant Pathol. 154, 919–930.

Carr, A., Diener, C., Baliga, N.S., Gibbons, S.M., 2019. Use and abuse of correlation analyses in microbial ecology. ISME J. 13, 2647–2655.

Carrere, C., Spilmont, A.-S., Loupit, G., Beutin, P., Stessels, C., Ollat, N., Cookson, S.-J., 2022. Evaluation of criteria to assist the selection of good quality grafted grapevines prior to their commercialisation. OENO One 56, 15–27.

Castañeda, L.E., Barbosa, O., 2017. Metagenomic analysis exploring taxonomic and functional diversity of soil microbial communities in Chilean vineyards and surrounding native forests. PeerJ 5, e3098.

Castañeda, L.E., Miura, T., Sánchez, R., Barbosa, O., 2018. Effects of agricultural management on phyllosphere fungal diversity in vineyards and the association with adjacent native forests. PeerJ 6, e5715.

Cavalieri, D., McGovern, P.E., Hartl, D.L., Mortimer, R., Polsinelli, M., 2003. Evidence for S. cerevisiae fermentation in ancient wine. J. Mol. Evol. 57, S226–S232.

Chandra, M., Mota, M., Silva, A.C., Malfeito-Ferreira, M., 2020. Forest oak woodlands and fruit tree soils are reservoirs of wine-related yeast species. Am. J. Enol. Vitic. 71, 191–197.

Chaparro, J.M., Badri, D.V., Vivanco, J.M., 2014. Rhizosphere microbiome assemblage is affected by plant development. ISME J. 8, 790–803.

Chaudhry, V., Runge, P., Sengupta, P., Doehlemann, G., Parker, J.E., Kemen, E., 2021. Shaping the leaf microbiota: plant–microbe–microbe interactions. J. Exp. Bot. 72, 36–56.

Chen, F., Guo, Y.B., Wang, J.H., Li, J.Y., Wang, H.M., 2007. Biological control of grape crown gall by *Rahnella aquatilis* HX2. Plant Dis. 91, 957–963.

Chen, M., Brun, F., Raynal, M., Makowski, D., 2019. Timing of grape downy mildew onset in bordeaux vineyards. Phytopathology 109, 787–795.

Chen, L.-X., Anantharaman, K., Shaiber, A., Eren, A.M., Banfield, J.F., 2020. Accurate and complete genomes from metagenomes. Genome Res. 30, 315–333.
Chen, Y.-H., Chiang, P.-W., Rogozin, D.Y., Degermendzhy, A.G., Chiu, H.-H., Tang, S.-L., 2021. Salvaging high-quality genomes of microbial species from a meromictic lake using a hybrid sequencing approach. Commun. Biol. 4, 996.
Chiapello, M., Rodríguez-Romero, J., Ayllón, M.A., Turina, M., 2020. Analysis of the virome associated to grapevine downy mildew lesions reveals new mycovirus lineages. Virus Evol. 6, veaa058.
Chiquet, J., Robin, S., Mariadassou, M., 2019. Variational inference for sparse network reconstruction from count data. In: Proceedings of the 36th International Conference on Machine Learning. Presented at the International Conference on Machine Learning. PMLR, pp. 1162–1171.
Chitwood, D.H., Klein, L.L., O'Hanlon, R., Chacko, S., Greg, M., Kitchen, C., Miller, A.J., Londo, J.P., 2016. Latent developmental and evolutionary shapes embedded within the grapevine leaf. New Phytol. 210, 343–355.
Claisse, O., Chaïb, A., Jaomanjaka, F., Philippe, C., Barchi, Y., Lucas, P.M., Le Marrec, C., 2021. Distribution of prophages in the *Oenococcus oeni* species. Microorganisms 9, 856.
Clavijo-Coppens, F., Ginet, N., Cesbron, S., Briand, M., Jacques, M.-A., Ansaldi, M., 2021. Novel virulent bacteriophages infecting mediterranean isolates of the plant pest *Xylella fastidiosa* and *Xanthomonas albilineans*. Viruses 13, 725.
Cobo-Díaz, J.F., Legrand, F., Le Floch, G., Picot, A., 2022. Influence of maize residues in shaping soil microbiota and *Fusarium* spp. communities. Microb. Ecol. 83, 702–713.
Cobos, R., Ibañez, A., Diez-Galán, A., Calvo-Peña, C., Ghoreshizadeh, S., Coque, J.J.R., 2022. The grapevine microbiome to the rescue: implications for the biocontrol of trunk diseases. Plan. Theory 11, 840.
Compant, S., Kaplan, H., Sessitsch, A., Nowak, J., Ait Barka, E., Clément, C., 2008. Endophytic colonization of *Vitis vinifera* L. by *Burkholderia phytofirmans* strain PsJN: from the rhizosphere to inflorescence tissues. FEMS Microbiol. Ecol. 63, 84–93.
Compant, S., Brader, G., Muzammil, S., Sessitsch, A., Lebrihi, A., Mathieu, F., 2013. Use of beneficial bacteria and their secondary metabolites to control grapevine pathogen diseases. BioControl 58, 435–455.
Compant, S., Samad, A., Faist, H., Sessitsch, A., 2019. A review on the plant microbiome: ecology, functions, and emerging trends in microbial application. J. Adv. Res. 19, 29–37.
Cookson, S.J., Clemente Moreno, M.J., Hevin, C., Nyamba Mendome, L.Z., Delrot, S., Trossat-Magnin, C., Ollat, N., 2013. Graft union formation in grapevine induces transcriptional changes related to cell wall modification, wounding, hormone signalling, and secondary metabolism. J. Exp. Bot. 64, 2997–3008.
Cordero-Bueso, G., Arroyo, T., Serrano, A., Tello, J., Aporta, I., Velez, M., Valero, E., 2011a. Influence of the farming system and vine variety on yeast communities associated with grape berries. Int. J. Food Microbiol. 145, 132–139.
Cordero-Bueso, G., Arroyo, T., Serrano, A., Valero, E., 2011b. Influence of different floor management strategies of the vineyard on the natural yeast population associated with grape berries. Int. J. Food Microbiol. 148, 23–29.
Cordero-Bueso, G., Arroyo, T., Valero, E., 2014. A long term field study of the effect of fungicides penconazole and sulfur on yeasts in the vineyard. Int. J. Food Microbiol. 189, 189–194.
Cordero-Bueso, G., Mangieri, N., Maghradze, D., Foschino, R., Valdetara, F., Cantoral, J.M., Vigentini, I., 2017. Wild grape-associated yeasts as promising biocontrol agents against *Vitis vinifera* fungal pathogens. Front. Microbiol. 8, 2025.
Cordero-Bueso, G., Vigentini, I., Foschino, R., Maghradze, D., Ruiz-Muñoz, M., Benitez-Trujillo, F., Cantoral, J.M., 2022. Culturable yeast diversity of grape berries from *Vitis vinifera* ssp. *sylvestris* (Gmelin) hegi. J. Fungi 8, 410.

Crane, S.L., van Dorst, J., Hose, G.C., King, C.K., Ferrari, B.C., 2018. Microfluidic qPCR enables high throughput quantification of microbial functional genes but requires strict curation of primers. Front. Environ. Sci. 6, 145.

Crisp, P., Wicks, T.J., Bruer, D., Scott, E.S., 2006. An evaluation of biological and abiotic controls for grapevine powdery mildew. 2. Vineyard trials. Aust. J. Grape Wine Res. 12, 203–211.

Cruz-Silva, A., Figueiredo, A., Sebastiana, M., 2021. First insights into the effect of mycorrhizae on the expression of pathogen effectors during the infection of grapevine with *Plasmopara viticola*. Sustainability 13, 1226.

Cureau, N., Threlfall, R., Marasini, D., Lavefve, L., Carbonero, F., 2021. Year, location, and variety impact on grape-associated mycobiota of arkansas-grown wine grapes for wine production. Microb. Ecol. 82, 845–858.

Dagostin, S., Schärer, H.-J., Pertot, I., Tamm, L., 2011. Are there alternatives to copper for controlling grapevine downy mildew in organic viticulture? Crop Prot. 30, 776–788.

Damme, R.V., Hölzer, M., Viehweger, A., Müller, B., Bongcam-Rudloff, E., Brandt, C., 2021. Metagenomics workflow for hybrid assembly, differential coverage binning, metatranscriptomics and pathway analysis (MUFFIN). PLoS Comput. Biol. 17, e1008716.

Darriaut, R., Martins, G., Dewasme, C., Mary, S., Darrieutort, G., Ballestra, P., Marguerit, E., Vivin, P., Ollat, N., Masneuf-Pomarède, I., Lauvergeat, V., 2021. Grapevine decline is associated with difference in soil microbial composition and activity. OENO One 55, 67–84.

Darriaut, R., Lailheugue, V., Masneuf-Pomarède, I., Marguerit, E., Martins, G., Compant, S., Ballestra, P., Upton, S., Ollat, N., Lauvergeat, V., 2022. Grapevine rootstock and soil microbiome interactions: keys for a resilient viticulture. Hortic. Res. 9, uhac019.

Das, M., Bhowmick, T.S., Ahern, S.J., Young, R., Gonzalez, C.F., 2015. Control of pierce's disease by phage. PLoS One 10, e0128902.

de Brito, L.M., Zanette, F., Biasi, L.A., Costa, M.D., dos Passos, J.F.M., 2019. Morphological identification of *Ilyonectria liriodendri* and its interaction with plant growth-promoting bacteria in grapevine rootstocks. Rev. Bras. Frutic. 41, e-140.

Del Frari, G., Gobbi, A., Aggerbeck, M.R., Oliveira, H., Hansen, L.H., Ferreira, R.B., 2019. Characterization of the wood mycobiome of vitis vinifera in a vineyard affected by esca. Spatial distribution of fungal communities and their putative relation with leaf symptoms. Front. Plant Sci. 10, 910.

del Martínez-Diz, M.P., Andrés-Sodupe, M., Bujanda, R., Díaz-Losada, E., Eichmeier, A., Gramaje, D., 2019. Soil-plant compartments affect fungal microbiome diversity and composition in grapevine. Fungal Ecol. 41, 234–244.

Derocles, S.A.P., Bohan, D.A., Dumbrell, A.J., Kitson, J.J.N., Massol, F., Pauvert, C., Plantegenest, M., Vacher, C., Evans, D.M., 2018. Biomonitoring for the 21st century: integrating next-generation sequencing into ecological network analysis. In: Advances in Ecological Research. Elsevier, pp. 1–62.

Deyett, E., Rolshausen, P.E., 2019. Temporal dynamics of the sap microbiome of grapevine under high pierce's disease pressure. Front. Plant Sci. 10, 1246.

Deyett, E., Rolshausen, P.E., 2020. Endophytic microbial assemblage in grapevine. FEMS Microbiol. Ecol. 96, fiaa053.

Di Vecchi-Staraz, M., Laucou, V., Bruno, G., Lacombe, T., Gerber, S., Bourse, T., Boselli, M., This, P., 2009. Low level of pollen-mediated gene flow from cultivated to wild grapevine: consequences for the evolution of the endangered subspecies *Vitis vinifera* L. subsp. *silvestris*. J. Hered. 100, 66–75.

Ding, Y., Wei, R., Wang, L., Yang, C., Li, H., Wang, H., 2021. Diversity and dynamics of microbial ecosystem on berry surface during the ripening of Ecolly (*Vitis vinifera* L.) grape in Wuhai, China. World J. Microbiol. Biotechnol. 37, 214.

Dohlman, A.B., Shen, X., 2019. Mapping the microbial interactome: statistical and experimental approaches for microbiome network inference. Exp. Biol. Med. 244, 445–458.

Douglas, G.M., Maffei, V.J., Zaneveld, J.R., Yurgel, S.N., Brown, J.R., Taylor, C.M., Huttenhower, C., Langille, M.G.I., 2020. PICRUSt2 for prediction of metagenome functions. Nat. Biotechnol. 38, 685–688.

Dreier, M., Berthoud, H., Shani, N., Wechsler, D., Junier, P., 2021. Development of a high-throughput microfluidic qPCR system for the quantitative determination of quality-relevant bacteria in cheese. Front. Microbiol. 11, 619166.

Dries, L., Bussotti, S., Pozzi, C., Kunz, R., Schnell, S., Löhnertz, O., Vortkamp, A., 2021. Rootstocks shape their microbiome—bacterial communities in the rhizosphere of different grapevine rootstocks. Microorganisms 9, 822.

Dufour, M.-C., Magnin, N., Dumas, B., Vergnes, S., Corio-Costet, M.-F., 2016. Highthroughput gene-expression quantification of grapevine defense responses in the field using microfluidic dynamic arrays. BMC Genomics 17, 957.

Ebenhöh, O., Handorf, T., Heinrich, R., 2004. Structural analysis of expanding metabolic networks. Genome Inform. Int. Conf. Genome Inform. 15, 35–45.

Eder, M., Nidelet, T., Sanchez, I., Camarasa, C., Legras, J.-L., Dequin, S., 2020. QTL mapping of modelled metabolic fluxes reveals gene variants impacting yeast central carbon metabolism. Sci. Rep. 10, 2162.

Eichmeier, A., Pečenka, J., Peňázová, E., Baránek, M., Català-García, S., León, M., Armengol, J., Gramaje, D., 2018. High-throughput amplicon sequencing-based analysis of active fungal communities inhabiting grapevine after hot-water treatments reveals unexpectedly high fungal diversity. Fungal Ecol. 36, 26–38.

Elena, G., Bruez, E., Rey, P., Luque, J., 2018. Microbiota of grapevine woody tissues with or without esca-foliar symptoms in northeast Spain. Phytopathol. Mediterr. 57, 425–438.

Esmaeel, Q., Jacquard, C., Sanchez, L., Clément, C., Ait Barka, E., 2020. The mode of action of plant associated Burkholderia against grey mould disease in grapevine revealed through traits and genomic analyses. Sci. Rep. 10, 19393.

Estaki, M., Jiang, L., Bokulich, N.A., McDonald, D., González, A., Kosciolek, T., Martino, C., Zhu, Q., Birmingham, A., Vázquez-Baeza, Y., Dillon, M.R., Bolyen, E., Caporaso, J.G., Knight, R., 2020. QIIME 2 enables comprehensive end-to-end analysis of diverse microbiome data and comparative studies with publicly available data. Curr. Protoc. Bioinformatics 70, 100.

Falk, S.P., 1995. Parasitism of *Uncinula necator* cleistothecia by the mycoparasite *Ampelomyces quisqualis*. Phytopathology 85, 794.

Falk, S., 1996. *Fusarium proliferatum* as a biocontrol agent against grape downy mildew. Phytopathology 86, 1010.

Fan, Y., Gao, L., Chang, P., Li, Z., 2020. Endophytic fungal community in grape is correlated to foliar age and domestication. Ann. Microbiol. 70, 30.

Fang, H., Huang, C., Zhao, H., Deng, M., 2015. CCLasso: correlation inference for compositional data through Lasso. Bioinformatics 31, 3172–3180.

Faust, K., Bauchinger, F., Laroche, B., de Buyl, S., Lahti, L., Washburne, A.D., Gonze, D., Widder, S., 2018. Signatures of ecological processes in microbial community time series. Microbiome 6, 120.

Fay, J.C., Benavides, J.A., 2005. Evidence for domesticated and wild populations of *Saccharomyces cerevisiae*. PLoS Genet. 1, e5.

Fay, J.C., McCullough, H.L., Sniegowski, P.D., Eisen, M.B., 2004. Population genetic variation in gene expression is associated with phenotypic variation in *Saccharomyces cerevisiae*. Genome Biol. 5, R26.

Feechan, A., Kocsis, M., Riaz, S., Zhang, W., Gadoury, D.M., Walker, M.A., Dry, I.B., Reisch, B., Cadle-Davidson, L., 2015. Strategies for *RUN1* deployment using *RUN2* and *REN2* to manage grapevine powdery mildew informed by studies of race specificity. Phytopathology 105, 1104–1113.

Felber, A.C., Orlandelli, R.C., Rhoden, S.A., Garcia, A., Costa, A.T., Azevedo, J.L., Pamphile, J.A., 2016. Bioprospecting foliar endophytic fungi of *Vitis labrusca* Linnaeus, Bordô and Concord cv. Ann. Microbiol. 66, 765–775.

Fleet, G.H., 1993. In: CRC Press (Ed.), Wine Microbiology and Biotechnology. Harwood Academic Publishers, Chur. Switzerland.

Fleet, G.H., 2003. Yeast interactions and wine flavour. Int. J. Food Microbiol. 86, 11–22.

Fontaine, M.C., Austerlitz, F., Giraud, T., Labbé, F., Papura, D., Richard-Cervera, S., Delmotte, F., 2013. Genetic signature of a range expansion and leap-frog event after the recent invasion of Europe by the grapevine downy mildew pathogen *Plasmopara viticola*. Mol. Ecol. 22, 2771–2786.

Fontaine, M.C., Labbé, F., Dussert, Y., Delière, L., Richart-Cervera, S., Giraud, T., Delmotte, F., 2021. Europe as a bridgehead in the worldwide invasion history of grapevine downy mildew, *Plasmopara viticola*. Curr. Biol. 31, 2155–2166.e4.

Fort, T., Robin, C., Capdevielle, X., Delière, L., Vacher, C., 2016. Foliar fungal communities strongly differ between habitat patches in a landscape mosaic. PeerJ 4, e2656.

Fort, T., Pauvert, C., Zanne, A.E., Ovaskainen, O., Caignard, T., Barret, M., Compant, S., Hampe, A., Delzon, S., Vacher, C., 2021. Maternal effects shape the seed mycobiome in *Quercus petraea*. New Phytol. 230, 1594–1608.

Friedman, J., Alm, E.J., 2012. Inferring correlation networks from genomic survey data. PLoS Comput. Biol. 8, e1002687.

Friedman, J., Higgins, L.M., Gore, J., 2017. Community structure follows simple assembly rules in microbial microcosms. Nat. Ecol. Evol. 1, 1–7.

Frioux, C., Fremy, E., Trottier, C., Siegel, A., 2018. Scalable and exhaustive screening of metabolic functions carried out by microbial consortia. Bioinformatics 34, i934–i943.

Frioux, C., Singh, D., Korcsmaros, T., Hildebrand, F., 2020. From bag-of-genes to bag-of-genomes: metabolic modelling of communities in the era of metagenome-assembled genomes. Comput. Struct. Biotechnol. J. 18, 1722–1734.

Gadoury, D.M., Cadle-Davidson, L., Wilcox, W.F., Dry, I.B., Seem, R.C., Milgroom, M.G., 2012. Grapevine powdery mildew (*Erysiphe necator*): a fascinating system for the study of the biology, ecology and epidemiology of an obligate biotroph: grapevine powdery mildew. Mol. Plant Pathol. 13, 1–16.

Gambetta, G.A., Herrera, J.C., Dayer, S., Feng, Q., Hochberg, U., Castellarin, S.D., 2020. The physiology of drought stress in grapevine: towards an integrative definition of drought tolerance. J. Exp. Bot. 71, 4658–4676.

Gan, H.M., Szegedi, E., Fersi, R., Chebil, S., Kovács, L., Kawaguchi, A., Hudson, A.O., Burr, T.J., Savka, M.A., 2019. Insight into the microbial co-occurrence and diversity of 73 grapevine (*Vitis vinifera*) crown galls collected across the northern hemisphere. Front. Microbiol. 10, 1896.

Gessler, C., Pertot, I., Perazzolli, M., 2011. *Plasmopara viticola*: a review of knowledge on downy mildew of grapevine and effective disease management. Phytopathol. Mediterr. 50, 3–44.

Ghule, S.B., Sawant, I.S., Sawant, S.D., Saha, S., Devarumath, R.M., 2019. Isolation and identification of three new mycoparasites of *Erysiphe necator* for biological control of grapevine powdery mildew. Australas. Plant Pathol. 48, 351–367.

Gloor, G.B., Macklaim, J.M., Pawlowsky-Glahn, V., Egozcue, J.J., 2017. Microbiome datasets are compositional: and this is not optional. Front. Microbiol. 8, 2224.

Gobbi, A., Kyrkou, I., Filippi, E., Ellegaard-Jensen, L., Hansen, L.H., 2020. Seasonal epiphytic microbial dynamics on grapevine leaves under biocontrol and copper fungicide treatments. Sci. Rep. 10, 681.

Gobbi, A., Acedo, A., Imam, N., Santini, R.G., Ortiz-Álvarez, R., Ellegaard-Jensen, L., Belda, I., Hansen, L.H., 2022. A global microbiome survey of vineyard soils highlights the microbial dimension of viticultural terroirs. Commun. Biol. 5, 241.

Gobbin, D., Rumbou, A., Linde, C.C., Gessler, C., 2006. Population genetic structure of *Plasmopara viticola* after 125 years of colonization in European vineyards: Population genetics of *P. viticola* in Europe. Mol. Plant Pathol. 7, 519–531.

González-García, S., Álvarez-Pérez, J.M., Sáenz de Miera, L.E., Cobos, R., Ibañez, A., Díez-Galán, A., Garzón-Jimeno, E., Coque, J.J.R., 2019. Developing tools for evaluating inoculation methods of biocontrol *Streptomyces* sp. strains into grapevine plants. PLoS One 14, e0211225.

Gonze, D., Coyte, K.Z., Lahti, L., Faust, K., 2018. Microbial communities as dynamical systems. Curr. Opin. Microbiol. 44, 41–49.

Gopal, M., Gupta, A., 2016. Microbiome selection could spur next-generation plant breeding strategies. Front. Microbiol. 7, 1971.

Gramaje, D., Armengol, J., 2011. Fungal trunk pathogens in the grapevine propagation process: potential inoculum sources, detection, identification, and management strategies. Plant Dis. 95, 1040–1055.

Gramaje, D., Úrbez-Torres, J.R., Sosnowski, M.R., 2018. Managing grapevine trunk diseases with respect to etiology and epidemiology: current strategies and future prospects. Plant Dis. 102, 12–39.

Gramaje, D., Eichmeier, A., Spetik, M., Carbone, M.J., Bujanda, R., Vallance, J., Rey, P., 2022. Exploring the temporal dynamics of the fungal microbiome in rootstocks, the lesser-known half of the grapevine crop. J. Fungi 8, 421.

Grassi, F., De Lorenzis, G., 2021. Back to the origins: background and perspectives of grapevine domestication. Int. J. Mol. Sci. 22, 4518.

Griggs, R.G., Steenwerth, K.L., Mills, D.A., Cantu, D., Bokulich, N.A., 2021. Sources and assembly of microbial communities in vineyards as a functional component of winegrowing. Front. Microbiol. 12, 673810.

Grube, M., Schmid, F., Berg, G., 2011. Black fungi and associated bacterial communities in the phyllosphere of grapevine. Fungal Biol. 115, 978–986.

Gu, C., Kim, G.B., Kim, W.J., Kim, H.U., Lee, S.Y., 2019. Current status and applications of genome-scale metabolic models. Genome Biol. 20, 121.

Guarnaccia, V., Groenewald, J.Z., Woodhall, J., Armengol, J., Cinelli, T., Eichmeier, A., Ezra, D., Fontaine, F., Gramaje, D., Gutierrez-Aguirregabiria, A., Kaliterna, J., Kiss, L., Larignon, P., Luque, J., Mugnai, L., Naor, V., Raposo, R., Sándor, E., Váczy, K.Z., Crous, P.W., 2018. *Diaporthe* diversity and pathogenicity revealed from a broad survey of grapevine diseases in Europe. Persoonia - Mol. Phylogeny Evol. Fungi 40, 135–153.

Gur, L., Reuveni, M., Cohen, Y., Cadle-Davidson, L., Kisselstein, B., Ovadia, S., Frenkel, O., 2021. Population structure of *Erysiphe necator* on domesticated and wild vines in the Middle East raises questions on the origin of the grapevine powdery mildew pathogen. Environ. Microbiol. 23, 6019–6037.

Gutierrez, A., Grillo, M.A., 2022. Effects of domestication on plant-microbiome interactions. Plant Cell Physiol., 1–13. online.

Habbadi, K., Duplay, Q., Chapulliot, D., Kerzaon, I., Benkirane, R., Benbouazza, A., Wisniewski-Dyé, F., Lavire, C., Achbani, E.H., Vial, L., 2020. Characterization and phylogenetic diversity of *Allorhizobium vitis* isolated from grapevine in Morocco. J. Appl. Microbiol. 128, 828–839.

Hacquard, S., Spaepen, S., Garrido-Oter, R., Schulze-Lefert, P., 2017. Interplay between innate immunity and the plant microbiota. Annu. Rev. Phytopathol. 55, 565–589.
Haidar, R., Yacoub, A., Vallance, J., Compant, S., Antonielli, L., Saad, A., Habenstein, B., Kauffmann, B., Grélard, A., Loquet, A., Attard, E., Guyoneaud, R., Rey, P., 2021. Bacteria associated with wood tissues of Esca-diseased grapevines: functional diversity and synergy with *Fomitiporia mediterranea* to degrade wood components. Environ. Microbiol. 23, 6104–6121.
Haidar R., Compant S., Antonielli L., Pandin C., Yacoub A., Chrysovergi M., Attard E., Rey P. and Guyoneaud R. Paenibacillus xylinteritus, a novel bacterial species isolated from the wood of the grapevine cultivar, Sauvignon Blanc. Int. J. Syst. Evol. Microbiol. (submitted).
Hao, Z., van Tuinen, D., Fayolle, L., Chatagnier, O., Li, X., Chen, B., Gianinazzi, S., Gianinazzi-Pearson, V., 2018. Arbuscular mycorrhiza affects grapevine fanleaf virus transmission by the nematode vector *Xiphinema index*. Appl. Soil Ecol. 129, 107–111.
Heid, C.A., Stevens, J., Livak, K.J., Williams, P.M., 1996. Real time quantitative PCR. Genome Res. 6, 986–994.
Héloir, M.-C., Adrian, M., Brulé, D., Claverie, J., Cordelier, S., Daire, X., Dorey, S., Gauthier, A., Lemaître-Guillier, C., Negrel, J., Trdá, L., Trouvelot, S., Vandelle, E., Poinssot, B., 2019. Recognition of elicitors in grapevine: from MAMP and DAMP perception to induced resistance. Front. Plant Sci. 10, 1117.
Hernandez, M.M., Menéndez, C.M., 2019. Influence of seasonality and management practices on diversity and composition of fungal communities in vineyard soils. Appl. Soil Ecol. 135, 113–119.
Hildebrand, F., Tadeo, R., Voigt, A.Y., Bork, P., Raes, J., 2014. LotuS: an efficient and user-friendly OTU processing pipeline. Microbiome 2, 30.
Hocking, A.D., Leong, S.L., Kazi, B.A., Emmett, R.W., Scott, E.S., 2007. Fungi and mycotoxins in vineyards and grape products. Int. J. Food Microbiol. 119, 84–88.
Hofstetter, V., Buyck, B., Croll, D., Viret, O., Couloux, A., Gindro, K., 2012. What if esca disease of grapevine were not a fungal disease? Fungal Divers. 54, 51–67.
Jaarsveld, W.J., Halleen, F., Bester, M.C., Pierron, R.J., Stempien, E., Mostert, L., 2021. Investigation of *Trichoderma* species colonization of nursery grapevines for improved management of black foot disease. Pest Manag. Sci. 77, 397–405.
Jacoby, R., Peukert, M., Succurro, A., Koprivova, A., Kopriva, S., 2017. The role of soil microorganisms in plant mineral nutrition—current knowledge and future directions. Front. Plant Sci. 8, 1617.
Jacquet, F., Jeuffroy, M.-H., Jouan, J., Le Cadre, E., Litrico, I., Malausa, T., Reboud, X., Huyghe, C., 2022. Pesticide-free agriculture as a new paradigm for research. Agron. Sustain. Dev. 42, 8.
Jakuschkin, B., Fievet, V., Schwaller, L., Fort, T., Robin, C., Vacher, C., 2016. Deciphering the pathobiome: intra- and interkingdom interactions involving the pathogen *Erysiphe alphitoides*. Microb. Ecol. 72, 870–880.
Jara, C., Laurie, V.F., Mas, A., Romero, J., 2016. Microbial terroir in chilean valleys: diversity of non-conventional yeast. Front. Microbiol. 7, 663.
Jassey, V.E.J., Reczuga, M.K., Zielińska, M., Słowińska, S., Robroek, B.J.M., Mariotte, P., Seppey, C.V.W., Lara, E., Barabach, J., Słowiński, M., Bragazza, L., Chojnicki, B.H., Lamentowicz, M., Mitchell, E.A.D., Buttler, A., 2018. Tipping point in plant–fungal interactions under severe drought causes abrupt rise in peatland ecosystem respiration. Glob. Chang. Biol. 24, 972–986.
Ji, W., Han, K., Cai, Y., Mu, Y., Zhao, L., Zhang, M., Hou, C., Gao, M., Zhao, Q., 2019. Characterization of rhizosphere bacterial community and berry quality of Hutai No.8 (*Vitis vinifera* L.) with different ages, and their relations. J. Sci. Food Agric. 99, 4532–4539.

Joseph, T.A., Shenhav, L., Xavier, J.B., Halperin, E., Peer, I., 2020. Compositional Lotka-Volterra describes microbial dynamics in the simplex. PLoS Comput. Biol. 16, e1007917.

Jürges, G., Kassemeyer, H.-H., Dürrenberger, M., Düggelin, M., Nick, P., 2009. The mode of interaction between *Vitis* and *Plasmopara viticola* Berk. & Curt. Ex de Bary depends on the host species. Plant Biol. 11, 886–898.

Kaminski, T.S., Scheler, O., Garstecki, P., 2016. Droplet microfluidics for microbiology: techniques, applications and challenges. Lab Chip 16, 2168–2187.

Kang, D.D., Li, F., Kirton, E., Thomas, A., Egan, R., An, H., Wang, Z., 2019. MetaBAT 2: an adaptive binning algorithm for robust and efficient genome reconstruction from metagenome assemblies. PeerJ 7, e7359.

Karp, P.D., Weaver, D., Latendresse, M., 2018. How accurate is automated gap filling of metabolic models? BMC Syst. Biol. 12, 73.

Karp, P.D., Midford, P.E., Billington, R., Kothari, A., Krummenacker, M., Latendresse, M., Ong, W.K., Subhraveti, P., Caspi, R., Fulcher, C., Keseler, I.M., Paley, S.M., 2019. Pathway Tools version 23.0 update: software for pathway/genome informatics and systems biology. Brief. Bioinform. 22, 109–126.

Kerdraon, L., Barret, M., Laval, V., Suffert, F., 2019. Differential dynamics of microbial community networks help identify microorganisms interacting with residue-borne pathogens: the case of *Zymoseptoria tritici* in wheat. Microbiome 7, 125.

Kernaghan, G., Mayerhofer, M., Griffin, A., 2017. Fungal endophytes of wild and hybrid *Vitis* leaves and their potential for vineyard biocontrol. Can. J. Microbiol. 63, 583–595.

Khalifa, M.E., MacDiarmid, R.M., 2021. A mechanically transmitted DNA mycovirus is targeted by the defence machinery of its host, *Botrytis cinerea*. Viruses 13, 1315.

Khodayari, A., Zomorrodi, A.R., Liao, J.C., Maranas, C.D., 2014. A kinetic model of Escherichia coli core metabolism satisfying multiple sets of mutant flux data. Metab. Eng. 25, 50–62.

Kiss, L., 2003. A review of fungal antagonists of powdery mildews and their potential as biocontrol agents. Pest Manag. Sci. 59, 475–483.

Kleyer, H., Tecon, R., Or, D., 2017. Resolving species level changes in a representative soil bacterial community using microfluidic quantitative PCR. Front. Microbiol. 8, 2017.

Knapp, D.G., Lázár, A., Molnár, A., Vajna, B., Karácsony, Z., Váczy, K.Z., Kovács, G.M., 2021. Above-ground parts of white grapevine *Vitis vinifera* cv. Furmint share core members of the fungal microbiome. Environ. Microbiol. Rep. 13, 509–520.

Kortekamp, A., 1997. *Epicoccum nigrum* Link: a biological control agent of *Plasmopara viticola* (Berk et Curt.) Berl. et De Toni? Vitis 36, 215–216.

Kraus, C., Voegele, R.T., Fischer, M., 2019. Temporal development of the culturable, endophytic fungal community in healthy grapevine branches and occurrence of GTD-associated fungi. Microb. Ecol. 77, 866–876.

Kuntal, B.K., Gadgil, C., Mande, S.S., 2019. Web-gLV: a web based platform for lotka-volterra based modeling and simulation of microbial populations. Front. Microbiol. 10, 288.

Kurtz, Z.D., Müller, C.L., Miraldi, E.R., Littman, D.R., Blaser, M.J., Bonneau, R.A., 2015. Sparse and compositionally robust inference of microbial ecological networks. PLoS Comput. Biol. 11, e1004226.

La Guerche, S., Chamont, S., Blancard, D., Dubourdieu, D., Darriet, P., 2005. Origin of (−)-geosmin on grapes: on the complementary action of two fungi, *Botrytis cinerea* and *Penicillium Expansum*. Antonie Van Leeuwenhoek 88, 131–139.

La Guerche, S., De Senneville, L., Blancard, D., Darriet, P., 2007. Impact of the *Botrytis cinerea* strain and metabolism on (−)-geosmin production by *Penicillium expansum* in grape juice. Antonie Van Leeuwenhoek 92, 331–341.

Labarthe, S., Polizzi, B., Phan, T., Goudon, T., Ribot, M., Laroche, B., 2019. A mathematical model to investigate the key drivers of the biogeography of the colon microbiota. J. Theor. Biol. 462, 552–581.

Lacombe, T., Boursiquot, J.-M., Laucou, V., Di Vecchi-Staraz, M., Péros, J.-P., This, P., 2013. Large-scale parentage analysis in an extended set of grapevine cultivars (*Vitis vinifera* L.). Theor. Appl. Genet. 126, 401–414.

Lade, S.B., Štraus, D., Oliva, J., 2022. Variation in fungal community in grapevine (*Vitis vinifera*) nursery stock depends on nursery, variety and rootstock. J. Fungi 8, 47.

Lagier, J.-C., Hugon, P., Khelaifia, S., Fournier, P.-E., La Scola, B., Raoult, D., 2015. The rebirth of culture in microbiology through the example of culturomics to study human gut microbiota. Clin. Microbiol. Rev. 28, 237–264.

Lagier, J.-C., Khelaifia, S., Alou, M.T., Ndongo, S., Dione, N., Hugon, P., Caputo, A., Cadoret, F., Traore, S.I., Seck, E.H., Dubourg, G., Durand, G., Mourembou, G., Guilhot, E., Togo, A., Bellali, S., Bachar, D., Cassir, N., Bittar, F., Delerce, J., Mailhe, M., Ricaboni, D., Bilen, M., Dangui Nieko, N.P.M., Dia Badiane, N.M., Valles, C., Mouelhi, D., Diop, K., Million, M., Musso, D., Abrahão, J., Azhar, E.I., Bibi, F., Yasir, M., Diallo, A., Sokhna, C., Djossou, F., Vitton, V., Robert, C., Rolain, J.M., La Scola, B., Fournier, P.-E., Levasseur, A., Raoult, D., 2016. Culture of previously uncultured members of the human gut microbiota by culturomics. Nat. Microbiol. 1, 1–8.

Lagier, J.-C., Dubourg, G., Million, M., Cadoret, F., Bilen, M., Fenollar, F., Levasseur, A., Rolain, J.-M., Fournier, P.-E., Raoult, D., 2018. Culturing the human microbiota and culturomics. Nat. Rev. Microbiol. 16, 540–550.

Laimer, M., Lemaire, O., Herrbach, E., Goldschmidt, V., Minafra, A., Bianco, P., Wetzel, T., 2009. Resistance to viruses, phytoplasmas and their vectors in the grapevine in Europe: a review. J. Plant Pathol. 91, 7–23.

Legras, J.-L., Merdinoglu, D., Cornuet, J.-M., Karst, F., 2007. Bread, beer and wine: *Saccharomyces cerevisiae* diversity reflects human history. Mol. Ecol. 16, 2091–2102.

Lemos, L.N., Mendes, L.W., Baldrian, P., Pylro, V.S., 2021. Genome-resolved metagenomics is essential for unlocking the microbial black box of the soil. Trends Microbiol. 29, 279–282.

Levy, A., Salas Gonzalez, I., Mittelviefhaus, M., Clingenpeel, S., Herrera Paredes, S., Miao, J., Wang, K., Devescovi, G., Stillman, K., Monteiro, F., Rangel Alvarez, B., Lundberg, D.S., Lu, T.-Y., Lebeis, S., Jin, Z., McDonald, M., Klein, A.P., Feltcher, M.E., Rio, T.G., Grant, S.R., Doty, S.L., Ley, R.E., Zhao, B., Venturi, V., Pelletier, D.A., Vorholt, J.A., Tringe, S.G., Woyke, T., Dangl, J.L., 2018. Genomic features of bacterial adaptation to plants. Nat. Genet. 50, 138–150.

Li, D., Liu, C.-M., Luo, R., Sadakane, K., Lam, T.-W., 2015. MEGAHIT: an ultra-fast single-node solution for large and complex metagenomics assembly via succinct de Bruijn graph. Bioinformatics 31, 1674–1676.

Li, R., Yang, S., Lin, M., Guo, S., Han, X., Ren, M., Du, L., Song, Y., You, Y., Zhan, J., Huang, W., 2022. The biogeography of fungal communities across different chinese wine-producing regions associated with environmental factors and spontaneous fermentation performance. Front. Microbiol. 12, 636639.

Likar, M., Stres, B., Rusjan, D., Potisek, M., Regvar, M., 2017. Ecological and conventional viticulture gives rise to distinct fungal and bacterial microbial communities in vineyard soils. Appl. Soil Ecol. 113, 86–95.

Liu, D., Howell, K., 2021. Community succession of the grapevine fungal microbiome in the annual growth cycle. Environ. Microbiol. 23, 1842–1857.

Liu, D., Zhang, P., Chen, D., Howell, K., 2019. From the vineyard to the winery: how microbial ecology drives regional distinctiveness of wine. Front. Microbiol. 10, 2679.

Liu, Z., Ma, A., Mathé, E., Merling, M., Ma, Q., Liu, B., 2021. Network analyses in microbiome based on high-throughput multi-omics data. Brief. Bioinform. 22 (2), 1639–1655.

Lleixa, J., Kioroglou, D., Mas, A., Portillo, M., 2018. Microbiome dynamics during spontaneous fermentations of sound grapes in comparison with sour rot and Botrytis infected grapes. Int. J. Food Microbiol. 281, 36–46.

Lo, C., Marculescu, R., 2017. MPLasso: Inferring microbial association networks using prior microbial knowledge. PLoS Comput. Biol. 13, e1005915.

Loqman, S., Barka, E.A., Clément, C., Ouhdouch, Y., 2009. Antagonistic actinomycetes from Moroccan soil to control the grapevine gray mold. World J. Microbiol. Biotechnol. 25, 81–91.

Lou, J., Yang, L., Wang, H., Wu, L., Xu, J., 2018. Assessing soil bacterial community and dynamics by integrated high-throughput absolute abundance quantification. PeerJ 6, e4514.

Louca, S., Parfrey, L.W., Doebeli, M., 2016. Decoupling function and taxonomy in the global ocean microbiome. Science 353, 1272–1277.

Lyu, D., Msimbira, L.A., Nazari, M., Antar, M., Pagé, A., Shah, A., Monjezi, N., Zajonc, J., Tanney, C.A.S., Backer, R., Smith, D.L., 2021. The coevolution of plants and microbes underpins sustainable agriculture. Microorganisms 9, 1036.

Magnin-Robert, M., Trotel-Aziz, P., Quantinet, D., Biagianti, S., Aziz, A., 2007. Biological control of *Botrytis cinerea* by selected grapevine-associated bacteria and stimulation of chitinase and β-1,3 glucanase activities under field conditions. Eur. J. Plant Pathol. 118, 43–57.

Mahadevan, R., Edwards, J.S., Doyle, F.J., 2002. Dynamic flux balance analysis of diauxic growth in *Escherichia coli*. Biophys. J. 83, 1331–1340.

Maher, N., Piot, J., Bastien, S., Vallance, J., Rey, P., Guérin-Dubrana, L., 2012. Wood necrosis in ESCA-affected vines: types, relationships and possible links with foliar symptom expression. J. Int. Sci. Vigne Vin 46, 15–27.

Majer, P., Hideg, É., 2012. Developmental stage is an important factor that determines the antioxidant responses of young and old grapevine leaves under UV irradiation in a green-house. Plant Physiol. Biochem. 50, 15–23.

Maliogka, V.I., Martelli, G.P., Fuchs, M., Katis, N.I., 2015. Control of viruses infecting grapevine. In: Advances in Virus Research. Elsevier, pp. 175–227.

Marais, A., Faure, C., Comont, G., Candresse, T., Stempien, E., Corio-Costet, M.-F., 2021. Characterization of the mycovirome of the phytopathogenic fungus, *Neofusicoccum parvum*. Viruses 13, 375.

Marasco, R., Rolli, E., Fusi, M., Michoud, G., Daffonchio, D., 2018. Grapevine rootstocks shape underground bacterial microbiome and networking but not potential functionality. Microbiome 6, 3.

Marchesi, J.R., Ravel, J., 2015. The vocabulary of microbiome research: a proposal. Microbiome 3, 31.

Mardanov, A.V., Chizhevskaya, E.P., Lazarev, A.M., Rakitin, A.L., Beletsky, A.V., Chebotar, V.K., Ravin, N.V., 2019. Complete genome sequences of endophytic bacilli isolated from grapevine plants. Microbiol. Resour. Announc. 8, e01265-19.

Marín, O., González, B., Poupin, M.J., 2021. From microbial dynamics to functionality in the rhizosphere: a systematic review of the opportunities with synthetic microbial communities. Front. Plant Sci. 12, 650609.

Marino, S., Baxter, N.T., Huffnagle, G.B., Petrosino, J.F., Schloss, P.D., 2014. Mathematical modeling of primary succession of murine intestinal microbiota. Proc. Natl. Acad. Sci. U. S. A. 111, 439–444.

Marsit, S., Dequin, S., 2015. Diversity and adaptive evolution of *Saccharomyces* wine yeast: a review. FEMS Yeast Res. 15, fov067.

Martelli, G.P., 1999. Proceedings of the Mediterranean Network on Grapevine Closteroviruses 1992-1997 and the Viroses and Virus-Like Diseases of the Grapevine. Centre international de hautes études agronomiques (méditerranéennes).

Martini, M., Musetti, R., Grisan, S., Polizzotto, R., Borselli, S., Pavan, F., Osler, R., 2009. DNA-dependent detection of the grapevine fungal endophytes *Aureobasidium pullulans* and *Epicoccum nigrum*. Plant Dis. 93, 993–998.

Martins, G., Miot-Sertier, C., Lauga, B., Claisse, O., Lonvaud-Funel, A., Soulas, G., Masneuf-Pomarède, I., 2012. Grape berry bacterial microbiota: impact of the ripening process and the farming system. Int. J. Food Microbiol. 158, 93–100.

Martins, G., Vallance, J., Mercier, A., Albertin, W., Stamatopoulos, P., Rey, P., Lonvaud, A., Masneuf-Pomarède, I., 2014. Influence of the farming system on the epiphytic yeasts and yeast-like fungi colonizing grape berries during the ripening process. Int. J. Food Microbiol. 177, 21–28.

Martins, G.E., Casini, C., Da Costa, J.-P., Geny, L., Lonvaud, A., Masneuf-Pomarède, I., 2020. Correlation between water activity (aw) and microbial epiphytic communities associated with grapes berries. OENO One 54, 49–61.

Mendoza, S.N., Cañón, P.M., Contreras, Á., Ribbeck, M., Agosín, E., 2017. Genome-scale reconstruction of the metabolic network in *Oenococcus oeni* to assess wine malolactic fermentation. Front. Microbiol. 8, 534.

Merdinoglu, D., Schneider, C., Prado, E., Wiedemann-Merdinoglu, S., Mestre, P., 2018. Breeding for durable resistance to downy and powdery mildew in grapevine. OENO One 52, 203–209.

Miotto-Vilanova, L., Jacquard, C., Courteaux, B., Wortham, L., Michel, J., Clément, C., Barka, E.A., Sanchez, L., 2016. *Burkholderia phytofirmans* PsJN confers grapevine resistance against *Botrytis cinerea* via a direct antimicrobial effect combined with a better resource mobilization. Front. Plant Sci. 7, 1236.

Miura, T., Sánchez, R., Castañeda, L.E., Godoy, K., Barbosa, O., 2017. Is microbial terroir related to geographic distance between vineyards? Microbial terroir related to geographic distance. Environ. Microbiol. Rep. 9, 742–749.

Miura, T., Sánchez, R., Castañeda, L.E., Godoy, K., Barbosa, O., 2019. Shared and unique features of bacterial communities in native forest and vineyard phyllosphere. Ecol. Evol. 9, 3295–3305.

Molitor, D., Beyer, M., 2014. Epidemiology, identification and disease management of grape black rot and potentially useful metabolites of black rot pathogens for industrial applications—a review. Ann. Appl. Biol. 165, 305–317.

Momeni, B., Xie, L., Shou, W., 2017. Lotka-Volterra pairwise modeling fails to capture diverse pairwise microbial interactions. Elife 6, e25051.

Mondello, V., Songy, A., Battiston, E., Pinto, C., Coppin, C., Trotel-Aziz, P., Clément, C., Mugnai, L., Fontaine, F., 2018. Grapevine trunk diseases: a review of fifteen years of trials for their control with chemicals and biocontrol agents. Plant Dis. 102, 1189–1217.

Morales-Cedeño, L.R., Ma del Orozco-Mosqueda, C., Loeza-Lara, P.D., Parra-Cota, F.I., de los Santos-Villalobos, S., Santoyo, G., 2021. Plant growth-promoting bacterial endophytes as biocontrol agents of pre- and post-harvest diseases: fundamentals, methods of application and future perspectives. Microbiol. Res. 242, 126612.

Morgan, H.H., du Toit, M., Setati, M.E., 2017. The grapevine and wine microbiome: insights from high-throughput amplicon sequencing. Front. Microbiol. 8, 820.

Mortimer, R., Polsinelli, M., 1999. On the origins of wine yeast. Res. Microbiol. 150 (3), 199–204.

Mourad, E.F., Sarhan, M.S., Daanaa, H.-S.A., Abdou, M., Morsi, A.T., Abdelfadeel, M.R., Elsawey, H., Nemr, R., El-Tahan, M., Hamza, M.A., Abbas, M., Youssef, H.H., Abdelhadi, A.A., Amer, W.M., Fayez, M., Ruppel, S., Hegazi, N.A., 2018. Plant materials are sustainable substrates supporting new technologies of plant-only-based culture media for *in vitro* Culturing of the plant microbiota. Microbes Environ. 33 (1), 40–49. advpub, ME17135.

Mugnai, L., Graniti, A., Surico, G., 1999. Esca (black measles) and brown wood-streaking: two old and elusive diseases of grapevines. Plant Dis. 83, 404–418.

Musetti, R., Vecchione, A., Stringher, L., Borselli, S., Zulini, L., Marzani, C., D'Ambrosio, M., di Toppi, L.S., Pertot, I., 2006. Inhibition of sporulation and ultrastructural alterations of grapevine downy mildew by the endophytic fungus alternaria alternata. Phytopathology 96, 689–698.

Myles, S., Boyko, A.R., Owens, C.L., Brown, P.J., Grassi, F., Aradhya, M.K., Prins, B., Reynolds, A., Chia, J.-M., Ware, D., Bustamante, C.D., Buckler, E.S., 2011. Genetic structure and domestication history of the grape. Proc. Natl. Acad. Sci. U. S. A. 108, 3530–3535.

Nai, C., Meyer, V., 2018. From axenic to mixed cultures: technological advances accelerating a paradigm shift in microbiology. Trends Microbiol. 26, 538–554.

Narduzzi-Wicht, B., Jermini, M., Gessler, C., Broggini, G.A.L., 2014. Microsatellite markers for population studies of the ascomycete *Phyllosticta ampelicida*, the pathogen causing grape black rot. Phytopathol. Mediterr. 53, 470–479.

Nayfach, S., Roux, S., Seshadri, R., Udwary, D., Varghese, N., Schulz, F., Wu, D., Paez-Espino, D., Chen, I.-M., Huntemann, M., Palaniappan, K., Ladau, J., Mukherjee, S., Reddy, T.B.K., Nielsen, T., Kirton, E., Faria, J.P., Edirisinghe, J.N., Henry, C.S., Jungbluth, S.P., Chivian, D., Dehal, P., Wood-Charlson, E.M., Arkin, A.P., Tringe, S.G., Visel, A., Abreu, H., Acinas, S.G., Allen, E., Allen, M.A., Andersen, G., Anesio, A.M., Attwood, G., Avila-Magaña, V., Badis, Y., Bailey, J., Baker, B., Baldrian, P., Barton, H.A., Beck, D.A.C., Becraft, E.D., Beller, H.R., Beman, J.M., Bernier-Latmani, R., Berry, T.D., Bertagnolli, A., Bertilsson, S., Bhatnagar, J.M., Bird, J.T., Blumer-Schuette, S.E., Bohannan, B., Borton, M.A., Brady, A., Brawley, S.H., Brodie, J., Brown, S., Brum, J.R., Brune, A., Bryant, D.A., Buchan, A., Buckley, D.H., Buongiorno, J., Cadillo-Quiroz, H., Caffrey, S.M., Campbell, A.N., Campbell, B., Carr, S., Carroll, J., Cary, S.C., Cates, A.M., Cattolico, R.A., Cavicchioli, R., Chistoserdova, L., Coleman, M.L., Constant, P., Conway, J.M., Cormack, W.P.M., Crowe, S., Crump, B., Currie, C., Daly, R., Denef, V., Denman, S.E., Desta, A., Dionisi, H., Dodsworth, J., Dombrowski, N., Donohue, T., Dopson, M., Driscoll, T., Dunfield, P., Dupont, C.L., Dynarski, K.A., Edgcomb, V., Edwards, E.A., Elshahed, M.S., Figueroa, I., Flood, B., Fortney, N., Fortunato, C.S., Francis, C., Gachon, C.M.M., Garcia, S.L., Gazitua, M.C., Gentry, T., Gerwick, L., Gharechahi, J., Girguis, P., Gladden, J., Gradoville, M., Grasby, S.E., Gravuer, K., Grettenberger, C.L., Gruninger, R.J., Guo, J., Habteselassie, M.Y., Hallam, S.J., Hatzenpichler, R., Hausmann, B., Hazen, T.C., Hedlund, B., Henny, C., Herfort, L., Hernandez, M., Hershey, O.S., Hess, M., Hollister, E.B., Hug, L.A., Hunt, D., Jansson, J., Jarett, J., Kadnikov, V.V., Kelly, C., Kelly, R., Kelly, W., Kerfeld, C.A., Kimbrel, J., Klassen, J.L., Konstantinidis, K.T., Lee, L.L., Li, W.-J., Loder, A.J., Loy, A., Lozada, M., MacGregor, B., Magnabosco, C., da Silva, A.M., McKay, R.M., McMahon, K., McSweeney, C.S., Medina, M., Meredith, L., Mizzi, J., Mock, T., Momper, L., Moran, M.A., Morgan-Lang, C., Moser, D., Muyzer, G., Myrold, D., Nash, M., Nesbø, C.L., Neumann, A.P., Neumann, R.B., Noguera, D., Northen, T., Norton, J., Nowinski, B., Nüsslein, K., O'Malley, M.A., Oliveira, R.S., de Oliveira, V.M., Onstott, T., Osvatic, J., Ouyang, Y., Pachiadaki, M., Parnell, J., Partida-Martinez, L.P., Peay, K.G., Pelletier, D., Peng, X., Pester, M., Pett-Ridge, J., Peura, S., Pjevac, P., Plominsky, A.M., Poehlein, A., Pope, P.B., Ravin, N., Redmond, M.C., Reiss, R., Rich, V.,

Rinke, C., Rodrigues, J.L.M., Rossmassler, K., Sackett, J., Salekdeh, G.H., Saleska, S., Scarborough, M., Schachtman, D., Schadt, C.W., Schrenk, M., Sczyrba, A., Sengupta, A., Setubal, J.C., Shade, A., Sharp, C., Sherman, D.H., Shubenkova, O.V., Sierra-Garcia, I.N., Simister, R., Simon, H., Sjöling, S., Slonczewski, J., de Souza, R.S.C., Spear, J.R., Stegen, J.C., Stepanauskas, R., Stewart, F., Suen, G., Sullivan, M., Sumner, D., Swan, B.K., Swingley, W., Tarn, J., Taylor, G.T., Teeling, H., Tekere, M., Teske, A., Thomas, T., Thrash, C., Tiedje, J., Ting, C.S., Tully, B., Tyson, G., Ulloa, O., Valentine, D.L., Goethem, M.W.V., VanderGheynst, J., Verbeke, T.J., Vollmers, J., Vuillemin, A., Waldo, N.B., Walsh, D.A., Weimer, B.C., Whitman, T., van der Wielen, P., Wilkins, M., Williams, T.J., Woodcroft, B., Woolet, J., Wrighton, K., Ye, J., Young, E.B., Youssef, N.H., Yu, F.B., Zemskaya, T.I., Ziels, R., Woyke, T., Mouncey, N.J., Ivanova, N.N., Kyrpides, N.C., Eloe-Fadrosh, E.A., 2020. A genomic catalog of Earth's microbiomes. Nat. Biotechnol. 39, 499–509.

Nearing, J.T., Douglas, G.M., Hayes, M.G., MacDonald, J., Desai, D.K., Allward, N., Jones, C.M.A., Wright, R.J., Dhanani, A.S., Comeau, A.M., Langille, M.G.I., 2022. Microbiome differential abundance methods produce different results across 38 datasets. Nat. Commun. 13, 342.

Nemergut, D.R., Schmidt, S.K., Fukami, T., O'Neill, S.P., Bilinski, T.M., Stanish, L.F., Knelman, J.E., Darcy, J.L., Lynch, R.C., Wickey, P., Ferrenberg, S., 2013. Patterns and processes of microbial community assembly. Microbiol. Mol. Biol. Rev. 77, 342–356.

Nerva, L., Turina, M., Zanzotto, A., Gardiman, M., Gaiotti, F., Gambino, G., Chitarra, W., 2019. Isolation, molecular characterization and virome analysis of culturable wood fungal endophytes in esca symptomatic and asymptomatic grapevine plants. Environ. Microbiol. 21, 2886–2904.

Nerva, L., Garcia, J.F., Favaretto, F., Giudice, G., Moffa, L., Sandrini, M., Cantu, D., Zanzotto, A., Gardiman, M., Velasco, R., Gambino, G., Chitarra, W., 2022. The hidden world within plants: metatranscriptomics unveils the complexity of wood microbiomes. J. Exp. Bot. 73, 2682–2697.

Nguyen, N.H., Song, Z., Bates, S.T., Branco, S., Tedersoo, L., Menke, J., Schilling, J.S., Kennedy, P.G., 2016. FUNGuild: an open annotation tool for parsing fungal community datasets by ecological guild. Fungal Ecol. 20, 241–248.

Nidelet, T., Brial, P., Camarasa, C., Dequin, S., 2016. Diversity of flux distribution in central carbon metabolism of *S. cerevisiae* strains from diverse environments. Microb. Cell Fact. 15, 58.

Ning, D., Deng, Y., Tiedje, J.M., Zhou, J., 2019. A general framework for quantitatively assessing ecological stochasticity. Proc. Natl. Acad. Sci. U. S. A. 116, 16892–16898.

Nisiotou, A.A., Nychas, G.J., 2007. Yeast populations residing on healthy or botrytis-infected grapes from a vineyard in Attica, Greece. Appl. Environ. Microbiol. 73 (8), 2765–2768.

Noman, M., Ahmed, T., Ijaz, U., Shahid, M., Azizullah, Li, D., Manzoor, I., Song, F., 2021. Plant–microbiome crosstalk: dawning from composition and assembly of microbial community to improvement of disease resilience in plants. Int. J. Mol. Sci. 22, 6852.

Novello, G., Gamalero, E., Bona, E., Boatti, L., Mignone, F., Massa, N., Cesaro, P., Lingua, G., Berta, G., 2017. The rhizosphere bacterial microbiota of *Vitis vinifera* cv. Pinot noir in an integrated pest management vineyard. Front. Microbiol. 8, 1528.

Nurk, S., Meleshko, D., Korobeynikov, A., Pevzner, P.A., 2017. metaSPAdes: a new versatile metagenomic assembler. Genome Res. 27, 824–834.

Ocete, R., Armendariz, I., Cantos, M., Álvarez, D., Azcón, R., 2015. Ecological characterization of wild grapevine habitats focused on arbuscular mycorrhizal symbiosis. Vitis 54, 207–211.

Ofek-Lalzar, M., Gur, Y., Ben-Moshe, S., Sharon, O., Kosman, E., Mochli, E., Sharon, A., 2016. Diversity of fungal endophytes in recent and ancient wheat ancestors *Triticum dicoccoides* and *Aegilops sharonensis*. FEMS Microbiol. Ecol. 92, fiw152.

Oliveira, M., Arenas, M., Lage, O., Cunha, M., Amorim, M.I., 2018. Epiphytic fungal community in *Vitis vinifera* of the Portuguese wine regions. Lett. Appl. Microbiol. 66, 93–102.

Oliver, J.E., Fuchs, M.F., 2011. Fanleaf Degeneration/Decline Disease of Grapevines. Cornell University and the New York State IPM Program, Ithaca, pp. 1–3.

Oren, A., Garrity, G.M., 2021. Valid publication of the names of forty-two phyla of prokaryotes. Int. J. Syst. Evol. Microbiol. 71 (10), 005056.

Orth, J.D., Thiele, I., Palsson, B.Ø., 2010. What is flux balance analysis? Nat. Biotechnol. 28, 245–248.

Ouadi, L., Bruez, E., Bastien, S., Vallance, J., Lecomte, P., Domec, J.-C., Rey, P., 2019. Ecophysiological impacts of Esca, a devastating grapevine trunk disease, on *Vitis vinifera* L. PLoS One 14, e0222586.

Ouadi, L., Bruez, E., Bastien, S., Yacoub, A., Coppin, C., Guérin-Dubrana, L., Fontaine, F., Domec, J.-C., Rey, P., 2021. Sap flow disruption in grapevine is the early signal predicting the structural, functional, and genetic responses to esca disease. Front. Plant Sci. 12, 695846.

Ovaskainen, O., Tikhonov, G., Norberg, A., Guillaume Blanchet, F., Duan, L., Dunson, D., Roslin, T., Abrego, N., 2017. How to make more out of community data? A conceptual framework and its implementation as models and software. Ecol. Lett. 20, 561–576.

Pacifico, D., Squartini, A., Crucitti, D., Barizza, E., Lo Schiavo, F., Muresu, R., Carimi, F., Zottini, M., 2019. The role of the endophytic microbiome in the grapevine response to environmental triggers. Front. Plant Sci. 10, 1256.

Pandey, B., Naidu, R.A., Grove, G.G., 2018. Detection and analysis of mycovirus-related RNA viruses from grape powdery mildew fungus *Erysiphe necator*. Arch. Virol. 163, 1019–1030.

Panstruga, R., Kuhn, H., 2019. Mutual interplay between phytopathogenic powdery mildew fungi and other microorganisms. Mol. Plant Pathol. 20, 463–470.

Paolinelli, M., Escoriaza, G., Cesari, C., Garcia-Lampasona, S., Hernandez-Martinez, R., 2022. Characterization of grapevine wood microbiome through a metatranscriptomic approach. Microb. Ecol. 83, 658–668.

Pauvert, C., Buée, M., Laval, V., Edel-Hermann, V., Fauchery, L., Gautier, A., Lesur, I., Vallance, J., Vacher, C., 2019. Bioinformatics matters: the accuracy of plant and soil fungal community data is highly dependent on the metabarcoding pipeline. Fungal Ecol. 41, 23–33.

Pauvert, C., Fort, T., Calonnec, A., Faivre d'Arcier, J., Chancerel, E., Massot, M., Chiquet, J., Robin, S., Bohan, D.A., Vallance, J., Vacher, C., 2020. Microbial association networks give relevant insights into plant pathobiomes. bioRxiv. 2020.02.21.958033.

Pearson, M.N., Bailey, A.M., 2013. Viruses of *Botrytis*. In: Advances in Virus Research. Elsevier, pp. 249–272.

Perazzolli, M., Antonielli, L., Storari, M., Puopolo, G., Pancher, M., Giovannini, O., Pindo, M., Pertot, I., 2014. Resilience of the natural phyllosphere microbiota of the grapevine to chemical and biological pesticides. Appl. Environ. Microbiol. 80, 3585–3596.

Perera, D., Savocchia, S., Prenzler, P.D., Thomson, P.C., Steel, C.C., 2021. Occurrence of fumonisin-producing black aspergilli in Australian wine grapes: effects of temperature and water activity on fumonisin production by *A. niger* and *A. welwitschiae*. Mycotoxin Res. 37, 327–339.

Peressotti, E., Wiedemann-Merdinoglu, S., Delmotte, F., Bellin, D., Di Gaspero, G., Testolin, R., Merdinoglu, D., Mestre, P., 2010. Breakdown of resistance to grapevine downy mildew upon limited deployment of a resistant variety. BMC Plant Biol. 10, 147.

Pérez-Valera, E., Verdú, M., Navarro-Cano, J.A., Goberna, M., 2020. Soil microbiome drives the recovery of ecosystem functions after fire. Soil Biol. Biochem. 149, 107948.

Pertot, I., Caffi, T., Rossi, V., Mugnai, L., Hoffmann, C., Grando, M.S., Gary, C., Lafond, D., Duso, C., Thiery, D., Mazzoni, V., Anfora, G., 2017. A critical review of plant protection tools for reducing pesticide use on grapevine and new perspectives for the implementation of IPM in viticulture. Crop Prot. 97, 70–84. Pesticide use and risk reduction with IPM.

Philippe, C., Krupovic, M., Jaomanjaka, F., Claisse, O., Petrel, M., le Marrec, C., 2018. Bacteriophage GC1, a novel tectivirus infecting *Gluconobacter Cerinus*, an acetic acid bacterium associated with wine-making. Viruses 10, 39.

Pilar Martínez-Diz, M., Díaz-Losada, E., Andrés-Sodupe, M., Bujanda, R., Maldonado-González, M.M., Ojeda, S., Yacoub, A., Rey, P., Gramaje, D., 2021. Field evaluation of biocontrol agents against black-foot and petri diseases of grapevine. Pest Manag. Sci. 77, 697–708.

Pinto, C., Pinho, D., Sousa, S., Pinheiro, M., Egas, C., Gomes, A.C., 2014. Unravelling the diversity of grapevine microbiome. PLoS One 9, e85622.

Pirrello, C., Mizzotti, C., Tomazetti, T.C., Colombo, M., Bettinelli, P., Prodorutti, D., Peressotti, E., Zulini, L., Stefanini, M., Angeli, G., Masiero, S., Welter, L.J., Hausmann, L., Vezzulli, S., 2019. Emergent ascomycetes in viticulture: an interdisciplinary overview. Front. Plant Sci. 10, 1394.

Poudel, R., Jumpponen, A., Schlatter, D.C., Paulitz, T.C., Gardener, B.B.M., Kinkel, L.L., Garrett, K.A., 2016. Microbiome networks: a systems framework for identifying candidate microbial assemblages for disease management. Phytopathology 106, 1083–1096.

Prodhomme, D., Valls Fonayet, J., Hévin, C., Franc, C., Hilbert, G., de Revel, G., Richard, T., Ollat, N., Cookson, S.J., 2019. Metabolite profiling during graft union formation reveals the reprogramming of primary metabolism and the induction of stilbene synthesis at the graft interface in grapevine. BMC Plant Biol. 19, 599.

Props, R., Kerckhof, F.-M., Rubbens, P., De Vrieze, J., Hernandez Sanabria, E., Waegeman, W., Monsieurs, P., Hammes, F., Boon, N., 2017. Absolute quantification of microbial taxon abundances. ISME J. 11, 584–587.

Prosser, J.I., Bohannan, B.J.M., Curtis, T.P., Ellis, R.J., Firestone, M.K., Freckleton, R.P., Green, J.L., Green, L.E., Killham, K., Lennon, J.J., Osborn, A.M., Solan, M., van der Gast, C.J., Young, J.P.W., 2007. The role of ecological theory in microbial ecology. Nat. Rev. Microbiol. 5, 384–392.

Puig-Pujol, A., Ferrando, N., Capdevila, F., Ocete, R., Revilla, E., 2016. Yeast biodiversity from *Vitis vinifera* L., subsp. sylvestris (Gmelin) Hegi to face up the oenological consequences of climate change. BIO Web Conf. 7, 02026.

Qiu, W., Feechan, A., Dry, I., 2015. Current understanding of grapevine defense mechanisms against the biotrophic fungus (*Erysiphe necator*), the causal agent of powdery mildew disease. Hortic. Res. 2, 15020.

Quince, C., Walker, A.W., Simpson, J.T., Loman, N.J., Segata, N., 2017. Shotgun metagenomics, from sampling to analysis. Nat. Biotechnol. 35, 833–844.

Radić, T., Likar, M., Hančević, K., Regvar, M., Čarija, M., Zdunić, G., 2021. Root-associated community composition and co-occurrence patterns of fungi in wild grapevine. Fungal Ecol. 50, 101034.

Raguideau, S., Plancade, S., Pons, N., Leclerc, M., Laroche, B., 2016. Inferring aggregated functional traits from metagenomic data using constrained non-negative matrix factorization: application to fiber degradation in the human gut microbiota. PLoS Comput. Biol. 12, e1005252.

Raspor, P., Miklič-Milek, D., Avbelj, M., Čadež, N., 2010. Food Technol. Biotechnol. 48, 336–343.

Rathnayake, R.M.S.P., Savocchia, S., Schmidtke, L.M., Steel, C.C., 2018. Characterisation of *Aureobasidium pullulans* isolates from *Vitis vinifera* and potential biocontrol activity for the management of bitter rot of grapes. Eur. J. Plant Pathol. 151, 593–611.

Remien, C.H., Eckwright, M.J., Ridenhour, B.J., 2021. Structural identifiability of the generalized Lotka–Volterra model for microbiome studies. R. Soc. Open Sci. 8, 201378.

Rezgui, A., Ben Ghnaya-Chakroun, A., Vallance, J., Bruez, E., Hajlaoui, M.R., Sadfi-Zouaoui, N., Rey, P., 2016. Endophytic bacteria with antagonistic traits inhabit the wood tissues of grapevines from Tunisian vineyards. Biol. Control 99, 28–37.

Rijavec, T., Lapanje, A., 2016. Hydrogen cyanide in the rhizosphere: not suppressing plant pathogens, but rather regulating availability of phosphate. Front. Microbiol. 7, 1785.

Rillig, M.C., Antonovics, J., Caruso, T., Lehmann, A., Powell, J.R., Veresoglou, S.D., Verbruggen, E., 2015. Interchange of entire communities: microbial community coalescence. Trends Ecol. Evol. 30, 470–476.

Rinaldi, P.A., Paffetti, D., Comparini, C., Broggini, G.A.L., Gessler, C., Mugnai, L., 2017. Genetic variability of *Phyllosticta ampelicida*, the agent of black rot disease of grapevine. Phytopathology 107, 1406–1416.

Riva, V., Mapelli, F., Bagnasco, A., Mengoni, A., Borin, S., 2022. A Meta-analysis approach to defining the culturable core of plant endophytic bacterial communities. Appl. Environ. Microbiol. 88, e02537-21.

Rivas, G.A., Semorile, L., Delfederico, L., 2022. Microbial diversity of the soil, rhizosphere and wine from an emerging wine-producing region of Argentina. LWT 153, 112429.

Rizaludin, M.S., Stopnisek, N., Raaijmakers, J.M., Garbeva, P., 2021. The chemistry of stress: understanding the 'cry for help' of plant roots. Metabolites 11, 357.

Rondot, Y., Reineke, A., 2019. Endophytic *Beauveria bassiana* activates expression of defence genes in grapevine and prevents infections by grapevine downy mildew *Plasmopara viticola*. Plant Pathol. 68, 1719–1731.

Roossinck, M.J., 2010. Lifestyles of plant viruses. Philos. Trans. R. Soc. B Biol. Sci. 365, 1899–1905.

Ross, T., McMeekin, T.A., 1994. Predictive microbiology. Int. J. Food Microbiol. 23, 241–264. Special Issue Predictive Modelling.

Röttjers, L., Faust, K., 2018. From hairballs to hypotheses–biological insights from microbial networks. FEMS Microbiol. Rev. 42, 761–780.

Rouxel, M., Mestre, P., Comont, G., Lehman, B.L., Schilder, A., Delmotte, F., 2013. Phylogenetic and experimental evidence for host-specialized cryptic species in a biotrophic oomycete. New Phytol. 197, 251–263.

Rouxel, M., Mestre, P., Baudoin, A., Carisse, O., Delière, L., Ellis, M.A., Gadoury, D., Lu, J., Nita, M., Richard-Cervera, S., Schilder, A., Wise, A., Delmotte, F., 2014. Geographic distribution of cryptic species of *Plasmopara viticola* causing downy mildew on wild and cultivated grape in eastern North America. Phytopathology 104, 692–701.

Rowhani, A., Uyemoto, J.K., Golino, D.A., Daubert, S.D., Al Rwahnih, M., 2017. Viruses involved in graft incompatibility and decline. In: Meng, B., Martelli, G.P., Golino, D.A., Fuchs, M. (Eds.), Grapevine Viruses: Molecular Biology, Diagnostics and Management. Springer International Publishing, Cham, pp. 289–302.

Ruiz-Padilla, A., Rodríguez-Romero, J., Gómez-Cid, I., Pacifico, D., Ayllón, M.A., 2021. Novel mycoviruses discovered in the mycovirome of a necrotrophic fungus. MBio 12, e03705–e03720.

Saikkonen, K., Nissinen, R., Helander, M., 2020. Toward comprehensive plant microbiome research. Front. Ecol. Evol. 8, 61.

Salvetti, E., Campanaro, S., Campedelli, I., Fracchetti, F., Gobbi, A., Tornielli, G.B., Torriani, S., Felis, G.E., 2016. Whole-metagenome-sequencing-based community profiles of Vitis vinifera L. cv. corvina berries withered in two post-harvest conditions. Front. Microbiol. 7, 937.

Santos, A., Marquina, D.Y., 2004. 2004. Killer toxin of *Pichia membranifaciens* and its possible use as a biocontrol agent against grey mould disease of grapevine. Microbiology 150, 2527–2534.

Saraf, M., Jha, C.K., Patel, D., 2010. The Role of ACC deaminase producing PGPR in sustainable agriculture. In: Maheshwari, D.K. (Ed.), Plant Growth and Health Promoting Bacteria. Plant Growth and Health Promoting Bacteria, Microbiology Monographs, Springer Berlin Heidelberg, Berlin, Heidelberg, pp. 251–272.

Sarhan, M.S., Patz, S., Hamza, M.A., Youssef, H.H., Mourad, E.F., Fayez, M., Murphy, B., Ruppel, S., Hegazi, N.A., 2018. G3 PhyloChip analysis confirms the promise of plant-based culture media for unlocking the composition and diversity of the maize root microbiome and for recovering unculturable candidate divisions/phyla. Microbes Environ. 33 (3), 317–325. advpub, ME18023.

Sarhan, M.S., Hamza, M.A., Youssef, H.H., Patz, S., Becker, M., ElSawey, H., Nemr, R., Daanaa, H.-S.A., Mourad, E.F., Morsi, A.T., Abdelfadeel, M.R., Abbas, M.T., Fayez, M., Ruppel, S., Hegazi, N.A., 2019. Culturomics of the plant prokaryotic microbiome and the dawn of plant-based culture media—a review. J. Adv. Res. 19, 15–27. Special Issue on Plant Microbiome.

Schacherer, J., Shapiro, J.A., Ruderfer, D.M., Kruglyak, L., 2009. Comprehensive polymorphism survey elucidates population structure of *Saccharomyces cerevisiae*. Nature 458, 342–345.

Schloss, P.D., Westcott, S.L., Ryabin, T., Hall, J.R., Hartmann, M., Hollister, E.B., Lesniewski, R.A., Oakley, B.B., Parks, D.H., Robinson, C.J., Sahl, J.W., Stres, B., Thallinger, G.G., Horn, D.J.V., Weber, C.F., 2009. Introducing mothur: open-source, platform-independent, community-supported software for describing and comparing microbial communities. Appl. Environ. Microbiol. 75, 7537–7541.

Seaver, S.M.D., Liu, F., Zhang, Q., Jeffryes, J., Faria, J.P., Edirisinghe, J.N., Mundy, M., Chia, N., Noor, E., Beber, M.E., Best, A.A., DeJongh, M., Kimbrel, J.A., D'haeseleer, P., McCorkle, S.R., Bolton, J.R., Pearson, E., Canon, S., Wood-Charlson, E.M., Cottingham, R.W., Arkin, A.P., Henry, C.S., 2021. The ModelSEED Biochemistry Database for the integration of metabolic annotations and the reconstruction, comparison and analysis of metabolic models for plants, fungi and microbes. Nucleic Acids Res. 49 (D1), D575–D588.

Seemann, T., 2014. Prokka: rapid prokaryotic genome annotation. Bioinformatics 30, 2068–2069.

Sereika, M., Kirkegaard, R.H., Karst, S.M., Michaelsen, T.Y., Sørensen, E.A., Wollenberg, R.D., Albertsen, M., 2022. Oxford Nanopore R10.4 long-read sequencing enables the generation of near-finished bacterial genomes from pure cultures and metagenomes without short-read or reference polishing. Nat. Methods 19, 823–826.

Shakir, S., Zaidi, S.S.-A., de Vries, F.T., Mansoor, S., 2021. Plant genetic networks shaping phyllosphere microbial community. Trends Genet. 37, 306–316.

Signorini, M., Borruso, L., Randall, K.C., Dumbrell, A.J., Pii, Y., Mimmo, T., Cesco, S., 2021. Soil heterogeneity within a vineyard impacts the beta but not the alpha microbial agro-diversity. Appl. Soil Ecol. 166, 104088.

Singh, P., Santoni, S., This, P., Péros, J.-P., 2018. Genotype-environment interaction shapes the microbial assemblage in grapevine's phyllosphere and carposphere: an NGS approach. Microorganisms 6, 96.

Singh, P., Santoni, S., Weber, A., This, P., Péros, J.-P., 2019. Understanding the phyllosphere microbiome assemblage in grape species (Vitaceae) with amplicon sequence data structures. Sci. Rep. 9, 14294.

Sivan, A., Rahimi, O., Lavi, B., Salmon-Divon, M., Weiss, E., Drori, E., Hübner, S., 2021. Genomic evidence supports an independent history of Levantine and Eurasian grapevines. Plants People Planet 3, 414–427.

Smith, N.A., Eamens, A.L., Wang, M.-B., 2011. Viral small interfering RNAs target host genes to mediate disease symptoms in plants. PLoS Pathog. 7, e1002022.

Staudt, G., Kassemeyer, H.H., 1995. Evaluation of downy mildew resistance in various accessions of wild *Vitis* species. Vitis 34, 225–228.

Stefanini, I., Cavalieri, D., 2018. Metagenomic approaches to investigate the contribution of the vineyard environment to the quality of wine fermentation: potentials and difficulties. Front. Microbiol. 9, 991.

Steimetz, E., Trouvelot, S., Gindro, K., Bordier, A., Poinssot, B., Adrian, M., Daire, X., 2012. Influence of leaf age on induced resistance in grapevine against *Plasmopara viticola*. Physiol. Mol. Plant Pathol. 79, 89–96.

Stein, R.R., Bucci, V., Toussaint, N.C., Buffie, C.G., Rätsch, G., Pamer, E.G., Sander, C., Xavier, J.B., 2013. Ecological modeling from time-series inference: insight into dynamics and stability of intestinal microbiota. PLoS Comput. Biol. 9, e1003388.

Stempien, E., Pierron, R.J., Adendorff, I., van Jaarsveld, W.J., Halleen, F., Mostert, L., 2020. Host defence activation and root colonization of grapevine rootstocks by the biological control fungus *Trichoderma atroviride*. Phytopathol. Mediterr. 59, 615–626.

Stranska, M., Dzuman, Z., Prusova, N., Behner, A., Kolouchova, I., Lovecka, P., Rezanka, T., Kolarik, M., Hajslova, J., 2022. Fungal endophytes of *Vitis vinifera*—plant growth promoters or potentially toxinogenic agents? Toxins 14, 66.

Sun, X., Kosman, E., Sharon, A., 2020. Stem endophytic mycobiota in wild and domesticated wheat: structural differences and hidden resources for wheat improvement. J. Fungi 6, 180.

Sun, D., Qu, J., Huang, Y., Lu, J., Yin, L., 2021. Analysis of microbial community diversity of muscadine grape skins. Food Res. Int. 145, 110417.

Swift, J.F., Hall, M.E., Harris, Z.N., Kwasniewski, M.T., Miller, A.J., 2021. Grapevine microbiota reflect diversity among compartments and complex interactions within and among root and shoot systems. Microorganisms 9, 92.

Taguiam, J.D., Evallo, E., Balendres, M.A., 2021. Epicoccum species: ubiquitous plant pathogens and effective biological control agents. Eur. J. Plant Pathol. 159, 713–725.

Tamaddoni-Nezhad, A., Bohan, D., Milani, G.A., Raybould, A., Muggleton, S., 2021. Human–machine scientific discovery. In: Human-Like Machine Intelligence. Oxford University Press, pp. 297–315.

Tedesco, S., Fevereiro, P., Kragler, F., Pina, A., 2022. Plant grafting and graft incompatibility: a review from the grapevine perspective. Sci. Hortic. 299, 111019.

Terral, J.-F., Tabard, E., Bouby, L., Ivorra, S., Pastor, T., Figueiral, I., Picq, S., Chevance, J.-B., Jung, C., Fabre, L., Tardy, C., Compan, M., Bacilieri, R., Lacombe, T., This, P., 2010. Evolution and history of grapevine (*Vitis vinifera*) under domestication: new morphometric perspectives to understand seed domestication syndrome and reveal origins of ancient European cultivars. Ann. Bot. 105, 443–455.

Theis, K.R., Dheilly, N.M., Klassen, J.L., Brucker, R.M., Baines, J.F., Bosch, T.C., Cryan, J.F., Gilbert, S.F., Goodnight, C.J., Lloyd, E.A., Sapp, J., Vandenkoornhuyse, P., Zilber-Rosenberg, I., Rosenberg, E., Bordenstein, S.R., 2016. Getting the hologenome concept right: an eco-evolutionary framework for hosts and their microbiomes. Msystems 1 (2), e00028-16.

Thibon, C., Shinkaruk, S., Jourdes, M., Bennetau, B., Dubourdieu, D., Tominaga, T., 2010. Aromatic potential of botrytized white wine grapes: Identification and quantification of new cysteine-S-conjugate flavor precursors. Anal. Chim. Acta 660, 190–196.

Thibon, C., Cluzet, S., Mérillon, J.M., Darriet, P., Dubourdieu, D., 2011. 3-sulfanylhexanol precursor biogenesis in grapevine cells: the stimulating effect of *Botrytis cinerea*. J. Agric. Food Chem. 59, 1344–1351.

This, P., Lacombe, T., Thomas, M., 2006. Historical origins and genetic diversity of wine grapes. Trends Genet. 22, 511–519.

Tikhonov, G., Opedal, Ø.H., Abrego, N., Lehikoinen, A., de Jonge, M.M.J., Oksanen, J., Ovaskainen, O., 2020. Joint species distribution modelling with the r-package Hmsc. Methods Ecol. Evol. 11, 442–447.

Toju, H., Peay, K.G., Yamamichi, M., Narisawa, K., Hiruma, K., Naito, K., Fukuda, S., Ushio, M., Nakaoka, S., Onoda, Y., Yoshida, K., Schlaeppi, K., Bai, Y., Sugiura, R., Ichihashi, Y., Minamisawa, K., Kiers, E.T., 2018. Core microbiomes for sustainable agroecosystems. Nat. Plants 4, 247–257.

Trivedi, P., Leach, J.E., Tringe, S.G., Sa, T., Singh, B.K., 2020. Plant–microbiome interactions: from community assembly to plant health. Nat. Rev. Microbiol. 18, 607–621.

Trivedi, P., Mattupalli, C., Eversole, K., Leach, J.E., 2021. Enabling sustainable agriculture through understanding and enhancement of microbiomes. New Phytol. 230, 2129–2147.

Trouvelot, S., Bonneau, L., Redecker, D., van Tuinen, D., Adrian, M., Wipf, D., 2015. Arbuscular mycorrhiza symbiosis in viticulture: a review. Agron. Sustain. Dev. 35, 1449–1467.

Tshikantwa, T.S., Ullah, M.W., He, F., Yang, G., 2018. Current trends and potential applications of microbial interactions for human welfare. Front. Microbiol. 9, 1156.

Uyemoto, J.K., Rowhani, A., Ludivisi, D., 2000. An association of rootstock stem lesions in Vitis species and different graft-transmissible agents. In: Proc. 13th ICVG Meet. Adel. Aust. vol. 83.

Uyemoto, J.K., Rowhani, A., Luvisi, D., Krag, C.R., 2001. New closterovirus in 'Redglobe' grape causes decline of grafted plants. Calif. Agric. 55, 28–31.

Vacher, C., Hampe, A., Porté, A.J., Sauer, U., Compant, S., Morris, C.E., 2016a. The phyllosphere: microbial jungle at the plant–climate interface. Annu. Rev. Ecol. Evol. Syst. 47, 1–24.

Vacher, C., Tamaddoni-Nezhad, A., Kamenova, S., Peyrard, N., Moalic, Y., Sabbadin, R., Schwaller, L., Chiquet, J., Smith, M.A., Vallance, J., Fievet, V., Jakuschkin, B., Bohan, D.A., 2016b. Learning ecological networks from next-generation sequencing data. In: Woodward, G., Bohan, D.A. (Eds.), Advances in Ecological Research, Ecosystem Services: From Biodiversity to Society, Part 2. Academic Press, pp. 1–39.

Vacher, C., Francioni, C., Michel, M., Faivre d'Arcier, J., Fort, T., Chancerel, E., Delmotte, F., Delmas, C.L., (2022), Fungal metabarcoding data for two grapevine varieties (Regent and *Vitis vinifera* L. cv. Cabernet–Sauvignon) inoculated with powdery mildew (*Erysiphe necator*) under drought conditions. Phytobiomes J. in press.

Vandenkoornhuyse, P., Quaiser, A., Duhamel, M., Le Van, A., Dufresne, A., 2015. The importance of the microbiome of the plant holobiont. New Phytol. 206 (4), 1196–1206.

Vass, M., Székely, A.J., Lindström, E.S., Langenheder, S., 2020. Using null models to compare bacterial and microeukaryotic metacommunity assembly under shifting environmental conditions. Sci. Rep. 10, 2455.

Vellend, M., 2010. Conceptual synthesis in community ecology. Q. Rev. Biol. 85, 183–206.

Villano, C., Aversano, R., 2020. Towards grapevine (*Vitis vinifera* L.) mildews resistance: molecular defence mechanisms and new breeding technologies. Italus Hortus 27, 1.

Vitulo, N., Lemos, W.J.F., Calgaro, M., Confalone, M., Felis, G.E., Zapparoli, G., Nardi, T., 2019. Bark and grape microbiome of *Vitis vinifera*: influence of geographic patterns and agronomic management on bacterial diversity. Front. Microbiol. 9, 3203.

Vukicevich, E., Thomas Lowery, D., Úrbez-Torres, J.R., Bowen, P., Hart, M., 2018. Groundcover management changes grapevine root fungal communities and plant-soil feedback. Plant and Soil 424, 419–433.

Waite, H., Whitelaw-Weckert, M., Torley, P., 2015. Grapevine propagation: principles and methods for the production of high-quality grapevine planting material. N. Z. J. Crop Hortic. Sci. 43, 144–161.

Wan, Y., Schwaninger, H.R., Baldo, A.M., Labate, J.A., Zhong, G.-Y., Simon, C.J., 2013. A phylogenetic analysis of the grape genus (*Vitis* L.) reveals broad reticulation and concurrent diversification during neogene and quaternary climate change. BMC Evol. Biol. 13, 141.

Warringer, J., Zörgö, E., Cubillos, F.A., Zia, A., Gjuvsland, A., Simpson, J.T., Forsmark, A., Durbin, R., Omholt, S.W., Louis, E.J., Liti, G., Moses, A., Blomberg, A., 2011. Trait variation in yeast is defined by population history. PLoS Genet. 7, e1002111.

Weiss, S., Van Treuren, W., Lozupone, C., Faust, K., Friedman, J., Deng, Y., Xia, L.C., Xu, Z.Z., Ursell, L., Alm, E.J., Birmingham, A., Cram, J.A., Fuhrman, J.A., Raes, J., Sun, F., Zhou, J., Knight, R., 2016. Correlation detection strategies in microbial data sets vary widely in sensitivity and precision. ISME J. 10, 1669–1681.

Weiss, S., Xu, Z.Z., Peddada, S., Amir, A., Bittinger, K., Gonzalez, A., Lozupone, C., Zaneveld, J.R., Vázquez-Baeza, Y., Birmingham, A., Hyde, E.R., Knight, R., 2017. Normalization and microbial differential abundance strategies depend upon data characteristics. Microbiome 5, 27.

Wemheuer, F., Taylor, J.A., Daniel, R., Johnston, E., Meinicke, P., Thomas, T., Wemheuer, B., 2020. Tax4Fun2: prediction of habitat-specific functional profiles and functional redundancy based on 16S rRNA gene sequences. Environ. Microbiome 15, 11.

Widder, S., Allen, R.J., Pfeiffer, T., Curtis, T.P., Wiuf, C., Sloan, W.T., Cordero, O.X., Brown, S.P., Momeni, B., Shou, W., Kettle, H., Flint, H.J., Haas, A.F., Laroche, B., Kreft, J.-U., Rainey, P.B., Freilich, S., Schuster, S., Milferstedt, K., van der Meer, J.R., Groβkopf, T., Huisman, J., Free, A., Picioreanu, C., Quince, C., Klapper, I., Labarthe, S., Smets, B.F., Wang, H., Soyer, O.S., 2016. Challenges in microbial ecology: building predictive understanding of community function and dynamics. ISME J. 10, 2557–2568.

Wu, M., Zhang, L., Li, G., Jiang, D., Ghabrial, S.A., 2010. Genome characterization of a debilitation-associated mitovirus infecting the phytopathogenic fungus *Botrytis cinerea*. Virology 406, 117–126.

Wu, Y.-W., Simmons, B.A., Singer, S.W., 2016. MaxBin 2.0: an automated binning algorithm to recover genomes from multiple metagenomic datasets. Bioinformatics 32, 605–607.

Wu, X., Cui, Z., Peng, J., Zhang, F., Liesack, W., 2022a. Genome-resolved metagenomics identifies the particular genetic traits of phosphate-solubilizing bacteria in agricultural soil. ISME Commun. 2, 17.

Wu, X., Rensing, C., Han, D., Xiao, K.-Q., Dai, Y., Tang, Z., Liesack, W., Peng, J., Cui, Z., Zhang, F., 2022b. Genome-resolved metagenomics reveals distinct phosphorus acquisition strategies between soil microbiomes. mSystems 7, e01107–e01121.

Xiao, Y., Angulo, M.T., Friedman, J., Waldor, M.K., Weiss, S.T., Liu, Y.-Y., 2017. Mapping the ecological networks of microbial communities. Nat. Commun. 8, 2042.

Xu, N., Zhang, Z., Shen, Y., Zhang, Q., Liu, Z., Yu, Y., Wang, Y., Lei, C., Ke, M., Qiu, D., Lu, T., Chen, Y., Xiong, J., Qian, H., 2022. Compare the performance of multiple binary classification models in microbial high-throughput sequencing datasets. Sci. Total Environ. 837, 155807.

Yacoub, A., Gerbore, J., Magnin, N., Chambon, P., Dufour, M.-C., Corio-Costet, M.-F., Guyoneaud, R., Rey, P., 2016. Ability of *Pythium oligandrum* strains to protect *Vitis vinifera* L., by inducing plant resistance against *Phaeomoniella chlamydospora*, a pathogen involved in Esca, a grapevine trunk disease. Biol. Control 92, 7–16.

Yergeau, É., Quiza, L., Tremblay, J., 2020. Microbial indicators are better predictors of wheat yield and quality than N fertilization. FEMS Microbiol. Ecol. 96, fiz205.

Yousaf, S., Bulgari, D., Bergna, A., Pancher, M., Quaglino, F., Casati, P., Campisano, A., 2014. Pyrosequencing detects human and animal pathogenic taxa in the grapevine endosphere. Front. Microbiol. 5, 327.

Yu, L., Sang, W., Wu, M.-D., Zhang, J., Yang, L., Zhou, Y.-J., Chen, W.-D., Li, G.-Q., 2015. Novel hypovirulence-associated RNA mycovirus in the plant-pathogenic fungus *Botrytis cinerea*: molecular and biological characterization. Appl. Environ. Microbiol. 81, 2299–2310.

Yue, Y., Huang, H., Qi, Z., Dou, H.-M., Liu, X.-Y., Han, T.-F., Chen, Y., Song, X.-J., Zhang, Y.-H., Tu, J., 2020. Evaluating metagenomics tools for genome binning with real metagenomic datasets and CAMI datasets. BMC Bioinf. 21, 334.

Zaller, J.G., Cantelmo, C., Santos, G.D., Muther, S., Gruber, E., Pallua, P., Mandl, K., Friedrich, B., Hofstetter, I., Schmuckenschlager, B., Faber, F., 2018. Herbicides in vineyards reduce grapevine root mycorrhization and alter soil microorganisms and the nutrient composition in grapevine roots, leaves, xylem sap and grape juice. Environ. Sci. Pollut. Res. 25, 23215–23226.

Zaneveld, J.R., McMinds, R., Vega Thurber, R., 2017. Stress and stability: applying the Anna Karenina principle to animal microbiomes. Nat. Microbiol. 2, 1–8.

Zanne, A.E., Abarenkov, K., Afkhami, M.E., Aguilar-Trigueros, C.A., Bates, S., Bhatnagar, J.M., Busby, P.E., Christian, N., Cornwell, W.K., Crowther, T.W., Flores-Moreno, H., Floudas, D., Gazis, R., Hibbett, D., Kennedy, P., Lindner, D.L., Maynard, D.S., Milo, A.M., Nilsson, R.H., Powell, J., Schildhauer, M., Schilling, J., Treseder, K.K., 2020. Fungal functional ecology: bringing a trait-based approach to plant-associated fungi. Biol. Rev. 95, 409–433.

Zanzotto, A., Morroni, M., 2016. Major biocontrol studies and measures against fungal and oomycete pathogens of grapevine. In: Compant, S., Mathieu, F. (Eds.), Biocontrol of Major Grapevine Diseases: Leading Research. CABI, Wallingford, pp. 1–34.

Zarraonaindia, I., Owens, S.M., Weisenhorn, P., West, K., Hampton-Marcell, J., Lax, S., Bokulich, N.A., Mills, D.A., Martin, G., Taghavi, S., van der Lelie, D., Gilbert, J.A., 2015. The soil microbiome influences grapevine-associated microbiota. MBio 6, e02527-14.

Zhang, X., Zhou, Y., Li, Y., Fu, X., Wang, Q., 2017. Screening and characterization of endophytic *Bacillus* for biocontrol of grapevine downy mildew. Crop Prot. 96, 173–179.

Zhang, S., Wang, Y., Chen, X., Cui, B., Bai, Z., Zhuang, G., 2020. Variety features differentiate microbiota in the grape leaves. Can. J. Microbiol. 66, 653–663.

Zhang, J., Liu, Y.-X., Guo, X., Qin, Y., Garrido-Oter, R., Schulze-Lefert, P., Bai, Y., 2021. High-throughput cultivation and identification of bacteria from the plant root microbiota. Nat. Protoc. 16, 988–1012.

Zherdev, A., Vinogradova, S., Byzova, N., Porotikova, E., Kamionskaya, A., Dzantiev, B., 2018. Methods for the diagnosis of grapevine viral infections: a review. Agriculture 8, 195.

Zhou, J., Ning, D., 2017. Stochastic community assembly: does it matter in microbial ecology? Microbiol. Mol. Biol. Rev. 81, e00002–e00017.

Zilber-Rosenberg, I., Rosenberg, E., 2008. Role of microorganisms in the evolution of animals and plants: the hologenome theory of evolution. FEMS Microbiol. Rev. 32 (5), 723–735.

Zohary, D., Hopf, M., 1993. Domestication of plants in the old world. In: Zohary, D., Hopf, M. (Eds.), The Origin and Spread of Cultivated Plants on West Asia, Europe and the Nile Valley. Clarendon Press, Oxford, England, pp. 143–150.

Zolti, A., Green, S.J., Sela, N., Hadar, Y., Minz, D., 2020. The microbiome as a biosensor: functional profiles elucidate hidden stress in hosts. Microbiome 8, 71.

Zorrilla, F., Buric, F., Patil, K.R., Zelezniak, A., 2021. metaGEM: reconstruction of genome scale metabolic models directly from metagenomes. Nucleic Acids Res. 49, e126.

CHAPTER THREE

Compendium of analytical methods for sampling, characterization and quantification of bioaerosols

Corinne Whitby[a,*], Robert M.W. Ferguson[a], Ian Colbeck[a], Alex J. Dumbrell[a], Zaheer A. Nasir[b], Emma Marczylo[c,d], Rob Kinnersley[e], Philippa Douglas[c,d], Gill Drew[b], Kam Bhui[f], Mark Lemon[g], Simon Jackson[h], Sean Tyrrel[b], and Frederic Coulon[b]

[a]School of Life Sciences, University of Essex, Colchester, Essex, United Kingdom
[b]School of Water, Energy and Environment, Cranfield University, Cranfield, United Kingdom
[c]UK Health Security Agency, Radiation, Chemical and Environmental Hazards Directorate, Oxfordshire, United Kingdom
[d]NIHR Health Protection Research Unit in Environmental Exposures and Health at the University of Leicester, Leicester, United Kingdom
[e]Environment Agency, Bristol, United Kingdom
[f]University of Oxford, World Psychiatric Association Collaborating Centre, Oxford, United Kingdom
[g]De Montfort University, Leicester, United Kingdom
[h]University of Plymouth, Plymouth, United Kingdom
*Corresponding author: e-mail address: cwhitby@essex.ac.uk

Contents

1. Introduction 104
2. Air sampling methods 109
   2.1 Active vs passive sampling 110
   2.2 Which air sampling method is best for cultivation or molecular analyses? 124
   2.3 Which air sampling method is best for which biological particle? 137
   2.4 Additional sampling considerations 143
   2.5 Summary of air sampling methods 146
   2.6 Recommendations and guidelines for standardization of air sampling methods 147
3. Bioaerosol analysis methods 149
   3.1 Culture-based methods 149
   3.2 Culture-independent methods 160
   3.3 Real-time (online) monitoring of bioaerosols 168
   3.4 Which bioaerosol analysis method for which taxa? 172
   3.5 Summary of bioaerosol analysis methods 175
   3.6 Recommendations and guidelines for bioaerosol analysis methods 176

Advances in Ecological Research, Volume 67
ISSN 0065-2504
https://doi.org/10.1016/bs.aecr.2022.09.004

4. Sampling and analysis methods to assess exposure and risk 177
4.1 Human exposure to bioaerosols 177
4.2 Standardizing "acceptable limits" for bioaerosols 185
4.3 How the various sampling and analysis methods would be useful for different environmental contexts for human exposure 190
4.4 Summary of sampling and analysis methods to assess exposure and risk 200
4.5 Recommendations and guidelines for assessment of exposure and human health risk 200
5. Conclusions and future perspectives 202
Acknowledgements 206
List of contributors 206
References 207

Abstract

Bioaerosols are suspensions of airborne particulate matter of biological origin (BioPM) which includes microorganisms and the products of these organisms. Bioaerosols are ubiquitous in indoor and outdoor environments and can become dispersed by attaching to other particles. Bioaerosols are diverse in terms of their size, composition and biological properties and are an important transmission route for infectious and sensitization agents. More recently, bioaerosols have received significant scientific and societal attention from industry, academia, government and the wider public due to the emergence and global spread of COVID-19 and the threat of bioterrorism. Yet despite their importance for human health, the microbiological components of aerosols and their species dispersal from various environments remains poorly understood. Moreover, there is a lack of understanding of the ecology and role that bioaerosols play in the environment.

As a result of these knowledge gaps, health officials and regulators have been hindered in their assessment of public and occupational health exposures and risk. For example, a better understanding of the concentrations and composition of bioaerosols in a particular environment, and the transmission dynamics of pathogens and their components, can inform on the appropriate ventilation rates and hygiene procedures to maintain good air quality and reduce human health risk. However, there are currently many uncertainties still remaining with regard to exposure assessment.

To better understand the impact of bioaerosol exposure on human health, comprehensive methods to detect, characterise and quantify bioaerosols are needed. Although significant advances in technologies for bioaerosol sampling and analysis have been achieved over the last two decades or so, a consensus on air sampling methods for a particular context or environment and a universal analysis method still does not exist. This makes it difficult for researchers to compare data across studies, and for regulators to set meaningful exposure limits.

BioAirNet is a UKRI NERC-funded project which acts as a leading voice for the UK BioPM science community and operates around four themes: Theme (1) BioPM sources and dynamics; Theme (2) BioPM sampling and characterisation; Theme (3) Human health, behaviour and wellbeing; and Theme (4) Policy and public engagement. As part of Theme 2, researchers, regulators, and public health officials have developed this compendium and Fig. 1 presents an overview of BioAirNet Theme 2.

This compendium aims to provide a comprehensive toolbox of current techniques, workflows, and technologies for bioaerosol sampling, characterisation, and monitoring across different environments for researchers, epidemiologists, regulators, public health officials and regulators involved in bioaerosols. *The overall goal* of this text is to support the development of useful standards to better regulate and monitor bioaerosols worldwide.

Abbreviations

AMV	Acceptable maximum value
BAMS	Bioaerosol mass spectrometry
CEN	European Committee for Standardization
CFU	Colony forming unit
DECOS	Dutch Expert Committee on Occupational Standards
DGGE	Denaturing gradient gel electrophoresis
EDC	Electrostatic dust cloth
ESP	Electrostatic precipitation
FISH	Fluorescent *in situ* hybridization
FLAPS	Fluorescence aerosol particle sensor
FTIR	Fourier transform infrared
GC	Gas chromatography
GTC	Growth tube collector
HPLC	High-performance liquid chromatography
HTS	High throughput sequencing
IMA	Index of microbial air contamination
IRMS	Isotope ratio mass spectrometry
ISO	International Standards Organization
LAMP	Loop-mediated isothermal amplification
LC	Liquid chromatography
LIF	Laser-induced fluorescence
LOAEL	Lowest-observed-adverse-effect level
MALDI-TOF	Matrix-assisted laser desorption/ionization mass spectrometry coupled with time-of-flight mass spectrometry
MCE	Mixed cellulose ester
MOUDI	Micro-orifice uniform deposition impactor
MPN	Most probable number
NOAEL	No-observed-adverse-effect level
ODTS	Organic dust toxic syndrome
OEL	Occupational exposure limit
PBS	Phosphate buffered saline
PAAS	Personal aeroallergen sampler
PCR	Polymerase chain reaction
PES	Polyethersulfone

PFU	Plaque forming units
$PM_{10}/PM_{2.5}$	Particulate matter (that are 10 or 2.5 μm and smaller)
PMT	Photo multiplier tube
PVDF	Polyvinylidene fluoride
PTFE filters	Polytetrafluoroethylene membrane filters
PY/GC/DMS	Pyrolysis gas chromatography differential mobility spectrometry
qPCR	Quantitative polymerase chain reaction
REPS	Rutgers electrostatic passive sampler
RSV	Respiratory syncytial virus
RT-qPCR	Reverse transcriptase quantitative polymerase chain reaction
SARS	Severe acute respiratory syndrome
SEM	Scanning electron microscopy
SIBS	Spectral intensity bioaerosol sensor
TGGE	Temperature gradient gel electrophoresis
TEM	Transmission electron microscopy
TLV	Threshold limit value
UV-APS	Ultraviolet aerodynamic particle sizer
VBNC	Viable but non-culturable
VOCs	Volatile organic compounds
WIBS	Waveband integrated bioaerosol sensor
WWTP	Wastewater treatment plant

1. Introduction

Bioaerosols are airborne particulate matter of biological origin (BioPM) with a particle diameter between ~10 nm and 100 μm in size with the majority mainly found in the range 0.1–10 μm. Because of their small size and mass, bioaerosols are easily transported from one place to another, facilitating the rapid spread of microorganisms and allowing for their genetic exchange (Fröhlich-Nowoisky et al., 2013). Bioaerosols can be generated from various natural and anthropogenic sources and are ubiquitous in indoor and outdoor environments, where they may remain airborne for long periods of time (Cox et al., 2020; Brown and Hovmøller, 2002; Coulon and Colbeck, 2017). Their residence time is influenced by a range of factors including particle size, meteorological conditions and source strength (Nazaroff, 2016; Kunkel et al., 2017; Kohanski et al., 2020).

Bioaerosols comprise of complex mixtures of airborne microorganisms (i.e. bacteria, fungi, archaea, viruses), the by-products of microbial metabolism, foreign proteins of microorganisms, fragments of cells, endotoxins, mycotoxins, enzymatic proteins, microbial volatile organic compounds (MVOCs) and allergens. Exposure to bioaerosols can lead to a range of deleterious health effects, including allergies, asthma, chronic obstructive pulmonary disease (COPD), and a range of respiratory and infectious diseases (Douwes et al., 2003; Douglas et al., 2017; Van Leuken et al., 2016; Bush and Portnoy, 2001; Dakhama et al., 1999; Gladding et al., 2003). Understanding and informing mitigations for the negative health impacts of occupational exposures is a major challenge for authorities and regulators. However, defining safe exposure limits for bioaerosols is difficult, because of their complexity, lack of dose-response relationships and multiple health outcomes. The health effects of bioaerosol components can also vary from person to person and from one environment to another.

Although there have been many studies to investigate the potential health effects of bioaerosols, there remains a lack of understanding of their composition, distribution, abundance, and dose-response relationships (O'Connor et al., 2015). Indeed, the risk of human exposure to bioaerosols is still difficult to quantify, particularly the temporal and spatial trends of infectious agents from different environments in real time. There is also a lack of information on how bioaerosols interact with other air pollutants in relation to human health effects. Such information is important for urban planning, and regulatory policy to better manage and monitor bioaerosol emissions from different environmental contexts. Before embarking on a bioaerosol study however, there are key decision events that need to be considered even before air sampling takes place and these are conceptualised in Figs. 1 and 2.

These decision events can be further broken down into some key questions that need to be considered before air sampling and analysis is undertaken (Box 1).

Once these questions have been addressed, this will facilitate the choice of air sampling method and downstream analysis to be performed. There are a variety of bioaerosol sampling methods in use today, (e.g. impactors, filters, and impingers). Each method has various advantages and disadvantages and no one method is favoured over another (see Section 2). Airborne microorganisms are difficult to assess accurately across different studies, due to differences in sample collection efficiency and variations in DNA extraction efficiencies (if molecular analysis approaches are being used).

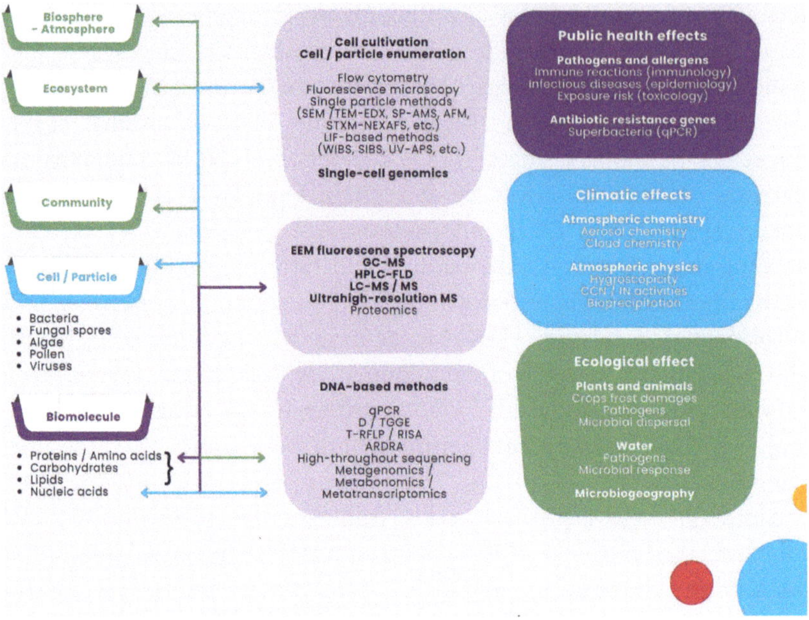

Fig. 1 Overview of Theme 2, BioAirNet.

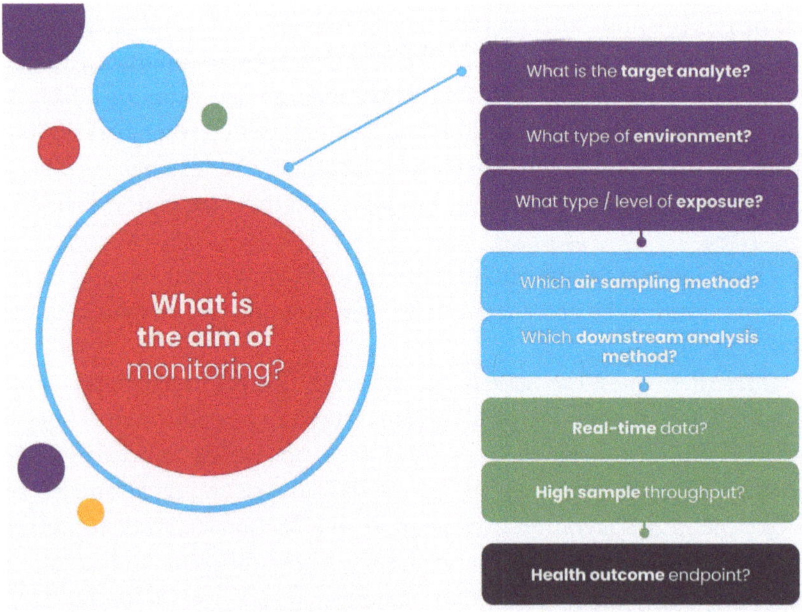

Fig. 2 Decision flow diagram conceptualizing the key decision events from target analytes to health outcome and endpoint.

BOX 1 Key questions to consider before air sampling

1. What are the research, regulatory or health impact questions to be addressed?
2. What are the objective and goals of monitoring (e.g. emission control efficacy, compliance monitoring, exploratory, exposure conditions)?
3. What information is needed (e.g. concentration, identification, worst-case concentration, particle size distribution)?
4. What is the relevance and relationship of average and worst-case concentrations?
5. Is personal or occupational exposure important?
6. What is the environment being sampled?—is it expected to have a high concentration of microorganisms or is this information unknown?
7. What is the anticipated nature and variability of the environment being sampled in time and space?
8. Are there logistical challenges to the sampling location (e.g. limited/no access to electricity supply)?
9. What human activity is occurring at the environment being sampled (e.g. office workers, hospital surgeries etc)?
10. What are the climatic conditions when the sample is being collected?
11. What are the optimal locations and time to collect samples (e.g. background, baseline, worst-case scenario, outdoors, indoors)?
12. Are microbiological taxa the target, or is it their specific components or products?
13. Are specific microbial groups or species (e.g. bacterial pathogens, viruses) important?
14. Are specific genes or biomarkers the target (e.g. pathogenicity genes, antimicrobial resistance genes) or are metagenomes needed?
15. What is the source of biological agent of interest?
16. Are viable microorganisms needed- for example for downstream culturing?
17. How will the collected air sample be analysed- for example culture-based or non-culture-based (molecular) methods?
18. What is the suitability, cost, availability, constraints, and requirements of sampling and downstream analysis methods?
19. What is the required sampling volume, the collection period (duration, frequency/interval) and the time pattern (e.g. continuous, periodic, random), including the levels of sensitivity for detection for the sampling/analysis method to be used?
20. Is real-time data important and/or high sample throughput required?

These challenges are exacerbated by the low biomass often recovered in air samples from certain environments (e.g. clinical settings), and thus obtaining a representative microbiological sample for downstream analyses can be very challenging. This is particularly crucial when investigating viral communities, given the extremely low biomass of viral particles in the air compared with other environments. Consequently, bioaerosol abundance and biodiversity can be underestimated, making it more difficult to link the specific bioaerosol components to particular diseases or human health effects. More efficient methods for air sample collection are therefore urgently required. Sampling method selection, however, is typically determined or dictated by the intended downstream analysis (and vice versa) and the key bioaerosol of interest. Consequently, differences in sampling strategies have hampered comparisons of bioaerosol measurements worldwide. Since there remains no consensus on which air sampling methods to use, no standardised protocol has been established. *Realistically, bespoke air sampling methods may need to be developed that are tailored to the environment being sampled and the specific research question being addressed, in order to set useful guidelines and recommendations for human health exposure limits.*

There are a variety of downstream methods that can be used to analyse air samples once they have been collected. These analysis methods can be broadly categorised into culture-dependent and culture-independent methods (see Section 3). Culture-based techniques are relatively simple and cost-effective and allow viable microorganisms to be isolated for further analyses. Often the regulator is only interested in viable microorganisms, by which they mean culturable. Yet it is a misconception that the culturable fraction of the community reflects the active/infective microorganisms. This because <1% of microorganisms are viable but non-culturable (VBNC) (Amann et al., 1995). There are also several inherent limitations with culture-based methods, which can lead to an underestimation or a misrepresentation of the communities present (Colbeck and Whitby, 2019).

Bioaerosol research however, has recently been enhanced by molecular tools, especially high throughput sequencing (HTS), which have allowed a rapid and more detailed characterisation of air samples, and overcomes the drawbacks of culturing. The large datasets obtained from HTS approaches also enables associations between microorganisms, microbial markers and human health effects to be determined. These vast datasets will further push epidemiological and toxicological studies to determine the health impacts on human exposure and will take bioaerosol research much closer towards establishing standards for acceptable bioaerosol exposure limits, that are applicable worldwide (see Section 4). However, these analysis methods are currently being used with no clear guidelines and recommendations are therefore needed for the most appropriate air sampling and analysis methods.

This compendium seeks to provide advice for bioaerosol researchers to help users decide what sampler to deploy (Section 2) and what bioaerosol analysis technique to use for a particular study or environment (Section 3).

This text also sets out how these sampling and analysis approaches can be used to assess human health exposure and risk that can benefit authorities and regulatory bodies so that meaningful guidelines for exposure limits may be set (Section 4).

2. Air sampling methods

There are a wide variety of bioaerosol sampling methods currently available, yet no standardised procedures have been firmly established. This is largely because there is still much debate on which sampling techniques to use (which is often determined by the downstream analyses to be performed, the type of sampler to use, the frequency, duration and location of sampling, and even whether checks, controls and recording sampler performance parameters are useful) (Napoli et al., 2012; Pasquarella et al., 2008). Since international standards leave the choice of system open, it is often very difficult to compare results from across different studies (Pasquarella et al., 2008). This is especially challenging when different samplers show a high variability in results and inherent biases can be introduced (Pasquarella et al., 2008; Ferguson et al., 2019). There is also lack of standardised metadata requirements to enable comparisons to be made across different studies. Furthermore, sampling design significantly affects the representativeness of the bioaerosol sample collected. Indeed, the same considerations one would undertake to collect a representative sample of aerosols in the environment also applies to bioaerosols (Kulkarni et al., 2011). At present, there is no consensus regarding the methods to be used and even how to interpret the data, especially in terms of health exposure (Viani et al., 2020), which makes bioaerosol research and monitoring challenging for the end-users, regulators and health officials involved.

2.1 Active vs passive sampling

Broadly speaking, there are two different types of sampling techniques to assess microbial air contamination—*active* and *passive* sampling—that are summarised in Fig. 3 and are detailed below.

Active sampling draws a known volume of air physically through or over a particle collection device (which can be a liquid or solid media). The sampler inlet must be designed to avoid disturbing the airflow as it enters the sampler, (i.e. air should enter the sampler isokinetically, to more accurately reflect the particle size of the sampled air (Delort and Amato, 2018)). Results are traditionally expressed as colony forming unit per cubic metre ($CFU\,m^{-3}$), although other metrics (such as particle number m^{-3} or relative abundance) are also used. Active sampling approaches can give high volumes and high biomass yield and so is more applicable to use when low concentrations of microorganisms are expected, such as hospital operating theatres (Napoli et al., 2012). Active methods also allow the quantification of bioaerosol concentration (Mainelis, 2020). Several types of devices are available, such as air impactors, impingers and filtration systems, which are described in more detail below.

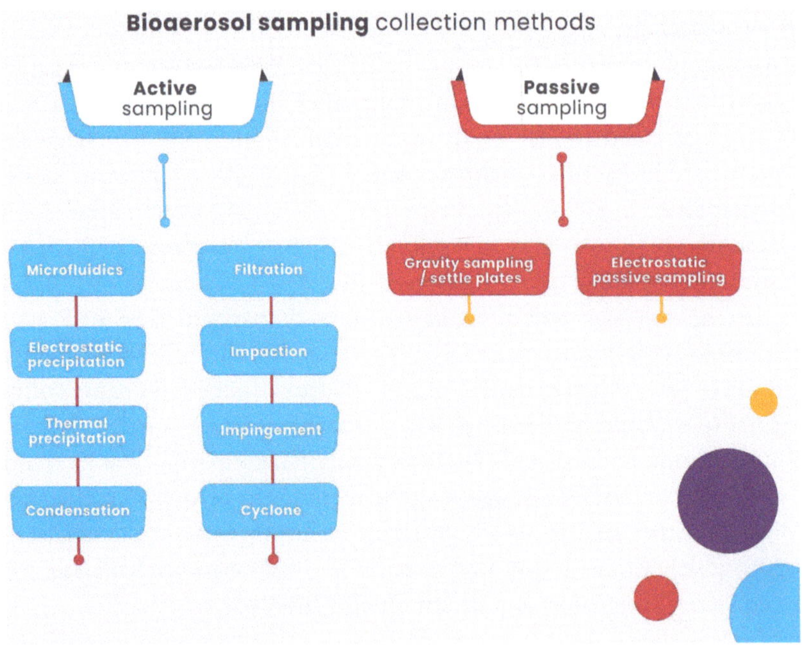

Fig. 3 Flow chart summarizing active and passive air sampling methods.

2.1.1 Overview of active sampling systems

(a) *Filtration:* Filtration systems are simple to use and typically have high collection efficiencies (>95%) for particles >0.5 μm in diameter (Miaskiewicz-Peska and Lebkowska, 2012). In general, filter pores sizes are between 0.2 and 0.45 μm (with 0.45 μm pore size the most commonly used) (Yoo et al., 2017). Filtration samplers use the mechanisms of interception, inertial impaction, diffusion, gravitational settling, and electrostatic attraction to collect biological particles onto a filter (Ghimire et al., 2019). This allows the collection of a wide range of particle sizes (Ghimire et al., 2019). The air flow rate is determined by the filter media and filter area. There are three main classes of filter materials: fibrous, membrane and flat filters, which include: cellulose nitrate or mixed cellulose ester (MCE), nylon, polycarbonate, gelatin and glass fibres. Fibrous filters trap particles within a fibre matrix (e.g. glass fibre, cellulose), whilst membrane filters have a complex structure of pores within which particles are deposited (e.g. gelatin, polyvinyl chloride) and flat filters pass air through the membrane pores and particles collect on the filter surface (e.g. polycarbonate) (Ferguson et al., 2019). For example, the Button Aerosol Sampler and IOM cassette (both by SKC Inc., Eighty Four, PA) are inhalable samplers, adapted for bioaerosol collection using various filters, including gelatin (Chang and Hung, 2012). Polyurethane foams (PUF) have also been used to collect thoracic and respirable bioaerosol fractions (Kenny et al., 1999).

(b) Impactors*:* There are several impactor systems and the most commonly used devices include jet-to-agar or slit-to-agar impactors, centrifugal agar impactors, multi-nozzle, sieve samplers, single and multi-stage impactors. They all require the use of an air pump to capture air which is then distributed evenly over the surface of agar plates for culture analyses (Macher, 1989; Willeke and Macher, 1999) or other solid collection surface such as a glass slide for microscopy analysis (Morris, 1995). Slits or holes control airflow, or in some cases, by varying the flow rate (typically between 10 and 700 L min^{-1}) (Yoo et al., 2017). To avoid overloading of the collection surface it is advisable to calculate the optimal sampling time (Reponen et al., 2011). Examples of impactors include spore traps, MB2 and N6 multi-hole impactors, and AirTrace slit-to-agar impactors. With the slit-to-agar impactor, a high-velocity stream of air is directed through a narrow slit and the particles impact on an agar plate by their inertia (Cooper et al., 2019; Calderon et al., 2002; Schafer et al., 2003; Willeke and Macher, 1999). There are also size-differentiating cascade impactors (e.g. Micro-Orifice Uniform Deposition Impactor (MOUDI), Andersen cascade impactor, Tisch cascade impactor), which have eight stages and distribute the captured particles into several size categories (such as cut-sizes down to 0.056 μm with the MOUDI) (Maenhaut et al., 1993). A number of hand-held Andersen-type single-stage impactors have also been developed that collect particles between 1 and 8 μm; with flow rates between 100 L min^{-1} and 180 L min^{-1} (Yao and Mainelis, 2006). Hand-held impactors are typically used in culture-based analyses, but some have been adapted for molecular analyses (Chang and Hung, 2012).

The collection efficiency of impactors depends on the diameter and density of the particle, diameter of the nozzle, and the air velocity of the jet (Haig et al., 2016). The d_{50}, (cut-off diameter) is defined as the diameter for which a collection efficiency of 50% is reached and is directly linked to the distance between the inlet and impaction plate and depends on the flow rate (Delort and Amato, 2018). Virtual impactors use the centrifugal force and inertia to separate particles, and have a collection probe, which carries the larger particles to a collection filter and the smaller particles are collected separately (Haig et al., 2016). Sticky plastic rods (e.g. Rotorods) or sticky glass plates may also act as impactors, where flow rate can be adjusted to vary the size of the particle collected (e.g. VersaTrap spore trap cassette) (Haig et al., 2016). There are also passive impactors where the particles are separated from the air either by displacement, natural wind, or turbulence.

(c) *Impingers:* Impingers use inertia to collect bioaerosols into a liquid and also by diffusion at the liquid interface (Cooper et al., 2019). Examples of impinger systems include SKC BioSampler (three tangential nozzles) and AGI-30 (single nozzle) impingers. The diameter of the particles collected is determined by the distance from the nozzle outlet to the surface of the liquid and the flow rate (Haig et al., 2016).

(d) *Cyclones:* The cyclone device creates a centrifuge or a cyclonic action to separate out particles which is captured on a filter paper mounted inside a cassette at the top of the sampler (e.g. two-stage cyclones). The particle-laden air is forced by the shape of the collection chamber into a spiral flow (Haig et al., 2016). Other cyclones use centrifugal force to carry particles towards the cyclone wall where they are separated from the air flow into a liquid (e.g. Coriolis micro) or wall/cone (e.g. Coriolis compact) (Haig et al., 2016). When the airflow reaches the bottom of the cyclone, the direction is reversed and smaller, uncollected particles go out of the cyclone (Haig et al., 2016).

(e) *Droplet-based microfluidic chips:* Droplet microfluidics is based on the creation of monodisperse microdroplets containing compartmentalised assay reagents (e.g. labelled antibodies, PCR reagents) created on a microfluidic droplet chip (Damit, 2017). The fluids are manipulated in channels with dimensions of tens of micrometres, and as a result, the bioaerosols are precisely focused into the droplet (Kaushik et al., 2020). The assay reagents react with the target species during an incubation step and the target is identified using appropriate downstream methods (Damit, 2017). Microfluidics has emerged as a distinct field with exciting potential for diagnostics.

(f) *Electrostatic precipitation (ESP):* This is a less widely used method as only a limited number of ESP samplers are commercially available. The bioaerosol particles are electrically charged at the inlet before progressing through an electric field, where they are separated from the air flow and deposited onto charged plates or collection medium by electrostatic attraction or repulsion (Haig et al., 2016). The collected particles are then removed from the collection medium for analysis. Here, the particle velocity is between 2 and 4 orders lower than with impactors or impingers (at comparable sampling flow rates) and so can be less damaging to microorganisms (Zhen et al., 2013).

(g) *Condensation* techniques (*Laminar-flow water-based condensational "growth tube" collector (GTC)*)*:* This method involves passing the air sample through a humidifier. The warm, humid air is rapidly cooled with the bioaerosol particles acting as condensation nuclei. This method has been used to improve the detection of smaller microbes and viruses (Degois et al., 2021; Ghosh et al., 2015; Park et al., 2011).

(h) *Thermal precipitation:* This technique relies on a process known as thermophoresis, whereby the bioaerosol particles pass through a narrow channel with a temperature gradient perpendicular to the flow.

The particles move away from the hot surface where they then deposit (Waldman and Schmitt, 1996).

(i) *Personal samplers:* These are small, portable devices that are attached to a persons' clothing and can be used to assess personal exposure, but typically have low flow rates (e.g. IOM 2L min^{-1}, Button sampler 4L min^{-1}) and so longer sampling times are often needed (Haig et al., 2016). With personal samplers, the particle fraction(s) that will be sampled: (i.e. total particulates, PM_{10}, $PM_{2.5}$, inhalable, thoracic, respirable, or combinations of these) has to be considered (Mainelis, 2020). Built-in pumps offer user mobility and convenience. The power requirements of personal samplers often dictates how long a sample could be collected, especially when battery-operated air movers are used. Here, filters with higher porosity are typically used to extend sampling time without a substantial loss of collection efficiency (Mainelis, 2020). Various filters including gelatin are also used in personal samplers (e.g. Button Aerosol Sampler and IOM cassette both by SKC Inc., Eighty Four, PA). Electrostatic collection is a low power method, that also can be applied to personal sampling (Mainelis, 2020; Frankel et al., 2012; Noss et al., 2010).

2.1.2 Overview of passive sampling systems

Unlike active samplers, passive samplers do not use air movers but instead particles are collected using gravitational settling and/or electrostatic attraction, or a combination of the two (Manibusen and Mainelis, 2022), onto a collection surface such as a Petri dish that contains culture media (known as "settle plates"). When using settle plates, the plates are exposed to the air for a given period of time and then incubated. Gravitational settling of particles however is affected by particle size, shape and air motion (Reponen et al., 2011), where larger particles are preferentially captured (Burge and Solomon, 1987). The smaller, lighter particles can remain airborne for longer than larger particles and may even remain suspended indefinitely (Haig et al., 2016). Careful decontamination is also required with passive sampling (Delort and Amato, 2018). Recently, the Rutgers Electrostatic Passive Sampler (REPS) was developed using a polyvinylidene fluoride (or PVDF) film, which improved passive particle capture as it works by both electrostatically attracting charged particles alongside gravitational capture (Therkorn et al., 2017).

The passive method been standardised, has to some extent, by the Index of Microbial Air contamination (IMA), which corresponds to the number of CFU counted on a Petri dish (9 cm in diameter) left open to the air according to the 1/1/1 scheme (for 1 h, 1 m above the floor and about 1 m away from walls and major obstacles) (Pasquarella et al., 2000). The IMA is expressed as CFU m^{-2}, dm^{-2} or cm^{-2} per unit time (Viani et al., 2020). Five classes of contamination and maximum acceptable levels of IMA have been proposed that are related to different infection or contamination risks: 0–5 very good; 6–25 good; 26–50 fair; 51–75 poor; >76 very poor (Pasquarella et al., 2000). The metrics used to quantify bioaerosols by passive sampling is CFU $plate^{-1}$ h^{-1} if using settle plates or the amount that has settled per surface area per unit time such as X cm^{-2} min^{-1} (Manibusen and Mainelis, 2022).

Whilst passive sampling is much less frequently used, the approach has various advantages and has been applied to both indoor and outdoor environments (Manibusen and Mainelis, 2022). For example, passive sampling can be used for extended sampling periods or even indefinitely and therefore is useful for determining exposures over longer time frames. It is a non-intrusive method with a simple design and that does not require any specialised mechanical equipment, making it inexpensive. However, passive sampling cannot easily be related to airborne concentrations, since the volume of air is not measured and therefore is not quantitative. Despite this, some studies suggest that passive sampling provides valid metrics for risk assessment, as it may capture the harmful part of bioaerosols which settle out onto a critical surface, such as surgical cuts, or on the instruments in operating theatres (French et al., 1980), and thereby identifies locations where more detailed sampling and analysis efforts need to be deployed. For example, studies using an Electrostatic Dustfall Collector (EDC), which collects airborne dust on a large scale, has been used in high-exposure occupational environments to measure endotoxins and allergens (Samadi et al., 2010; Kilburg-Basnyat et al., 2014; Noss et al., 2010). Passive sampling can therefore complement active sampling methods, particularly for characterising bioaerosol sources and their relative contributions (Duchaine and Mériaux, 2001). However, each air sampling method described above has various advantages and disadvantage which are summarised in Table 1. Moreover, passive and active sampling methods can yield very different results in terms of species diversity due to size differences of the captured microbes, deposition rates, and other factors (Cox et al., 2017).

Table 1 Advantages and disadvantages of active and passive air sampling methods to characterize and monitor microorganisms in bioaerosols.

Air sampling methods	Advantages	Disadvantages	References
Active Sampling Methods	• In general, can be used for both culture-based and molecular analyses (see below) • High volume and high biomass. (Large volumes of air can be sampled within short periods (hours-days) or more extended sampling, and so suitable for capturing temporal variability as well as microorganisms present in lower abundances (e.g. hospitals)	• Sample collection is dependent on the size selectivity of the sampler • Expensive	(Cox et al., 2020) Napoli et al. (2012)
Filtration Systems (e.g. cellulose nitrate or mixed cellulose ester (MCE), nylon, polycarbonate, gelatine, glass fibres)	• Simple, relatively inexpensive and widely used. High collection efficiencies (>95% for particles >0.5 μm in diameter) making them suitable for collecting bacteria, fungi and viruses • Give best recovery of molecular material for molecular analysis. DNA may be directly extracted from the filters for downstream molecular analyses • High sampling volume can be achieved as can be sampled over longer sampling time periods (For example, high volume air sampling using a military Dry Filter Unit 1000 DFU-1000 produces significantly higher relative sampling efficiency compared	• For culture-based analyses, particle elution from the filter is an extra step, that can lead to particle losses. This is especially the case with fibrous filters • Type of filter and extraction buffer can influence endotoxin extraction efficiencies (with increased efficiencies with glass fibre filters and 0.05% Tween. • Potential for sampling bias whereby spore forming microorganisms may be preferentially recovered, depending on filtration time, pore size and filter type • Gelatin filters demonstrate some bacterial contamination and so not suitable for molecular analyses. Potential for filter overloading and contamination by	Casey (2019) Chang and Hung (2012) Cooper et al. (2019) Cox et al. (2020) Ferguson et al. (2019) Haig et al. (2016) Miaskiewicz-Peska and Lebkowska (2012) Gandolfi et al. (2013) Haig et al. (2016) Olenchock et al. (1989) Spaan et al. (2008) Yoo et al. (2017)

	to the commercial Biocapture-650, XMX2L-MIL, and Microbio MB2) • Captured bioaerosols remain viable and can be resuspended back into the air using reverse air flow. However, this is not always the case as samples can become desiccated • Can be used for both culture-based (e.g. gelatine) and molecular (e.g. polycarbonate) analyses. Also amenable for direct microscope analysis • Can be used to analyse background microbial communities and provides an accurate assessment of total coverage • Can be used with personal samplers attached to workers' clothing to provide an assessment of human exposure	non-biological particles when using microscopy • Longer air sampling times are required to obtain sufficient DNA yield for molecular analyses (e.g. for metagenomics/metabarcoding approaches, and pooling of samples may still be required for an amplification-free metagenome analysis) • Longer sampling times could lead to sample desiccation (especially with Gram negative bacteria) although Gelatin filters or wetted porous foams can be used to reduce desiccation effects • Can suffer from sample overloading • Filters with smaller pore sizes typically have a higher air flow resistance, which reduces the maximum air flowrate • Personal samplers have low flow rates (some as low as 2 L min^{-1}) and so longer sampling times are needed	
Impactors (e.g. jet-to-agar or slit-to-agar impactors. Various single-stage and multi-stage Anderson impactors, Aerotech N-6 impactor, Air Test	• Convenient and easy to use, portable and relatively inexpensive • Highly suitable as particle size classifiers, with particles greater than a given size being separated from the air flow, whilst smaller particles remain airborne. (e.g. Cascade impactors is based on the particle-laden air	• Recovery efficiency is often low as low sample volumes due to low flow rates, shear forces may cause additional losses and the captured microorganisms may not always adhere to the agar surface (particle bounce)	Alonso et al. (2017) Cox et al. (2020) Cooper et al. (2019) Delort and Amato (2018) Griffin (2007) Pankhurst et al. (2012)

Continued

Table 1 Advantages and disadvantages of active and passive air sampling methods to characterize and monitor microorganisms in bioaerosols.—cont'd

Air sampling methods	Advantages	Disadvantages	References
Omega, Air Samplair Mas-100, BioImpactor 100-08, BioStage impactor)	flow being passed through successive tiers of nozzles and impaction plates. Each tier (or stage) collects particles of a specific size. The fraction of the total mass in each particle size range can be established) • Allows direct collection of microorganisms • Allows the total sample to be examined without any need for extraction or sub-sampling • Reduces post sampling efforts and errors • Reduce in between sterilization • Allows collection of smaller particles. (Both Andersen and Tisch cascade impactors perform equally well for detection of total virus concentration) • Can be used for both culture-based and molecular analyses • Collection efficiencies can be increased by the use of multi-jet impactors (e.g. with >100 nozzles) • The use of a collection probe with virtual impactors rather than an agar plate avoids issues of particle bounce and deposition build-up • Virtual impactors useful as particle concentrators	• Particles with sufficiently low inertia will be carried by the streamlines and escape capture. Often have low efficiencies with small particles • Airflow over the impaction plates may also be disturbed by the build-up of deposited particles, leading to altered collection efficiencies • Cell viability may be lost due to impact stress • Less efficient for culture based methods compared to other methods (e.g. gelatine filters) • Cascade impactor does not necessarily behave like a sieve as some particles are deposited between stages or bounce off the impaction plates and avoid capture • Virtual impactors can have collection losses near to the size of the cut-off diameter at the inlet of the probe • Possibility of microbial overload on plates resulting in inaccurate enumeration and/or difficulty in obtaining isolates • With passive impactors the airflow is not precisely controlled, so the cut-off diameter remains unknown and varies with the environmental and meteorological conditions	Haig et al. (2016) Willeke and Baron (1993)

Impingers (e.g. SKC BioSampler and AGI-30 impingers, SAS sampler)	• Relatively low cost, widely used • Shorter air sampling time than filtration needed to obtain sufficient biomass and DNA yield for molecular studies • Moderate to high collection efficiency (~70% at 0.5 μm and ~90% at 1 μm). The diffusion process enhances the collection efficiency especially of small particles especially viruses • Liquid collection matrix can be distributed between various different analyses (e.g. culture and non-culture) • Angled nozzles may reduce shear stress on captured microorganisms and minimize re-aerosolization of captured organisms from the collection fluid • Airflow rates for impingers are generally higher (300–600 L min^{-1}, compared to 2–300 L min^{-1} for filtration), permitting shorter sampling periods • Collection on to liquid prevents desiccation of the collected particles and overcomes enumeration issues due to sample overloading • Can be used for a snapshot assessment, or spatial/temporal studies, may also detect the rare members of a community • Often used to collect airborne viruses for culture-based analysis as the liquid collection media preserves the viability of sensitive viruses much better than dry collection	• Inlet characteristics of the sampler influence collection and recovery efficiency. Some low flow rate impingers have low capture rates, particularly of virus-sized particles and a loss of viability (e.g. the liquid impingers AGI-30 and BioSamplerTM both frequently failed to detect the presence of nebulizer-generated *C. difficile* spore aerosols in a lab environment), so may not be useful for culture-based analyses • Sample loss, lower cell numbers recovered-Evaporation of impinger liquid occurs if sampling is >30 mins results in potential sample loss and a decline in microbial viability due to desiccation. (e.g. impinger operation produced significant culturability losses and failed to retain *C. difficile* endospores). However, it is possible to continue collection following refreshing of the collection fluid. Further sample losses may also occur with re-aerosolization of previously collected particles and adherence of particles to the internal walls of the collection chamber. However, viscous fluid e.g. mineral oil can be used to reduce evaporation losses • Potential for cell viability to be lost due to impact stress	Agranovski et al. (2004) Casey (2019) Cooper et al. (2019) Ghimire et al. (2019) Haig et al. (2016) Lin et al. (1999) Verreault et al. (2008)

Continued

Table 1 Advantages and disadvantages of active and passive air sampling methods to characterize and monitor microorganisms in bioaerosols.—cont'd

Air sampling methods	Advantages	Disadvantages	References
		• May be less efficient for culture-based methods compared to other methods (e.g. gelatine filters) as shear forces in the jet in conjunction with the turbulence caused by the air being forced into the chamber may result in loss of viability • Glass tubes can get easily broken with field sampling and so their robustness is a concern • In between sampling sterilisation required	
Cyclones (e.g. Coriolis® μ/Compact SASS 2300, Burkard Cyclone sampler)	• Can be used for collecting large volumes of air and operates at high flow rates • Gentle, low-stress sampling method • Smaller cyclones can be worn on a person's clothing and so has the potential to sample in potentially hazardous environments to assess exposure • Efficiency can be increased by injecting a film of liquid near the cyclone's inlet to wet the cyclone walls and the liquid can then be collected for analysis • Frequently used as a pre-screen i.e. removing larger particles from an airflow before further size classification is deployed with other types of the sampler	• Generally larger particles are more likely to be collected than smaller particles, causing an under-sampling of smaller bioaerosols • A calibrated airflow is essential for accurate collection efficiency • Shear forces may reduce viability and reduce collection efficiency. Collection losses may also occur via evaporation of the collection liquid, resulting in re-aerosolization of previously collected material • Liquid carryover may also occur causing sample losses whereby the liquid injected into the cyclone travels over the cyclone roof and vortex before escaping from the system, carrying with it collected particles	Griffiths and Boysan (1996) Ferguson et al. (2019) Haig et al. (2016) Krames and Buttner (1994)

Droplet-based Microfluidic chips	• Can map single aerosols to individual droplets • Precise aerosol loading • Immediate initiation of assay upon collection • A series of intermediate droplet manipulations can be performed to enable multiple assays • Allows microelectronics-based droplet manipulations during collection	• Not widely used	Damit (2017)
Electrostatic Precipitation	• Less damaging to the microorganisms than impactors and impingers due to lower particle velocity • Most are open-channel devices thus allowing the use of batteries to operate the device • Compatible with multiple collection media types, (e.g. agar, liquid, and solid surfaces) and can be analysed by a variety of downstream methods • Can be incorporated into personal samplers and thus allows concentrating of personal samples improving the detection limit and is especially useful in low concentration environments • Avoids unwanted stresses on samples • Very low power is needed • Collection efficiency is mainly controlled by the self-parameters such as voltage, flow rate, dimensions of the precipitators and initial particle charging level	• Not widely used. Limited number commercially available • Production of ozone during particle charging may affect microorganism viability and integrity as well as potential health effects of personnel with the generation of secondary by-products. Viability dependent on electric charge • Desiccation of microorganisms is a concern when sampling over longer time periods • limiting initial charge at inlet may result in loss of microorganisms	Ghosh et al. (2015) Han et al. (2018) Mainelis (2020) Mainelis et al. (2002) Gollakota et al. (2021)

Continued

Table 1 Advantages and disadvantages of active and passive air sampling methods to characterize and monitor microorganisms in bioaerosols.—cont'd

Air sampling methods	Advantages	Disadvantages	References
Thermal Precipitation	• High collection efficacy for particle sizes ranging from 0.01 μm to 5 μm	Low collection rate ranging between $7\,cm^3\,min^{-1}$ and $1\,L\,min^{-1}$	Watson (1958) Cox and Wathes (1995)
Condensation (e.g. BioSpot-VIVAs™, The Spot Sampler™ BC, BioSpot-Gem™)	• Quick method (entire process takes less than 1 second), increasing detection efficiency • Can be used effectively for the detection of small microbes including viruses. High efficiency across 5 nm to 10 μm particle sizes • Collection is gentle and does not suffer from desiccation, thermal/mechanical stress and re-aerosolization • Samples can be concentrated into various media	• Not widely used • Complex to use • Heat transfer to the microbes may result in a loss of viability • Not portable	Ghosh et al. (2015) Park et al. (2011) Gollakota et al. (2021)
Microfluidics	• Rapid capture (e.g. shown to distinguish between *E. coli* and non-biological aerosols within 20 s)	• Variable recovery efficiency (e.g. between 27% and 100%)	Damit (2017)
Passive Sampling Methods Generally low volume and low biomass (e.g. Agar settle plates, settling filters, Electrostatic dust fall collectors (EDCs))	• Can be used for extended sampling periods • Does not require any mechanical equipment so easy operation and portable without the need for a power source. As no mechanical equipment is required, passive sampling does not disturb the surrounding air	• Can cause the microorganisms to desiccate and lose viability • As samples are impacted directly onto nutrient agar, it is limited to culturing each sample on one growth medium at a time • Passive sampling rates are generally low $<5\,m^3\,day^{-1}$	Cox et al. (2020) French et al. (1980) Haig et al. (2016) Hayward et al. (2010) Napoli et al. (2012) Noss et al. (2010) Smith et al. (2018)

	• Non-intrusive. Relatively low cost, easy to use. Does not require prior optimisation of flow rates • Useful when determining the contamination of surfaces by settled particles, such as the contamination of wounds during surgery or food in production lines • Can provide a valid risk assessment as it measures the harmful viable components of bioaerosols that fall onto a critical surface (e.g. surgical cuts) • Allows the total sample to be examined without any need for extraction or sub-sampling • Generally used for culture-based analyses though nucleic acids may be extracted directly from the collected dust	• Data is treated generally as qualitative since the sampled air volume is unknown • Generally low biomass so not usually used for molecular analyses • The source volume of air collected sample is usually unknown • Only collects particles large enough to deposit	Manibusen and Mainelis (2022)
Gravity sampling/Settle plates	• Provides a picture of the total number of collected bioaerosols • Settled dust can be used as a means of identifying bioaerosols in buildings, especially allergens and mould • Easy to use for the participants of a study to be sent the plates, adhesive tape etc by mail	• Non-quantitative and non-qualitative. • Settled dust can also contain particles from non-aerosol sources which can make data interpretation difficult.	Cox et al. (2020) Burge and Solomon 1987
Electrostatic passive samplers (e.g. Electrostatic Dustfall Collector (EDC)	• Can be been used in high-exposure occupational environments • Can be used to measure endotoxins and allergens		Samadi et al. (2010) Noss et al. (2010)

In addition to the active and passive air sampling methods described above, there has been recent developments in the real-time measurement and identification of aerosol particles directly in the ambient air within seconds. These on-line methods include Light-Induced Fluorescence (LIF), ultraviolet aerodynamic particle sizer (UV-APS), Spectral Intensity Bioaerosol Sensor (SIBS) and Wideband Integrated Bioaerosol Sensors (WIBS), which have been utilised in continuous measurement of bioaerosols (O'Connor et al., 2015; Könemann et al., 2019; Nasir et al., 2019; Tian et al., 2020). These systems provide real-time data on local concentrations of airborne biological aerosol particles (pollen, spores etc.) and other particles and are described more fully in Section 3. These air sampling devices also differ in terms of their biological and physical efficiencies, making comparisons between different studies difficult (Viani et al., 2020). Table 2 compares some specific samplers in terms of their technical specifications and efficiencies.

2.2 Which air sampling method is best for cultivation or molecular analyses?

The choice of air sampler and air sampling method comes with certain caveats. One of these considerations is what the downstream analyses of the bioaerosols will involve. For example, whether this analysis will involve cultivation into media (i.e. culture-based analysis) or the extraction of nucleic acids for molecular analyses (i.e. non-culture-based analysis). Fig. 4 summarises the most appropriate type of air sampling methods for either cultivation or culture-independent downstream analysis methods.

In general, active sampling devices can be used for both culture and molecular analyses. The three most common methods for collecting bioaerosol environmental DNA for molecular analyses are air filtration, liquid impingement and impaction (Ferguson et al., 2019; Pankhurst et al., 2012). Impingers and impactors are also commonly used for culture-based analysis, especially of airborne indoor bacteria (Lindsley et al., 2017). Filters have also been widely used for both molecular (by extraction into a liquid) and for culture-based analyses (by collecting on a gelatin filter, an agar surface or into liquid media) (Lindsley et al., 2017; Ghimire et al., 2019). Filter selection is largely dependent on the aim of the research (Ferguson et al., 2019). For example, glass fibre or polyvinyl chloride membrane have been used for gravimetric analysis (Ghimire et al., 2019); whereas polycarbonate or cellulose ester are better for microscopy (Humbal et al., 2018). However, there is the potential for filter overloading and contamination by non-biological

Table 2 Key features of some example air samplers for the characterization of bioaerosols

Air sampler	Key features[a]
Laminar-flow water condensation growth tube (CGT) method	
BioSpot-VIVAS™ (Aerosol Devices Inc)	• High efficiency >90% <10 nm to 10 μm • Collects viruses, bacteria, fungal spores, toxins, proteins, allergens • Collects at 8 L/min flow rate (approximate to human lung breathing rate) • Collects into a Petri dish for particle into liquid sampling • High sensitivity as samples concentrated into a liquid • Gentle collection so cell viability is maintained • Can instantly preserve DNA/RNA so suitable for downstream molecular analyses
Series 110A Liquid Spot Sampler™ (Aerosol Devices Inc)	• Versatile for wide range of different applications • Collects a concentrated suspension of particles into a small liquid-filled vial that can be configured for on-line batch sampling • Gentle collection, no turbulence so no particle bounce or re-aerosolization • Captures both soluble and insoluble particles • High collection efficiency >95% for particles from 5 to 2.5 μm for dry collection and >90% for particles up to 10 μm for liquid collection • Collects into both liquid and as a "spot" deposit on solid substrates for concentrated sample collection for higher sensitivity • Moderate temperatures required for condensational growth (25–30 °C) • No steam injection required • Particles are concentrated in a 0.5 mL volume of liquid • Uninterrupted sampling from minutes to hours • Automated sample handling eliminates need for sample preparation
BC, BioSpot-Gem™ (Aerosol Devices Inc)	• Fast and easy to use, weight 4.5 kg • High efficiency (>95% effective for particles <10 nm to 10 μm) using a sterile swab • Collects viruses, fungal spores, toxins, proteins, allergens, particles from exhaled breath

Continued

Table 2 Key features of some example air samplers for the characterization of bioaerosols—cont'd

Air sampler	Key features[a]
	• Suitable for field applications (e.g. public, industrial, medical, agricultural environments), infectious disease surveillance, IAQ mitigation assurance • Maximum particle collection of 10^5 particles cm^{-3} • Compact, lightweight, and quiet operation • Flow rate 1.2 L min^{-1} • Can sample from between 1 min up to 24 h • Samples onto a stabilizer for instant DNA/RNA preservation for downstream molecular analyses • Compatible with virus/microbial testing methods (e.g. PCR, sequencing) • Environmental temperature range of 10–35 °C • Decontamination with hydrogen peroxide, ozone, or alcohol rinse
Impactors	
Next Generation Impactor (NGI) (Copley)	• User friendly with easy automation • Flow rate 28.3 L min^{-1} • 60–90 L min^{-1} for high flow rate testing whilst retaining the 28.3 L min^{-1} cut off diameters • Seven stages (five with cut-offs between 0.54 and 6.12 μm at flow rates of 30 to 100 L min^{-1} • Particle size range 0.4–9.0 μm (28.3 L min^{-1}), 0.3–8.6 μm (60 L min^{-1}), 0.2–8.0 μm (90 L min^{-1}) • Low inter stage losses (~5%) for good recovery
Andersen Cascade Impactor (ACI) (Copley)	• User friendly with easy automation • Flow rate 15–100 L min^{-1} (usually run at 28 L min^{-1}) • Seven stages (five with cut-offs between 0.54 and 6.12 μm at flow rates from 30 to 100 L min^{-1}) • Particle size range 0.24–14.1 μm (depending on flow rate) • Low to high inter stage losses • Reduced stack option for work with nasal aerosols and sprays • Low flow resistance at high flow rates when stages 6 and 7 are removed
Reuter Centrifugal agar impactor (Biotest AG, Germany)	• Flow rates ranging from 40 to 100 L min^{-1} (Grinshpun et al., 2015)

Table 2 Key features of some example air samplers for the characterization of bioaerosols—cont'd

Air sampler	Key features[a]
SpinAir sampler (IUL micro, Barcelona, Spain)	• A multi-nozzle portable impactor with a Petri dish rotating at 1–4 rpm to increase the sampling surface (Grinshpun et al., 2015)
BIAP Slit Sampler (Scantago, ApS, Denmark)	• Allows time-resolved sampling as long as 3 h (Grinshpun et al., 2015)
Spore Traps: Single use: Air-O-Cell cassettes (Zefon International, Ocala, FL)	• Easy to use • Flow rates ranging from 5 to 30 L min^{-1} with particles at 2 μm and higher (Grinshpun et al., 2015) • Can be applied with automated microscopy (Wagner and Macher, 2012)
Spore Trap:Multi use: BioSIS Slit Impaction Air Sampler (EMS Inc., Charleston, SC); Personal Volumetric Air Sampler (Burkard Manufacturing Co., Ltd.). Hirst spore traps; Burkard 7-day Volumetric Spore Trap (Burkard Manufacturing Company Ltd., Hertfordshire, UK)	• Multi-use e.g. Hirst spore traps can also be used for the collection of pollen (Hirst, 1952). • Personal Volumetric Air Sampler has a built-in pump • Burkard 7-day Volumetric Spore Trap can be used to determine temporal variability of mould spore or pollen concentrations over seven days (Li et al., 2019); can be used for microscopy or molecular methods; widely used for pollen and fungal spore monitoring • Performance of the above samplers for pollen capture was orientation-dependent (Michel et al., 2012)
Impingers	
All Glass Impinger (AGI) (Ace Glass, Inc., Vineland, NJ),	• Operate at 12.5 L min^{-1} and collect particles into 20 mL of liquid • Has either 4- or 3-mm distances between the sampling nozzle and the bottom of the collection vessel, with the nozzle being submerged into the collection liquid • Long, curved inlet designed to simulate the nasal passage and its capture of inhaled particles (Grinshpun et al., 1994) • Efficient for particles of 0.3 μm, but the collection liquid evaporates in about 90 min and high impingement velocities lead to microorganism damage and re-aerosolization (Lin et al., 1997)

Continued

Table 2 Key features of some example air samplers for the characterization of bioaerosols—cont'd

Air sampler	Key features[a]
Multi-stage Liquid Impinger (MSLI) (Copley)	• Quick and easy to use • Flow rate 30–100 L min^{-1} • Particle size range 17–13.0 μm • Four stages with a final filter stage • Virtually no inter stage loss • Eliminates particle bounce and re-entrainment problems
Burkard multistage liquid impinger (Burkard Manufacturing Co. Limited, England)	• Capable of collecting particles in three size fractions
Glass Twin Impinger (GTI) (Copley)	• Flow rate 60 L min^{-1} • Particle size up to 6.4 μm • One stage • Virtually no inter stage loss • Eliminates particle bounce and re-entrainment problems
BioSampler (SKC Inc., Eighty Four, PA)	• Combines impingement with centrifugal motion (Willeke et al., 1998) and has become the *de facto* reference sampler in bioaerosol studies (Mainelis, 2020) • Operates at 12.5 L min^{-1} and can be used with either 5- or 20mL sampling cups • Reduces microorganism stress (Tseng et al., 2014)
Cyclone Samplers	
Coriolis μ (Bertin Instruments)	• Liquid output (up to 15 mL), compatible for culture and molecular based analyses • High flow rate 100–300 L min^{-1} • Particle size from 0.5 μm • Collection time between 10 mins and 6 h • Can be used indoors and outdoors for a range of environments • Can detect viruses as low as 100 genome copies m^{-3} (in a few hours) • Small, lightweight (2.8 kg) and portable with a 1-h battery life • Decontamination with hydrogen peroxide

Table 2 Key features of some example air samplers for the characterization of bioaerosols—cont'd

Air sampler	Key features[a]
Coriolis Compact (Bertin Instruments)	• Easy to use and set up • Dry cyclonic sampler, compatible for culture and molecular based analyses • Operates up to 8 h • Flow rate 50 L min^{-1} • Particle size from 500 nm to 10 μm • Can be applied to both indoor and outdoor environments • Environmental temperature range of 5–45 °C with between 10% and 90% moisture content • Collects bacteria, fungi, viruses • Small, portable, lightweight (1.2 kg) and with an 8-h battery life
Burkard Multi-Vial Cyclone Sampler (Burkard Manufacturing Co)	• Primarily for field operation but can be used within (suitable) indoor environments • Flow rate of 16.5 L min^{-1}. • High efficiency sampling in the 1 μm range. • Particles are collected directly into a 1.5 mL eppendorf and can be analysed by molecular or culture-based methods • Using the cyclone system, continuous sampling can be carried out over long periods without overloading, loss or change of sampling efficiency • Minimum operator interaction over long sampling campaigns (requires tube collection and change once a week) • Rapid preparation of each sample
High volume cyclone air sampler (Agri Samplers Ltd)	• Faster sampling rate with a flow rate of 270 L min^{-1} • Semi-automated and increased sensitivity • Can be used for detecting pollen, fungal spores • Collected particles can be analysed by molecular or culture-based methods
Portable dry air samplers	
SASS 3100 (Research International Inc., WA)	• Operates for long periods (up to several days) enabling continuous sampling • High efficiency for sub-micron sized particles • Flow rate 50–300 L min^{-1} • Wireless controlled • Easy decontamination • Portable and light weight

Continued

Table 2 Key features of some example air samplers for the characterization of bioaerosols—cont'd

Air sampler	Key features[a]
SASS 2300 (Research International Inc., WA)	• Operates for long periods (up to several days) enabling continuous sampling • High efficiency for sub-micron sized particles • Flow rate 50–300 L min^{-1} • Wireless controlled • Long-life primary and rechargeable battery options • Easy decontamination, including fan rotor • Portable, compact and light weight
MD8, MD8 Airscan, Airport MD8 (Sartorius)	• Portable and lightweight • Widely used
MicroBio MB2 (220/400 hole) (Cantium Scientific)	• Suitable for culture-based methods • Flow rate 100 L min^{-1} • Lightweight
AirPrep™ Cub Sampler	• Operates for long periods (up to several days) enabling continuous sampling • High efficiency for sub-micron sized particles • Flow rate 50–300 L min^{-1} • Wireless controlled • Easy decontamination • Portable and light weight
Burkard volumetric air sampler (Burkard Manufacturing Co Ltd)	• Collects directly onto glass slides • Designed for short-term sampling in domestic or industrial environments particularly where no power supplies are available • Air throughput of 10 L min^{-1}
Portable liquid air sampler	
BioCapture 650 (Thermo Fisher Scientific, Inc.)	• For field operation • Uses disposable collection cartridges—developed for emergency response detection • 2 h battery life • Captures particles between 0.5 and 10 μm
Passive samplers	
Air Germ (Technology Care LLC)	• Easily measures bioaerosols, fibres including asbestos, PM_{10} fine dust and other airborne particles in various environmental settings • Requires lab processing

Table 2 Key features of some example air samplers for the characterization of bioaerosols—cont'd

Air sampler	Key features[a]
Electrostatic Dustfall Collector (EDC) (Noss et al., 2010)	• Reliable method to collect airborne dust on a large scale • Easy to use for the participants of a study to be sent the collector by mail • Useful in high-exposure environments for both endotoxin and allergen measurements • Particles deposited at lower velocities and thus less microorganism damage compared to inertia-based methods (Zhen et al., 2013) • Difficult to elute particles from the collection material and may lose charge over time (Manibusen and Mainelis, 2022)
Rutgers Electrostatic Passive Sampler (REPS) (Therkorn et al., 2017)	• Substantially improves passive capture of airborne biological particles by electrostatically attracting charged particles in addition to gravitational capture • Uses a spiral film shape to increase total surface area for collection • Uses 50 mL conical centrifuge tubes so easy to collect, transport and perform downstream analyses. • Requires large collection volume (35 mL) which reduces levels of sensitivity (Manibusen and Mainelis, 2022). • Particles deposited at lower velocities and thus less microorganism damage compared to inertia-based methods (Zhen et al., 2013)
Automated electrostatic sampler (AES) (Tan et al., 2011)	• Collects 70–90% of bioaerosol particles sized 0.3–2 μm • Low sampling flow rate of $1.2\,\mathrm{L\,min^{-1}}$ • Promising tool for real-time detection of bioaerosols
HVAC Air filters	• Uses existing filtration system to collect bioaerosol samples. • Airflow can be determined • Unmaintained filters can be a source of microorganisms and cause sample bias
Personal samplers	
Button Aerosol Sampler and IOM cassette (both by SKC Inc., Eighty Four, PA)	• Used with various filters, including gelatin, to capture inhalable biological particles with low flow rates (e.g. IOM $2\,\mathrm{L\,min^{-1}}$, Button sampler $4\,\mathrm{L\,min^{-1}}$) (Mainelis, 2020)

Continued

Table 2 Key features of some example air samplers for the characterization of bioaerosols—cont'd

Air sampler	Key features[a]
NIOSH BS 251 (NIOSH)	• Operates at 3.5 L min^{-1} • Includes a two-stage cyclone • Cut-off sizes of 4 μm and 1 μm plus a final 37 mm filter enable collection of three bioaerosol size fractions • Can be used for molecular analysis
Personal Electrostatic Bioaerosol Sampler (PEBS) (Han et al., 2017)	• Operates up to 4 h at 10 L min^{-1} on a single battery charge • Reduced ozone production (<10 ppb) due to the sampler's unique charger design • Good biological sample recovery, including of culturable microorganisms (>80% collection efficiency for bacteria and fungi) • Collected samples can be analysed by multiple methods • Feasible for personal bioaerosol exposure assessment

[a]Technical specifications obtained from respective manufacturer product briefing documents unless otherwise referenced.

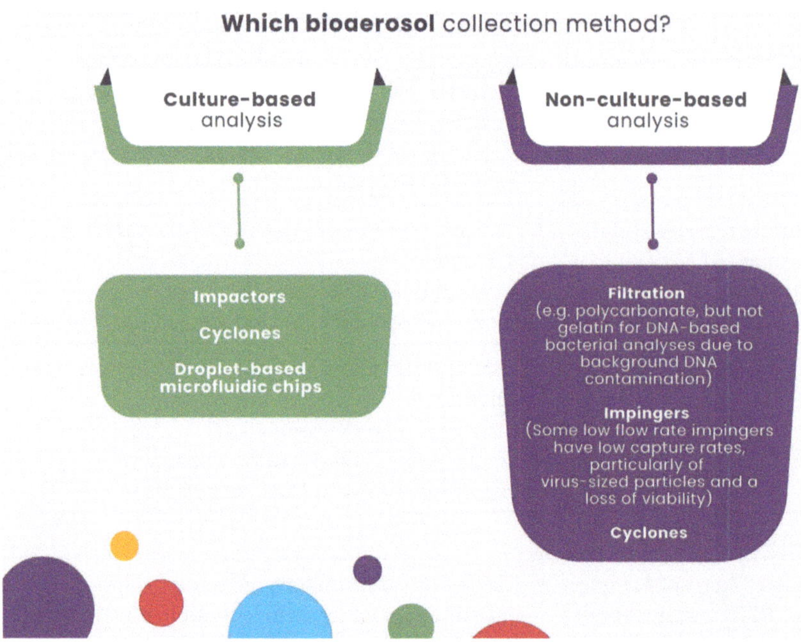

Fig. 4 Flow chart summarizing air sampling methods for downstream culture-based and non-culture-based analysis of bioaerosols.

particles when using microscopy (Ferguson et al., 2019). Gelatin filters are easy to use for cultivation as they readily dissolve directly into the agar, whereas mixed cellulose ester (MCE) filters must absorb agar media in sufficient amounts to support colony development (Cooper et al., 2019). However, gelatin filters have been found to contain contaminating bacterial DNA and thus are unsuitable for molecular analyses (Ferguson et al., 2019; Grydaki et al., 2021). Furthermore, gelatin filters may inhibit PCR analysis (Yamamoto et al., 2010).

For immunological assays, biomarkers (e.g. Endotoxin, (1,3)-β-D-glucan, Protein, Ergosterol, Mannitol, Arabitol, 3-hydroxy fatty acids), or PCR-based methods filters have been used (Georgakopoulos et al., 2009). There are several advantages of using filters. Notably, filter-based high-flow rate air samplers have demonstrated a higher microbial diversity coverage than liquid-based high-flow rate air samplers when characterising bacterial and fungal communities by amplicon-based, high-throughput sequencing (Lemieux et al., 2019; Mbareche et al., 2018). However, the way the filter traps the particle also influences both what is collected (e.g. spores or cells) and how easily it is released for downstream analysis (Duquenne et al., 2018; Dybwad et al., 2014; Yoo et al., 2017), leading to lower recoveries and inherent biases. Filter samplers can be used for culture-based analyses, but particle elution from the filter can lead to particle losses (Ferguson et al., 2019).

Indeed, depending on the air sampling method used, some microorganisms are preferentially recovered, or maintain viability over others. For example, passive samplers can cause the individual microorganisms to desiccate and consequently lose viability (Cox et al., 2020). Some active sampling devices such as impactors or impingers may also cause a loss in cell viability as a result of impact stress as the microorganism contacts the solid surface or liquid. With filtration, desiccation can affect cell integrity and reduce bacterial viability resulting in a preferential recovery of more resilient spore forming taxa (e.g. *Bacillus*), depending on filtration time, pore size and filter type (Yoo et al., 2017; Dybwad et al., 2014; Zhen et al., 2013). Impingement overcomes the problems associated with organism desiccation that can occur with filters and impactors. However, depending on the liquid matrix used, storage times and temperatures, the effects on cell counts and composition for culture-based studies can vary (Chang and Wang, 2015). Moreover, since different media will preferentially support the growth and expansion of different microbes, the numbers of individual species and thus overall composition of the sample is dependent on the type of media, temperature and time stored prior to processing. Thus, the composition of the analysed sample can be very different from the point at which it was sampled. Furthermore, preserving the viability of an airborne virus both during and after bioaerosol collection, is also often more difficult than preserving the viability of bacteria or fungi (Cox et al., 2020).

However, if molecular methods are being applied (e.g. metagenomics), then maintaining cell viability is not essential unless metatranscriptomics is required. As with cultivation, however, for metatranscriptomics it is the RNA present at the time of sampling that is required. This too will change depending on the sample collection and storage conditions so that the RNAs at the point of analysis may be very different to that at the point of sampling. Yet, it may be the case that the microorganisms only need to be viable at the point of collection and the snapshot of nucleic acids are preserved at the point of sampling. Whilst RNA is less robust than DNA, it is possible to immediately preserve it at the point of sampling. So, samples for metatranscriptomics (like those for metagenomics) do not require cell viability to be maintained—in fact this is more likely to cause the problems described above. If the loss of viability caused by devices such as impactors or impingers is not an issue, those samplers can be readily deployed. However, where the research poses questions that addresses the viable members of a population (analysed by either cultivation or transcriptomics), then impactors, impingers and passive sampling may not necessarily be a good option.

With molecular approaches, maintaining nucleic acid integrity and DNA yield is essential, especially if metagenomics is being performed and/or RNA viruses are the target. High fidelity recovery of genomic material has been reported when using condensational capture (Nieto-Caballero et al., 2019). Collecting sufficient genetic material is also often problematic, especially in environments such as a hospital setting, where there is low sample biomass present. Moreover, cross contamination may occur due to the high sensitivity of the molecular approaches. Thus, the DNA extraction method applied needs to have high recovery efficiency when working with low-biomass bioaerosol samples (Morgan et al., 2010). Using filters for molecular analyses have the advantage that the nucleic acids can be extracted directly from the filter (Yoo et al., 2017). Although elution from filters can cause sample losses, efficiencies can be increased with properly designed protocols (Ferguson et al., 2019). Ferguson et al. (2019) showed that air filtration using polycarbonate filters with a phenol chloroform DNA extraction method gave the highest bacterial DNA yield recovery.

The polyethersulfone (PES) membrane filters have also been used as they dissolve directly into buffer, removing the elution step and making it amenable to molecular techniques (Rahav et al., 2019; Zhen et al., 2018). Some studies have increased recovery rates from filters (especially for more robust organisms) by cutting them up and bead beating or vortexing the filters to resuspend the bacteria into a liquid for subsequent DNA extraction (Be et al., 2014; Madsen et al., 2015; Pankhurst et al., 2012; Ferguson et al., 2019). However, using filters only provides a single snapshot of information whereas, sampling with impingement may be more practical for multiple sampling within shorter periods of time, since they generally run at higher flow rates and less likely to get overloaded, thus provide more fine-scale temporal/spatial data (Ferguson et al., 2019).

With liquid samplers, the type of matrix used is important to consider as this may affect DNA yield and can lead to inherent biases in recovery (Ferguson et al., 2019). Specifically, Ferguson et al. (2019) showed that choice of liquid impingement matrix can lead to inherent biases against Gram negative bacteria with mixed environmental communities. These authors found that the optimum impingement matrix to circumvent these biases was phosphate buffered saline (PBS) (Ferguson et al., 2019). Another potential issue with liquid samplers is sample matrix loss through evaporation, which has to be replenished, especially if sampling occurs over longer time periods. Automated systems (e.g. SASS-2400 Research International, WA) that maintain collection fluid levels are available, but are more expensive (Mainelis, 2020). With liquid-based samplers, sample losses may also occur due to particle bounce, re-aerosolization or as a result of remaining on internal sampler surfaces (Grinshpun et al., 1997; Han and Mainelis, 2012). Sample losses from evaporation, bounce and re-aerosolization losses may be reduced by the use of viscous liquids such as mineral oil (Simon et al., 2016), which is compatible with downstream molecular methods (He and Yao, 2011). With impactors, recovery efficiency is also reduced due to particle bounce as well as low flow rates (Griffin, 2007), which in turn can reduce DNA yields and cause biases in the recovery of certain taxa (Ferguson et al., 2019). Furthermore, the amount of agar used in impactors is not standardised and could be anything between 25 and 40 mL in typical Petri dishes (Mainelis, 2020). Since the amount of agar affects the jet-to-plate distance and the strength of air jets when they reach the agar plate, agar amount can affect sampler performance (Therkorn and Mainelis, 2013). The type of agar selected by the user also depends on the organisms being cultured (see Section 3).

Another important consideration is ensuring a representative sample of the microbial diversity present in a given environment is obtained (but without exceeding the upper quantification limit of the sampler) (Ferguson et al., 2019). Thus, the duration of sample collection and the volume of air to be sampled is important. Studies vary in terms of sampling time periods used from as low as 3 min to several hours, and this is typically dictated by the study (Haig et al., 2016). Differences in sampling methods (e.g. high vs low volume rates), also mean that sampling times will vary depending on the air sampling method used. It is also important to note that the total microbial counts obtained with an impinger are generally not linear with sampling time, as microbial stress, deagglomeration and reaerosolization of microorganisms may occur which can affect the viable count in a time-dependent manner (Willeke et al., 1995).

When using samplers with different flow rates deployed at the same time, one option is to calculate the sampling time for each device so that the volume of air captured is comparable (Haig et al., 2016). However, caution should be taken when comparing air samples of varying duration and flow rate, as collecting 300 L over 10 min may not be the same as collecting 300 L over 2 h and could represent fundamentally different microbial communities. When a large amount of material is collected, the rare members of a community might be overlooked (Ferguson et al., 2019). Furthermore, simply increasing sampling time or flow rates may not necessarily increase DNA yield. For example, doubling the flow rate of impingers from 300 L min^{-1} to 600 L min^{-1} does not significantly increase DNA yields (Ferguson et al., 2019). It may also not be appropriate to increase the sampling time with filters due to sample desiccation. Luhung et al. (2015) found that bacterial DNA yield did not increase with time, and the sample was only representative of the bacteria collected during the latter stages of sampling. This is particularly important when determining microbial function such as with metatranscriptomic analyses, whereby RNA can be degraded quickly (Ferguson et al., 2019). Optimal sampling times for molecular analyses, was evaluated by Ferguson et al. (2019), who showed that sampling for at least 120 min (3.36 m^3 air) was required for air filtration, whilst only 20 min (6 m^3 air) was required for impingement when coupled with metagenomic/metabarcoding approaches. The same study highlighted that pooling of samples may still be required for amplification-free metagenome analyses. Thus, sampling duration should be of sufficient length to result in a representative sample.

The main limitation in molecular analysis of bioaerosol samples is the difficulty in concentrating samples with low biomass and/or the potential carryover of PCR inhibitors. Although collection devices with high flow rates or high collection efficiencies may be useful (Morgan et al., 2010), under certain instances such as in low concentration environments or when sampling duration is limited, even high flow rates might not yield sufficiently high sample concentrations. Thus, it may be necessary to concentrate samples post-collection. Several approaches to address this include centrifugation, accelerated evaporation, (e.g. Smart Evaporator (BioChromato, SanDiego, CA)), or filtration followed by reconstitution. Ferguson et al. (2019) demonstrated that the optimal method for bioaerosol concentration to recover bacteria from the liquid impingement matrix was centrifugation. However, this method of sample concentration may cause biases due to a loss of fungal cells, caused by the hydrophobicity and polarity of their cell membranes (Mbareche et al., 2019). Alternatively, if access to a centrifuge is limited (e.g. in certain outdoor environments), then filters can be used for concentrating samples (Ferguson et al., 2019). Therefore, if molecular analyses are to be performed, it is important to consider the effect of different sampling matrices, sample concentration strategies, sampling duration, sample storage conditions and the time between sampling and analysis has on DNA yield, which in turn may affect the microbial diversity recovered (Ferguson et al., 2019).

2.3 Which air sampling method is best for which biological particle?

The tree of life consists of three domains: Bacteria; Archaea; and, Eukarya, to which the fungi belong. Both active and passive sampling have been used to assess microbial communities across all domains of life (Delort and Amato, 2018). However, it is well known that sampler selection and sampling method can affect the type of microbiome as well as the size of the microorganisms recovered (Haig et al., 2016; Hoisington et al., 2014; Ferguson et al., 2019). Indeed, a preferential collection of some microbial species by particular samplers has been found (Serrano-Silva and Calderón-Ezquerro, 2018). The filter-based Button Aerosol Sampler and the liquid-based BioSampler have been found to yield different microbial communities (Zhen et al., 2018). A field study using a filter sampler, impactor, and impinger showed good agreement for the dominant species but not so for the less abundant species (Fahlgren et al., 2011). Filter-based high-flow rate air samplers produce a higher diversity coverage than liquid-based high-flow rate air samplers for bacterial and fungal amplicon-based high-throughput sequencing approaches (Ferguson et al., 2019; Lemieux et al., 2019; Mbareche et al., 2018). This is possibly because the collection media in the liquid cyclonic impactors may evaporate depending on the environmental conditions (e.g. relative humidity etc) resulting in differences in the microbial diversity recovered (Cox et al., 2020; Lemieux et al., 2019). Another important consideration is that filter pore size does not indicate the minimum aerosol particle size that will be collected by that filter and so shouldn't be used for size-selection. Most filters will efficiently collect aerosol particles that are much smaller than the nominal pore size. For collecting across all domains, filters typically have high collection efficiencies (>95% for particles ca. 0.035 μm in diameter), making them suitable for collecting bacteria, fungi and viruses (Gandolfi et al., 2013). However, it has more recently been demonstrated that depending on pore size, filter type and collection time, spore forming microorganisms may be preferentially recovered (Yoo et al., 2017). So, a key question for end-users is, which sampler is optimal for which taxa?

2.3.1 Fungi

Most airborne fungi are within the breathable fraction of bioaerosols (e.g. <8 μm) (Jones and Cookson, 1983) and between 30% and 35% of fungal species are estimated to be potentially allergenic to humans (Adhikari et al., 2004). If fungal organisms in the environment (mycobiome) are of interest, then these can be readily collected with impactors, impingers, cyclones or filters (Haig et al., 2016; Lindsley et al., 2017). Culture-based methods are often used to analyse fungal spores. However, many fungal species are non-culturable (Bridge and Spooner, 2001), and so molecular methods are useful for characterising a greater diversity of environmental fungi. For sampling total fungal spores, slit-to-slide samplers are commonly used for microscopic analysis, whilst Andersen cascade impactors collect directly onto culture plates (Cox et al., 2020). Airborne fungi can also be detected using other fungal components and metabolites such as ergosterol, arabitol mannitol or volatile organic compounds (VOCs) (Polizzi et al., 2012; Elbert et al., 2007). More recently, a review of outdoor fungal spore seasonality across Europe found that Hirst type samplers (such as the Burkard cyclonic samplers) were the most commonly used coupled with microscopy (Anees-Hill et al., 2021).

A particular class of bioaerosol impactor known as spore traps, which generally have a single nozzle or a slit directing airborne particles onto a sticky glass slide are widely used to assess indoor air quality, especially where mould is a concern (Godish and Godish, 2008). However, sampling duration is a potential issue for spore traps, where the impaction area is limited and can be easily overloaded and therefore spore traps are typically limited to 15 mins sampling time (Mainelis, 2020; Codina et al., 2008). When sampling for fungi, the collection efficiency of the Coriolis is lower than the BioSampler which combines impingement with centrifugal motion (Chang et al., 2019). A personal electrostatic sampler that operates at a flow rate of 5 L min^{-1} has also been used to capture fungal spores (Roux et al., 2016). Passive bioaerosol sampling using settle plates or EDC has also been used for fungi (Cox et al., 2017; Normand et al., 2009; Frankel et al., 2012). Filtration often results in higher yields both qualitatively and quantitatively for fungal exposure, especially in indoor environments (Mbareche et al., 2019).

Although fungi are ubiquitous organisms, concentrations can vary by location and season (Anees-Hill et al., 2021). For example, the abundance of fungal spores typically ranges from $\sim 10^3$ to $\sim 10^6$ m^{-3} near-ground level (Delort and Amato, 2018). Large spatial and temporal variations in fungal abundance have been found that are linked to environmental and meteorological conditions (Burch and Levetin, 2002; Jones and Cookson, 1983; Adhikari et al., 2004; Anees-Hill et al., 2021; Delort and Amato, 2018). Human activity and higher relative humidity may also result in significantly higher concentrations of airborne fungal spores (Lehtonen et al., 1993; Scheff et al. 2000). Extremely high concentrations of up to $\sim 10^9$ spores m^{-3} of air have been found in specific indoor environments (e.g. mouldy/damp buildings, waste processing buildings) as well as near strong sources of bioaerosols such as composting facilities, farms, or greenhouses (Durand et al., 2002; Ferguson et al., 2021; Radon et al., 2002). In contrast, lower concentrations ($<10^3$ spores m^{-3}) have been found at high altitudes and in polar areas (Duncan et al., 2010; Marshall, 1997). Some fungal species are known to be more abundant indoors, (e.g. *Aspergillus*), whilst others are rarely found indoors, (e.g. basidiomycetes) (National Research Council, 2005). In general, most indoor fungi originate from the outdoor air and are present in lower concentrations indoors compared to levels outdoors (Burge et al., 2000; Grydaki et al., 2021). Where indoor fungal levels are higher than the outdoor levels, an indoor fungal source could be indicated (Rao et al. 1996).

2.3.2 Bacteria

Understanding bacteria in the environment (bacteriome) is important, especially pathogenic microorganisms or their components that may have detrimental effects on health. Typical concentrations of airborne bacteria near the ground ranges from 10^2 to 10^6 cells m^{-3} but large spatial and temporal variability occurs (Bowers et al., 2013). For collecting bacterial aerosols, filters are commonly used (Lindsley et al., 2017, Ferguson et al., 2019; Grydaki et al., 2021). Ferguson et al. (2019) showed that polycarbonate filters significantly outperformed glass fibre filters, with recoveries of approximately three orders of magnitude more 16S rRNA gene copies for Gram-positive and one order of magnitude for Gram-negative bacteria. In another study, MCE filters showed significantly higher sampling efficiencies compared to the slit-to-agar impactor AirTraceTM and liquid impinger air samplers AGI-30 and BioSamplerTM of nebuliser-generated *C. difficile* spore aerosols in a lab environment (Cooper et al., 2019). Passive sampling (e.g. EDC) has also been used for bacteria (Normand et al., 2009). In addition, GTC has been used to collect viable *E. coli* cells and may be more suitable for sampling viable aerosolised Gram negative bacteria compared to the BioSampler (Pan et al., 2018).

2.3.3 Archaea

Archaea are currently understudied in bioaerosol research, despite possible human health implications. In particular, the presence of archaea in indoor air and the factors affecting their abundance are not well characterised. For example, exposure to archaea can induce a full immune response in a murine model of airway exposure (Blais Lecours et al., 2011). In one study, three samplers were compared: IOM sampler with a gelatin membrane, NIOSH two-stage cyclone, and the Coriolis liquid impinger. All gave comparable results in terms of archaeal gene copy number m^{-3} of air in dairy farms (Blais Lecours et al., 2012).

2.3.4 Viruses

Understanding viruses in the environment (viriome) and their routes of transmission is challenging. Yet such knowledge is crucial to understand the sources of transmission of viruses, characterise the viriome, and enable more efficient virucidal air treatments to be deployed. Improved sampling methods are particularly crucial especially for investigating viral communities, due to an extremely low biomass of viral particles in the air compared with other environments. Thus, to characterise viruses, sampling methods need to have high collection efficiencies, (especially for viral metagenome studies) (Prussin et al., 2014), and preserve virus integrity (Degois et al., 2021). Alternatively, filters may be used as they would allow sampling of a larger volume of air, and collection of viable viruses is not necessary for viral metagenomics (Prussin et al., 2014). Otherwise infectious viral concentrations could be greatly underestimated due to sampling stress (Verreault et al., 2008). With impingers, the viral bioaerosols are deposited into a liquid media, which preserves their viability (Colbeck and Whitby, 2019). Sensitivity to aerosolization and sampling conditions (e.g. temperature, relative humidity, chemical composition of the air, and time spent in aerosol state) also affect viral integrity and can be strain-specific (Degois et al., 2021).

Particle size is an important consideration regarding the viability of aerosolized viruses (Anderson et al., 2017). Collection efficiencies are typically <1% for particles smaller than 100 nm (Tseng and Li, 2005; Dart and Thornburg, 2008). Viruses are the smallest class of bioaerosols and can be found associated with particles of all sizes (Yang et al., 2011). In general, larger particle sizes (300–450 nm) have higher survivability compared to particles measuring closer to the actual size of the virions (100–200 nm) (Zuo et al., 2013, Alonso et al., 2015). Thus, methods for characterising viruses in bioaerosols would benefit from the routine inclusion of methods to evaluate particle size. Yet, many commercially available air samplers are optimally designed for collecting micrometre-sized particles (e.g. fungal spores, bacteria), and so the collection efficiencies of submicron particles (<1 μm) tend to be low (Hogan et al., 2005). Capture efficiency of viruses is also affected by virus morphology and hydrophobicity (Mainelis, 2020). Hydrophilic viruses are captured 10–100-fold more efficiently than hydrophobic viruses using samplers such as the Andersen impactor, AGI-30 impinger, and filters (e.g. gelatin, nucleopore) (Tseng and Li, 2005). The BioSampler is often used as a reference tool when investigating other samplers for collection of viruses (Mainelis, 2020).

Impingers are commonly used to collect airborne viruses for culture-based analysis as the liquid collection media preserves the viability of sensitive viruses much better than dry collection, and this factor usually outweighs the drawbacks of impingers (Verreault et al., 2008). Liquid and dry impaction are also commonly used to study viruses in bioaerosols (Verreault et al., 2008). Indeed, both Andersen and Tisch cascade impactors have been shown to perform equally well for the detection of viral loadings (Alonso et al. 2017). Both the AGI-30 and the BioSampler collectors have been used to collect viruses, but with low capture efficiency (Hogan et al., 2005).

The G-II uses a slit impactor to remove particles >5 μm and then grows the remaining particles <1 μm to be captured by another slit impactor, with collection efficiencies for influenza virus similar to that of the BioSampler (McDevitt et al., 2013).

Electrostatic Precipitation (ESP) has also been used for the efficient capture of viruses (Hogan et al., 2004) including influenza, MS2 and T3, with relatively high virus recovery rates (Hong et al., 2016; Ladhani et al., 2017). Although viral particles are usually smaller than the typical filter pore sizes, they have been collected using filters (Burton et al., 2007) such as gelatin filters, but their performance can be affected by humidity (Verreault et al., 2008). A recently-developed aerosol sampler using growth tube collector (GTC) has been developed to collect airborne viruses with minimal damage and maintains viability (Pan et al., 2016). This was successfully used to capture MS2 bacteriophage and was 20X more effective for capturing infective virions compared to the AGI-4 impinger (Walls et al., 2016). It has also been used to capture MS2 coliphages, influenza H1N1 virus (Lednicky et al., 2016; Pan et al., 2016) and determine the size distribution of infectious viruses (Pan et al., 2019). The Spot Sampler™ also showed good recovery of bacteriophage and better preservation of virus infectivity (e.g. with influenza) (Degois et al., 2021), and has recently been used in the hospital rooms of COVID-19 patients to collect viable SARS-CoV-2 (for genome sequencing) (Lednicky et al., 2020). Filter and cyclone-based aerosol samplers are often used to collect viral bioaerosols for culture-independent analysis because of their simplicity and they are effective at collecting aerosol particles of all sizes (Lindsley et al., 2017). An aerosol sampler using a condensation-based growth tube collector is also reported to collect airborne viruses with minimal damage, and maintains viability (Pan et al., 2016). The methods described for viral aerosol collection also applies to bacteriophages, which can be a problem in indoor industrial environments where bacteria are used (e.g. dairy production plants) (Cox et al., 2020; Verreault et al., 2011).

2.3.5 The air resistome

Antimicrobial resistance (AMR) is a global public health issue causing >700,000 deaths every year (Williams et al., 2016). Although the environment (soil, water, air) acts as a reservoir of AMR (i.e. the resistome) (Surette and Wright, 2017), most AMR studies have focused on aquatic and soil environments with very little attention towards the transmission of antimicrobial resistance in the air (Williams et al., 2016; Jones, 2020).

Antimicrobial resistance genes (ARGs) contained in the bacteria circulating in air can invade other pathogenic bacteria through mobile genetic elements (e.g. integrons, transposons, and plasmids) (D'Costa et al., 2006; Van Hoek et al., 2011). Furthermore, a large number of ARGs are unknown or uncharacterised and the abundance of organisms and diversity of ARGs will vary depending on the environment and season (Asaduzzaman et al., 2019).

The Andersen two-stage sampler has been applied to swine production facilities, and high rates of antibiotic-resistant bacteria that could pose a risk to individuals in close proximity were found both within, and downwind of, the facility (Gibbs et al., 2004, 2006). Devices for sampling particulate matter (PM), typically at 2.5 and 10 μm aerodynamic diameter (Gandolfi et al., 2013) have been used to study AMR in air samples (Xie et al., 2019). Although, PM does not differentiate between organic and inorganic matter (Morawska et al., 2001), $PM_{2.5}$ has been recently shown to affect the proliferation and transfer of ARGs in urban air (Xie et al., 2019). Analysis of cultivable bacteria within PM_{10} sampled from milking parlours demonstrated that opportunistic pathogens (e.g. *Staphylococcus*, *Aerococcus*, *Bacillus*, *Pseudomonas*, among others) contained low levels of acquired resistance to antibiotics commonly used to treat human or animal infections (Bayle et al., 2021).

2.3.6 Pollen, endotoxins, and other allergens

Pollen grains can range from 7 to 100 μm depending on plant species, but on average are between 30 and 40 μm (Delort and Amato, 2018). The personal aeroallergen sampler (PAAS) is designed for coarse particles and has been successfully applied to measure exposures to pollen (Yamamoto et al., 2006, 2007). EDCs (Noss et al., 2008) and endotoxins (DECOS, 2010; Liebers et al., 2012). However, heating for sterilisation of EDC cloths can diminish its capabilities and the location of the EDC cloth significantly affects the endotoxin levels detected (Kilburg-Basnyat et al., 2016), most likely due to the loss of charge (Brown et al., 1995).

2.4 Additional sampling considerations

Bioaerosol concentrations vary over space and time and sampling at multiple time points or for long durations is frequently needed, especially in indoor environments (LeBouf et al., 2008; Lindsley et al., 2010). Often the bioaerosol source is intermittent or seasonal (Emerson et al., 2017). Several studies have assessed seasonal variations in bioaerosols and seasonal shifts in concentrations demonstrated (Kim et al., 2007; Nehmé et al., 2008; Kumari and Choi, 2014; Anees-Hill et al., 2021). For example, Kim et al. (2007) found that fungal spores were more detectable during the summer months, whilst higher concentrations of bacteria were found during the winter. Seasonal variations have also been shown with viruses, where for example the influenza A virus (IAV) RNA increased during the winter when temperatures were below 20 °C (Anderson et al., 2016).

2.4.1 *Spatial and temporal variations—When and where to deploy samplers?*

To address the question of when and where to deploy samplers to assess seasonal and spatial variations is not simply a case of increasing the sampling time, as sampling times are often constrained by the sampling device used. For example, indoor bioaerosols are often collected for short time periods (typically 15–30 min with impactors and impingers), but this only provides the average bioaerosol concentration during the collection time ((Cox et al., 2020). For assessing indoor bioaerosols over extended periods, building HVAC filters have been analysed as a simple method of collection (Haaland and Siegel, 2017; Maestre et al., 2018). Similarly, car filters have been used for investigating mobile environments (Hurley et al., 2019; Viegas et al., 2018; Mainelis, 2020). Although such approaches provide long-term integration of samples which improves detection limits, the longer collection durations (e.g. which can be up to several months), can affect efficiency, making comparison between filters difficult. The higher sampling rates achievable with (some) impingement systems (>100 L min^{-1}), (compared to filtration (2–30 L min^{-1})), makes impingement suitable for snapshot sampling (Ferguson et al., 2019; Pankhurst et al., 2012). However, in terms of human exposure it is necessary to consider both the bioaerosol concentration and the exposure time (Cox et al., 2020), whereby any fluctuations in bioaerosol concentration can greatly affect overall exposures, which can be missed if sampling occurs over a few short-term collections (Cox et al., 2020). Instead, continuous on-site detection of bioaerosols in real-time would be of greater benefit, and techniques such as LIF and SIBs are continually being developed and deployed (O'Connor et al., 2015; Könemann et al., 2019; Tian et al., 2020) (see Section 3). However, it should be noted that these approaches do not resolve biological from non-biological particulate matter.

In addition to temporal variability, spatial variations also occur, and sampling location is also an important consideration, especially in terms of risk of human exposure (see Section 4). Despite this, it is not uncommon for studies to sample in only one location. Sampling location is especially important for sampling indoor bioaerosols where bioaerosol concentrations can vary within a building or room for example, due to changes in building airflow patterns or ventilation systems (Crawford et al., 2009; Luongo et al., 2016; Morey, 2007). When sampling indoors, samplers are typically placed approximately 1.1 m above the floor (USEPA, 1994). Indoor and outdoor air samples should be collected simultaneously with additional outdoor sites included near to, and/or far from, potential sources of bioaerosols. Other best practices include avoidance of strong airflows at the point of sampling, and the sampler's inlet should be placed horizontally to minimise sampling bias (Haig et al., 2016). Systematic grid sampling approaches can be used for finding hotspots for target organisms/genes over a specific time period, as samples are taken at regular intervals (Keith, 1991). If the spatial variations in bioaerosol concentrations are not of interest or the bioaerosol concentrations are low, then samples can be pooled to get an average concentration, which also reduces sampling and analysis costs (Cox et al., 2020; Ferguson et al., 2019). Sampling outdoors has additional challenges (e.g. wind speed and direction), which may affect collection efficiencies (Li et al., 2000). Thus, the performance of bioaerosol samplers should include tests with wind speed (Su et al., 2012b) and direction (Su et al., 2012a). Certain samplers (e.g. Button Aerosol Sampler) are designed to minimise sensitivity to the wind (Kalatoor et al., 1995). Sampler choice is also determined by whether it is an environmental area or a personal exposure that is to be measured.

2.4.2 Quality control and other environmental considerations

Here we ask: *what is deemed an acceptable sample for the downstream analysis approaches*? Such quality control decisions are method-specific (i.e. the type of quality control checks required are dependent on the particular analysis method to be used (Macher et al., 1995). Specifically, the use of blanks, controls and field blanks (whereby the collection media is removed from the sampler just prior to sample collection) is essential. Blank filters and blank agar plates should also be processed and any observed microbial growth (with culture-based analyses) or background nucleic acids (with molecular analyses) should be subtracted from the sample datasets. Contamination is generally a big quality control issue in bioaerosol sampling as biomass concentrations may be low in certain environments, so any contamination can drown out the real signal. This is especially challenging with molecular methods (see Section 3).

Environmental conditions at each sampling location should always be recorded, including temperature, relative humidity, air movement (e.g. wind direction/speed for outdoor environments and if heating, ventilation and air conditioning was on/off for indoor environments), and carbon dioxide levels. The occupancy and activity in the room/building or outdoor environment and conditions during sampling also need to be considered when evaluating bioaerosols, particularly when comparing studies. For example, if assessing a ventilation system for microbial contamination, samples should be collected at different times during the unit's cycle, including when it has been turned on and off. Such information is useful as it allows variations in microbial diversity to be linked to specific environmental conditions and human activities during sampling (Cox et al., 2020).

2.5 Summary of air sampling methods

No single universal bioaerosol sampler exists and there are no standardised bioaerosol sampling protocols. This is largely due to the fact that there are a range of air samplers, that differ in their operation and performance characteristics and efficacy with downstream analysis methods. This lack of consensus makes it difficult to determine dose-effect relationships. Furthermore, given that certain environments (e.g. clinical settings) where low biomass may occur, samplers with low sampling flow rates or short sampling times may miss potential fluctuations/variations. Consequently, bioaerosol composition and concentration is not being reliably captured resulting in incomplete characterisation. The vast majority of samplers rely on filtration, impaction, impingement, and electrostatic precipitation, or combinations of these. For each device, there are various pros and cons. It therefore comes down to what is being measured and sampling device/method selection depends on the objectives of the study (see Fig. 7). For example, if exposure to an individual is important then, personal samplers with the inhalable fractions combined with size fractionation samplers (e.g. Andersen) could be used. Sample method should also be appropriate for the particle size of interest (e.g. bacteria, fungi, viruses, pollen etc) and standardized methods need to be developed for each particle being assessed. If culture-based analysis is used, then the sampler and collection medium best suited for cultivation should be chosen. For combined molecular and cultivation analyses, samplers with liquid collection medium is recommended so samples can readily be split between the different methods. Molecular methods are widely applied to characterise bioaerosols, yet such methods are not standardised (unlike culture-based methods where there is some degree of standardisation) (see Section 3).

2.6 Recommendations and guidelines for standardization of air sampling methods

Recently a set of guidelines and recommendations applicable for different environmental scenarios was described (Ferguson et al., 2019; Fig. 5). For example, whilst air filtration using polycarbonate filters may give the highest DNA yield recovery, the faster sampling rates that are possible with impingement, mean that this method is recommended for fine-scale temporal/spatial ecological studies (Ferguson et al., 2019). It is therefore important to understand that for each study, sampling optimisation is required, which may depend on the environment (i.e. indoor vs. outdoor), regulatory context (e.g. hospital operating theatre vs. poultry farm) and the specific questions being addressed (e.g. quantification, taxonomic identity, pathogen exposure).

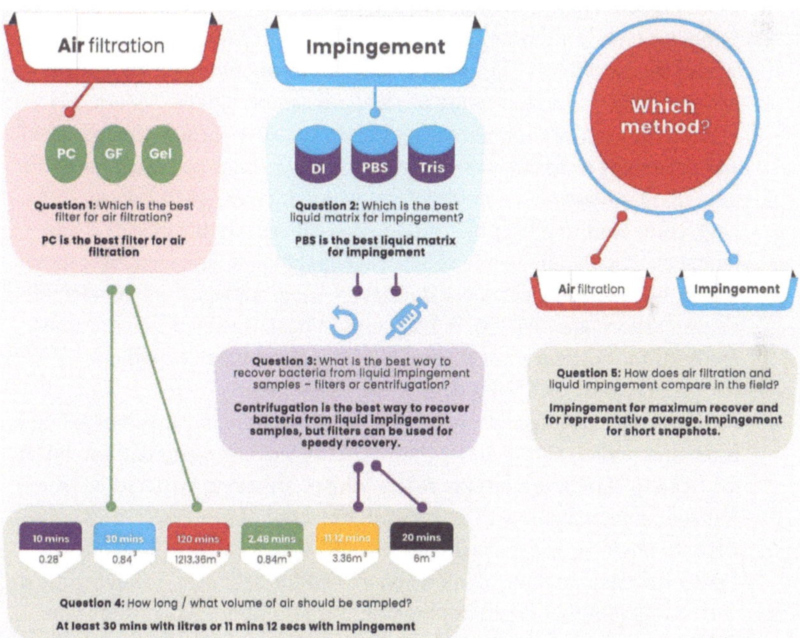

Fig. 5 Flow diagram of recommendations/guidelines for air sampling. *Modified from Ferguson et al. (2019).*

In conclusion, although there is no one optimal air sampling method, any given sampler selected must obtain a representative sample of the environment being investigated. Sample integrity must also be preserved for the downstream analysis method to be performed. To achieve this, bespoke combinations of methods may be what is needed, that are tailored to the environment, context, downstream analysis and research questions under investigation (Box 2).

BOX 2 Recommendations for standardization of sampling methods

1. Although harmonized sampling protocols would improve the reproducibility of bioaerosol studies, one overarching standardised method is not practical (i.e. one size does not fit all). Bespoke standardised methods are needed for each environmental context, target microorganism(s) (i.e. whether concentration, cell size, cell viability, a range of taxa or specific species/strain) are needed to be retained to assess exposure and ensure a representative sample of the specific environment is collected.
2. Sample preservation and storage conditions prior to downstream analysis should be considered to maintain sample integrity.
3. Sampling time is important, where long-term sampling at selected sites should be considered. This is dependent on location and on flow rates. To address Point 1 above, a range of standard sampling times/volumes is needed for each environmental setting, to obtain sufficient biomass yield and determine exposure risk. (For example, lower volumes could be set for targeting pathogens in a poultry farm where biomass yield is likely to be high, compared to a hospital setting where lower biomass yields mean higher air volumes are needed.) Since active sampling gives higher volumes and higher biomass yield, these devices are more applicable when microbial concentrations are low (e.g. in hospital settings). However, standardizing methods for hospital settings should also include passive methods (i.e. settle plates), as a valid patient risk assessment.
4. For downstream molecular analyses, air filtration using polycarbonate filters is recommended, as it gives the highest DNA yield recovery and should be included in standard methods where viruses are being targeted or where metagenome analysis such as for AMR detection is required.
5. Centrifugation should be included to concentrate bacteria from liquid matrix but, syringe filters can be considered for field use. However, this is with the caveat that liquid impingement is also useful when short time periods are needed (i.e. filters for highest recovery, impingement for short samples).

BOX 2 Recommendations for standardization of sampling methods—cont'd

6. Ideally an air sampler (especially for indoors) should be portable, be able to sample large volumes of air, fractionate by particle size, and allow 100% recovery of DNA.
7. Standardized methods for characterising viruses in bioaerosols should routinely include methods to evaluate particle size, given the link to particle size and viability of viruses.
8. Standardized methods should include sampling at multiple time points or for long durations at multiple locations for improved assessment of spatial and temporal bioaerosol dynamics and transmission routes. For long-term integrated samples, longer sampling times or integration of several samples are preferred. Sampling duration should be sufficiently long to result in a representative sample. Thus impingement, or faster filters are recommended.
9. Sampling frequency should account for how rapidly the bioaerosol concentration is anticipated to change. For fast changing levels, multiple samples should be collected, especially when the peak exposures are a concern.
10. Operators should take steps to minimize particle losses during sampling and take losses into account during data analysis.

3. Bioaerosol analysis methods

There are a variety of methods for analysing the biological components of bioaerosols, some rely on culturing the microorganisms, whilst others are culture independent. Selection of the most appropriate method often depends upon the taxa that needs to be characterized and the purpose of the study. Fig. 6 summarises some of the various culture-dependent and culture-independent methods currently used to analyse collected bioaerosols in relation to some typical questions or the study being addressed. However, each method has various advantages and disadvantages, which are detailed below and summarised in Table 3.

3.1 Culture-based methods

With culture-based methods, the collected airborne microorganisms are grown using liquid or solid media, which is incubated for a specific period.

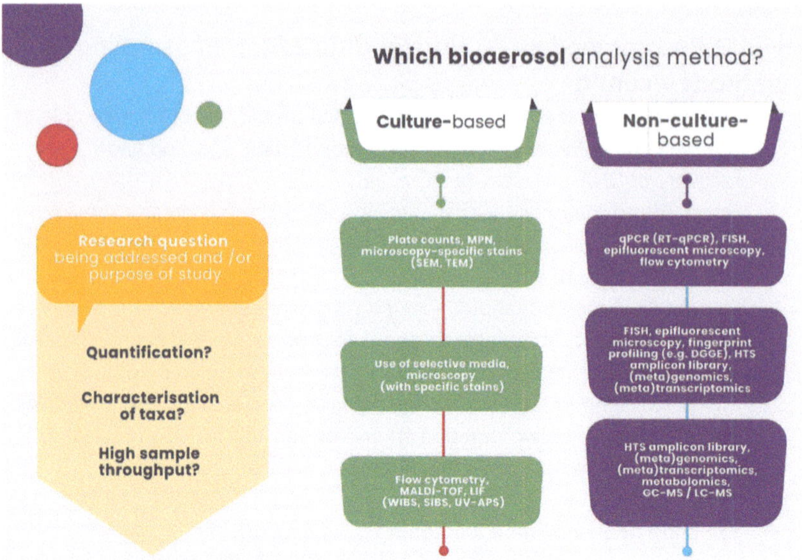

Fig. 6 Flow chart summarizing the different culture-based strategies that can be applied for bioaerosol analysis depending on the research question being addressed and/or purpose of the study.

The results are expressed as colony-forming units (CFU), or most probable number (MPN) (Heidelberg et al., 1997). Virus particles can also be enumerated using plaque assays and expressed as Plaque Forming Units (PFU). However, for many viruses this approach cannot be used and alternative approaches are needed (see Section 3.3). Culture-based techniques are relatively simple and cost-effective and allow microorganisms to be isolated for further analyses, such as for physiological studies. Impactors are commonly used with culture-based methods to enumerate and identify viable microbes in aerosols (Ghosh et al., 2015; Heidelberg et al., 1997; Nonnenmann et al., 2010; Pillai and Ricke, 2002). Bacteria and fungi are often collected directly onto culture plates or impinged into liquid media that is then inoculated onto agar plates (Cox et al., 2020).

Culture-based methods however have several drawbacks. They are often time-consuming and labour intensive (Table 3). In addition, since different microorganisms have different growth requirements, it is often necessary to use a range of media, nutrients and growth conditions in order to capture the wider picture of microbial diversity present (Heidelberg et al., 1997; Nonnenmann et al., 2010; Pillai and Ricke, 2002). Thus, repeated cultivations using a range of different culture media may be needed to identify the

Table 3 Advantages and disadvantages of culture and non-culture based approaches to analyse microorganisms in bioaerosols.

	Advantages	Disadvantages	References
Culture-dependent approaches			
Plate counts	• Easy and cheap • Can use selective or non-selective media • Useful for assessing the viable fraction	• Requires use of several different media types to assess the diversity of culturable organisms present (could require prior knowledge of taxa present) • Cells may be aggregated together and may be indistinguishable as individual colonies • Overload of microorganisms may affect the result • Likely underestimate true numbers and low taxonomic resolution • Time consuming	Ghosh et al. (2015)
Most Probable Number (MPN)	• Easy and cheap • Can be used for viable counts	• As above • For more accurate results, a large number of replicates are required	Ghosh et al. (2015)
Classical microscopy (i.e. analysis of cultures using light microscopy)	• Easy and cheap • Can be used to quantify total and viable counts and identify different taxa depending on stains used	• Time consuming, many fields of view need to be analysed to get statistically robust results • Generally not reliable as may get under or over estimation of counts e.g. due to aggregation • Stains can fade (although anti-fade can be used) • May require specialized knowledge and expertise during sample processing and staining	Ghosh et al. (2015) Terzieva et al. (1996)

Continued

Table 3 Advantages and disadvantages of culture and non-culture based approaches to analyse microorganisms in bioaerosols.—cont'd

	Advantages	Disadvantages	References
Culturomics	• Allows high sample throughput • Can identify previously unculturable species	• High capital cost for robotics, and downstream sequencing (although can be combined with spectroscopic approaches)	Lagier et al. (2016) Azhar et al. (2017)
Culture-independent approaches			
Microfluidics	• Rapid reaction kinetic effect		Whitesides (2006)
Spectroscopy (e.g. MALDI-TOF), dynamic particle-size spectrometry	• High sensitivity • Used for measuring the total and size-selective counts whilst microorganisms are still in the airborne state	• Requires specific databases • Compounds less than 600 Da in size cannot be detected due to intense matrix signal	Qian et al. 1995 Madsen et al. (2015)
Microscopy (e.g. Fluorescence in situ hybridization (FISH, CARD-FISH), Epifluorescence Immunofluorescence, radioactive probes, coulter counter)	• Relatively cheap • Gives quantitative data on both culturable and non-culturable cells giving a more representative enumeration of the microorganisms present • Depending on stains used both live (viable) and dead cells can be enumerated (e.g. acridine orange-total counts, DAPI) • High throughput of samples possible if image analysis software implemented	• Restricted ability to identify specific microbial taxa but can be overcome if combined with a range of FISH and taxa-specific probes • Fluorochromes may bind to abiotic particles giving "false positives" and overestimation of counts • Not reliable for counting cell aggregates	Terzieva et al. (1996)

	• Very sensitive technique and high throughput of samples if combined with flow cytometry • Gives quantitative data on cell size and cell numbers	• Frequently show discrepancies in counts, which are caused by their complexity and limited performance of the staining procedure. Stains may fade although can be overcome with use of anti-fade methods • Image analysis software may count abiotic particles within the size range parameters specified • Difficult to differentiate cell numbers within an aggregate or aggregates may obscure single cells resulting in a misrepresentation of the microorganisms present
Microarrays	• Allows many genes to be targeted simultaneously • The genes examined are not influenced by a preselection of genes • cDNA microarrays can be used to measure the expression profiles of many genes simultaneously in response to a particular treatment or condition	• The signal is linear only over a limited range of concentrations • It is often difficult to design arrays in which multiple related DNA/RNA sequences will not bind to the same probe • DNA arrays are limited by the sequences that the array was designed to detect and where there is no complimentary sequence; those species will be overlooked • Genomes from reference strain are often used which may not contain many of the genes found in other isolates

Continued

Table 3 Advantages and disadvantages of culture and non-culture based approaches to analyse microorganisms in bioaerosols.—cont'd

	Advantages	Disadvantages	References
Biomarkers (biochemical) markers	• Specific taxa of microorganisms can be identified • Can give information of microbial activity and metabolisms • Provide biochemical results • Relatively cheap	• Whole cells are not measured so not quantitative in terms of cell numbers • Phylogenetic resolution is not possible • Some biomarker assays are affected by non-biological particles (e.g. dust or other microbial cell components) • Cannot differentiate between viable and nonviable • Microorganisms • Amount of biomarker per cell may vary with species • Microbial volatile organic compounds (MVOCs) are unsuitable to quantify fungi, but can serve as good indicators of fungal presence • Chemical markers may reflect environment rather than specific taxa (e.g. MVOCs might reflect the current substrates available to the microbes)	Ghosh et al. (2015) Garcia-Alcega et al. (2017)
PCR, q-PCR, ddPCR based detection	• Relatively fast, reliable, and sensitive method • If q-PCR used real-time quantification of total microorganisms across domains or specific taxonomic groups or species can be performed. Can target and quantify specific microbial taxa	• PCR inhibitors may be co-extracted which impedes amplification of samples • Biases may be introduced by the nucleic acid extraction method. Yield and microbial diversity will depend on the	Ghosh et al. (2015)

	• Targets only biological material containing nucleic acids. • Can quantify even agglomerated organisms	nucleic acid extraction protocol being used. The DNA extraction method needs to have high recovery efficiencies especially when used for low biomass bioaerosol samples. Bioaerosol quantification may be affected by contamination during DNA extractions • Sufficient biomass needs to be collected from bioaerosol sampling in order to produce enough DNA for amplification • Often only the dominating sequences are detected meaning that the "rare biosphere" may be missed • There are no truly "universal" PCR primers and this is particularly pertinent for viruses • Bioaerosol quantification may be highly affected by contamination	
Loop-mediated isothermal amplification (LAMP)	• Highly selective and sensitive • Relatively quick and easy to perform with the results visible to the naked eyes • Requires no sophisticated instrument, thus convenient to perform in the field	• A period of method development and optimization is required for each target organism	Jiang et al. (2016a)

Continued

Table 3 Advantages and disadvantages of culture and non-culture based approaches to analyse microorganisms in bioaerosols.—cont'd

	Advantages	Disadvantages	References
DNA Fingerprinting or DNA profiling methods (e.g. DGGE, TGGE, RFLP)	• Can be used for broad community characterization • Samples can be multiplexed so can be analysed simultaneously • As for NGS- Can analyse any microbial community without prior knowledge of species present	• Have limited sensitivity, often with poor detection limits and can be time consuming • Limits phylogenetic resolution as only short fragments (ca. 250 bp) are used • Issues with multiple bands from single species due to microheterogeneity of rRNA • Is only semi-quantitative due to variation in 16S rRNA copy numbers in different species	Pankhurst et al. (2012)
High Throughput Sequencing (metagenetic, metagenomic, meta-transcriptomic)	• Targets only biological material containing nucleic acids • Very sensitive and gives fine-scale resolution of the microbial communities • Depth of sequencing means that "rare" biosphere may potentially be detected and characterized • Use of Nanopore based technology (e.g. MinION) is a portable, real-time device for DNA and RNA sequencing which can be used to sequence samples directly in the field • Can analyse any community (metagenomics) and activity (metatranscriptomics) of members without prior knowledge of species present	• Has relatively high capital and running costs • Can be time consuming but advances in technology means that both costs and run times are rapidly decreasing (e.g. *MinION* generates sequence data in real time) • Works on relatively short read lengths (~<600 bp) (depending on technology, see Nanopore and PacBio, which may limit phylogenetic characterization (although read lengths are continually increasing as technology advances)	Ferguson et al. (2019)

	• Samples can be multiplexed so can be analysed simultaneously • Can determine the relative abundance of specific microbial taxa within a community • Metagenomics used to study viral communities in bioaerosols	• Sufficient biomass needs to be collected from bioaerosol sampling in order to produce enough DNA/RNA for analysis • Sequencing artifacts (e.g. chimeric sequences, sequencing errors and homopolymer sequences) may lead to errors, but can be removed or corrected for using advanced bioinformatic pipelines • Metagenomics does not distinguish between viable and inactivated viruses, as genomic material from both is sequenced • Sequence databases need improvement especially with regard to viruses where only a very small fraction of viral genomes are available • With bacterial transcriptomics, ribosomal RNA, needs to be removed and the bacterial mRNA enriched for which can be a challenge especially with low biomass systems such as many bioaerosols • Analysis requires significant computational resources and produces large amounts of data which needs to be stored

Continued

Table 3 Advantages and disadvantages of culture and non-culture based approaches to analyse microorganisms in bioaerosols.—cont'd

	Advantages	Disadvantages	References
Real-time monitoring approaches			
Flow cytometry	• Requires low volumes (cells can be quantified in <1 mL of sample) • Allows rapid throughput of samples ($\sim$50 μL min^{-1})	• Relatively time consuming	Ghosh et al. (2015)
Raman Spectroscopy	• Fast and non-invasive method	• Spectra can be considerably affected by proteins and amino acids present in aerosol samples • Low sensitivity for detecting low concentrations of biological agents in aerosols as Intensities of spectra can be low	Sengupta et al. (2005)
Terahertz spectroscopy	• Detects biological particles including bacterial cells/spores in atmospheric samples	• Low sensitivity for detecting low concentrations of biological agents in aerosols as Intensities of spectra can be low • Multiple resonances may be produced by biological samples	Globus and Gelmont (2014)
LIF, WIBS, SIBS, UV-APS	• Allows *long-term, real-time monitoring of* particle classes (0.5–300 μm which includes pollen) • High flow rates air sampling volume (e.g. SwisensPoleno Mars 40 L min^{-1}) • Low maintenance and is compact • High detection sensitivity as the signals are observed against dark background • 2D or 3D images can be obtained	• High initial capital costs	Nasir et al. (2019) Zare (2012) Ghosh et al. (2015) Huffman et al. (2020)

range of specific taxa present in a sample. It also requires prior knowledge of what you would expect to find, leading to biases in future studies/monitoring. There are other inherent biases with culture-based methods, where for example, the fastest growing or most abundant microorganisms are often selected, leading to a misrepresentation of the communities present (Colbeck and Whitby, 2019). Overlapping colonies or cell aggregates can also occur, which lead to an underestimation of counts (Chang et al., 1994, 1995). Furthermore, the so called *viable but non-culturable* (*VBNC*) organisms whereby <1% of bacteria and up to 40% fungi are culturable, meaning that many microorganisms although present, may be overlooked (Li et al., 2014 Amann et al., 1995; Bridge and Spooner, 2001). Moreover, these VBNC microorganisms, include many pathogenic bacteria that can potentially influence human health (Li et al., 2014). In addition, the non-viable, non-cultivable components of bioaerosols can also be harmful, since several health issues such as allergies related to bioaerosol exposure are not linked to microorganism viability or infectious potential (Cox et al., 2020). Furthermore, cell debris, dead microbial cells or other microbial components, which may also have toxic or allergenic properties are not detected by cultivation (Griffin, 2007). These problems are further exacerbated when analysing airborne microorganisms, where the sampling method used can lead to cell stress or desiccation, causing a further loss of cell viability and underestimation of the microorganisms present (see Section 2).

3.1.1 Plate counts

Here, serially-diluted cultures are plated onto solid agar and incubated. Following incubation, colonies are counted and expressed as CFU per volume of sample spread on the agar. These counts give an approximation, since several cells may aggregate together and become indistinguishable as individual colonies. It is generally accepted that plates must contain between 30 and 300 CFUs for a reliable count, to avoid colony overlap. This, however, is especially problematic when analysing bioaerosols, where the microbiological components may be present in low abundance, and thus the resultant plate counts may have too few colonies to count.

3.1.2 Most probable number (MPN)

Here, the sample is serially diluted (usually by 10-fold), and known volumes are inoculated into replicate tubes containing liquid medium and incubated. Following incubation, the tubes are scored for growth, usually visually by turbidity or a colour change in the media used. The MPN is calculated as

1 multiplied by the dilution factor in the tube containing the most diluted inoculum where microbial growth occurred (Ghosh et al., 2015). This assumes that the most diluted volume of sample contained one to 10 (in the case of decimal dilutions) culturable cells. For more accurate results, MPN requires a large number of replicates, and is usually performed in 96-well plates (Ghosh et al., 2015).

3.1.3 Culturomics

Lately, culture-based methods have gained in popularity due to culturomics. Culturomics uses multiple high-throughput culture conditions combined with either 16S rRNA amplification and sequencing or matrix-assisted laser desorption/ionization-time of flight (MALDI-TOF) (see Section 3.3) to identify previously unculturable species (Dickson, 2017). Culturomics has been successfully used to analyse the human gut microbiome, where organisms corresponding to sequences previously not assigned were cultured, and application of this method resulted in double the number of previously isolated species (Lagier et al., 2016). Culturomics combined with metagenomics has also been used to analyse air samples in an urban environment, where a new *Pseudomonas* species was identified (Azhar et al., 2017).

3.2 Culture-independent methods

To circumvent the inherent limitations of culturing (Section 3.1), culture-independent methods provide an alternative approach. These can be considered as non-nucleic acid based (e.g. spectroscopic, spectrometric, fluorescent methods), or nucleic acid based (i.e. molecular methods) or combinations of the two.

3.2.1 Light and epifluorescence microscopy

Microscopic analysis of airborne biological particles involves the air samples to be drawn onto glass slides, semi-solid media or filters. Impaction or liquid impingement onto a glass slide has been used to quantify airborne bacteria and fungal spores by microscopy (Yoo et al., 2017). Bacteria from bioaerosols can be classically Gram stained and observed via light microscopy (usually at 400X or 1000X magnification), whereas fungi are generally viewed under phase contrast (Delort and Amato, 2018). Electron microscopy (e.g. transmission (TEM) and scanning (SEM)) have also been used to analyse aerosol particles at high magnification (up to 10,000,000X) (Delort and Amato, 2018). When coupled with energy-dispersive X-ray (EDX) spectrometers, the shape and elemental composition of biological particles in aerosols may also be determined (Matthias-Maser et al., 2000). Microscopy has also been combined with microfluidics (see below) where monodisperse microdroplets are generated from aerosols with compartmentalized reagents for specific assays (Kamperman et al., 2018). Although this technique is highly sensitive, it requires accurate and valid automation, and prevention of contamination in the microfluidic channels (Ghimire et al., 2019).

A number of different stains or dyes are commercially available that stain viable and non-viable cells. A frequently used stain is DAPI (4′,6-diamidino-2-phenylindole), which stains both viable or dead cells and has been used to count total bacteria and fungi in aerosols (Paez-Rubio et al., 2005). Other stains include CTC (5-cyano-2,3-ditolyl tetrazolium chloride) which targets living and respiring cells, and Acridine orange (AO), which targets both DNA and RNA and differentiates viable and non-viable cells. Both have been used to quantify viable cells in bioaerosols (Hill et al., 2007). Live/DeadR (Baclight™) combines two dyes which bind to DNA and distinguishes between viable and dead bacteria, based on cell membrane integrity. Alternatively, propidium monoazide (PMA) staining, can be used to distinguish between live/dead cells (Nocker et al., 2007). Other widely used stains include SYTO9, Sytox© Green, SYBR Green, SYBR GOLD, SYBR SAFE (Colbeck and Whitby, 2019; Delort and Amato, 2018).

Microscopy has the advantage that it quantifies both culturable and non-culturable microorganisms. Microorganisms can also be analysed in terms of morphology, counts, and phenotype. However, this method is time-consuming and requires specialized knowledge and expertise during sample processing, staining and visualisation (Ghosh et al., 2015). Furthermore, identification of microorganisms to the species level is usually not possible, unless taxon-specific, immune-specific or fluorescent staining methods (e.g. fluorescence *in situ* hybridisation (FISH)) are applied (see below).

3.2.2 Fluorescence in situ hybridization (FISH) and flow cytometry

FISH allows the detection, identification, and enumeration of microorganisms from different environments (Kubota, 2013). FISH uses specific molecular probes that target either the ribosomal RNA (rRNA), messenger RNA (mRNA), or functional genes (Kubota, 2013). However, FISH has several drawbacks including insufficient sensitivity, due to the low number of target molecules in cells, low probe permeability of cells, and poor probe hybridization efficiency (Amann et al., 1995). Often the fluorophores used in FISH bind to abiotic material, making it difficult to differentiate between microbial cells and abiotic particles resulting in errors in enumeration (Colbeck and Whitby, 2019). Baclight^{TM} staining however, may overcome this issue as it is less susceptible to binding abiotic material Terzieva et al., 1996; Colbeck and Whitby, 2019; Pillai and Ricke, 2002). CARD-FISH (also known as tyramide signal amplification (TSA)-FISH) is an advancement that allows both rRNA-targeted phylogenetic identification and phylogeny to be linked to physiology and metabolic activity (Kubota, 2013).

FISH is also often combined with flow cytometry to give higher sample throughput and has been used to detect airborne bacteria and fungi (Chen and Li, 2005; Rule et al., 2007; Hernlem and Ravva, 2007; Lange et al., 1997). Flow cytometry requires the cells to be in aqueous solution and hence is typically performed after air collection usually by impingement, cyclone, or filtration (Colbeck and Whitby, 2019). The particles are separated in a sheath fluid that is passed through a laser beam. The fluorescence emitted and the light dispersed by each particle is measured with photon detectors. Upon light emission, samples can be analysed for several parameters simultaneously (i.e. counts, cell size and shape, biological and chemical properties, and bacterial ice nucleation) (Ghosh et al., 2015). Flow cytometry can also detect several taxon-specific targets simultaneously to categorize taxa, or even species, and viruses as well as discriminating larger cells (e.g. fungi) from smaller cells (e.g. bacteria) (Ghosh et al., 2015).

3.2.3 Molecular methods

Molecular-approaches targeting nucleic acids are gaining wide acceptance in bioaerosol studies (Colbeck and Whitby, 2019) and offer a broader view of the microbial diversity present, especially the use of amplicon-based sequencing or metagenomics (Cox et al., 2020). However, there are several considerations when applying molecular techniques to analysing bioaerosols. For example, yield and microbial diversity obtained depends on the nucleic extraction method and DNA biases may be introduced (Ferguson et al., 2019). Specifically, DNA based methods (via PCR) do not discriminate between intact cells and cell debris and both types of material can be amplified by PCR (Reponen et al., 2007). Thus, there is the issue of whether a microbial DNA signature comes from a viable organism. As most health effects are caused from exposure to the microbial products (e.g. endotoxins, mycotoxins, etc.) instead of the viable microorganism themselves, methods to measure these compounds in relation to human response are important (Castellan et al., 1987) (See Section 4). In addition, since air samples generally have a low biomass and may come from environments with high concentrations of PCR inhibitors, the DNA extraction method needs to have high recovery efficiencies, especially when used for metagenomics (Ferguson et al., 2019). Characterizing microorganisms from bioaerosols can be hindered by PCR inhibitors (e.g. humic acids and inorganic particles), which may impede downstream molecular analysis (Luhung et al., 2015; Peccia & Hernandez, 2006). Furthermore, given that many of the air sampling methods only take a "snap-shot" sample in time, generally the most abundant members of the bioaerosol community are recovered, rather than the rare biosphere, leading to biases in the observed microbial diversity (Ferguson et al., 2019) (see Section 2). Molecular methods are also liable to false positives as they have low detection limits and collect genetic material from dead cells (Ferguson et al., 2019). However, with the continued development of sequence databases, molecular tools have great potential for providing new information on bioaerosol diversity.

3.2.3.1 Quantitative (qPCR) and reverse transcriptase qPCR

qPCR uses fluorescent dyes (e.g. SYBR) or TaqMan probes for a rapid and accurate, real-time quantification of microorganisms and is expressed as the number of gene copies or target organism (genome) in a sample (Colbeck and Whitby, 2019). One advantage of qPCR is that it can quantify even agglomerated organisms (Colbeck and Whitby, 2019). qPCR has been used to characterize bioaerosols across all domains of life (Pankhurst et al., 2012; Ferguson et al., 2019, 2021; Unterwurzacher et al., 2018). Although there are no truly "universal" PCR primers, the most commonly used gene targets are the 16S rRNA gene for bacteria and archaea and the internal transcribed spacer region (ITS) for fungi. With viruses, the specific virus has to be targeted, which may lead to an underestimation of the viriome (Carducci et al., 2013). Using the 16S RNA or ITS genes however is problematic, as the number of 16S RNA gene copies varies with species (Klappenbach et al., 2001), and there is inherent variability in the ITS region, even within species (Haugland et al., 2004). The 16S rRNA gene can also be restricted in its taxonomic resolution and does not give information on the evolutionary and functional mechanisms at the species level (Shah et al., 2011). Thus, the use of specific functional gene targets or antimicrobial resistance (AMR) genes might be a better approach. Another consideration is that PCR inhibitors may be co-extracted during DNA/RNA extraction which would need removing prior to amplification. Despite these drawbacks, detection of a high abundance of a microbial pathogen is still a good indicator of a potential health risk (Colbeck and Whitby, 2019).

3.2.3.2 Loop-mediated isothermal amplification (LAMP, RT-LAMP)

LAMP is an isothermal nucleic acid analysis technique (Notomi et al., 2000, Fang et al., 2010). The method has the advantages that it is highly sensitive and selective, relatively quick to perform and the results are easily visualised via a colour assay. Thus, LAMP has great potential in clinical diagnostics and in-field detection of microorganisms such as targeted pathogens. One study combined LAMP with a microfluidic system that could capture and enrich airborne pathogens with good detection limits down to approximately 24 cells per reaction for *Staphylococcus aureus* (Jiang et al., 2016a,b). Reverse transcription loop-mediated isothermal amplification (RT-LAMP) has also been developed that utilises a hand-held device that was successfully used to detect H1N1 influenza viruses (Jiang et al., 2021).

3.2.3.3 High throughput sequencing (HTS)

For many years, molecular fingerprinting approaches, such as denaturing gradient gel electrophoresis (DGGE), temperature gradient gel electrophoresis (TGGE), single-strand conformational polymorphism (SSCP), terminal restriction fragment length polymorphism (T-RFLP), and automated ribosomal intergenic spacer analysis (ARISA) were used to assess microbial biodiversity (Colbeck and Whitby, 2019). In general, these methods had limited sensitivity, poor detection limits, were time consuming and labour intensive and often missed the rare biosphere (Colbeck and Whitby, 2019). Previously, cloning and (Sanger) sequencing was also used to analyse microbial diversity, but this approach was also labour and time intensive. These methods have since been superseded by the advancement of High Throughput Sequencing (HTS) technologies, which have completely revolutionized microbial ecology studies.

HTS produces large sequence datasets and enables the spatial and temporal distribution of microorganisms in the air to be characterised at the fine-scale (Colbeck and Whitby, 2019). These approaches are now routinely used to target phylogenetic and functional marker genes in bioaerosols (e.g. Pankhurst et al., 2012; Ferguson et al., 2019, 2021; Franchitti et al., 2020; Grydaki et al., 2021; Mbareche et al., 2017a,b). HTS has the advantage that it gives a greater phylogenetic resolution. However, amplicon sequence libraries may be influenced by sequencing artifacts such as chimeric sequences, sequencing errors and the presence of homopolymer sequences, although such errors can be removed or corrected for using advanced bioinformatic pipelines (Colbeck and Whitby, 2019). In addition, HTS is only as good as the sequence databases available for bioinformatics analysis and whilst new sequences are continually being added to the databases, analysis is only as good as the reference database that is being used for comparison. This is especially the case for genome sequence databases (see below).

3.2.3.4 Metagenomics (whole-genome shotgun sequencing)

Here DNA is extracted from an environmental sample, and either cloned into a suitable vector, and transformed into a host bacterium or random fragments are generated. Transformants are screened or randomly sequenced. The sequenced fragments are assembled into larger pieces by identifying overlaps using specific assembly packages (e.g. CLC assembly, MegaHIT, metaSPAdes) and reconstructed whole genomes identified by comparison with databases. Thus, metagenomic analyses requires very complex, high-resolution assembly, binning, and sequence annotation. In theory, with metagenomics every DNA sequence present in the particular sample will be sequenced. This allows the taxonomic composition and metabolic potential of samples to be determined. However, in practice most of the DNA sequences recovered will come from the most dominant members of the community (and/or those with the largest genomes). Further, as there is no targeted PCR step to select for taxa of interest, a large amount of non-target DNA will be sequenced (animal, plant etc.). As such, although the taxonomic resolution is higher with metagenomics, the total diversity recovered can be lower than for amplicon sequencing.

Data analysis and storage also remains challenging due to a lack of best practice methods and reference sequences in the database to compare genome sequence data to. Although metagenomics is not generally subject to PCR biases and errors associated with amplicon sequencing, the relative abundances of entire microorganisms can vary significantly depending on the DNA extraction and sequencing protocol used (Shah et al., 2011; Ferguson et al., 2019). Furthermore, low biomass in air samples means that generally a random PCR step is required during library preparation (e.g. Nextera XT). However, this does not introduce the same level of PCR bias as an amplicon-based method. Thus, a major drawback is that metagenomics requires a good nucleic acid concentration, and with the low air biomass, the limitation is often the sample processing rather than the analytical method.

Metagenomics has been widely used in the characterization of airborne viral diversity (Posada-Cespedes et al., 2017; Prussin et al., 2014; Rosario et al., 2018). However, a major drawback with this approach is that it is unable to distinguish between viable and inactivated viruses, as genomic material from both is sequenced (Colbeck and Whitby, 2019).

Current sequence databases also need improvement, especially with regard to viruses, where only a very small fraction of viral genomes are available in reference databases. Currently the application of metagenomic sequencing to study bioaerosols has been limited, and largely only been applied to the indoors, such as hospital environments (Tong et al., 2017; King et al., 2016). Despite this, metagenomics is a powerful tool for the study of complex microbial communities, and in particular for identifying exposure risk to human health to bioaerosols.

3.2.3.5 Metatranscriptomics

Gene activity can be analysed at the transcriptome level, by analysing the gene transcripts rather than the genes themselves. However, this requires the metabolic state of the microorganisms to be frozen immediately upon sampling in order to reflect their state whilst airborne. This can be done by direct freezing in liquid nitrogen or using fixative agents (e.g. RNA Later®). RNA is then extracted, reverse-transcribed into cDNA and processed (e.g. by RNASeq or TruSeq sequencing). This approach also gives a unique insight into the transcriptional/metabolic activity of the microbes at the time of collection (i.e. it provides more information about how they were functioning immediately prior to collection). This technique is particularly useful for understanding the effects of gene expression in relation to human health (Hegarty et al., 2018). With eukaryotic systems, it is relatively straightforward to extract and enrich mRNA, however with bacterial transcriptomics, rRNA, (which comprises 90% to 95% of the total RNA) needs to be removed first and the bacterial mRNA enriched. This can be especially challenging when analysing bioaerosol transcriptomes, which may have low biomass. However, novel extraction methods and library preparation can overcome these challenges (Colbeck and Whitby, 2019). Despite this, metatranscriptomics opens up new avenues to bioaerosol health-related studies.

3.2.3.6 Microarrays

Microarrays (DNA chips/gene chips) uses a solid matrix (typically a glass slide) to simultaneously measure DNA or RNA sequences. There are different types of microarrays including cDNA, oligonucleotide, BAC and SNP microarrays, MMChips (for microRNA populations) as well as protein, peptide and tissue microarrays. Phylogenetic and functional microarrays such as GeoChips and PhyloChips have also been used to analyse microbial community structure and functional genes. Labelling is usually performed by incorporating fluorescently labelled nucleotides into the hybridization probes. Sample probes are hybridized to the microarray slide and washed to remove any unbound probe. The fluorescent signal is detected and measured using a scanner with signal intensity taken as a measure of gene expression. cDNA microarrays can also be used to determine expression profiles of many genes simultaneously in response to a particular treatment or condition.

Microarrays have the advantage that many genes may be targeted simultaneously. However, they only provide an indirect measure of relative concentration and are limited by the fact that the fluorescent signal is linear only over a limited range of concentrations. Thus, at high concentrations the array becomes saturated whilst at low concentrations, equilibrium favours no binding. It is also often difficult to design arrays where related DNA/RNA sequences do not bind to the same probe. DNA arrays are also limited by the sequences that the array was designed to detect. Thus, genes that have not yet been annotated in a genome will be missed. Furthermore, for highly variable genomes (e.g. bacteria), the arrays are typically designed using genome information from a reference strain, which may not contain some of the genes found in other isolates and therefore become overlooked.

3.2.3.7 Proteomics and biomarkers

Liquid and gas chromatography coupled with mass spectrometry (LC/GC-MS) and spectroscopy (nuclear magnetic resonance (NMR), Fourier transform infrared (FTIR), matrix- assisted laser desorption/ionization mass spectrometry coupled with time-of-flight mass spectrometry (MALDI-TOF) have all been used to quantity various biomarkers in bioaerosols (Georgakopoulos et al., 2009; Ghimire et al., 2019). High-performance liquid chromatography (HPLC), liquid chromatography (LC) and gas chromatography (GC) can also be coupled with isotope ratio mass spectrometry (IRMS) to analyse changes in the natural variation in isotope ratios (Hayes, 2001; Liang et al., 2012). Ayer et al. (2008) also used pyrolysis gas chromatography differential mobility spectrometry (PY/GC/DMS) to differentiate proteins and viruses.

Examples of specific biomarker targets used to analyse bioaerosols include: endotoxins, mycotoxins, (1→3) β-D-glucan, beta-*N*-acetylhexosaminidase assays, ergosterol, extracellular polysaccharides, DNA/RNA and their sequences, amino acids, adenosine triphosphate (ATP), fatty acids, phospholipids, glycolipids, lipopeptides, volatile metabolites and specific microorganisms (bacteria, archaea, fungi, viruses) (Bloom et al., 2009; Colbeck and Whitby, 2019; Cox et al., 2020; Ghimire et al., 2019; Elbert et al., 2007). Different molecules have also been used to assess different microbial groups.

For example, protein markers for total bacteria, 3-hydroxy fatty acids (3-OH FAs) for Gram-negative bacteria, and ergosterol for fungal biomass (Delort and Amato, 2018). Microbial volatile organic compounds (MVOC's), which are products of fungal and bacterial metabolism, have also been used to characterize the microorganisms in bioaerosols (Garcia-Alcega et al., 2017, 2018). Although MVOCs are unsuitable to quantify fungi, they can serve as good indicators of fungal presence (Korpi et al., 1998). Alternatively, arabitol and mannitol have been used as targets for quantifying fungal spores in PM_{10} (Bauer et al., 2008).

Biomarker analysis, however, has several drawbacks. For example, it cannot differentiate between viable and nonviable microorganisms; often specific genera/species cannot be identified and the amounts of biomarker per cell may vary with species (Chen and Hildemann, 2009). Immunoassays which bind antibodies to a specific target antigen (e.g. fluorescence immunoassay, monoclonal antibodies, enzyme-linked immunoassay and radioimmunoassay) have also been used in bioaerosol analysis to detect and enumerate components in bioaerosols based on antibodies specific for antigenic components of microorganisms (Burge and Solomon, 1987; Speight et al., 1997). Other approaches include, ATP bioluminescence, which is a relatively cheap and easy to use optical label-free detection method, that allows for the analysis of microbe metabolism (Moon et al., 2012). These biomarkers and molecular tracers may offer a means for assessing bioaerosols, particularly their metabolic and physiological characteristics and their contributions to PM mass.

3.3 Real-time (online) monitoring of bioaerosols

The popularity of real-time analysis of bioaerosols is growing (Fennelly et al., 2017; Huffman et al., 2010, 2012, 2020). These approaches allow samples to be detected very rapidly (often within milliseconds), are generally nondestructive to the target microorganisms, and usually have less chance of contamination during handling. Real-time analysis methods can be broadly split into those methods that utilise fluorescence (e.g. flow cytometry), molecular approaches that target DNA or RNA (e.g. minion), spectroscopic (e.g. Raman spectroscopy) or spectrometric (e.g. whole-cell mass spectrometry) methods. However, more information is required about all of these real-time methods in terms of their strengths, and limitations in detecting the biological components of air.

3.3.1 *Real-time fluorescence approaches (culture-independent)*

Biological cells comprise fluorophores, (e.g. nicotinamide adenine dinucleotide (phosphate), flavin, and riboflavin), which autofluorescence at different wavelengths (Brosseau et al., 2000), and several fluorescence detectors with automated biosensors have been developed and applied to monitor bioaerosols in real-time (Hairston et al., 1997). These include Ultraviolet Aerodynamic Particle Sizer (UV-APS) which uses PM_{10} samplers (Agranovski et al., 2003); UV light-emitting diode (UV-LED) based fluorescence sensor for real-time bioparticle detection (Ryskevic et al., 2010); and Fluorescence aerosol particle sensor (FLAPS) (Cox and Wathes, 1995).

Laser-induced fluorescence (LIF) which combines fluorescence with spectroscopy is based on the principle that specific biological molecules (e.g. proteins, coenzymes, structural compounds, pigments) exhibit fluorescence (or autofluorescence) and has been mostly used for studying the structure of molecules and detection of selective species (especially bacteria) (Ghosh et al., 2015; Pan, 2015; Tian et al., 2020). The air samples are excited by a laser, within a few nanoseconds the species de-excite and emit light (with wavelength longer than the excitation wavelength) (Zare, 2012). This emitted fluorescent light is then recorded by a photomultiplier tube (PMT) or filtered photo diode. The data is expressed as individual "events" which contains the measurement data describing the properties of a single particle. Ultraviolet laser-induced fluorescence (UV-LIF) has been used with filter samplers and applies a chamber-based library approach to detect fluorescent bioaerosol particles (Ghimire et al., 2019). UV-LIF works on the basis of the excitation and emission properties of bioaerosols together with phenotypic characteristics and high throughput sequences (Government Office for Science, 2014).

Waveband Integrated Bioaerosol Sensor (WIBS) (Droplet Measurement Technologies, Longmont, CO, USA) is another instrument whereby multichannel fluorescence analyses are merged with optical diameter measurement and fluorescence spectra are produced from the airborne particles (i.e. fungal spores, pollens, bacteria) (Nasir et al., 2019). WIBS can also be used for quantitative, qualitative, and phenotypic observation of bioaerosols (Nasir et al., 2019). WIBS-3 and UV-APS have also been compared to characterize airborne PMs both in the laboratory and field (Gosselin et al., 2016). Spectral intensity bioaerosol sensor (SIBS), which is an extension of WIBS, differs in terms of the channels for fluorescence intensity measurement across 298–735 nm wavelengths (Ghimire et al., 2019). SIBS is dual wavelength excitation multiple fluorescence band system, providing information on size, shape and generates highly resolved spectral information (over 14 channels) of single particles in real time (Nasir et al., 2019).

SIBS coupled with MALDI–TOF, qPCR and size-fractionated endotoxins has also been recently used to characterize bioaerosols emission from compost material (Nasir et al., 2018; Ferguson et al., 2021; Nasir et al., 2019; Tian et al., 2020). Some instruments primarily focus on real time measurements of pollens, such as, Rapid-E (Kawashima et al., 2017; Šikoparija, 2020), and wisens Poleno (Sauvageat et al., 2020). Other commercially available fluorescence bioaerosol detection systems include ENVI BioScout, IMD-A, Intascope signature library for mould, spores, fungi, bacteria and viruses, and FLIR IBAC systems (formerly known as FIDO B2 (Jonsson and Kullander, 2014; Hernandez et al., 2016; https://instascope.co/science/). These systems possess a single excitation wavelength with fluorescence detection, which can differentiate unknown types of harmful microorganisms, allergens, toxins, or any particles from biological origin from the natural background (Ghimire et al., 2019). Whilst real-time UV-LIF based measurements have the potential to provide profile, size distribution and composition of bioaerosols there are still challenges to advance the real time detection of bioaerosols. There also remains a need for developing standard reference materials for fluorescence measurements to improve the certainty in the assignment of the fluorescence to potential biofluorophores; alongside the intercomparison of UV-LIF and advanced data analytics for discriminating the components of BioPM (Huffman et al., 2020; Nasir et al., 2019; Tian et al., 2020).

3.3.2 Real-time spectrometry approaches

Whole-cell mass spectrometry has been used to identify a wide range of bacteria and fungi (Welker and Moore, 2011). For example, MALDI-TOF, can enable the identification of bacteria and fungi (Delort and Amato, 2018). However, this approach is limited in that it can only analyse compounds greater than 600 Da (Ghosh et al., 2015). Bioaerosol mass spectrometry (BAMS) is another tool that has been used to detect biological agents in airborne samples (e.g. vegetative bacteria, bacterial spores, viruses, and biological toxins) (Ghimire et al., 2019). BAMS has the advantage that airborne particles can be sampled directly from the air in a size-dependent manner (Ghimire et al., 2019). This approach provides an accurate real-time measurement of bioaerosols, but it is complex, large in size, and costly; and thus, not widely applicable (Ghimire et al., 2019). Furthermore, the use of BAMS for the selective detection and discrimination of target microorganisms with specific metabolic/molecular markers requires further development (Szponar and Larsson, 2001).

3.3.3 Real-time spectroscopy approaches

Raman spectroscopy detects the chemical composition of samples, which in turn provides a fingerprint that can be related to the microorganisms present. This method is fast, specific, reliable, and non-invasive (Bowley et al., 2012; Sengupta et al., 2005). Several models of Raman spectroscopy have been developed such as surface-enhanced Raman spectroscopy-SERS, which have been applied for bioaerosol detection (Sengupta et al., 2005). However, applying Raman spectroscopy to analyse biological aerosol samples has inherent problems, whereby the spectra can be considerably affected by the proteins and amino acids present (Sengupta et al., 2005). Another spectroscopic approach is terahertz spectroscopy which works using a terahertz frequency range (i.e., 0.1–10 THz) or in the sub-millimetre wave range and can detect biological particles including bacterial cells/spores in air samples (Ghimire et al., 2019; Globus and Gelmont, 2014). However, both Raman spectrometry and terahertz spectroscopy have shortcomings, whereby intensities of spectra can be low, leading to reduced sensitivity Ghimire et al., 2019). This is especially problematic when detecting low concentrations of biological agents present in aerosols (Ghimire et al., 2019). Furthermore, when analysing the biological components of aerosol samples, multiple resonances may be produced by terahertz spectroscopy (Globus and Gelmont, 2014). In another study, Parker et al. (2000) developed laser ablation mass spectroscopy and multivariate analysis to detect and analyse pollen, bacteria, and other aerosol particles.

3.3.4 Real-time molecular approaches

It has recently become possible to sequence samples directly in the field with single molecule sequencing based technologies (e.g. Minion, Oxford Nanopore Technologies). This is a portable, real-time device for DNA and RNA sequencing. The device has the potential for bioaerosol studies allowing air samples to be sequenced in real time in the field (provided sufficient DNA/RNA yields are obtained from the air sampling method used). The MinION generates between 10 and 20 Gb of DNA sequence data with read lengths of hundreds of kb. Since the MinION streams data in real time, data analysis can be performed almost immediately. Devices such as the PromethION, GridION and the portable MinION can be used for DNA and RNA sequencing (Oxford Nanopore Technologies). These devices could become powerful tools for real time monitoring of biological hazards in the air.

3.4 Which bioaerosol analysis method for which taxa?

Culture-based methods are often used to determine if the microorganisms are viable and capable of causing an infection, or to study the performance of indoor air disinfection techniques for example. Thus, cultivation methods tend to be more widely used to characterize indoor airborne microbial communities. However, non-culturable and non-viable bacterial aerosols can carry endotoxins and other toxic components that can also have adverse health effects. Therefore, both culture-based and culture-independent methods may be needed. Both approaches have been widely used to characterise various taxa from both indoor and outdoor environments (e.g. Lindsley et al., 2017; Pankhurst et al., 2012; Ferguson et al., 2021; Grydaki et al., 2021).

As discussed in Section 2, the efficiency of sampling and nucleic acid recovery for molecular approaches can vary depending on the air sampling technique used. With HTS the issue of nucleic acid recovery leads to variations in the number of operational taxonomic units (OTUs) or amplicon sequencing variants (ASVs) identified, which in turn affects the microbial diversity observed (Mbareche et al., 2018). For example, one study found that only 14% of bacterial OTUs and 44% of fungal OTUs were shared in four co-located samplers (Hoisington et al., 2014). Furthermore, using an electrostatic dry filter consistently yielded more fungal and bacterial OTUs compared to two liquid cyclonic samplers (Mbareche et al., 2018). In addition, data processing techniques, and bioinformatic analyses for HTS approaches have yet to be standardized, particularly in aerosol studies and thus cross-comparison among studies is difficult and must be done with caution (Pauvert et al., 2019). Analysis of the airborne microbial diversity is a rapidly emerging area and we detail below the various methods used for targeting different taxa.

3.4.1 Bacteria

For bacterial analysis, air samples are often collected directly onto culture plates or impinged into liquid media that is then applied to plates. One of the most widely used non-selective media used in culture-based aerosol studies is trypticase soy agar (TSA), a nutrient-rich medium known to grow a large spectrum of organisms, and like nutrient broth or agar is used for growing Gram-positive bacteria (e.g. *Staphylococcus* spp., *Micrococcus* spp., *Bacillus* spp. etc) (Delort and Amato, 2018). Luria–Bertani (LB) medium is a nutrient-rich medium, generally used to grow *Escherichia coli* and other bacterial species (e.g. *Bacillus* spp., *Curtobacterium* spp., *Clavibacter* spp., *Arthrobacter* spp., *Psychrobacter* spp.) (Delort and Amato, 2018). Other media such as R2A, have been used to grow *Pseudomonas* spp., *Sphingomonas* spp., *Methylobacterium* spp., *Bacillus* spp., among others, and Zobell's medium has been used to grow *Pseudomonas* spp. and *Sphingomonas* spp. from aerosols (Delort and Amato, 2018). Drigalski and MacConkey media has been used for Gram-negative rods and enteric bacteria, and King's B medium for Pseudomonads (Delort and Amato, 2018). To make culture-based approaches more selective, antibiotics such as cycloheximide, actidione plus others may also be added.

Culture-independent methods (e.g. HiSeq) have been used to analyse bacteria in aerosols and provide important information on the air quality control. For example, HTS was used to analyse caged poultry houses and showed seasonal changes in microbial community composition, especially with opportunistic pathogens that had relatively high abundances during the winter (Jiang et al., 2018). Real-time approaches such as terahertz frequency and sub-millimetre-wave Fourier transform have also been used to detect bacterial spores (e.g. *Bacillus* spp.) (Globus and Gelmont, 2014). Endospores from Gram positive *Bacillus* spp. have also been detected using Raman spectroscopy, but this was suitable only for high-concentration aerosols and more research is still required in this area, especially for measuring low concentrations of bacteria (Sengupta et al., 2005).

3.4.2 Fungi

Many fungi can be easily cultured, and much of the bioaerosol data obtained for indoor fungi have been obtained using culture or microscopy methods (Nazaroff, 2016). To culture fungi from atmospheric samples, Sabouraud's medium, potato dextrose agar, and malt extract agar have all been used, most notably for the recovery of *Aspergillus* spp., *Cladosporium* spp., and *Penicillium* spp. (Delort and Amato, 2018). However, some fungi grow more slowly than others, and competition can occur when multiple species are present. Non-viable spore traps are commonly used to collect airborne fungi for analysis by microscopy in combination with specific stains to measure certain parameters (see Section 3.3). However, this approach is labour intensive, needs specialist expertise in fungal identification, and identification can be limited to spore types or groups of genera.

As is the case with other taxa, information on the culturable fungi only reflects the viable organisms that were successfully sampled and grown, rather than all the fungal species present. Q-PCR has also been used to quantify both total fungi and specific fungal species (Haugland et al., 2004). Based on 36 specific mould species obtained from settled dust sampling in the United States, the Environmental Relative Mouldiness Index (ERMI) established protocols including species-specific primers and probes for fungal qPCR analysis (Cox et al., 2017). The ERMI however, is not dependent on microbial viability, and is limited to the 36 known species it was developed for (Vesper et al., 2007; Cox et al., 2017). Transcriptome analysis targeting ribosomal genes may be used to identify active fungal groups and was successfully achieved for fungal communities by impingement, using a preservation solution as the collection liquid (Womack et al., 2015). Biochemical markers have been used to measure airborne fungal exposure including: ergosterol, extracellular polysaccharides, ß-1,3-glucan and beta-*N*-acetylhexosaminidase assays (Miller et al., 1988; Miller and Young, 1997). Sugar alcohols, such as mannitol and arabitol, have also been used for fungi as well as the biochemical tracers for fungal spores including mannitol and glucans (Delort and Amato, 2018).

3.4.3 Archaea

Although Archaea are ubiquitous in the environment, they are largely understudied in regard to bioaerosols. Their presence in the air and the factors affecting their abundance are not well characterized. Culture-independent techniques have allowed the detection and identification of archaea. Specifically, qPCR has been used to measure total archaea in indoor environments (Just et al., 2013; Nehmé et al., 2009). To study archaeal diversity, amplicon-based HTS methods targeting the prokaryotic 16S rRNA gene have been used (Cox et al., 2020). Specific primers targeting different sub-regions of the 16S rRNA gene have also been shown to offer a better coverage of archaeal diversity (Bahram et al., 2018) and thus primer selection for targeting archaea needs to be carefully considered.

3.4.4 Viruses

Airborne viruses are most often studied because of concerns about infectious disease transmission, such as the recent pandemic with COVID-19. As with bacterial aerosols, analysis of airborne viruses can be broadly divided into culture-dependent and culture-independent methods, and many of the same considerations apply (Lindsley et al., 2017; Verreault et al., 2008). For culture-based methods, viability needs to be preserved during and after bioaerosol collection, which is more difficult than preserving the viability of bacteria or fungi. In addition, because viruses require host cells to reproduce, viral assays are considerably more complex and difficult than bacterial or fungal assays, and many viruses currently cannot be cultured. Culture-independent methods of virus detection allow the use of a broader range of aerosol sampling equipment. Thus, PCR and other culture-independent methods are more widely used than culture-based methods, but they do not determine if the airborne virus is potentially infectious or not (Cox et al., 2020).

3.5 Summary of bioaerosol analysis methods

Although advances in molecular methods have expanded our understanding of what microorganisms are found in bioaerosols and how they function and interact with humans, surfaces and other environments; large knowledge gaps still remain. For example, current recommendations of acceptable exposure limits for fungal and bacterial bioaerosols still rely on culturable and microscopic total count determination as this approach is simple and low cost. Yet a more extensive characterisation and quantification of bioaerosol components (especially in indoor environments) is necessary to better understand their health impacts on humans.

Standardisation of bioaerosol analysis methods is urgently required especially given the rapid development in analytical and molecular methods. However, airborne microorganisms are very difficult to assess accurately, especially under field conditions, due to differences in sample collection and nucleic acid extraction efficiencies. Consequently, bioaerosol abundance and biodiversity can be underestimated, making it difficult to link specific bioaerosol components to disease and exposure. Of particular relevance, especially for indoor air, is whether the microbial DNA signature comes from a viable organism, and we should not confuse viability with infectivity. It is a common misconception that microorganisms identified with molecular methods, but do not grow in culture, are assumed to be derived from dead material and we caution the use of viability as the "*gold standard*" to what is infectious in the air. We should be clear that where DNA is recovered and identified by molecular methods, but the organism is not grown in culture, does not conclusively prove that the organism is dead and non-infectious. Indeed, given that <1% of bacteria and ~40% of fungi cannot be cultured, it is likely that a vast majority of pathogens in a bioaerosol sample will not be detectable with culture based methods. Given the low biomass in most air samples, it remains a challenge to recover sufficient DNA to obtain a representative sample to assess exposure levels. Thus, methods that characterize the entire population of bioaerosol communities should be considered.

Bioaerosol research has been enhanced by the recent advancements in molecular tools, particularly HTS which have allowed a faster characterisation of the microbial components in air over spatial and temporal scales. Knowledge of the advantages and disadvantages for each method should be considered before application and studies combining techniques to overcome the inherent limitations of each may provide a more fuller image of indoor and outdoor bioaerosols. Although there are still many issues that need to be addressed, careful selection and optimisation of the most appropriate bioaerosol analysis techniques will ensure more information can be gleaned on the role of microbes in bioaerosols and improve the management of air quality. Indeed, as costs of HTS methods decreases, more-detailed characterisation of the rate of transmission of microbes between the environment and humans will become possible in the future.

3.6 Recommendations and guidelines for bioaerosol analysis methods

See Box 3.

BOX 3 Recommendations for standardisation of bioaerosol analysis methods

1. Harmonized protocols for bioaerosol analysis would improve the reproducibility of bioaerosol studies. With culture-based analyses, for example, agar volume should be standardized and reported.
2. Sampler efficiency and ability to cover microbial diversity needs to be considered when selecting methods to investigate microbial diversity.
3. Combining culture-based with culture-independent methods and a move away from culture-based as a sacred *"gold standard"* would provide a fuller image of the microbial components and their activities in bioaerosols but should be treated with caution as they combine biases inherent to both.
4. Most health effects are caused due to exposure to the microbial products (e.g. endotoxins, mycotoxins, etc); instead of the viable microorganism themselves (i.e. we should not confuse viability with infectivity). Efforts therefore should include measuring these compounds in relation to exposure and the human response.
5. There is a need for long-term integrated analysis of bioaerosols e.g. with real-time analysis methods.
6. Current sequence databases also need improvement, especially with regard to viruses.

BOX 3 Recommendations for standardisation of bioaerosol analysis methods—cont'd

7. Best practice methods are required for sequence data analysis and storage alongside improved reference sequences in the database to compare genome sequence data to.
8. Simply characterizing the aerosol microbiome is only the first step to understanding the microbial processes occurring in bioaerosols in different environments. A combination of metagenetics (amplicon sequencing) and metatranscriptomics will provide a better understanding of how communities respond both functionally and taxonomically to changes in their environment.

4. Sampling and analysis methods to assess exposure and risk

Bioaerosols represent a significant human health concern. For example, they play an important role in the distribution of infection, allergic, and carcinogenic compounds that impact on human health leading to infectious diseases, acute toxic effects, allergies, communicable diseases, respiratory diseases, cardiovascular diseases, and cancer (Douglas et al., 2017). Bioaerosols may also cause occupational health effects which may lead to respiratory symptoms, which can range from acute to chronic and even death. In contrast, many microbial exposures are beneficial and essential in human development (Sharma and Gilbert, 2018). Section 4 presents some of the components of bioaerosols that impact on human health. We also discuss how sampling and bioaerosol analysis approaches may be used to assess human exposure and risk in various environmental and occupational contexts.

4.1 Human exposure to bioaerosols

Particle size, agglomeration and dispersion play an important role in determining human exposure, as these factors determine the penetration of the human respiratory system (Wery, 2014). During inhalation, larger particles present in bioaerosols become trapped in the upper respiratory tract, whilst smaller particles ($<6\,\mu m$ in diameter) enter the lung and alveoli (Stetzenbach, 2009; Ferguson et al., 2021). Following inhalation, infectious viruses, bacteria, and fungi can establish in the respiratory tract whilst others may be translocated to other parts of the body, such as the gastrointestinal tract and other tissues (Stetzenbach, 2009). One way of assessing the health relevance of air pollution is to collect material from the inhalable/respirable size range (e.g. ISO 7708:1995; International Standards Organisation (ISO) (Ferguson et al., 2019). For example, the use of polycarbonate filters, as recommended by Ferguson et al. (2019), could be used with an IOM Multidust sampler head to select for health-relevant fractions (e.g. inhalable, thoracic or respirable).

However, the health impacts of bioaerosols occur not just from the microorganisms themselves (i.e. bacteria, fungi, archaea, viruses) but also from their components. For example, endotoxins, mycotoxins, enzymatic proteins, volatile organic compounds (VOCs) and allergens. These by-products of microbial metabolism, foreign proteins of microorganisms and fragments of cells can result in allergic reactions, irritation, inflammation, infection, and respiratory disease (Stetzenbach, 2009). Yet defining acceptable exposure levels is extremely challenging. This is largely due to the complexity of bioaerosols as they contain both biological and non-biological components, and the health effects of the various components varies from species to species and person to person. Health impacts are dependent on several factors such as exposure, concentration, and the general health of the individual. Whilst healthy persons are usually adapted to bioaerosols in the environment, bioaerosol exposure may pose a health risk, especially for vulnerable groups such as immunosuppressed persons, persons suffering from allergies or respiratory diseases (Biermann et al., 2013; Walser et al., 2015). Some examples of airborne pathogenic microorganisms are shown in Table 4 and described below.

4.1.1 Bacteria

Airborne bacteria are derived from a variety of natural and anthropogenic environments. When pathogenic bacteria are released into the air, they can be readily transported via bioaerosols to humans. Diseases caused by airborne bacteria include Legionnaires' disease, caused by *Legionella pneumophila*. This disease arises when infective aerosols are generated from contaminated water sprays (e.g. showers, fountains, humidifiers, and whirlpools). There are also *Legionella* bacteria that cannot replicate in the lung and are linked to Pontiac fever, with mild flu-like symptoms (Stetzenbach, 2009). Other bacterial diseases include tuberculosis, and non-tubercule mycobacterial disease (Stetzenbach, 2009). Some Gram-negative bacteria such as *Escherichia coli, Pantoea* (*Enterobacter*) *agglomerans, Pseudomonas* spp., Q-fever (*Coxiella burnetii*) and *Acinetobacter* spp. also cause health problems, especially to young, elderly,

Table 4 Examples of airborne pathogenic microorganisms.

Pathogen	Approximate size	Disease/source of airborne transmission	References
Bacteria			
Legionella pneumophila	Length 2 μm; width 0.5–1.0 μm	Legionnaires disease	Kim et al. (2018)
		Inhalation of water aerosol containing the bacteria	Stetzenbach (2009)
Mycobacterium tuberculosis	Length 1–10 μm; width 0.2–0.6 μm	Tuberculosis	Kim et al. (2018)
		Person to person via the air	Stetzenbach (2009)
Bordetella pertussis	Length 40–100 nm; width 2 nm	Whooping cough	Kim et al. (2018)
		Inhalation of airborne droplets	
Bacillus anthracis	Length 3–5 μm; width 1.0–1.2 μm	Anthrax	Kim et al. (2018)
		Inhalation of airborne anthrax spores	
Neisseria meningitidis	0.6–1.0 μm	Meningitis	Caugant and Brynildsrud (2020)
Corynebacterium diphtheriae	~2 μm in length	Diphtheria	Hoskisson (2018)
Coxiella burnetii	0.2–0.4 μm wide, 0.4–1 μm length	Q Fever	Eldin et al. (2017)
Klebsiella pneumoniae	1–2 μm by 0.5 μm	*Klebsiella* pneumonia (environmental)	Effah et al. (2020)
Fungi			
Aspergillus spp. e.g. *A. fumigatus, A. flavus, A. nidulans, A. terrus, A. niger*	~1 μm in diameter	Aspergillosis, allergic sinusitis, allergic bronchopulmonary disease	Stetzenbach (2009)
		Inhalation of spores, compost, soil, plant material, dust, building materials	

Continued

Table 4 Examples of airborne pathogenic microorganisms.—cont'd

Pathogen	Approximate size	Disease/source of airborne transmission	References
Blastomyces dermatitidis	8–15 μm in diameter	Chronic lung disease	Stetzenbach (2009)
		Moist soil, decomposing organic debris	
Coccidioides immitis	5 μm in diameter	Coccidioidomycosis/San Joaquin Valley fever	Stetzenbach (2009)
		Desert soils	
Cryptococcus neoformans	10 μm in diameter	Cryptococcosis	Stetzenbach (2009)
		Bird droppings, soils associated with birds	
Histoplasma capsulatum	2–5 μm in diameter	Chronic lung disease	Stetzenbach (2009)
		Environment	
Viruses			
Severe Acute Respiratory syndrome (SARS) and Coronavirus SARS-CoV-2	RNA	Acute respiratory distress	Stetzenbach (2009)
		Infected persons	Marjomäki et al. (2021) Stetzenbach (2009)
Middle East Respiratory Syndrome (MERS)	Positive-sense RNA	lethal zoonotic disease of humans endemic to The Middle East	Azhar et al. (2017, 2019)
			Gordon et al. (2020)
Influenza A virus	80–100 nm, ssRNA	Influenza, pneumonia, acute respiratory distress	Stetzenbach (2009)
		Human respiratory secretions, avian flu (associated with birds)	
Adenovirus	70 and 90 nm diameter, DNA	mild infections of the upper or lower respiratory tract, gastrointestinal (GI) tract, or conjunctiva, haemorrhagic cystitis, hepatitis, haemorrhagic colitis, pancreatitis, nephritis, or encephalitis	Besson et al. (2020)
			Lynch et al. (2011)

Table 4 Examples of airborne pathogenic microorganisms.—cont'd

Pathogen	Approximate size	Disease/source of airborne transmission	References
Enteroviruses	30 nm, RNA	Acute and chronic infections	Marjomäki et al. (2021)
Rotavirus	70 nm diameter	Severe, dehydrating gastroenteritis	Crawford et al. (2017)
Variola vera	Length 220–450 nm; width 140–260 nm	Smallpox	Kim et al. (2018)
		Inhalation of airborne virus	
Varicella zoster	150–200 nm, dsDNA	Chickenpox	Stetzenbach (2009)
		Human reservoir	
Measles/Rubeola	100–250 nm, RNA	Acute, febrile disease with an exanthematous rash, dry cough, sore throat, headache	Stetzenbach (2009)
		Human reservoir	
Norwalk-like virus (*norovirus*)	~30 nm, RNA	Diarrhoea, sudden vomiting	Stetzenbach (2009)
		Infected persons, surfaces	
Sin Nombre virus	ssRNA	Respiratory distress	Stetzenbach (2009)
		Rodent urine, saliva	

and immunocompromised individuals (Stetzenbach, 2009). These microorganisms are commonly found in animal housing (e.g. cow barns, pig houses, poultry barns) as well as sewage treatment facilities. Some airborne Gram-positive bacteria such as *Staphylococcus* spp., which are commonly found in indoor air, have also been shown to increase the likelihood of nosocomial infections, especially in healthcare settings (Stetzenbach, 2009). In addition, inhalation of the Gram positive actinomycetes, (usually associated with water damaged buildings) can also result in disease (Stetzenbach, 2009).

4.1.2 Archaea

Archaea have generally been ignored as agents detrimental to human health. Yet, Archaea are major components of bioaerosols particularly in agricultural environments and subcomponents of archaea have been found to be immunogenic (Blais Lecours et al., 2011). Using a murine model to study the pulmonary immune responses, one study showed that specific archaeal species caused a pulmonary immune response and that dose and archaeal species determined the magnitude and nature of the immune response (Blais Lecours et al., 2011). Thus, members of the Archaeal communities should not be ignored in future bioaerosol research studies.

4.1.3 Fungi

Fungal spores make up a significant proportion of organic matter within the air and may result in allergic sensitisation to fungi leading to conditions such as allergic fungal airway disease (Anees-Hill et al., 2021). Fungal concentrations may be influenced by meteorological conditions (e.g. temperature, precipitation, relative humidity), land-use such as proximity to agricultural and coastal areas (Anees-Hill et al., 2021). One study found higher bacterial concentrations in the severe colder zones compared to higher fungal concentrations in hot summer and warm winter zones (Guo et al., 2020). Bacterial and fungal levels may also vary among different buildings, with classrooms having the highest average levels (Guo et al., 2020). One study found concentration of indoor bacteria and fungi varied (with between 72.5 and 7500 CFU m^{-3} for bacteria; and between 12 and 9730 CFU m^{-3} for fungi) (Guo et al., 2020). Another study found that specifically, fungal spore concentrations of *Alternaria* spp. and *Cladosporium* spp. exceeded clinical thresholds, with concentrations of 665 and 18,827 CFU m^{-3}, respectively (Anees-Hill et al., 2021).

Most indoor fungi originate from the outdoor air but are generally present in lower concentrations (Burge et al., 2000; Grydaki et al., 2021). When higher indoor fungal levels are recorded, this is generally indicative of an indoor fungal source (Rao et al., 1996); although, some species of fungus are abundant in indoor air (e.g. *Aspergillus*) (Cox et al., 2019). Several *Aspergillus* species (e.g. *A. fumigatus*, *A. flavus*, *A. terreus, A. nidulans*, and *Aspergillus niger*) are linked to a pulmonary infection known as Aspergillosis (Stetzenbach, 2009). Exposure occurs through inhalation of the spores from soil, compost, decomposing organic matter and dust. Exposure may result in allergic sinusitis, allergic bronchopulmonary disease or translocate to other organs, including brain, skin, and bone (Stetzenbach, 2009). Histoplasmosis, which is caused by the inhalation of *Histoplasma capsulatum* spores mainly affects the lungs, but symptoms vary from asymptomatic to flu-like symptoms or chronic lung disease which may be fatal (Stetzenbach, 2009). Another disease iscoccidiomycosis (or San Joaquin Valley fever), is caused by the inhalation of airborne *Coccidioides immitis* arthroconidia (Stetzenbach, 2009). Whilst most infected people develop flu-like illness, others may suffer chronic pulmonary infection or widespread infection and immunocompromised persons may present severe pulmonary conditions and lung disease (Stetzenbach, 2009). Blastomycosis, caused by *B. dermatitidis,* grows in moist soil and exposure occurs via inhalation of airborne spores, following disturbance of the contaminated soil. Infection results in a flu-like illness, but chronic pulmonary infection, lung damage or infections of the skin, bones, genitourinary tract, and occasionally the brain may occur (Stetzenbach, 2009). Cryptococcal infections are caused by *C. neoformans* var. *neoformans* which are found in soil contaminated with decaying pigeon or chicken droppings. Exposure generally occurs when dust containing *Cryptococcus* disperses in the air (Stetzenbach, 2009). Whilst most people do not get sick with cryptococcosis, it can cause disease in immunocompetent and immunocompromised persons including brain and spinal cord diseases (e.g. meningoencephalitis) (Stetzenbach, 2009).

4.1.4 Viruses

Following inhalation, infectious viruses can establish in the respiratory tract, where they can cause a variety of respiratory syndromes (e.g. common cold, croup, cough, fever, bronchiolitis, and pneumonitis); whilst others are translocated to the gastrointestinal tract and other tissues (Stetzenbach, 2009). Respiratory viruses can also be expelled from the respiratory tract in exhaled breath as virus aerosols (e.g. influenza, respiratory syncytial virus (RSV), coronaviruses, and rhinoviruses (Tang et al., 2021). Viral diseases include influenza caused by flu viruses A, B, and C. Whilst symptoms can be mild, influenza may result in severe illness and death especially the elderly, people with chronic medical conditions, and young children. Transmission is from person to person in respiratory droplets expelled during coughing and sneezing (Stetzenbach, 2009). Human infection with avian influenza A viruses (e.g. H5N1, H7N2, H9N2, H7N3) may also occur via direct contact with infected poultry or contaminated surfaces and ranges from mild flu-like symptoms to pneumonia and acute respiratory distress (Stetzenbach, 2009). Severe acute respiratory syndrome (SARS) is a viral respiratory illness caused by a coronavirus, termed SARS-associated coronavirus (SARS-CoV) whereby exposure occurs from close personal contact with infected individuals (Stetzenbach, 2009). Recently, the airborne, aerosol transmission of SARS-CoV-2 led to the resulting pandemic disease, COVID-19 (Anderson et al., 2020; Morawska and Cao, 2020; Rahmani et al., 2020).

Recently, a microfluidics device consisting of an aerosol collecting and mixing chamber was applied for the rapid detection of SARS-CoV-2 nucleic acid and was highly effective in clinical samples (Xiong et al., 2021). The active virus becomes enriched on the filter membrane and is later turned inactive where the nucleic acid of SARSCoV-2 is extracted and loaded into the sample chamber of the microfluidic chip. Other airborne viral diseases include Norwalk-like viruses (NLVs), hantavirus pulmonary syndrome, measles, and varicella (chickenpox).

4.1.5 Endotoxins, mycotoxins, β-glucans and allergens

Microbial metabolites present in bioaerosols also play a major role in inflammatory reactions, lung deterioration and other health effects (Heedrik et al., 1991, Ghio and Roggli, 1995; Douwes et al., 2003). Endotoxins are biomolecules made up of lipopolysaccharides from the outer membranes of Gram-negative bacteria with very high pro-inflammatory properties (Armstrong et al., 2013). Since endotoxins easily bind to dust, they are readily inhaled (Wallace et al., 2016). Exposure to endotoxins may cause a range of health impacts including a decrease in lung function (Hoppin et al., 2014; Thilsing et al., 2015). For example, inhalation of >80 mg of endotoxins in healthy individuals and >20 mg of endotoxins in asthmatics decreased lung function (Kharitonov and Sjöbring, 2007). The Dutch Expert Committee on Occupational Standards, previously set an occupational health-based exposure limit of 50 EU m^{-3} but this was later increased to 90 EU m^{-3} in 2010, based on acute respiratory effects (Samadi et al., 2013). Endotoxins are also the main contributor to occupational lung diseases and organic dust toxic syndrome that is largely reported in textile, dairy, animal feed and sewage treatment workers (Paudyal et al., 2011; Mitchell et al., 2015; Cyprowski et al., 2015).

Mycotoxins are secondary biomolecules of low molecular weight produced by fungi and moulds (Andretta et al., 2011). Mycotoxins demonstrate a range of health impacts including weakened immune systems, allergies or irritations, and even death, depending on the mycotoxin type, exposure, concentration, and the individual (e.g. age, health, gender) (Marin et al., 2013). Pulmonary mycotoxicosis from mycotoxin exposure has also been reported in workers in agriculture, textile, and composting sectors (Ghio and Roggli, 1995; Pankhurst et al., 2012).

β-glucans are glucose polymers that occur in the cell walls of many microorganisms (e.g. bacteria, fungi, algae, yeast). Exposure to β glucans may cause inflammatory responses and directly attack the lungs causing respiratory symptoms depending on the type of glucan and exposure (Kurek et al. 2016; Richter et al., 2015). Concentrations of β glucans have been found to range from 0.02 ng m^{-3}, with higher levels (e.g. 538 ng m^{-3}) found in animal feeding facilities (Crawford et al., 2009; Dong and Yao, 2010; Yang et al., 2013). Allergens are derived from a variety of organisms including fungi (e.g. spores, hyphae) and plants (e.g. pollen) (Peden and Reed, 2010). Exposure to allergens can trigger allergic reactions (e.g. sneezing, runny nose and itchy eyes) as well as asthma, decreased lung activity and pulmonary inflammation (Peden and Reed, 2010; Baxi Phipatanakul, 2010; Baldacci et al., 2015). Many other factors influence allergens released by pollen and spores such as air pollution, and climatic conditions (Behrendt and Becker, 2001).

4.2 Standardizing "acceptable limits" for bioaerosols

Standardization of bioaerosols is becoming increasingly more urgent. Several technical organizations have focused on this for bioaerosols such as ASTM International which has a subcommittee on microorganisms, the International Organization for Standardization (ISO) and the European Committee for Standardization (CEN). Although, standards and recommendations for sampling for bioaerosols have been created by several organisations (e.g. ASTM, 2014; CEN, 2011; ISO, 2008); these are primarily aimed at airborne fungi and specific organisms such as *Penicillium* (Eduard, 2009) or Gram negative bacteria (SUVA, 2007). Guidelines for fungi in indoor environments have also been proposed by several organisations. However, these criteria have been developed for the assessment of indoor mould problems and are not health-based (Rao et al., 1996; WHO, 2009). Indeed, few exposure–response relationships have been proposed for fungal constituents like glucans and extracellular polysaccharides (EPS), which further complicates the standard setting process (Eduard et al., 2012).

There is currently limited information regarding the "*acceptable exposure limits*" for airborne fungi and bacteria, and standards do not exist at all for viruses, protozoa, or archaea (Cox et al., 2020). Moreover, "*acceptable concentrations*" of bioaerosols have been found to differ from country to country (Table 5). Exposure limits of 10^5 fungal spores m^{-3} of air has also been suggested, but this was limited to non-pathogenic fungal spores that did not produce mycotoxins (Eduard, 2009). "*Acceptable concentrations*" for bacteria also differ between different countries (Table 5).

Table 5 Examples of international exposure limits.[a]

Country	Exposure limit	Reference
Total bioaerosol		
Korea	800 CFU m^{-3}	Ghosh et al. (2015)
Netherlands	10,000 CFU m^{-3}	Ghosh et al. (2015)
Fungi		
Brazil	750 CFU m^{-3}	De Aquino Neto and de Góes (2004)
Germany	10,000 CFU m^{-3} Workplace limit of 50,000 CFU m^{-3} for mesophilic fungi (BAuA, 2013)	Institut für Arbeitsschutz der Deutschen Gesetzlichen Unfallversicherung (IFA), (2004) The Federal Institute for Occupational Safety and Health in Germany BAuA (2013)
Portugal	500 CFU m^{-3}	Pegas et al. (2010)
Switzerland	1000 CFU m^{-3}	SUVA (2007)
UK and Ireland	Green waste composting sites with 'sensitive receptors' (i.e. a place where people live or work for more than 6 h at a time): Ambient environmental compliance parameters at 250 m downwind are 500 CFU m^{-3} (+30%) for *A. fumigatus* (not epidemiologically based)	"Technical Guidance Note M9: Monitoring Methods for Ambient Air" UK Environment Agency
ACGIH	1000 CFU m^{-3}	ACGIH "Guidelines for the Assessment of Bioaerosols in the Indoor Environment" American Conference of Governmental Industrial Hygienists (ACGIH), Cincinnati, Ohio (1989)
Bacteria		
Finland	4500 CFU m^{-3}	Reviewed in Ghosh et al. (2015); Eduard (2009)
Germany	10,000 CFU m^{-3}	Institut für Arbeitsschutz der Deutschen Gesetzlichen Unfallversicherung (IFA) (2004)

Table 5 Examples of international exposure limits.[a]—cont'd

Country	Exposure limit	Reference
Netherlands	10,000 CFU m^{-3}	Reviewed in Ghosh et al. (2015); Eduard (2009)
UK and Ireland	Green waste composting sites with 'sensitive receptors' (i.e. a place where people live or work for more than 6 h at a time): Ambient environmental compliance parameters at 250 m downwind are 1000 (+30%) CFU m^{-3} for total bacteria (not epidemiologically based)	"Technical Guidance Note M9: Monitoring Methods for Ambient Air" UK Environment Agency
Endotoxins		
Netherlands	Dutch exposure standard limit of 90 Endotoxin Units (EU) m^{-3} for air-borne endotoxin	Dutch Expert Committee on Occupational Safety (DECOS) (DECOS, 2010). The Hague, The Netherlands: Health Council of the Netherlands Publication No 2010/040SH 2010
Subtilisin (Protease from *Bacillus subtilis*)		
USA, Norway	60 ng m^{-3} total wood dust	ACGIH, *The Labour Inspection Authority*. Norway: Trondheim; 2011. Aug 16, Guidance for administrative standards for contamination of the work environment (in Norwegian) www.arbeidstilsynet.no/artikkel.html?tid=78880
Dust		
Canada	50 CFU m^{-3} (normal), and 500 CFU m^{-3} (accepted) if dominated by *Cladosporium*	The Health and Welfare Department of Canada
		WHO (2009)
UK, EU	Hardwood 3 mg m^{-3}; softwood 5 mg m^{-3}	https://www.hse.gov.uk/woodworking/wooddust.htm
USA	0.5–1 mg m^{-3} total wood dust	ACGIH

Continued

Table 5 Examples of international exposure limits.[a]—cont'd

Country	Exposure limit	Reference
Norway	1–2 $mg\,m^{-3}$ total wood dust	*The Labour Inspection Authority.* Norway: Trondheim; 2011. Aug 16, Guidance for administrative standards for contamination of the work environment (in Norwegian) www.arbeidstilsynet.no/artikkel.html?tid=78880
USA, Norway	0.2 $mg\,m^{-3}$ total cotton dust	ACGIH, *The Labour Inspection Authority.* Norway: Trondheim; 2011. Aug 16, Guidance for administrative standards for contamination of the work environment (in Norwegian) www.arbeidstilsynet.no/artikkel.html?tid=78880
USA	4 $mg\,m^{-3}$ total grain dust	ACGIH
USA	0.5 $mg\,m^{-3}$ total flour dust	ACGIH
Norway	3 $mg\,m^{-3}$ total flour dust	*The Labour Inspection Authority.* Norway: Trondheim; 2011. Aug 16, Guidance for administrative standards for contamination of the work environment (in Norwegian) www.arbeidstilsynet.no/artikkel.html?tid=78880

[a]Also refer to the database: https://ieqguidelines.org/table.html (Toyinbo et al., 2022).

Toxicological and epidemiological data does allow for exposure levels to be categorised. For example, where adverse effects no longer appear, (i.e. No-Observed-Adverse-Effect Level-NOAEL) or Lowest-Observed-Adverse-Effect Level (LOAEL) (Walser et al., 2015). Epidemiological studies support LOELs of $\sim 10^5$ spores m^{-3} for diverse fungal species in non-sensitised populations (i.e. individuals who are used to exposure to high levels of bioaerosol exposures such as through occupational exposure rather than non-sensitised population (in the sense that it is anyone without a respiratory disease) (Eduard, 2009). Exposure limits for endotoxins and fungal spores have also been proposed (Eduard et al., 2012). For example, a lowest observed effect level (LOEL) of 100,000 spores m^{-3} for non-pathogenic and non-mycotoxin producing fungal species relating to inflammatory respiratory effects (Eduard, 2006). Yet, more specific OELs are required, but given the complexity of bioaerosols assessing such risks is a major challenge.

For comprehensive bioaerosol exposures, a NOAEL or LOAEL have yet to be developed (Walser et al., 2015; Swan et al., 2003). This is largely due to a lack of detailed toxicological, epidemiological data and quantitative dose-response studies in bioaerosol research. Certain terminologies are also in use that are different worldwide. For example, "*maximum acceptable values*" (De Aquino Neto and de Góes, 2004), "*orientation values*" (SUVA, 2007), "*acceptable maximum value, AMV*" (Becher et al., 2000, Jo and Seo, 2005), and "*Threshold limit value, TLV*" (American Conference of Governmental Industrial Hygienists (ACGIH), 2009). In conclusion, a LOEL of 10^5 spores m^{-3} seems appropriate as a basis for an OEL for spores from diverse fungi. This LOEL is probably too high if spores from mycotoxin-producing and/or opportunistic pathogenic species are prevalent. However, there are not sufficient data to support LOELs for spores from such species. Individuals with asthma and sensitised to fungal allergens are more sensitive than working populations in general. The available evidence suggests a safety factor of 10 (Marchand et al., 1995).

In general, existing standards and guidelines are reliant on short-term "snap-shot" sampling, often using culture-based analyses that give limited data and large variability regarding the potential health effects between different types of bioaerosols (see Section 3 for limitations of culture-based methods). This issue is exacerbated by a lack of consensus for air sampling and analysis protocols (See Sections 2 and 3), and thus exposure limits are often not used by bioaerosol researchers. There is also a lack of evidence regarding human dose-response relationships for establishing exposure limits (Walser et al., 2015). It could also be argued that workplace environments select for people resilient to sensitisation, whereas it is possible that the limit for the general population would likely be lower. Moreover, as novel technologies are developed, new approaches are not covered by the standards (Hart et al., 2020). Yet there remains a need to encourage consistency within the bioaerosol research sector for more efficient interpretation and cross-comparison of bioaerosol data. Furthermore, without scientifically proven standards and guidelines, inappropriate testing and data interpretation may occur, with significant consequences for human health and society (WHO, 2009). Thus, more data is required to fill these knowledge gaps. Recent methodological advancements will aid the identification of new biomarkers of exposure and improve our understanding of bioaerosol mediated disease to improve the assessment of bioaerosol exposure in future studies.

4.3 How the various sampling and analysis methods would be useful for different environmental contexts for human exposure

4.3.1 Workplace exposure

Exposure to bioaerosols in the workplace is of great concern because of the adverse health effects of bioaerosols and their prevalence in many occupational environments. Occupational exposure may lead to respiratory diseases, allergies, severe infections, diseases and even death (Arendrup et al., 2006; Douwes et al., 2003; Herr et al., 1999; Walser et al., 2014, 2015; Wery, 2014). For example, extrinsic allergic alveolitis (commonly known as farmers lung) is caused by inhalation of spores including *Aspergillus fumigatus* from sources such as composting facilities (Poole and Wong, 2013). For industries that utilize or manufacture enzymes that are known skin or respiratory sensitizing agents, in-house occupational exposure guidelines derived from minimal effect levels have been established (Basketter et al., 2010). Under EU legislation (REACH), there is a requirement to define a derived no effect level (DNEL) or a derived minimal effect level (DMEL) (Basketter et al., 2010). For bacterial and fungal enzymes which are recognized respiratory sensitizers, a DMEL can be established for example by, setting the employee medical surveillance data against exposure records over extended periods of time (Basketter et al., 2010). In the detergent industry, a long-established limit of $60\,\mathrm{ng\,m^{-3}}$ for pure enzyme protein has been set as the occupational health limits for sensitization (Basketter et al., 2010). This limit has enabled an appropriate level of security for new enzymes whose potency has not been fully characterized. For example, the general use of occupational exposure limits 3–10 times lower than the $60\,\mathrm{ng\,m^{-3}}$ limit (Basketter et al., 2010). For other agents, risk assessments may be based on exposure–response associations from epidemiological studies (e.g. $\beta(1\rightarrow3)$-glucans and allergens (Eduard et al., 2012). To better understand the relationship between an individual's exposure to bioaerosols and their connection with long-term health effects, sampling tools are needed that allow us to differentiate between personal bioaerosol exposures and ambient bioaerosol concentrations.

To assess workplace exposure, personal sampling devices (ISO standard EN 689 ISO, 2018) have an advantage over static sampling as these samplers can be worn on the person's clothing and have been shown to successfully capture fungi, bacteria and viruses (Borodulin et al., 2006; Kenny et al., 1998; Aizenberg et al., 2000; Wang et al., 2015). A personal bioaerosol sampler has to be lightweight, unobtrusive, easy to wear, capable of sampling for prolonged time periods, and have a robust and convenient sample removal. Application of personal bioaerosol samplers will help us link individual exposures with health effects. For example, Organic Dust Toxic Syndrome (ODTS) is a flu-like illness following respiratory exposure to organic dusts (Seifert et al., 2003). One study used GSP inhalable samplers attached to workers' clothing to collect bioaerosol samples to investigate organic dust toxic syndrome (ODTS) in a seed handling factory and calculated the concentration of airborne microbes the workers were exposed to from the dust (Madsen et al., 2015). However, one drawback with personal samplers to examine worker exposure is that it is dependent on the position of the sampler on the individual, whereby exhaled air as well as inhaled air may also be captured (Hall et al., 2013; Ghosh et al., 2015). Furthermore, as personal samplers have low flow rates; sometimes as low as $2\,\mathrm{L\,min^{-1}}$ (see Section 2), sample viability may be lost through desiccation (Hall et al., 2013). Yet to assess inhalation risk, viability needs to be determined, as non-viable microorganisms cannot cause infectious diseases (Pan et al., 2018). The limitations of personal samplers may be overcome however, by using larger static samplers with higher flow rates placed away from workers' immediate environment to sample a larger background air volume and obtain more representative air samples over the same period. *Thus, a combination of personal and static samplers may provide a more complete picture of the airborne microbiome to determine exposure and risk more accurately.*

Spatial variability in the workplace is also crucial, particularly in clinical settings to assess patient exposure. Seasonal variation also needs to be considered in the context of human exposure. For example, one study of a hospital in Taiwan showed seasonal shifts in the airborne adenovirus and *Mycoplasma pneumoniae,* with adenovirus peaking in the summer, and *M. pneumoniae* increasing in the autumn and winter (Wan et al., 2012).

Thus, identifying seasonal shifts in bioaerosol contamination is crucial in order to direct effective ventilation and control airflow to affected areas and reduce patient risk (Wan et al., 2012). The prevailing environmental conditions should also be considered for sampling the different workplace settings. For example, gelatin filters perform poorly in high-humidity environments (e.g. shower rooms), thus other approaches should be used in these scenarios (Chang and Hung, 2012).

Humans are also a common source of airborne bacteria and there is a relationship between human activity and increased concentrations of airborne microorganisms in indoor environments (Woodhead et al., 2002; Meadow et al., 2014; Lehtonen et al., 1993; Scheff et al., 2000). Hence, increasing ventilation, cleaning and maintenance activities can play a vital role in improving indoor air quality. However, even under ideal conditions for preventative maintenance, risk of exposure to bioaerosols persist. This is important especially in certain environmental contexts. For example, in healthcare settings, an increase in the number of people and door openings in an operating theatre, increased bacterial counts (Agodi et al., 2015). Similarly, human occupancy in a classroom greatly increased bacterial bioaerosols (Hospodsky et al., 2012). The use of certain aerosol generating equipment such as for certain medical procedures, humidifiers, water sprays etc. can also increase risk of exposure to pathogens in indoor air (Sattar, 2016; Best et al., 2012; Springston and Yocavitch, 2017). However, since outdoor air may influence the composition of indoor air (Grydaki et al., 2021; Adams et al., 2015), caution should be taken when investigating the relationship between human occupancy and levels of airborne microorganisms (Haig et al., 2016). For indoor environments, the American Conference of Governmental Industrial Hygienists (ACGIH) recommends that sampling be conducted in both expected high and low exposure areas, and an outdoor area adjacent to the air intake for the building (Macher, 1999).

Animals are also important sources of indoor airborne bacteria, particularly in indoor agricultural environments (Blais Lecours et al., 2012; Just et al., 2011, 2013) and the incidence of zoonotic diseases is likely to increase with climate change (Baker et al., 2022). Bioaerosols from intensive livestock farming cause symptoms and/or illnesses and consequently there is concern about the potential health effects on people who work or live near intensive farms (Douglas et al., 2017). Indoor and outdoor air samples should thus be collected simultaneously with the general assumption that indoor levels of airborne microorganisms should be lower than outdoor levels.

The difficulties of assessing the environmental health risks in the workplace, particularly for agriculture and waste processing industries also makes it difficult to assess the exposure limits for local residents living near these facilities (Walser et al., 2015). For example, some indoor locations such as those near farms, waste treatment facilities and composting sites, transport of outdoor bioaerosols into the indoor environment is a concern (Douglas et al., 2018; Pearson et al., 2015). This is exacerbated by factors that may affect particle dispersion rates (Douglas et al., 2016). Although, there is a general agreement that bioaerosol concentrations tend to reach background levels within 250 m of their source (Environment Agency, 2001, 2010); there is evidence, that bioaerosols may travel significantly further in elevated concentrations (Recer et al., 2001; Fischer et al., 2008; Schlosser et al., 2009; Ferguson et al., 2021; Pankhurst et al., 2012). Modelling will become an important tool for understanding the interconnection of air pollutants, their dispersal and health effects across different environmental contexts (Gollakota et al., 2021).

Despite the challenges in assessing exposure and risk, this general lack of guidance places tremendous limitations when applying methods to measure and analyse bioaerosols for human health exposure and risk. One approach is to establish defined methods and recommendations for each specific environmental or workplace context. Below we describe some examples of different workplace environments where higher emissions of bioaerosols may occur (e.g. waste processing facilities) or where human exposure involves vulnerable groups (e.g. clinical settings). We have broadly categorised these into waste processing, food production (agricultural/farming), and clinical settings.

4.3.2 Waste-processing environments

4.3.2.1 Wastewater treatment plant (WWTP)

Wastewater treatment systems may spread microorganisms via bioaerosols, resulting in adverse health effects for plant workers (O'Hara and Rubin, 2005; Fracchia et al., 2006). Previous studies have found high levels of viable airborne bacteria and fungi present in these facilities and aeration systems (Brandi et al., 2000, Sánchez-Monedero et al., 2008, Li et al., 2011). Personal samplers have been used to investigate worker exposure at wastewater treatment plants (WWTP) (Uhrbrand et al., 2011). One study used inhalable GSP samplers (CIS; BGI Inc, Waltham, MA (Madsen, 2006); fitted with Teflon filters (1 μm; Millipore) or polycarbonate filters (1 μm; GE Water & Process Technologies, USA) with an average sampling period of 242 min to collect either endotoxin or bacteria, moulds and viruses, respectively from airborne dust. The study found high concentrations of Noroviruses (NoVs) and endotoxins that could pose an occupational health risk (Uhrbrand et al., 2011), with similar findings at other WWTPs (Rylander, 1999; Thorn et al., 2002). Several measures have been implemented in WWTP facilities to regulate and control the impact of bioaerosols such as floating balls (Hung et al., 2010); air purification systems (e.g. UV germicidal irradiation) (Green and Scarpino, 2002). One study showed that intensive solar radiation, together with high ambient temperatures, alongside optimal wastewater treatment are the most important factors for lower numbers of airborne microbes (Karra and Katsivela, 2007).

4.3.2.2 Composting facilities

Composting facilities are one of the biggest anthropogenic sources of bioaerosols, with potential negative health impacts following exposure (Pankhurst et al., 2012; Pearson et al., 2015; Robertson et al., 2019). Bioaerosol emissions largely occur as a result of vigorous compost agitation activities such as shredding, turning and screening which release airborne infectious microorganisms (e.g. *Aspergillus fumigatus*, *Penicillium* sp., *Saccharopolyspora rectivirgula*, actinomycetes), and allergenic particles (e.g. endotoxins, 1-3 β-glucans) (Pankhurst et al., 2011, Wery, 2014). Exposure may result in a range of respiratory illnesses and infections including extrinsic allergic alveolitis (e.g. Farmers' lung), hypersensitivity pneumonitis and aspergillosis (Pearson et al., 2015; Robertson et al., 2019). Given this complex mixture of microorganisms, constituents and metabolites and the variety of symptoms and diseases associated with exposure at composting facilities it is often not clear which agents are involved in the health outcomes of exposed workers (Schlosser, 2019).

Several knowledge gaps remain regarding exposure levels in composting and solid waste facilities (Schlosser, 2019). Exposure assessment can range from very simple distance from a source as a proxy for exposure (e.g. location, duration, sample size, stationary and personal sampling) (Douwes et al., 2003), to more complicated individual-level biomonitoring, or dispersion modelling. Threshold limits for UK bioaerosol emissions were previously set as follows: Gram-negative bacteria (300 CFU m^{-3}), total bacteria (1000 CFU m^{-3}) and total fungi (1000 CFU m^{-3}) (Wheeler et al., 2001). Since 2010 however, these levels were re-defined as not exceeding: (i) those before the start of composting processes or (ii) $<$ 1000 CFU m^{-3} total bacteria, 500 CFU m^{-3} *A. fumigatus* (which replaces the previous 1000 CFU m^{-3} total fungi limit) and 300 CFU m^{-3} Gram-negative bacteria (Pearson et al., 2015; Robertson et al., 2019; Environment Agency, 2001, 2010). Furthermore, reports of exposure to viruses in solid waste processing facilities is limited (Schlosser, 2019). To date, there are no long-term epidemiological studies which have been able to help determine a "*safe*" advisory level for bioaerosol concentrations (in CFU m^{-3}) for occupational or ambient settings (Pearson et al., 2015; Robertson et al., 2019). Instead, the compliance parameters are based on an arbitrary $\times 10$ factor to allow for comparison with ambient background levels (Douglas et al., 2018).

These levels are for fungal spores (50 CFU m^{-3} for *A. fumigatus*) and bacteria (100 CFU m^{-3} for mesophilic bacteria) (Pearson et al., 2015; Robertson et al., 2019). However, whether concentrations exceeding above background level is a health concern is currently unknown (Pearson et al., 2015; Robertson et al., 2019). This is contrary to the environmental monitoring of particulate matter ($PM_{2.5}$ and PM_{10}) where limit values and target values are based on epidemiological data (UK Air Quality Limits-Defra UK).

To detect fungal spores, a range of sampling and analysis methods have been used, such as impaction and Scanning Electron Microscopy (Vestlund et al., 2014); the IOM (inhalable dust) filtration sampler (Kenny et al., 1999) and culture-independent analyses (Pankhurst et al. 2012). To overcome the issues of obtaining "snap-shot" samples, long-term monitoring using on-line approaches such as Light Induced Fluorescence (LIF) and laser scattering and Wideband Integrated Bioaerosol Sensor (WIBS) have also been used with relatively good levels of detection (Eversole et al., 2001; Kaye et al., 2005; Nasir et al., 2019, Ferguson et al., 2020). Thus, for assessing exposures from waste management facilities like WWTPs and composting facilities a combination of personal, static sampling and online monitoring of bioaerosol microorganisms and their components (e.g. MVOCs) is important for improving health risk assessments for workers (Garcia-Alcega et al., 2017, 2018; Persoons et al., 2010). Another approach is to apply dispersion modelling in association with health. This has been used by regulators of the waste and recycling industry (Pearson et al., 2015; Robertson et al., 2019). Here, size is an important factor in parameterising dispersion models for assessing human health risk (Douglas et al., 2017). For example, Williams et al. (2019) modelled exposure risk of *A. fumigatus* from outdoor composting activities and found that exposure was driven primarily by wind speed, direction and time-varying emissions factors including seasonal fluctuations. Specifically, *A. fumigatus* concentrations decreased rapidly from the facility boundary and plateaued beyond 1.5–2.0 km. Where multiple composting facilities were within 4 km of each other, complex exposure risk patterns were evident (Williams et al., 2019). However, more research is required in this area including more long-term bioaerosol monitoring in close proximity to facilities to more accurately assess exposure and health risks.

4.3.3 Agriculture/farming/food production

4.3.3.1 Poultry farms

Poultry facilities are associated with a high production of dust, gases and odours (Douglas et al., 2018; Whyte, 1993). There are two common types of poultry housing facilities: cage-housed operations, where birds are housed in cages for egg production, and floor-housed operations, where birds are housed on the floor for meat production, and thus the type of housing may influence the levels of environmental contaminants. It has been shown previously that floor-housed poultry operations have higher concentrations of dust, endotoxin, and bacteria than the bioaerosols from cage-housed (Just et al., 2011). Indeed, workers from these facilities report a greater prevalence of respiratory symptoms, and clinical diseases (e.g. allergic and non-allergic rhinitis, organic dust toxic syndrome (ODTS), chronic bronchitis, hypersensitivity pneumonitis (farmers lung) and asthma) (Iversen et al., 2000; Just et al. 2011, 2012; Radon et al., 2001; Redente and Massengale, 2006). Previous work has shown that the microbial diversity on chickens depends on the bird's age and use of antibiotics (Pedroso et al., 2006). Indeed, the use of antimicrobial feed additives in food animal production may select for drug resistance in bacterial pathogens, which can then be released into the environment through occupational exposures, high volume ventilation of animal houses, and land application of animal wastes (Rule et al., 2008). Furthermore, airborne bacteria, endotoxin and inhalable dust levels increase during the chicken growth cycle (Oppliger et al., 2008).

4.3.3.2 Dairy farms

In dairy production facilities, general handling of ingredients, spray drying, and milling operations can generate bioaerosols (Dungan et al., 2010). Wet and dry cleaning operations often result in the formation of bioaerosols as water droplets or dry dust originating from sweeping or directly from the exhaust of vacuum cleaners (Abt et al., 2000). Exposure to *Saccharopolyspora rectivirgula* (causal agent of farmer's lung) is a health risk to dairy farmers (Blais Lecours et al., 2012). It has also been reported that higher levels of antigen-specific IgG in the plasma of workers was related to high concentrations of *S. rectivirgula* cells and that IgG could be used as a diagnostic tools for exposure (Blais Lecours et al., 2012). To reduce bioaerosols in milk processing production high-efficiency particulate air (HEPA) filters are used (Jacobs et al., 2011; Kandhai et al., 2004; Mullane et al., 2008).

4.3.3.3 Slaughterhouse

It has been shown previously that airborne bacteria have an important role in carcass contamination in slaughterhouses (Rahkio and Korkeala, 1997). A study which examined the exposure risk of workers in a slaughterhouse used personal samplers consisting of PAS-6 sampling heads with polytetrafluoroethylene (PTFE) filters (1 μm pore size) (Merck Millipore SAS, Molsheim, France) (Hall et al., 2013). The PTFE filters successfully captured the viruses WU polyomavirus and human papillomavirus 120 along with other pathogens (Hall et al., 2013).

4.3.4 Clinical settings

4.3.4.1 Hospitals and healthcare settings

Sampling bioaerosol in hospitals is essential, especially in operating theatres to reduce the incidence of infections. For example, the toilet flushing of patient faecal waste has been shown to be a source of bioaerosols (Knowlton et al., 2018). Furthermore, poor indoor air quality in healthcare settings has been linked to an increase in hospital-acquired infections (Abu-Rub et al., 2022). One study, evaluated the relationship between heating, ventilation, air conditioning (HVAC) systems an laminar air flow (LAF) and bioaerosol concentrations in hospital rooms, showed that HVAC systems in hospitals can effectively remove bioaerosols in hospitals and HEPA filters are an effective option for areas that are under-ventilated or require additional protection whilst the LAF system (other than the HEPA filter) was not effective in removing airborne bacteria and fungi (Dai et al., 2021). Sensitive hospital areas such as operating rooms and intensive care units had mean indoor bacterial concentrations of 180.3 $CFU\,m^{-3}$ and 204.6 $CFU\,m^{-3}$ respectively (i.e. below WHO recommendations of 300 $CFU\,m^{-3}$ for airborne bacteria in areas occupied by immunosuppressed people), with *Staphylococcaceae*, *Enterobacteriaceae*, *Pseudomonadaceae* and *Bacillaceae* commonly identified (Abu-Rub et al., 2022).

Spatial variability is crucial in clinical settings for assessing human exposure and therefore sampling devices should be placed at the same height as the patient (Tang and Wan, 2013). Most healthcare settings are fitted with HEPA filters and particulate counts can be routinely monitored from these filters. Both passive and active sampling methods have been applied in clinical settings (Pasquarella et al., 2012; Agodi et al., 2015). Pasquarella et al. (2012) found a moderately strong correlation between the active and passive sampling methods. The differences obtained were attributed to the relatively short sampling period and limited spatial collection zone of the active sampler compared with the longer exposure time of the settle plates (Pasquarella et al., 2012). Another study found a positive correlation between airborne bacterial concentration and particulate matter (Shintani et al., 2004).

Larger samplers have also been used extensively in healthcare environments to successfully capture viruses, bacteria, and spores (Tang and Wan, 2013; Jung et al., 2015; Wan et al., 2012). Midget impingers (SKC, Inc.) have been used successfully to assess surgical masks against aerosolized influenza (Booth et al., 2013). There is also growing interest in the correlation between microbiological and dust contamination. It has been suggested to standardize dust collection in operating theatres in relation to the surgical activity to better define risk factors affecting air quality (Scaltriti et al., 2007).

4.3.4.2 Dentists

Bioaerosols may also be generated in dental clinics during certain dental procedures such as ultrasonic scaling, or high-speed drilling (Leggat and Kedjarune, 2001). Consequently, a number of sources of bacterial aerosols exist inside and outside dental clinics, where staff and patients may be exposed to various infectious agents and increasing the risk of cross-infection between patients (Mirhoseini et al., 2021; Leggat and Kedjarune, 2001). One study used an active single-stage Andersen sampler to collect air samples during patient treatment and found the concentrations of airborne bacteria and fungi varied with distances from the patient (Mirhoseini et al., 2021). To reduce exposure, dental staff should continue to use personal protective measures and air-conditioning and ventilation systems should be regularly maintained to reduce contaminants and prevent recirculation of bioaerosols (Leggat and Kedjarune, 2001). Procedural rinsing by patients with mouthwashes as well as vacuum and electrostatic extraction of aerosols during dental procedures could also be applied (Leggat and Kedjarune, 2001).

4.3.4.3 Podiatry

Studies have also reported the presence of fungal bioaerosols in the indoor air of podiatrists, especially during nail drilling procedures (Alija-Martínez et al., 2021). In one study, the most prevalent filamentous fungi detected were *Cladosporium* spp. and *Penicillium* spp. with *Trichophyton rubrum* and *Trichophyton interdigitale* reported in samples during the drilling procedures (Alija-Martínez et al., 2021).

For health officials and regulators, it would be useful to have a better understanding of specific target organisms for each specific environmental context or setting and thus bespoke guidelines for exposure limits may be more informative. However, this information is currently lacking and more studies using HTS methods are still needed to establish which microbial components predominate under certain environmental conditions. Current knowledge is predominantly based on traditional analyses methods that are either biased towards those microorganisms that are easier to grow or identify and/or are targeted towards known pathogenic organisms. Some examples of target organisms for specific environmental contexts are presented (Table 6). HTS and other techniques offer the opportunity to characterise a wider diversity of environmental organisms in a non-targeted way and provide more information in the future.

Table 6 Examples of target organisms for specific environmental contexts.

Target organism	Environmental setting	Reference
Aspergillus fumigatus, Zygomycete spp., Legionella spp., *enteric pathogens, Gram negative bacteria* (*e.g. Pseudomonas* spp.), *Endotoxin, Gram positive bacteria* (*e.g. Bacilli, Listeria* spp., *Corynebacterium* spp., *actinomycetes*), *beta*(*1→3*) *glucans*	Composting and industrial waste, municipal solid waste facilities	Robertson et al. (2019), Pearson et al. (2015), Schlosser (2019)
Aspergillus fumigatus, thermophilic ascomycetes spp.	Residential areas surrounding composting facilities	Schlosser (2019)
Endotoxin, fungal spores, bacteria, total microorganisms, viable microorganisms, fungal antigens, actinomycetes, glucans, total bacteria	Intensive farms	Douglas et al. (2018)
Cladosporium spp., *Alternaria* spp.	Ambient air	Anees-Hill (2021)
Aspergillus spp., *Penicillium* spp., *Stachybotrys* spp.	Indoor air, damp environments	Guo et al. (2020)
		Fakunle et al. (2021)
		Dai et al. (2021)
Legionella spp., *Staphylococcaceae, Enterobacteriaceae, Pseudomonadaceae, Bacillaceae*	Hospitals, dentists	Abu-Rub et al. (2022)
		Zemouri et al. (2017)

4.4 Summary of sampling and analysis methods to assess exposure and risk

Exposure-response relationships for most agents in bioaerosols is currently lacking (Eduard, 2009; Schlosser, 2019). One of the major challenges is establishing the definition of exposure (e.g. what indicator parameter or what exposure time-scale?) and the definition of response (e.g. what critical effect end-point?) (Schlosser, 2019). This is further complicated by the complex nature of bioaerosols and that combined effects may occur with the different components associated with bioaerosols, alongside variations between and within individuals and interactions between the bioaerosols and the environmental conditions (Schlosser, 2019). However despite these challenges, there remains an urgent need for health-related exposure limits and assessment tools that can be utilised by authorities, regulatory bodies and bioaerosol researchers for both indoor and outdoor air settings.

More comprehensive epidemiological approaches are needed to determine the association between humans, animals, and environmental risk. These epidemiological approaches provide observational results and risk measurement estimates (Schlosser, 2019). These epidemiological studies should also consider both exposure level and individual risk factors as well as include larger sample sizes, and appropriate controls (Anderson et al., 2017; Schlosser, 2019). High-throughput sequencing approaches should also be used for future epidemiological studies to identify long-term associations between the core microbiome of a specific environment, the microbial markers, and the health effects following exposure. More sampling data is also needed to validate dispersion modelling with enough confidence to allow for a more robust, rapid and cheaper assessment of bioaerosol exposure (reducing spatial/temporal limitations) (Douglas et al., 2017).

Whilst static sampling devices can measure bioaerosol concentration at a specific location, personal samplers can be deployed for human exposure assessments (Cox et al., 2020). Thus both types of samplers may be needed to characterise human exposure (Toivola et al., 2002). Sampling the inhalable fraction (including both microorganisms and their components) is also crucial for determining bioaerosol exposures and understanding their health impacts. Such approaches could take bioaerosol studies further towards determining standards for acceptable bioaerosol exposure limits (Cox et al., 2020).

4.5 Recommendations and guidelines for assessment of exposure and human health risk

See Box 4.

BOX 4 Recommendations

1. Sampling the inhalable fraction (and over longer time-scales) is important for determining actual bioaerosol exposures and understanding their health effects. Thus, combine both personal sampling with static sampling for more accurate assessment of exposure and risk.
2. Sample both indoor and outdoor air simultaneously to assess exposure levels.
3. Increase the sample size with sufficient power.
4. The activity in the room and conditions during sampling should be considered when evaluating air quality especially in indoor air environments when making recommended values. Awareness of airflow in and between areas under investigation, in addition to patterns of human activity, ventilation are also important.
5. The prevailing environmental conditions should be taken into account for different workplace settings including seasonality and background concentrations in the air. [Avoid gelatin filters for examining exposure in high humidity environments.]
6. Ensure occupational studies include a diversity of individuals (e.g. gender, age, health status and socio-economic factors) to help identify more representative (and indeed susceptible populations).
7. To assess exposure in certain contexts (e.g. composting facilities), simultaneous measuring the levels of compounds (e.g. MVOCs) and bioaerosols is important for improving health risk assessments. As different components of bioaerosols may give rise different health effects, biochemical measurements alongside health measurements should be performed.
8. The transmission dynamics of some pathogens is still not fully understood. More comprehensive epidemiological and toxicological data and exposure-response data (e.g. using animal models) is needed to better ascertain the association between human, animal, and environmental risk factors. This data will enable a NOAEL (No-Observed-Adverse-Effect Level) or LOAEL (Lowest-Observed-Adverse-Effect Level) to be developed (Walser et al., 2015) for bioaerosol exposures.
9. Not just the viability of airborne microorganisms and the risks that they pose that are important but non-viable cells and their products can cause adverse health effects- so guidelines need to consider all elements for assessing human health exposures.
10. Consensus in the sampling methods used to quantify and characterise bioaerosols to allow easier comparison between studies, and across different countries.

Continued

BOX 4 Recommendations—cont'd

11. Increased use of model systems to help link exposure and health effects enabling causal associations to be established.
12. Clearer definitions and or terminology of the health outcome/effect are required.
13. Exposure limits need to also address time-scale limits as well as numbers.

5. Conclusions and future perspectives

In conclusion, bioaerosols are found in varying concentrations in indoor and outdoor environments and may be hazardous to human health causing distinct infectious and non-infectious diseases. However, more research is still needed to unravel the link between epidemiology, exposure levels, and characterisation of the specific microorganisms involved. There are many techniques and sampling devices available for collecting and analysing bioaerosols, but their selection remains a challenge and there is currently no standardised approach that is widely acceptable. When choosing a bioaerosol sampler, multiple factors should be considered, including sampler efficiency, sensitivity and recovery, the type of bioaerosol of concern, downstream analysis methods, sampling environment (whether indoor or outdoor), sampling duration, convenience, and cost. For example, whilst active sampling of bioaerosols offer many advantages and there are various active sampling devices commercially available, the added benefits of passive sampling should not be overlooked and could serve as a valuable complementary tool, especially if long sampling times are needed or if multiple samples are required to cover wide areas.

Recent advances in analytical and molecular approaches to analyse the collected bioaerosols, including on-line methods are starting to reveal a near-real scenario on a molecular scale. Selection of sampling and downstream analysis methods should tie into the overall research question being addressed alongside the taxa being targeted for investigation. It therefore comes down to what is being measured and sampling device/method selection depends on the objectives of the study (Fig. 7). For example, if exposure to an individual is important then, a personal sampler with the inhalable

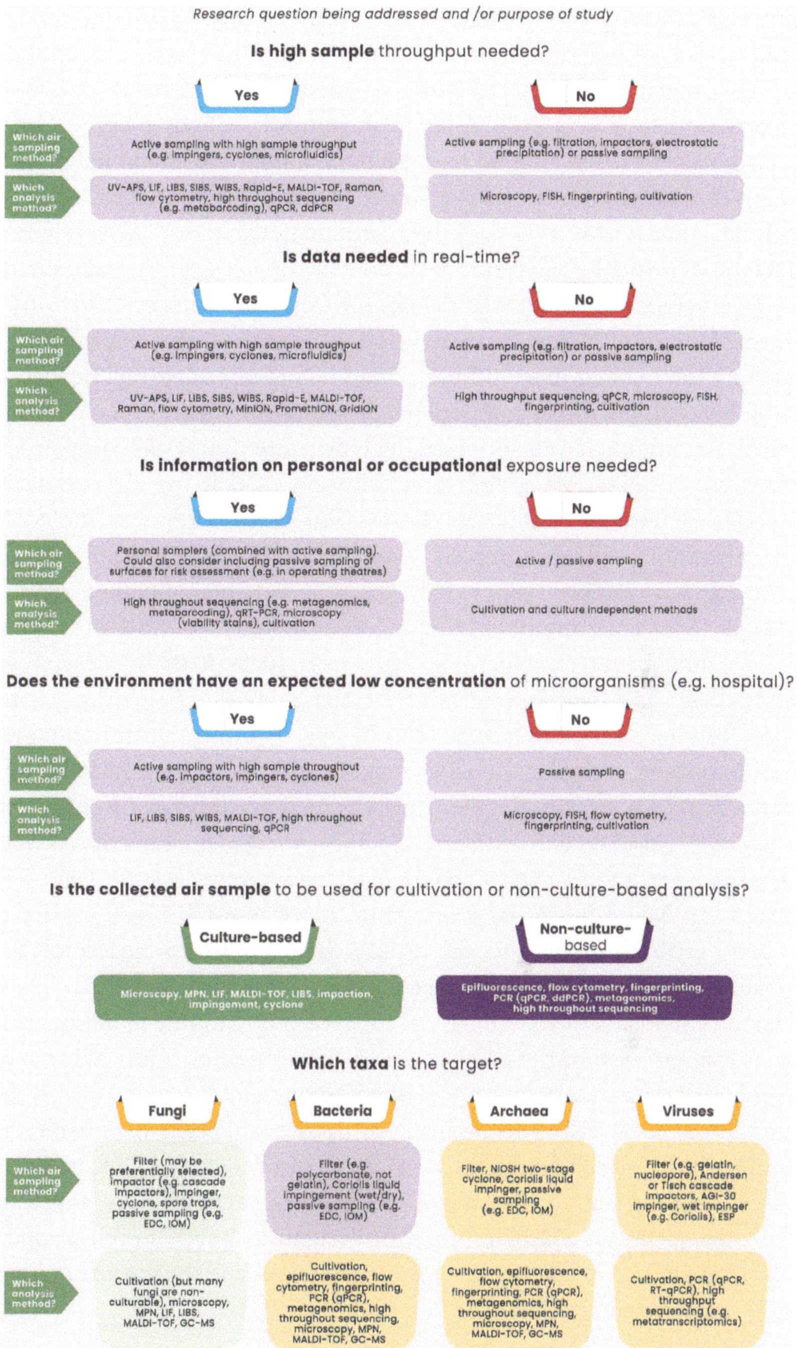

Fig. 7 Flow chart broadly summarizing the sampling and analysis strategies.

fractions combined with size fractionation samplers (e.g. Andersen) could be used. Sample method should also be appropriate for the particle size of interest. To facilitate the standardisation of methods to enable comparisons between studies, we thus recommend standardized methods to be developed for each particle being assessed (e.g. bacteria, fungi, viruses and pollen etc). To this end, more information is still needed on the specific target organisms (including non-viable cells and their products) that cause adverse health effects in a particular environmental context or setting. This will enable the development of bespoke guidelines for exposure limits and allow more informed risk assessments to be developed. However, this information is currently lacking and more studies using HTS methods are still needed to establish which microbial components predominate under certain environmental conditions. If culture-based analysis is used, then the sampler and collection medium best suited for cultivation of that specific organism/pathogen of interest should be chosen. For combinations of molecular and cultivation analyses, samplers with liquid collection medium is recommended so samples can readily be split between different analyses methods. Molecular methods are a major component of how bioaerosols are currently being characterised. However, such methods are not yet standardised/consistent (unlike culture-based methods where there is some degree of standardisation) (see Section 3). Fig. 7 summarises this compendium into some of the key decision-making questions that need to be considered in relation to sampling and analysis method that may be employed. *Realistically, guidelines and recommendations comprising of bespoke exposure limits may need to be developed that are tailored to the environment, biological target being sampled and human health end point.* Thus, more research is still needed to properly establish better assessment tools for exposure to bioaerosols and its validation. This may have wider implications for our understanding of bioaerosol mediated disease in the occupational environment and improve the assessment of bioaerosol exposure in future studies. Thus, there is an urgent need for a common protocol that is universally accepted, reliable that allows cross-comparisons to be made of datasets collected worldwide. To achieve this, more comprehensive studies are needed that link pollution sources, exposure risk and control to health impacts in space and time. Such advancements will help bioaerosol scientists, public health planners, policymakers, and regulators to improve air quality and the future of human health on a global scale (Box 5).

BOX 5 Summary of compendium recommendations, solutions and guidelines

Section 1: Considerations prior to air sampling:

- What are the objective and goals of monitoring (e.g. emission control efficacy, compliance monitoring, exploratory, exposure conditions, research, regulatory or health impact)?

Section 2: Sampling Methods:

- An overarching standardised method is not practical and instead bespoke methods are needed for each environment, target microorganism(s) being sampled to ensure a representative sample of the specific environment is collected and that sample integrity is maintained.
- Choice of sampling method should also consider downstream analysis method.

Section 3: Analysis Methods:

- Combining culture-based with culture-independent methods would provide a fuller image of the microbial components and their activities in bioaerosols but with a caveat of their inherent limitations.
- Current sequence databases also need improvement, especially with regard to viruses.
- Most of the health effects are caused due to exposure to the instead of the viable microorganism themselves. Efforts therefore should include measuring these microbial components (e.g. microbial products, endotoxins, mycotoxins, etc) as well as the microorganism itself to more fully assess environmental exposure.
- More long-term integrated analyses of bioaerosols is needed with real-time analysis methods.

Section 4: Human exposure and risk:

- Sampling the inhalable fraction (and over longer time-scales) as well as both indoor and outdoor air simultaneously is important for determining exposures and health effects.
- Combined personal sampling with static sampling may provide a more accurate assessment of exposure and risk.
- Prevailing environmental conditions, human activities, diversity of individuals should be reported when making recommended values.
- Simultaneous measuring the levels of compounds (e.g. MVOCs, biochemical measurements) and bioaerosols is important for improving health risk assessments as different components of bioaerosols may give rise different health effects.
- More comprehensive epidemiological and toxicological data and exposure-response data is needed.

Continued

BOX 5 Summary of compendium recommendations, solutions and guidelines—cont'd

- Increased use of validated model systems to help link exposure and health effects.
- Clearer definitions and or terminology of the health outcome/effect are required.
- Exposure limits need to address time-scale limits as well as numbers.

Acknowledgements

This work was supported by UKRI Natural Environment Research Council (NERC) funded projects: BioAirNet (NE/V002171/1), RAMBIE (NE/M010813/1) and BioSkyNet (NE/V008293/1). The research was supported by National Institute for Health Research (NIHR) Health Protection Research Unit in Environmental Exposures and Health, a partnership between UK Health Security Agency, the Health and Safety Executive and the University of Leicester. The views expressed are those of the author(s) and not necessarily those of the NHS, the NIHR, the Department of Health and Social Care or UK Health Security Agency. We thank Futurum for technical assistance with the graphic designs.

List of contributors

See Fig. 8 for list of contributors of the compendium of analytical methods for sampling, characterization and quantification of bioaerosols.

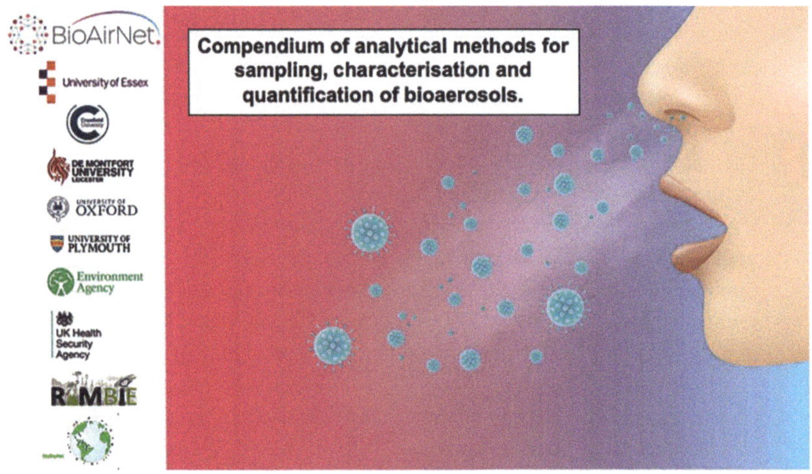

Fig. 8 Contributors of the compendium of analytical methods for sampling, characterization and quantification of bioaerosols.

References

Abt, E., Suh, H.H., Allen, G., Koutrakis, P., 2000. Characterization of indoor particle sources: a study conducted in the metropolitan Boston area. Environ. Health Perspect. 108 (1), 35–44. https://doi.org/10.1289/ehp.0010835.

Abu-Rub, L.I., Johar, A.A., Al Mana, H., Abdelrahman, H.A., Althani, A.A., Qotba, H., Yassine, H.M., Eltai, N.O., 2022. Bacterial indoor air contaminations in hospitals in MENA region: a systematic review. Int. J. Environ. Health Res. 6, 1–15.

Adams, R.I., Bhangar, S., Pasut, W., 2015. Chamber bioaerosol study: outdoor air and human occupants as sources of indoor airborne microbes. PLoS One 10, e0128022.

Adhikari, A., Sen, M.M., Gupta-Bhattacharya, S., Chanda, S., 2004. Airborne viable, non-viable, and allergenic fungi in a rural agricultural area of India: a 2-year study at five outdoor sampling stations. Sci. Total Environ. 326 (1–3), 123–141.

Agodi, A., Auxilia, F., Barchitta, M., 2015. Operating theatre ventilation systems and microbial air contamination in total joint replacement surgery: results of the GISIO-ISChIA study. J. Hosp. Infect. 90, 213–219.

Agranovski, V., Ristovski, Z., Hargreaves, M., Blackall, P.J., Morawska, L., 2003. Real-time measurement of bacterial aerosols with the UVAPS: performance evaluation. J. Aerosol Sci. 34 (3), 301–317.

Agranovski, I.E., Safatov, A.S., Borodulin, A.I., Pyankov, O.V., Petrishchenko, V.A., Sergeev, A.N., Agafonov, A.P., Ignatiev, G.M., Sergeev, A.A., Agranovski, V., 2004. Inactivation of viruses in bubbling processes utilized for personal bioaerosol monitoring. Appl. Environ. Microbiol. 70 (12), 6963–6967.

Aizenberg, V., Grinshpun, S.A., Willeke, K., Smith, J., Baron, P.A., 2000. Performance characteristics of the button personal inhalable aerosol sampler. Am. Ind. Hyg. Assoc. J. 61, 398–404.

Alija-Martínez, B., de Bengoa, B., Vallejo, R., Sevillano Fernández, D., González, N., Losa Iglesias, M.E., Collado, L., Espinosa-Rubio, R., Alou, L., 2021. Fungal bioaerosol as an occupational hazard in the podiatrist's workplace. Int. J. Environ. Health Res., 1–12. https://doi.org/10.1080/09603123.2021.2013451.

Alonso, C., Raynor, P.C., Davies, P.R., Torremorell, M., 2015. Concentration, size distribution, and infectivity of airborne particles carrying swine viruses. PLoS One 10 (8), e0135675.

Alonso, C., Raynor, P.C., Goyal, S., Olson, B.A., Alba, A., Davies, P.R., Torremorell, M., 2017. Assessment of air sampling methods and size distribution of virus-laden aerosols in outbreaks in swine and poultry farms. J. Vet. Diagn. Invest. 29 (3), 298–304.

Amann, R.I., Ludwig, W., Schleifer, K.H., 1995. Phylogenetic identification and in situ detection of individual microbial cells without cultivation. Microbiol. Rev. 59, 143–169.

American Conference of Governmental Industrial Hygienists (ACGIH), 2009. Threshold Limit Values (TLVs) for Chemical Substances and Physical Agents and Biological Exposure Indices (BEIs). ACGIH, Cincinnati, USA, pp. 223–226. Available at www.acgih.org/TLV/stu-dies.htm.

Anderson, B.D., Ma, M., Xia, Y., Wang, T., Shu, B., Lednicky, J.A., et al., 2016. Bioaerosol sampling in modern agriculture: a novel approach for emerging pathogen surveillance? J. Infect. Dis. 214 (4), 537–545.

Anderson, B.D., Lednicky, J.A., Montserrat, T., Gray, G.C., 2017. The use of bioaerosol sampling for airborne virus surveillance in swine production facilities: a mini review. Front. Vet. Sci. 4, 121.

Anderson, E.L., Turnham, P., Griffin, J.R., Clarke, C.C., 2020. Consideration of the aerosol transmission for COVID-19 and public health. Risk Anal. 40 (5), 902–907.

Andretta, I., Kipper, M., Lehnen, C.R., Hauschild, L., Vale, M.M., Lovatto, P.A., 2011. Meta-analytical study on productive and nutritional interactions of mycotoxins in broilers. Poult. Sci. 90, 1934–1940.

Anees-Hill, S., Douglas, P., Pashley, C.H., Hansell, A., Marczylo, E.L., 2021. A systematic review of outdoor airborne fungal spore seasonality across Europe and the implications for health. Sci. Total Environ. 151716.

Arendrup, M., Driscoll, B., Petersen, E., Denning, D., 2006. Acute pulmonary aspergillosis in immunocompetent subjects after exposure to bark chippings. Scand. J. Infect. Dis. 38, 945–949.

Armstrong, M.T., Rickles, F.R., Armstrong, P.B., 2013. Capture of lipopolysaccharide (endotoxin) by the blood clot: a comparative study. PLoS One 8, e80192.

Asaduzzaman, M., Hossain, M.I., Saha, S.R., Islam, M.R., Ahmed, N., Islam, M.A., 2019. Quantification of airborne resistant organisms with temporal and spatial diversity in Bangladesh: protocol for a cross-sectional study. JMIR Res. Protoc. 8 (12), e14574.

ASTM, 2014. Standard Practice for Collection of Total Airborne Fungal Structures via Inertial Impaction Methodology. ASTM d7788-14. 4, ASTM International, West Conshohocken, PA.

Ayer, S., Zhao, W.X., Davis, C.E., 2008. Differentiation of proteins and viruses using pyrolysis gas chromatography differential mobility spectrometry (PY/GC/DMS) and pattern recognition. IEEE Sens. J. 8 (9–10), 1586–1592.

Azhar, E.I., Papadioti, A., Bibi, F., Ashshi, A.M., Raoult, D., Angelakis, E., 2017. Pseudomonas saudimassiliensis' sp. nov. a new bacterial species isolated from air samples in the urban environment of Makkah, Saudi Arabia. New Microbes New Infect. 16, 43–44.

Azhar, E.I., Hui, D.S., Memish, Z.A., Drosten, C., Zumla, A., 2019. The middle east respiratory syndrome (MERS). Infect. Dis. Clin. 33 (4), 891–905.

Bahram, M., Anslan, S., Hildebrand, F., Bork, P., Tedersoo, L., 2018. Newly designed 16S rRNA metabarcoding primers amplify diverse and novel archaeal taxa from the environment. Environ. Microbiol. Rep. 11 (4), 487–494.

Baker, R.E., Mahmud, A.S., Miller, I.F., Rajeev, M., Rasambainarivo, F., Rice, B.L., Takahashi, S., Tatem, A.J., Wagner, C.E., Wang, L.F., Wesolowski, A., 2022. Infectious disease in an era of global change. Nat. Rev. Microbiol. 20 (4), 193–205.

Baldacci, S., Maio, S., Cerrai, S., Sarno, G., Baïz, N., Simoni, M., et al., 2015. Allergy and asthma: effects of the exposure to particulate matter and biological allergens. Respir. Med. 109 (9), 1089–1104.

Basketter, D.A., Broekhuizen, C., Fieldsend, M., Kirkwood, S., Mascarenhas, R., Maurer, K., Pedersen, C., Rodriguez, C., Schiff, H.E., 2010. Defining occupational and consumer exposure limits for enzyme protein respiratory allergens under REACH. Toxicology 268 (3), 165–170.

BAuA, 2013. Technische Regel für Biologische Arbeitsstoffe 214 (Technical Rules for Biological Agents 214), Bundesanstalt für Arbeitsschutz und Arbeitsmedizin (Federal Institute for Occupational Safety and Health). http://www.baua.de/de/Themenvon-A-Z/Biologische-Arbeitsstoffe/TRBA/TRBA-405.html.

Bauer, H., Claeys, M., Vermeylen, R., Schueller, E., Weinke, G., Berger, A., Puxbaum, H., 2008. Arabitol and mannitol as tracers for the quantification of airborne fungal spores. Atmos. Environ. 42, 588–593.

Baxi, S.N., Phipatanakul, W., 2010. The role of allergen exposure and avoidance in asthma. Adolesc. Med. State Art Rev. 21 (1), 57–63.

Bayle, S., Drapeau, A., Rocher, J., et al., 2021. Characterization of cultivable airborne bacteria and their antimicrobial resistance pattern in French milking parlour. Environ. Sci. Pollut. Res. 28, 11689–11696.

Be, N.A., Thissen, J.B., Fofanov, V.Y., Allen, J.E., Rojas, M., Golovko, G., Jaing, C.J., 2014. Metagenomic analysis of the airborne environment in urban spaces. Microb. Ecol. 69 (2), 346–355.

Becher, R., Honglso, J.K., Dybing, E., 2000. Guidelines for indoor air in Norway—a practical approach. Pollut. Atmos. 166, 245–246.

Behrendt, H., Becker, W.M., 2001. Localization, release and bioavailability of pollen allergens: The influence of environmental factors. Curr. Opin. Immunol. 13, 709–715.

Besson, S., Vragniau, C., Vassal-Stermann, E., Dagher, M.C., Fender, P., 2020. The adenovirus dodecahedron: beyond the platonic story. Viruses 12 (7), 718.

Best, E., Sandoe, J., Wilcox, M., 2012. Potential for aerosolization of Clostridium difficile after flushing toilets: the role of toilet lids in reducing environmental contamination risk. J. Hosp. Infect. 80 (1), 1–5.

Biermann, J., Merk, H., Wehrmann, W., Klimek, L., Wasem, J., 2013. Allergic disorders of the respiratory tract—findings from a large patient sample in the German statutory health insurance system. Allerg. J. 22, 366–373.

Blais Lecours, P., Duchaine, C., Taillefer, M., Tremblay, C., Veillette, M., Cormier, Y., Marsolais, D., 2011. Immunogenic properties of archaeal species found in bioaerosols. PLoS One 6 (8), e23326.

Blais Lecours, P., Veillette, M., Marsolais, D., Duchaine, C., 2012. Characterization of bioaerosols from dairy barns: reconstructing the puzzle of occupational respiratory diseases by using molecular approaches. Appl. Environ. Microbiol. 78, 3242–3248. https://doi.org/10.1128/AEM.07661-11.

Bloom, E., Nyman, E., Must, A., Pehrson, C., Larsson, L., 2009. Molds and mycotoxins in indoor environments—a survey in water-damaged buildings. J. Occup. Environ. Hyg. 6 (11), 671–678.

Booth, C.M., et al., 2013. Effectiveness of surgical masks against influenza bioaerosols. J. Hosp. Infect. 84 (1), 22–26.

Borodulin, A.I., Desyatkov, B.M., Lapteva, N.A., Sergeev, A.N., Agranovski, I.E., 2006. Personal sampler for monitoring of viable viruses; modelling of outdoor sampling conditions. Atmos. Environ. 40, 6687–6695.

Bowers, R.M., Clements, N., Emerson, J.B., Wiedinmyer, C., Hannigan, M.P., Fierer, N., 2013. Seasonal variability in bacterial and fungal diversity of the near-surface atmosphere. Environ. Sci. Technol. 47 (21), 12097–12106.

Bowley, H.J., Gerrard, D.L., Louden, J.D., Turrell, G., 2012. Practical Raman Spectroscopy. Springer, Berlin.

Brandi, G., Sisti, M., Amagliani, G., 2000. Evaluation of the environmental impact of microbial aerosols generated by wastewater treatment plants utilizing different aeration systems. J. Appl. Microbiol. 88 (5), 845–852.

Bridge, P., Spooner, B., 2001. Soil fungi: diversity and detection. Plant Soil 232 (1-2), 147–154.

Brosseau, L.M., et al., 2000. Differences in detected fluorescence among several bacterial species measured with a direct- reading particle sizer and fluorescence detector. Aerosol Sci. Tech. 32 (6), 545–558.

Brown, J.K., Hovmøller, M.S., 2002. Aerial dispersal of pathogens on the global and continental scales and its impact on plant disease. Science 297 (5581), 537–541.

Brown, R.C., Hemingway, M.A., Wake, D., Thompson, J., 1995. Field trials of an electret-based passive dust sampler in metal-processing industries. Ann. Occup. Hyg. 39 (5), 603–622.

Burch, M., Levetin, E., 2002. Effects of meteorological conditions on spore plumes. Int. J. Biometeorol. 46 (3), 107–117.

Burge, H.A., Solomon, W.R., 1987. Sampling and analysis of biological aerosols. Atmos. Environ. 21 (2), 451–456.

Burge, H.A., Pierson, D.L., Groves, T.O., Strawn, K.F., Mishra, S.K., 2000. Dynamics of airborne fungal populations in a large office building. Curr. Microbiol. 40 (1), 10–16.

Burton, N.C., Grinshpun, S.A., Reponen, T., 2007. Physical collection efficiency of filter materials for bacteria and viruses. Ann. Occup. Hyg. 51, 143–151.

Bush, R.K., Portnoy, J.M., 2001. The role and abatement of fungal allergens in allergic diseases. J. Allergy Clin. Immunol. 107, S430–S440.

Calderon, C., Ward, E., Freeman, J., McCartney, A., 2002. Detection of airborne fungal spores sampled by rotating-arm and Hirst-type spore traps using polymerase chain reaction assays. J. Aerosol Sci. 33 (2), 283–296.

Carducci, A., Federigi, I., Verani, M., 2013. Virus occupational exposure in solid waste processing facilities. Ann. Occup. Hyg. 57 (9), 1115e1127.

Casey, C., 2019. Development, Evaluation, and Demonstration of Air Sampling Methods to Assess Airborne Transport of Clostridium Difficile Spores in the Healthcare Environment. Technical Report, Air Force Institute of Technology Wright Patterson AFB United States.

Castellan, R.M., Olenchock, S.A., Kinsley, K.B., Hankinson, J.L., 1987. Inhaled endotoxin and decreased spirometric values: an exposure-response relation for cotton dust. N. Engl. J. Med. 317, 605–610.

Caugant, D.A., Brynildsrud, O.B., 2020. *Neisseria meningitidis*: using genomics to understand diversity, evolution and pathogenesis. Nat. Rev. Microbiol. 18, 84–96.

CEN, 2011. Ambient Air Quality—Measurement of Bioaerosols—Part 1: Determination of Moulds Using Filter Sampling Systems and Culture-Based Analyses, in Cen/TS 16115-1. 2011. European Committee for Standardization, Brussels, Belgium.

Chang, C.W., Hung, P.Y., 2012. Evaluation of sampling techniques for detection and quantification of airborne legionellae at biological aeration basins and shower rooms. J. Aerosol Sci. 48, 63–74.

Chang, C., Hwang, Y., Grinshpun, S., Macher, J., Willeke, K., 1994. Evaluation of counting error due to colony masking in bioaerosol sampling. Appl. Environ. Microbiol. 60, 3732–3738.

Chang, C., Grinshpun, S., Willeke, K., Macher, J., 1995. Factors affecting microbiological colony count accuracy for bioaerosol sampling and analysis. Am. Ind. Hyg. Assoc. J. 56, 979–986.

Chang, C.W., Ting, Y.T., Horng, Y.J., 2019. Collection efficiency of liquid-based samplers for fungi in indoor air. Indoor Air 29 (3), 380–389.

Chang, C.-W., Wang, L.-J., 2015. Methods for quantifying *Staphylococcus aureus* in indoor air. Indoor Air 25 (1), 59–67. https://doi.org/10.1111/ina.12124.

Chen, Q., Hildemann, L.M., 2009. Size-resolved concentrations of particulate matter and bioaerosols inside versus outside of homes. Aerosol Sci. Tech. 43, 699–713.

Chen, P.S., Li, C.S., 2005. Bioaerosol characterization by flow cytometry with fluorochrome. J. Environ. Monit. 7, 950–959.

Codina, R., Fox, R., Lockey, R., DeMarco, P., Bagg, A., 2008. Typical levels of airborne fungal spores in houses without obvious moisture problems during a rainy season in Florida, USA. J. Investig. Allergol. Clin. Immunol. 18, 156.

Colbeck, I., Whitby, C., 2019. In: Harrison, E.R.M., Hester, R.E. (Eds.), Biological Particles in the Indoor Environment. Issues in Environmental Science and Technology No. 48 Indoor Air Pollution. Royal Society of Chemistry.

Cooper, C.W., Aithinne, K.A.N., Floyd, E.L., Stevenson, B., Johnson, D.L., 2019. A comparison of air sampling methods for Clostridium difficile endospore aerosol. Aerobiologia 35, 411–420.

Cox, C., Wathes, C., 1995. Bioaerosols Handbook. Lewis Publisher, United States of America.

Coulon, F., Colbeck, I., 2017. RAMBIE, rapid monitoring of bioaerosols in urban, agricultural and industrial environments. NERC. Impact 11, 12–14.

Cox, J., Indugula, R., Vesper, S., Zhu, Z., Jandarov, R., Reponen, T., 2017. Comparison of indoor air sampling and dust collection methods for fungal exposure assessment using quantitative PCR. Environ. Sci.: Processes Impacts 19, 1312–1319.

Cox, J., Mbareche, H., Lindsley, W.G., Duchaine, C., 2020. Field sampling of indoor bioaerosols. Aerosol Sci. Tech. 54 (5), 572–584.
Crawford, C., Reponen, T., Lee, L.Y., Levin, L., Adhikari, A., Grinshpun, S.A., 2009. Temporal and spatial variation of indoor and outdoor airborne fungal spores, pollen, and (1→3)-β-D-glucan. Aerobiologia 25 (3), 147–158.
Crawford, S.E., Ramani, S., Tate, J.E., Parashar, U.D., Svensson, L., Hagbom, M., Franco, M.A., Greenberg, H.B., O'Ryan, M., Kang, G., Desselberger, U., 2017. Rotavirus infection. Nat. Rev. Dis. Primers. 3 (1), 1–6.
Cyprowski, M., Sobala, W., Buczyńska, A., 2015. Szadkowska-Stańczyk I. Endotoxin exposure and changes in short-term pulmonary function among sewage workers. Int. J. Occup. Med. Environ. Health 28 (5), 803–811.
Dai, R., Liu, S., Li, Q., Wu, H., Wu, L., Ji, C., 2021. A systematic review and meta-analysis of indoor bioaerosols in hospitals: the influence of heating, ventilation, and air conditioning. PLoS One 16 (12), e0259996.
Dakhama, A., Hegele, R.G., Laflamme, G., Israel-Assayag, E., Cormier, Y., 1999. Common respiratory viruses in lower airways of patients with acute hypersensitivity pneumonitis. Am. J. Respir. Crit. Care Med. 159, 1316–1322.
Damit, B., 2017. Droplet-based microfluidics detector for bioaerosol detection. Aerosol Sci. Tech. 51 (4), 488–500.
Dart, A., Thornburg, J., 2008. Collection efficiencies of bioaerosol impingers for virus-containing aerosols. Atmos. Environ. 42, 828–832.
D'Costa, V.M., McGrann, K.M., Hughes, D.W., Wright, G.D., 2006. Sampling the antibiotic resistome. Science 311 (5759), 374–377.
De Aquino Neto, F.R., de Góes, S.L.F., 2004. Guidelines for indoor air quality in offices in Brazil. Proc. Healthy Build. 4, 549–554.
DECOS, 2010. Endotoxins: Health Based Recommended Exposure Limit. A Report of the Health Council of the Netherlands. Publication No 2010/040SH. Health Council of The Netherlands, The Hague, The Netherlands.
Degois, J., Dubuis, M.E., Turgeon, N., Veillette, M., Duchaine, C., 2021. Condensation sampler efficiency for the recovery and infectivity preservation of viral bioaerosols. Taylor Francis J. https://doi.org/10.1080/02786826.2021.1889960.
Delort, A.M., Amato, P. (Eds.), 2018. Microbiology of Aerosols. Wiley Blackwell.
Dickson, I., 2017. Culturomics: illuminating microbial dark matter. Nat. Rev. Gastroenterol. Hepatol. 14, 3.
Dong, S., Yao, M., 2010. Exposure assessment in Beijing, China: biological agents, ultrafine particles, and lead. Environ. Monit. Assess. 170 (1–4), 331–343.
Douglas, P., Hayes, E.T., Williams, W.B., Tyrrel, S.F., Kinnersley, R.P., Walsh, K.O., Driscoll, M., Longhurst, P.J., Pollard, S.J.T., Drew, G.H., 2017. Use of dispersion modelling for environmental impact assessment of biological air pollution from composting: progress, problems and prospects. Waste Manag. 70, 22–29.
Douglas, P., Robertson, S., Gay, R., Hansell, A.L., Gant, T.W., 2018. A systematic review of the public health risks of bioaerosols from intensive farming. Int. J. Hyg. Environ. Health 221 (2), 134–173.
Douglas, P., Tyrrel, S.F., Kinnersley, R.P., Whelan, M., Longhurst, P.J., Walsh, K., Pollard, S.J., Drew, G.H., 2016. Sensitivity of predicted bioaerosol exposure from open windrow composting facilities to ADMS dispersion model parameters. J. Environ. Manage. 184, 448–455.
Douwes, J., Thorne, P., Pearce, N., Heederik, D., 2003. Bioaerosol health effects and exposure assessment: progress and prospects. Ann. Occup. Hyg. 47, 187–200.
Duchaine, C., Mériaux, A., 2001. The importance of combining air sampling and surface analysis when studying problematic houses for mold biodiversity determination. Aerobiologia 17 (2), 121–125.

Duncan, S.M., Farrell, R.L., Jordan, N., Jurgens, J.A., Blanchette, R.A., 2010. Monitoring and identification of airborne fungi at historic locations on Ross Island, Antarctica. Pol. Sci. 4 (2). 275–283. 53.

Dungan, R.S., Leytem, A.B., Verwey, S.A., Bjorneberg, D.L., 2010. Assessment of bioaerosols at a concentrated dairy operation. Aerobiologia 26, 171–184.

Duquenne, P., Coulais, C., Bau, S., Simon, X., 2018. Performances of the BC-112 NIOSH cyclone for the measurement of endotoxins in bioaerosols: a study in laboratory conditions. J. Aerosol Sci. 116, 92–105.

Durand, K.T.H., Muilenberg, M.L., Burge, H.A., Seixas, N.S., 2002. Effect of sampling time on the culturability of airborne fungi and bacteria sampled by filtration. Ann. Occup. Hyg. 46 (1). 113–118 51.

Dybwad, M., Skogan, G., Blatny, J.M., 2014. Comparative testing and evaluation of nine different air samplers: end-to-end sampling efficiencies as specific performance measurements for bioaerosol applications. Aerosol Sci. Tech. 48 (3), 282–295.

Eduard, W., 2006. The Nordic Expert Group for Criteria Documentation of Health Risks from Chemicals: 139. Fungal spores. National Institute for Working Life. Report nr 2006:21. Available at https://gupea.ub.gu.se/bitstream/handle/2077/4359/ah2006_21.pdf.

Eduard, W., 2009. Fungal spores: a critical review of the toxicological and epidemiological evidence as a basis for occupational exposure limit setting. Crit. Rev. Toxicol. 39, 799–864.

Eduard, W., Heederik, D., Duchaine, C., Green, B.J., 2012. Bioaerosol exposure assessment in the workplace: the past, present and recent advances. J. Environ. Monit. 14 (2), 334–339.

Effah, C.Y., Sun, T., Liu, S., et al., 2020. *Klebsiella pneumoniae*: an increasing threat to public health. Ann. Clin. Microbiol. Antimicrob. 19, 1.

Elbert, W., Taylor, P.E., Andreae, M.O., Pöschl, U., 2007. Contribution of fungi to primary biogenic aerosols in the atmosphere: wet and dry discharged spores, carbohydrates, and inorganic ions. Atmos Chem. Phys. 7 (17), 4569–4588.

Eldin, C., Mélenotte, C., Mediannikov, O., Ghigo, E., Million, M., Edouard, S., Mege, J.L., Maurin, M., Raoult, D., 2017. From Q fever to Coxiella burnetii infection: a paradigm change. Clin. Microbiol. Rev. 30 (1), 115–190.

Emerson, J.P., Keady, P., Clements, N., Morgan, E., Awerbuch, J., Miller, S., Fierer, N., 2017. High temporal variability in airborne bacterial diversity and abundance inside single-family residences. Indoor Air 27 (3), 576–586.

Environment Agency, 2001. Health Effects of Composting. A Study of Three Compost Sites and Review of Past Data. Environment Agency, Bristol, UK.

Environment Agency, 2010. Composting and Potential Health Effects From Bioaerosols: Our Interim Guidance for Permit Applicants. Position Statement 031, Version 1.0. Environment Agency, Bristol, UK.

Eversole, J.D., Cary, W.K., Scotto, C.S., Pierson, R., Spence, M., Campillo, A.J., 2001. Continuous bioaerosol monitoring using UV excitation fluorescence: outdoor test results. Field Anal. Chem. Technol. 5, 205–212.

Fahlgren, C., Bratbak, G., Sandaa, R.A., Thyrhaug, R., Zweifel, U.L., 2011. Diversity of airborne bacteria in samples collected using different devices for aerosol collection. Aerobiologia 27 (2), 107–120.

Fakunle, A.G., Jafta, N., Naidoo, R.N., Smit, L.A.M., 2021. Association of indoor microbial aerosols with respiratory symptoms among under-five children: a systematic review and meta-analysis. Environ. Health 20 (1), 77.

Fang, X., Liu, Y., Kong, J., Jiang, X., 2010. Loop-mediated isothermal amplification integrated on microfluidic chips for point-of-care quantitative detection of pathogens. Anal. Chem. 82 (7), 3002–3006.

Fennelly, M., Sewell, G., Prentice, M., O'Connor, D., Sodeau, J., 2017. The use of real-time fluorescence instrumentation to monitor ambient primary biological aerosol particles (PBAP). Atmosphere 9 (1), 1.

Ferguson, R.M.W., Garcia-Alcega, S., Coulon, F., Dumbrell, A.J., Whitby, C., Colbeck, I., 2019. Bioaerosol biomonitoring: sampling optimisation for molecular microbial ecology. Mol. Ecol. Resour. 19, 672–690.

Ferguson, R.M.W., Neath, C.E.E., Nasir, Z.A., Garcia-Alcega, S., Tyrrel, S., Coulon, F., Dumbrell, A.J., Colbeck, I., Whitby, C., 2021. Size fractionation of bioaerosol emissions from green-waste composting. Environ. Int. 147, 106327.

Fischer, G., Albrecht, A., Jackel, U., Kampfer, P., 2008. Analysis of airborne microorganisms, MVOC and odour in the surrounding of composting facilities and implications for future investigations. Int. J. Hyg. Environ. Health 211, 132–142.

Fracchia, L., Pietronave, S., Rinaldi, M., Martinotti, M.G., 2006. Site-related airborne biological hazard and seasonal variations in two wastewater treatment plants. Water Res. 40 (10), 1985–1994.

Franchitti, E., Pascale, E., Fea, E., Anedda, E., Traversi, D., 2020. Methods for bioaerosol characterization: limits and perspectives for human health risk assessment in organic waste treatment. Atmos. 11 (5), 452.

Frankel, M., Timm, M., Hansen, E., Madsen, A., 2012. Comparison of sampling methods for the assessment of indoor microbial exposure. Indoor Air 22 (5), 405–414.

French, M.L.V., Eitzen, H.E., Ritter, M.A., Leland, D.S., 1980. Environmental control of microbial contamination in the operating room. In: Hunt, T.K. (Ed.), Wound Healing and Wound Infection. Appleton-Century Crofis, New York, pp. 254–261.

Fröhlich-Nowoisky, J., Kampf, C.J., Weber, B., Huffman, J.A., Pöhlker, C., Andreae, M.O., Lang-Yona, N., Burrows, S.M., Gunthe, S.S., Elbert, W., Su, H., Hoor, P., Thines, E., Gandolfi, I., Bertolini, V., Ambrosini, R., Bestetti, G., Franzetti, A., 2013. Unravelling the bacterial diversity in the atmosphere. Appl. Microbiol. Biotechnol. 97 (11), 4727–4736.

Gandolfi, I., Bertolini, V., Ambrosini, R., Bestetti, G., Franzetti, A., 2013. Unravelling the bacterial diversity in the atmosphere. Appl. Microbiol. Biotechnol. 97, 4727–4736. https://doi.org/10.1007/s00253-013-4901-2.

Garcia-Alcega, S., Nasir, Z.A., Ferguson, R.M.W., Noel, C., Cravo-Laureau, C., Whitby, C., Dumbrell, A.J., Colbeck, I., Tyrell, S., Coulon, F., 2018. Can chemical and molecular biomarkers help discriminate between industrial, rural and urban environments? Sci. Tot. Environ. 631-632, 1059–1069. https://doi.org/10.1016/j.scitotenv.2018.03.062.

Garcia-Alcega, S., Nasir, Z.A., Ferguson, R.M.W., Whitby, C., Dumbrell, A.J., Colbeck, I., Gomes, D., Tyrell, S., Coulon, F., 2017. Fingerprinting outdoor air environment using microbial volatile organic compounds (MVOCs)—a review. Trends Anal. Chem. 86, 75–83.

Georgakopoulos, D., et al., 2009. Microbiology and atmospheric processes: biological, physical and chemical characterization of aerosol particles. Biogeosciences 6 (4), 721–737.

Ghimire, P.S., Tripathee, L., Chen, P., Kang, S., 2019. Linking the conventional and emerging detection techniques for ambient bioaerosols: a review. Rev. Environ. Sci. Biotechnol. 18, 495–523.

Ghio, A.J., Roggli, V.L., 1995. Mycotoxins and interstitial lung disease. Chest 108, 1185–1196.

Ghosh, B., Lal, H., Srivastava, A., 2015. Review of bioaerosols in indoor environment with special reference to sampling, analysis and control mechanisms. Environ. Int. 85, 254–272.

Gibbs, S.G., Green, C.F., Tarwater, P.M., Scarpino, P.V., 2004. Airborne antibiotic resistant and nonresistant bacteria and fungi recovered from two swine herd confined animal feeding operations. J. Occup. Environ. Hyg. 1 (11), 699–706.

Gibbs, S.G., Green, C.F., Tarwater, P.M., Mota, L.C., Mena, K.D., Scarpino, P.V., 2006. Isolation of antibiotic-resistant bacteria from the air plume downwind of a swine confined or concentrated animal feeding operation. Environ. Health Perspect. 114 (7), 1032–1037.

Gladding, T.L., Thorn, J., Stott, D., 2003. Organic dust exposure and work-related effects among recycling workers. Am. J. Ind. Med. 43, 584–591.

Globus, T., Gelmont, B., 2014. Biological Detection With Terahertz Spectroscopy, Bioaerosol Detection Technologies. Springer, Berlin, pp. 241–264.

Godish, D., Godish, T., 2008. Total airborne mold particle sampling: evaluation of sample collection, preparation and counting procedures, and collection devices. J. Occup. Environ. Hyg. 5 (2), 100–106.

Gollakota, A.R.K., Gautam, S., Santosh, M., Sudan, H.R., Gandhi, R., Jebadurai, V.S., Shu, C.M., 2021. Bioaerosols: Characterization, Pathways, Sampling Strategies, and Challenges to Geo-Environment and Health. Gondw. Res. 99, 178–203.

Gordon, C.J., Tchesnokov, E.P., Feng, J.Y., Porter, D.P., Götte, M., 2020. The antiviral compound remdesivir potently inhibits RNA-dependent RNA polymerase from Middle East respiratory syndrome coronavirus. J. Biol. Chem. 295 (15), 4773–4779.

Gosselin, M.I., et al., 2016. Fluorescent bioaerosol particle, molecular tracer, and fungal spore concentrations during dry and rainy periods in a semi-arid forest. Atmos. Chem. Phys. 16 (23), 15165–15184.

Government Office for Science, 2014. Blackett Review on Wide-Area Biological Detection. Available at https://www.gov.uk/government/publications/wide-area-biological-detection-blackett-review.

Green, F.C., Scarpino, P.V., 2002. The use of ultraviolet germicidal irradiation (UVGI) in disinfection of airborne bacteria. Environ. Eng. Policy 3 (1), 101–107.

Griffin, D.W., 2007. Atmospheric movement of microorganisms in clouds of desert dust and implications for human health. Clin. Microbiol. Rev. 20 (3), 459–477.

Griffiths, W.D., Boysan, F., 1996. Computational fluid dynamics (CFD) and empirical modelling of the performance of a number of cyclone samplers. J. Aerosol Sci. 27, 281–304.

Grinshpun, S.A., Chang, C.W., Nevalainen, A., Willeke, K., 1994. Inlet characteristics of bioaerosol samplers. J. Aerosol Sci. 25 (8), 1503–1522.

Grinshpun, S.A., Willeke, K., Ulevicius, V., Juozaitis, A., Terzieva, S., Donnelly, J., Stelma, G.N., Brenner, K.P., 1997. Effect of impaction, bounce and reaerosolization on the collection efficiency of impingers. Aerosol Sci. Tech. 26 (4), 326–342.

Grinshpun, S.A., Buttner, M.P., Mainelis, G., Willeke, K., 2015. Sampling for airborne microorganisms. In: Yates, M.V., Nakatsu, C.H., Miller, R.V., Pillai, S.D. (Eds.), Manual of Environmental Microbiology. American Society for Microbiology, Washington, DC, pp. 3.2.2-1–3.2.2-17.

Grydaki, N., Colbeck, I., Mendes, L., Eleftheriadis, K., Whitby, C., 2021. Bioaerosols in the Athens metro: metagenetic insights into the PM10 microbiome in a naturally ventilated subway station. Environ. Int. 146, 106186.

Guo, K., Qian, H., Zhao, D., Ye, J., Zhang, Y., Kan, H., Zhao, Z., Deng, F., Huang, C., Zhao, B., Zeng, X., Sun, Y., Liu, W., Mo, J., Sun, C., Guo, J., Zheng, X., 2020. Indoor exposure levels of bacteria and fungi in residences, schools, and offices in China: a systematic review. Indoor Air 30 (6), 1147–1165.

Haaland, D., Siegel, J.A., 2017. Quantitative filter forensics for indoor particle sampling. Indoor Air 27 (2), 364–376.

Haig, C.W., Mackay, W.G., Walker, J.T., Williams, C., 2016. Bioaerosol sampling: sampling mechanisms, bioefficiency and field studies. J. Hosp. Infect. 93 (3), 242–255.

Hairston, P.P., Ho, J., Quant, F.R., 1997. Design of an instrument for real-time detection of bioaerosols using simultaneous measurement of particle aerodynamic size and intrinsic fluorescence. J. Aerosol Sci. 28, 471–482.

Hall, R.J., Leblanc-Maridor, M., Wang, J., 2013. Metagenomic detection of viruses in aerosol samples from workers in animal slaughterhouses. PLoS One 8 (8), e72226.

Han, T.W., Mainelis, G., 2012. Investigation of inherent and latent internal losses in liquid-based bioaerosol samplers. J. Aerosol Sci. 45, 58–68.

Han, T.T., Thomas, N.M., Mainelis, G., 2017. Design and development of a self-contained personal electrostatic bioaerosol sampler (PEBS) with a wire-to-wire charger. Aerosol Sci. Tech. 51 (8), 903–915.

Han, T.T., Thomas, N.M., Mainelis, G., 2018. Performance of personal electrostatic bioaerosol sampler (PEBS) when collecting airborne microorganisms. J. Aerosol Sci. 124, 54–67.

Hart, M.B., Scotto, C.S., Tucker, J.E., Mcpherson, D.C., Minter, Z., Kesavan, J., Silcott, D., Lin, H.-B., Eversole, J.D., 2020. Toward biological aerosol reference standards. Aerosol Sci. Tech. 54 (5), 601–610.

Haugland, R.A., Varma, M., Wymer, L.J., Vesper, S.J., 2004. Quantitative PCR analysis of selected Aspergillus, Penicillium and Paecilomyces species. Syst. Appl. Microbiol. 27 (2), 198–210.

Hayes, J.M., 2001. Fractionation of carbon and hydrogen isotopes in biosynthetic processes. Rev. Minerol. Geochem. 43 (1), 225–277.

Hayward, S.J., Gouin, T., Wania, F., 2010. Comparison of four active and passive sampling techniques for pesticides in air. Environ. Sci. Technol. 44 (9), 3410–3416.

He, Q., Yao, M., 2011. Integration of high volume portable aerosol-to-hydrosol sampling and qPCR in monitoring bioaerosols. J. Environ. Monit. 13 (3), 706–712.

Heedrik, D., Brouwer, R., Biersteker, K., Boleij, J.S.M., 1991. Relationship of airborne endotoxins and bacteria levels in pig farms with the lung function and respiratory symptoms of farmers. Int. Arch. Occup. Environ. Health 62, 595–601.

Hegarty, B., Dannemiller, K., Peccia, J., 2018. Gene expression of indoor fungal communities under damp building conditions: implications for human health. Indoor Air 28 (4), 548–558.

Heidelberg, J., et al., 1997. Effect of aerosolization on culturability and viability of gram-negative bacteria. Appl. Environ. Microbiol. 63 (9), 3585–3588.

Hernandez, M., Perring, A.E., McCabe, K., Kok, G., Granger, G., Baumgardner, D., 2016. Chamber catalogues of optical and fluorescent signatures distinguish bioaerosol classes. Atmos. Meas. Tech. 9, 3283–3292.

Hernlem, B.J., Ravva, S.V., 2007. Application of flow cytometry and fluorescent in situ hybridisation for assessment of exposures to airborne bacteria. J. Environ. Monit. 9, 1317–1322.

Herr, C., Bittighofer, P.M., Bünger, J., Eikmann, T., Fischer, A.B., Grüner, C., Idel, H., Zur Nieden, A., Palmgren, U., Seidel, H.J., Velcovsky, H.G., 1999. Effect of microbial aerosols on the human. Schriftenr. Ver. Wasser Boden Lufthyg. 59, 229–239.

Hill, K.A., Shepson, P.B., Galbavy, E.S., Anastasio, C., Kourtev, P.S., Konopka, A., et al., 2007. Processing of atmospheric nitrogen by clouds above a forest environment. J. Geophys. Res. Atmos. 112 (D11), D11301.

Hirst, J.M., 1952. An automatic volumetric spore trap. Ann. Appl. Biol. 39 (2), 257–265.

Hogan Jr., C.J., Kettleson, E.M., Lee, M.H., Ramaswami, B., Angenent, L.T., Biswas, P., 2005. Sampling methodologies and dosage assessment techniques for submicrometre and ultrafine virus aerosol particles. J. Appl. Microbiol. 99 (6), 1422–1434.

Hogan, C.J., Lee, M.H., Biswas, P., 2004. Capture of viral particles in soft X-ray–enhanced corona systems: Charge distribution and transport characteristics. Aerosol Sci. Tech. 38 (5), 475–486.

Hoisington, A.J., Maestre, J.P., King, M.D., Siegel, J.A., Kinney, K.A., 2014. Impact of sampler selection on the characterization of the indoor microbiome via high-throughput sequencing. Build. Environ. 80, 274–282.

Hong, S., Bhardwaj, J., Han, C.-H., Jang, J., 2016. Gentle sampling of submicrometer airborne virus particles using a personal electrostatic particle concentrator. Environ. Sci. Technol. 50 (22), 12365–12372.

Hoppin, J.A., Umbach, D.M., Long, S., Rinsky, J.L., Henneberger, P.K., Salo, P.M., et al., 2014. Respiratory disease in United States farmers. Occup. Environ. Med. 71 (7), 484–491.

Hoskisson, P.A., 2018. Microbe profile: Corynebacterium diphtheriae—an old foe always ready to seize opportunity. Microbiology (Reading) 164 (6), 865–867.

Hospodsky, D., Qian, J., Nazaroff, W.W., Yamamoto, N., Bibby, K., Rismani-Yazdi, H., Peccia, J., 2012. Human occupancy as a source of indoor airborne bacteria. PLoS One 7 (4), e34867.

Huffman, J., Treutlein, B., Poschl, U., 2010. Fluorescent biological aerosol particle concentrations and size distributions measured with an ultraviolet aerodynamic particle sizer (UV-APS) in Central Europe. Atmos. Chem. Phys. 10 (7), 3215–3233.

Huffman, J., et al., 2012. Size distributions and temporal variations of biological aerosol particles in the Amazon rain- forest characterized by microscopy and real-time UV-APS fluorescence techniques during AMAZE-08. Atmos. Chem. Phys. 12 (24), 11997–12019.

Huffman, J.A., Perring, A.E., Savage, N.J., Clot, B., Crouzy, B., Tummon, F., Shoshanim, O., Damit, B., Schneider, J., Sivaprakasam, V., Zawadowicz, M.A., 2020. Real-time sensing of bioaerosols: review and current perspectives. Aerosol Sci. Tech. 54 (5), 465–495.

Humbal, C., Gautam, S., Trivedi, U., 2018. A review on recent progress in observations, and health effects of bioaerosols. Environ. Int. 118, 189–193.

Hung, H.F., Kuo, Y.M., Chien, C.C., Chen, C.C., 2010. Use of floating balls for reducing bacterial aerosol emissions from aeration in wastewater treatment processes. J. Hazard. Mater. 175 (1–3), 866–871.

Hurley, K.V., Wharton, L., Wheeler, M.J., Skjøth, C.A., Niles, C., Hanson, M.C., 2019. Car cabin filters as sampling devices to study bioaerosols using eDNA and microbiological methods. Aerobiologia 35 (2), 215–225.

Institut für Arbeitsschutz der Deutschen Gesetzlichen Unfallversicherung (IFA), 2004. Verfahren zur Bestimmung der Bakterienkonzentration in der Luft am Arbeitsplatz (#9430). Erich Schmidt Verlag, Berlin, Germany.

ISO, 2008. Indoor air—Part 16: Detection and enumeration of moulds—Sampling by filtration. ISO 16000-16:2008(en).

ISO, 2018. Workplace Exposure—Measurement of Exposure by Inhalation to Chemical Agents—Strategy for Testing Compliance With Occupational Exposure Limit Values, Standard: CSN EN 689, Brussels, Belgium.

Iversen, M., et al., 2000. Human health effects of dust exposure in animal confinement buildings. J. Agric. Saf. Health 6, 283–288.

Jacobs, C., Braun, P, Hammer, P., 2011. Reservoir and routes of transmission of Enterobacter sakazakii (*Cronobacter* spp.) in a milk powder-producing plant. J. Dairy Sci. 9 (8), 3801–3810. https://doi.org/10.3168/jds.2011-4318.

Jiang, X., Jing, W., Sun, X., Liu, Q., Yang, C., Liu, S., Qin, K., Sui, G., 2016a. High-throughput microfluidic device for LAMP analysis of airborne bacteria. ACS Sens. 1, 958–962. https://doi.org/10.1021/acssensors.6b00282.

Jiang, Y., Zou, S., Cao, X., 2016b. Rapid and ultra-sensitive detection of foodborne pathogens by using miniaturized microfluidic devices: a review. Anal. Methods 8 (37), 6668–6681.

Jiang, L., Zhang, J., Tang, J., Li, M., Zhao, X., Zhu, H., Yu, X., Li, Y., Feng, T., Zhang, X., 2018. Analyses of aerosol concentrations and bacterial community structures for closed cage broiler houses at different broiler growth stages in winter. J. Food Prot. 81 (9), 1557–1564. https://doi.org/10.4315/0362-028X.JFP-17-524.

Jiang, X., Loeb, J.C., Pan, M., Tilly, T.B., Eiguren-Fernandez, A., Lednicky, J.A., Wu, C.Y., Fan, Z.H., 2021. Integration of sample preparation with RNA-Amplification in a hand-held device for airborne virus detection. Anal. Chim. Acta 1165, 338542.

Jo, W.K., Seo, Y.J., 2005. Indoor and outdoor bioaerosol levels at recreation facilities, elementary schools, and homes. Chemosphere 61, 1570–1579.

Jones, 2020. Environment Agency. available at https://www.gov.uk/government/publications/review-of-airborne-antimicrobial-resistance.

Jones, B.L., Cookson, J.T., 1983. Natural atmospheric microbial conditions in a typical suburban area. Appl. Environ. Microbiol. 45 (3), 919–934.

Jonsson, P., Kullander, F., 2014. Bioaerosol Detection With Fluorescence Spectroscopy, Bioaerosol Detection Technologies. Springer, Berlin, pp. 111–141.

Jung, C.-C., Wu, P.-C., Tseng, C.-H., Su, H.-J., 2015. Indoor air quality varies with ventilation types and working areas in hospitals. Build. Environ. 85, 190–195.

Just, N., Kirychuk, S., Gilbert, Y., Letourneau, V., Veillette, M., Singh, B., Duchaine, C., 2011. Bacterial diversity characterization of bioaerosols from cage-housed and floor-housed poultry operations. Environ. Res. 4, 492–498.

Just, N., Lecours, P.B., Marcoux-Voiselle, M., Kirychuk, S., Veillette, M., Singh, B., Duchaine, C., 2013. Archaeal characterization of bioaerosols from cage-housed and floor-housed poultry operations. Can. J. Microbiol. 59 (1), 46–50.

Kalatoor, S., Grinshpun, S.A., Willeke, K., Baron, P., 1995. New aerosol sampler with low wind sensitivity and good filter collection uniformity. Atmos. Environ. 29 (10), 1105–1112.

Kamperman, T., Trikalitis, V.D., Karperien, M., Visser, C.W., Leijten, J., 2018. Ultrahigh-throughput production of monodisperse and multifunctional janus microparticles using in-air microfluidics. ACS Appl. Mater. Interfaces 10 (28), 23433–23438.

Kandhai, M.C., Reij, M.W., Gorris, L.G., Guillaume-Gentil, O., van Schothorst, M., 2004. Occurrence of Enterobacter sakazakii in food production environments and households. The Lancet 363 (9402), 39–40. https://doi.org/10.1016/S0140-6736(03)15169-0.

Karra, S., Katsivela, E., 2007. Microorganisms in bioaerosol emissions from wastewater treatment plants during summer at a Mediterranean site. Water Res. 41, 1355–1365.

Kaushik, A.K., Dhau, J.S., Gohel, H., Mishra, Y.K., Kateb, B., Kim, N.-Y., Goswami, D.Y., 2020. Electrochemical SARS-CoV-2 sensing at point-of-care and artificial intelligence for intelligent COVID-19 management. ACS Appl. Bio. Mater. 3, 7306–7325.

Kawashima, S, Thibaudon, M., Matsuda, S., Fujita, T., Lemonis, N., Clot, B., Oliver, G., 2017. Automated pollen monitoring system using laser optics for observing seasonal changes in the concentration of total airborne pollen. Aerobiologia 33, 351–362. https://doi.org/10.1007/s10453-017-9474-6.

Kaye, P., Stanley, W., Hirst, E., Foot, E., Baxter, K., Barrington, S., 2005. Single particle multichannel bio-aerosol fluorescence sensor. Opt. Exp. 13, 3583–3593.

Keith, L., 1991. Environmental Sampling and Analysis: A Practical Guide, first ed. Taylor & Francis Group, New York.

Kenny, L.C., Stancliffe, J.D., Crook, B., 1998. The adaptation of existing personal inhalable aerosol samplers for bioaerosol sampling. Am. Ind. Hyg. Assoc. J. 59, 831–841.

Kenny, L.C., Bowry, A., Crook, B., Stancliffe, J.D., 1999. Field testing of a personal size-selective bioaerosol sampler. Ann. Occup. Hyg. 43 (6), 393–404.

Kharitonov, S.A., Sjöbring, U., 2007. Lipopolysaccharide challenge of humans as a model for chronic obstructive lung disease exacerbations. Contrib. Microbiol. 14, 83–100.

Kilburg-Basnyat, B., Metwali, N., Thorne, P.S., 2014. Effect of deployment time on endotoxin and allergen exposure assessment using electrostatic dust collectors. Ann. Work Expo. Health 59, 104–115.

Kilburg-Basnyat, B., Metwali, N., Thorne, P.S., 2016. Performance of electrostatic dust collectors (EDCs) for endotoxin assessment in homes: effect of mailing, placement, heating, and electrostatic charge. J. Occup. Environ. Hyg. 13 (2), 85–93.

Kim, K.Y., Ko, H.J., Kim, H.T., Kim, Y.S., Roh, Y.M., Lee, C.M., et al., 2007. Influence of extreme seasons on airborne pollutant levels in a pig-confinement building. Arch. Environ. Occup. Health 62 (1), 27–32.

Kim, K.Y., Kabir, E., Jahan, S.A., 2018. Airborne bioaerosols and their impact on human health. J. Environ. Sci. 67, 23–35.

King, P., Pham, L.K., Waltz, S., Sphar, D., Yamamoto, R.T., Conrad, D., Taplitz, R., Torriani, F., Forsyth, R.A., 2016. Longitudinal metagenomic analysis of hospital air identifies clinically relevant microbes. PLoS One 11 (8), e0160124.

Klappenbach, J.A., Saxman, P.R., Cole, J.R., Schmidt, T.M., 2001. RRNDB: the ribosomal RNA operon copy number database. Nucleic Acids Res., 29181–29184.

Knowlton, S.D., Boles, C.L., Perencevich, E.N., Diekema, D.J., Nonnenmann, M.W., 2018. Bioaerosol concentrations generated from toilet flushing in a hospital-based patient care setting. Antimicrob. Resist. Infect. Control 7 (1), 1–8.

Kohanski, M.A., Lo, L.J., Waring, M.S., 2020. Review of indoor aerosol generation, transport, and control in the context of COVID-19. Int. Forum Allergy Rhinol. 10 (10), 1173–1179.

Könemann, T., Savage, N., Klimach, T., Walter, D., Fröhlich-Nowoisky, J., Su, H., Pöschl, U., Huffman, J.A., Pöhlker, C., 2019. Spectral intensity bioaerosol sensor (SIBS): an instrument for spectrally resolved fluorescence detection of single particles in real time. Atmos. Meas. Tech. 12, 1337–1363.

Korpi, A., Pasanen, A.-L., Pasanen, P., 1998. Volatile compounds originating from mixed microbial cultures on building materials under various humidity conditions. Appl. Environ. Microbiol. 64, 2914–2919.

Krames, J., Buttner, H., 1994. The cyclone scrubber—a high efficiency wet separator. Chem. Eng. Technol. 17, 73–80.

Kubota, K., 2013. CARD-FISH for environmental microorganisms: technical advancement and future applications. Microb. Environ. 28 (1), 3–12.

Kulkarni, P., Baron, P.A., Willeke, K. (Eds.), 2011. Aerosol Measurement: Principles, Techniques, and Applications. John Wiley & Sons.

Kumari, P., Choi, H.L., 2014. Seasonal variability in airborne biotic contaminants in swine confinement buildings. PLoS One 9 (11), e112897.

Kunkel, S.A., Azimi, P., Zhao, H., Stark, B.C., Stephens, B., 2017. Quantifying the size-resolved dynamics of indoor bioaerosol transport and control. Indoor Air 27 (5), 977–987.

Kurek, M.A., Wyrwisz, J., Wierzbicka, A., 2016. Effect of β-glucan particle size on the properties of fortified wheat rolls CyTA. J. Food 14, 124–130.

Ladhani, L., Pardon, G., Meeuws, H., Van Wesenbeeck, L., Schmidt, K., Stuyver, L., Van Der Wijngaart, W., 2017. Sampling and detection of airborne influenza virus towards point-of-care applications. PLoS One 12 (3), e0174314.

Lagier, J.C., Khelaifia, S., Alou, M., et al., 2016. Culture of previously uncultured members of the human gut microbiota by culturomics. Nat. Microbiol. 1, 16203.

Lange, J.L., Thorne, P.S., Lynch, N., 1997. Application of flow cytometry and fluorescent *in situ* hybridisation for assessment of exposures to airborne bacteria. Appl. Environ. Microbiol. 63, 1557–1563.

LeBouf, R., Yesse, L., Rossner, A., 2008. Seasonal and diurnal variability in airborne mold from an indoor residential environment in Northern New York. J. Air Waste Manage. Assoc. 58 (5), 684–692.

Lednicky, J., Pan, M., Loeb, J., Hsieh, H., Eiguren-Fernandez, A., Hering, S., Fan, Z.H., Wu, C.-Y., 2016. Highly efficient collection of infectious pandemic influenza H1N1 virus (2009) through laminar-flow water based condensation. Aerosol Sci. Tech. 50 (7), i–iv.

Lednicky, J.A., Lauzardo, M., Hugh Fan, Z., Cherabuddi, K., Glenn Morris Jr., J., Wu, C.-Y., 2020. Viable SARS-CoV-2 in the air of a hospital room with COVID-19 patients. Int. J. Infect. Dis. 100, 476–482.

Leggat, P.A., Kedjarune, U., 2001. Bacterial aerosols in the dental clinic: a review. Int. Dent. J. 51 (1), 39–44.

Lehtonen, M., Reponen, T., Nevalainen, A., 1993. Everyday activities and variation of fungal spore concentrations in indoor air. Int. Biodeterior. Biodegrad. 31, 25–39.

Lemieux, J., Veillette, M., Mbareche, H., Duchaine, C., 2019. Re-aerosolization in liquid-based air samplers induces bias in bacterial diversity. Aerosol Sci. Tech. 53, 1244–1260.

Li, S.-N., Lundgren, D.A., Rovell-Rixx, D., 2000. Evaluation of six inhalable aerosol samplers. Am. Ind. Hyg. Assoc. 61 (4), 506–516.

Li, L., Gao, M., Liu, J.X., 2011. Distribution characterization of microbial aerosols emitted from a wastewater treatment plant using the orbal oxidation ditch process. Process Biochem. 46 (4), 910–915.

Li, L., Mendis, N., Oliver, T.H., JD., Faucher SP., 2014. The importance of the viable but non-culturable state in human bacterial pathogens. Front. Microbiol. 5, 258.

Li, Z., Xu, X., Thompson, L.A., Gross, H.E., Shenkman, E.A., DeWalt, D.A., Huang, I.C., 2019. Longitudinal effect of ambient air pollution and pollen exposure on asthma control: The patient-reported outcomes measurement information system (PROMIS) pediatric asthma study. Acad. Pediatr. 19 (6), 615.

Liang, L., He, K., Duan, F., 2012. Progress on quantitative assessment methods of biological aerosols in the atmosphere. Biogeosci. Discuss. 9 (2), 1511–1528.

Liebers, V., van Kampen, V., Bünger, J., Düser, M., Stubel, H., Brüning, T., Raulf-Heimsoth, M., 2012. Assessment of airborne exposure to endotoxin and pyrogenic active dust using electrostatic dustfall collectors (EDCs). J. Toxicol. Environ. Health A 75 (8–10), 501–507.

Lin, X., Willeke, K., Ulevicius, V., Grinshpun, S.A., 1997. Effect of sampling time on the collection efficiency of all-glass impingers. Am. Ind. Hyg. Assoc. J. 58 (7), 480–488.

Lin, X.J., Reponen, T.A., Willeke, K., Grinshpun, S.A., Foarde, K.K., Ensor, D.S., 1999. Long-term sampling of airborne bacteria and fungi into a non-evaporating liquid. Atmos. Environ. 33 (26), 4291–4298.

Lindsley, W.G., Blachere, F.M., Davis, K.A., Pearce, T.A., Fisher, M.A., Khakoo, R., Davis, S.M., Rogers, M.E., Thewlis, R.E., Posada, J.A., et al., 2010. Distribution of airborne influenza virus and respiratory syncytial virus in an urgent care medical clinic. Clin. Infect. Dis. 50 (5), 693–698.

Lindsley, W.G., Green, B.J., Blachere, F.M., Martin, S.B., Law, B.F., Jensen, P.A., Schafer, M.P., 2017. Sampling and characterization of bioaerosols. In: Ashley, K., O'Connor, P.F. (Eds.), NIOSH Manual of Analytical Methods. National Institute for Occupational Safety and Health, Cincinnati, OH, pp. BA1–115.

Luhung, I., Wu, Y., Ng, C.K., Miller, D., Cao, B., Chang, V.-W.-C., 2015. Protocol improvements for low concentration DNA-based bioaerosol sampling and analysis. PLoS One 10 (11), e0141158.

Luongo, J.C., Fennelly, K.P., Keen, J.A., Zhai, Z.J., Jones, B.W., Miller, S.L., 2016. Role of mechanical ventilation in the airborne transmission of infectious agents in buildings. Indoor Air 26 (5), 666–678.

Lynch, J.P., Fishbein, M., Echavarria, M., 2011. Adenovirus. In: Seminars in Respiratory and Critical Care Medicine. vol. 32. Thieme Medical Publishers, pp. 494–511.

Macher, J.M., 1989. Positive-hole correction of multiple-jet impactors for collecting viable microorganisms. Am. Ind. Hyg. Assoc. J. 50 (11), 561–568.

Macher, J.M., 1999. Bioaerosols: Assessment and Control. American Conference of Governmental Industrial Hygienists (ACGIH), Cincinnati, OH.

Macher, J.M., Chatigny, M.A., Burge, H.A., 1995. Sampling airborne microorganisms and aeroallergens. In: Cohen, B.S., Hering, S.V. (Eds.), Air Sampling Instruments for Evaluations of Atmospheric Contaminants, eighth ed. American Conference of Governmental Industrial Hygienists, Inc., Cincinnati, OH, pp. 589–617.

Madsen, A.M., 2006. Exposure to airborne microbial components in autumn and spring during work at Danish biofuel plants. Ann. Occup. Hyg. 50, 821–831.

Madsen, A.M., Zervas, A., Tendal, K., Nielsen, J.L., 2015. Microbial diversity in bioaerosol samples causing ODTS compared to reference bioaerosol samples as measured using Illumina sequencing and MALDI-TOF. Environ. Res. 140, 255–267.

Maenhaut, W., Ducastel, G., Hillamo, R.E., Pakkanen, T.A., 1993. Evaluation of the applicability of the MOUDI impactor for aerosol collections with subsequent multielement analysis by PIXE. Nucl. Instrum. Methods Phys. Res. Sect. B 75 (1-4), 249–256.

Maestre, J.P., Jennings, W., Wylie, D., Horner, S.D., Siegel, J., Kinney, K.A., 2018. Filter forensics: microbiota recovery from residential HVAC filters. Microbiome 6 (1), 22.

Mainelis, G., 2020. Bioaerosol sampling: classical approaches, advances, and perspectives. Aerosol Sci. Tech. 54 (5), 496–519.

Mainelis, G., Willeke, K., Adhikari, A., Reponen, T., Grinshpun, S.A., 2002. Design and collection efficiency of a new electrostatic precipitator for bioaerosol collection. Aerosol Sci. Tech. 36, 1073–1085.

Manibusen, S., Mainelis, G., 2022. Passive bioaerosol samplers: a complementary tool for bioaerosol research. A review. J. Aerosol Sci., 105992.

Marchand, G., Lavoie, J., Lazure, L., 1995. Evaluation of bioaerosols in a municipal solid waste recycling and composting plant. J. Air Waste Manage. Assoc. 45, 778–781.

Marin, S., Ramos, A.J., Cano-Sancho, G., Sanchis, V., 2013. Mycotoxins: occurrence, toxicology, and exposure assessment. Food Chem. Toxicol. 60, 218–237.

Marjomäki, V., Kalander, K., Hellman, M., Permi, P., 2021. Enteroviruses and coronaviruses: similarities and therapeutic targets. Expert Opin. Ther. Targets 25 (6), 479–489. https://doi.org/10.1080/14728222.2021.1952985.

Marshall, W.A., 1997. Seasonality in Antarctic airborne fungal spores. Appl. Environ. Microbiol. 63 (6), 2240–2245.

Matthias-Maser, S., Bogs, B., Jaenicke, R., 2000. The size distribution of primary biological aerosol particles in cloud water on the mountain Kleiner Feldberg/Taunus (FRG). Atmos. Res. 54 (1), 1–13.

Mbareche, H., Brisebois, E., Veillette, M., Duchaine, C., 2017a. Bioaerosol sampling and detection methods based on molecular approaches: no pain no gain. Sci. Total Environ. 599, 2095–2104.

Mbareche, H., Veillette, M., Bonifait, L., Dubuis, M.E., Benard, Y., Marchand, G., Bilodeau, G.J., Duchaine, C., 2017b. A next generation sequencing approach with a suitable bioinformatics workflow to study fungal diversity in bioaerosols released from two different types of composting plants. Sci. Total Environ. 1 (601), 1306–1314.

Mbareche, H., Veillette, M., Bilodeau, G.J., Duchaine, C., 2018. Bioaerosol sampler choice should consider efficiency and ability of samplers to cover microbial diversity. Appl. Environ. Microbiol. 84. e01589–01518.

Mbareche, H., Veillette, M., Teertstra, W., Kegel, W., Bilodeau, G.J., Wösten, H.A.B., Duchaine, C., 2019. Recovery of fungal cells from air samples: a tale of loss and gain. Appl. Environ. Microbiol. 85 (9), e02941-18.

McDevitt, J.J., Koutrakis, P., Ferguson, S.T., Wolfson, J.M., Fabian, M.P., Martins, M., Pantelic, J., Milton, D.K., 2013. Development and performance evaluation of an exhaled-breath bioaerosol collector for influenza virus. Aerosol Sci. Tech. 47 (4), 444–451.

Meadow, J.F., Altrichter, A.E., Kembel, S.W., Kline, J., Mhuireach, G., Moriyama, M., Northcutt, D., O'Connor, T.K., Womack, A.M., Brown, G.Z., et al., 2014. Indoor airborne bacterial communities are influenced by ventilation, occupancy, and outdoor air source. Indoor Air 24 (1), 41–48.

Miaskiewicz-Peska, E., Lebkowska, M., 2012. Comparison of aerosol and bioaerosol collection on air filters. Aerobiologia 28 (2), 185–193.
Michel, D., Rotach, M.W., Gehrig, R., Vogt, R., 2012. On the efficiency and correction of vertically oriented blunt bioaerosol samplers in moving air. Int. J. Biometeorol. 56 (6), 1113.
Miller, J.D., Young, Y.C., 1997. The use of ergosterol to measure exposure to fungal propagules in indoor air. Am. Ind. Hyg. Assoc. J. 58, 39–43.
Miller, J.D., Laflamme, A.M., Sobol, Y., Lafontaine, P., Greenhalgh, R., 1988. Fungi and fungal products in some Canadian houses. Intl Biodet 24, 103–120.
Mirhoseini, S.H., Koolivand, A., Bayani, M., et al., 2021. Quantitative and qualitative assessment of microbial aerosols in different indoor environments of a dental school clinic. Aerobiologia 37, 217–224.
Mitchell, D.C., Armitage, T.L., Schenker, M.B., Bennett, D.H., Tancredi, D.J., Langer, C.E., et al., 2015. Particulate matter, endotoxin, and worker respiratory health on large Californian dairies. J. Occup. Environ. Med. 57 (1), 79–87.
Moon, H.-S., Lee, J.-H., Kwon, K., Jung, H.-I., 2012. Review of recent progress in micro-systems for the detection and analysis of airborne microorganisms. Anal. Lett. 45 (2–3), 113–129.
Morawska, L., Cao, J., 2020. Airborne transmission of SARS-CoV-2: the world should face the reality. Environ. Int. 139, 105730.
Morawska, L., He, C., Hitchins, J., Gilbert, D., Parappukkaran, S., 2001. The relationship between indoor and outdoor airborne particles in the residential environment. Atmos. Environ. 35 (20), 3463–3473.
Morey, P.R., 2007. Microbiological sampling strategies in indoor environments. In: Sampling and Analysis of Indoor Microorganisms, pp. 51–74.
Morgan, J.L., Darling, A.E., Eisen, J.A., 2010. Metagenomic sequencing of an in vitro-simulated microbial community. PLoS One 5 (4), e10209.
Morris, K.J., 1995. Modern microscopic methods of bioaerosol analysis. In: Cox, C.S., Wathes, C.M. (Eds.), Bioaerosols Handbook. CRC Press, pp. 285–316.
Mullane, N., Healy, B., Meade, J., Whyte, P., Wall, P.G., Fanning, S., 2008. Dissemination of Cronobacter spp. (Enterobacter sakazakii) in a powdered milk protein manufacturing facility. Appl. Environ. Microbiol. 74, 5913–5917. https://doi.org/10.1128/AEM.00745-08.
Napoli, C., Marcotrigiano, V., Montagna, M.T., 2012. Air sampling procedures to evaluate microbial contamination: a comparison between active and passive methods in operating theatres. BMC Public Health 12, 594.
Nasir, Z.A., et al., 2018. A controlled study on the characterisation of bioaerosols emissions from compost. Atmosphere 9 (10), 379.
Nasir, Z.A., et al., 2019. Scoping studies to establish the capability and utility of a real-time bioaerosol sensor to characterise emissions from environmental sources. Sci. Total Environ. 648, 25–32.
National Research Council, 2005. Section 3 Indoor and Outdoor Bioaerosol Backgrounds and Sampling Strategies in Sensor Systems for Biological Agent Attacks: Protecting Buildings and Military Bases. The National Academies Press, Washington, DC, pp. 1–45.
Nazaroff, W.W., 2016. Indoor bioaerosol dynamics. Indoor Air 26 (1), 61–78.
Nehmé, B., Letourneau, V., Forster, R.J., Veillette, M., Duchaine, C., 2008. Culture-independent approach of the bacterial bioaerosol diversity in the standard swine confinement buildings, and assessment of the seasonal effect. Environ. Microbiol. 10 (3), 665–675.
Nehmé, B., Gilbert, Y., Létourneau, V., Forster, R.J., Veillette, M., Villemur, R., Duchaine, C., 2009. Culture-independent characterization of archaeal biodiversity in swine confinement building bioaerosols. Appl. Environ. Microbiol. 75, 5445–5450.

Nieto-Caballero, M., Savage, N., Keady, P., Hernandez, M., 2019. High fidelity recovery of airborne microbial genetic materials by direct condensation capture into genomic preservatives. J. Microbiol. Methods 157, 1–3.
Nocker, A., Sossa-Fernandez, P., Burr, M.D., Camper, A.K., 2007. Use of propidium monoazide for live/dead distinction in microbial ecology. Appl. Environ. Microbiol. 73 (16), 5111–5117.
Nonnenmann, M., Bextine, B., Dowd, S., Gilmore, K., Levin, J., 2010. Culture-independent characterization of bacteria and fungi in a poultry bioaerosol using pyrosequencing: a new approach. J. Occup. Environ. Hyg. 7 (12), 693–699.
Normand, A.-C., Vacheyrou, M., Sudre, B., Heedrik, D.J.J., Piarroux, R., 2009. Assessment of dust sampling methods for the study of cultivable-microorganism exposure in stables. Appl. Environ. Microbiol. 75 (24), 7617–7623.
Noss, I., Wouters, I.M., Visser, M., Heederik, D.J., Thorne, P.S., Brunekreef, B., Doekes, G., 2008. Evaluation of a low-cost electrostatic dust fall collector for indoor air endotoxin exposure assessment. Appl. Environ. Microbiol. 74 (18), 5621–5627.
Noss, I., Doekes, G., Sander, I., Heederik, D.J., Thorne, P.S., Wouters, I.M., 2010. Passive airborne dust sampling with the electrostatic dustfall collector: optimization of storage and extraction procedures for endotoxin and glucan measurement. Ann. Occup. Hyg. 54, 651–658.
Notomi, T., Okayama, H., Masubuchi, H., Yonekawa, T., Watanabe, K., Amino, N., Hase, T., 2000. Loop-mediated isothermal amplification of DNA. Nucleic Acids Res. 28, E63.
O'Connor, D.J., Daly, S.M., Sodeau, J.R., 2015. On-line monitoring of airborne bioaerosols released from a composting/green waste site. Waste Manag. 42, 23–30.
O'Hara, R.E., Rubin, R., 2005. Reducing bioaerosol dispersion from wastewater treatment and its land application: a review and analysis. J. Environ. Health 68, 24–29.
Olenchock, S.A., Lewis, D.M., Mull, J.C., 1989. Effects of different extraction protocols on endotoxin analyses of airborne grain dusts. Scand. J. Work Environ. Health 15, 430–435.
Oppliger, A., et al., 2008. Exposure to bioaerosols in poultry houses at different stages of fattening; use of real-time PCR for airborne bacterial quantification. Ann. Occup. Hyg. 52, 405–412.
Paez-Rubio, T., Viau, E., Romero-Hernandez, S., Peccia, J., 2005. Source bioaerosol concentration and rRNA gene-based identification of microorganisms aerosolized at a flood irrigation wastewater reuse site. Appl. Environ. Microbiol. 71 (2), 804–810.
Pan, Y.L., 2015. Detection and characterization of biological and other organic-carbon aerosol particles in atmosphere using fluorescence. J. Quant. Spectrosc. Radiat. Transf. 150, 12–35.
Pan, M., Eiguren-Fernandez, A., Hsieh, H., Afshar-Mohajer, N., Hering, S.V., Lednicky, J., Hugh Fan, Z., Wu, C.-Y., 2016. Efficient collection of viable virus aerosol through laminar-flow, water-based condensational particle growth. J. Appl. Microbiol. 120 (3), 805–815.
Pan, M., Carol, L., Lednicky, J.A., et al., 2018. Collection of airborne bacteria and yeast through water-based condensational growth. Aerobiologia 34, 337–348.
Pan, M., Carol, L., Lednicky, J.A., Eiguren-Fernandez, A., Hering, S., Fan, Z.H., Wu, C.-Y., 2019. Determination of the distribution of infectious viruses in aerosol particles using water-based condensational growth technology and a bacteriophage MS2 model. Aerosol Sci. Tech. 53 (5), 583–593.
Pankhurst, L.J., Whitby, C., Pawlett, M., Larcombe, L.D., McKew, B., Deacon, L.J., Coulon, F., 2012. Temporal and spatial changes in the microbial bioaerosol communities in green-waste composting. FEMS Microbiol. Ecol. 79 (1), 229–239.
Park, C.W., Yoon, K.Y., Kim, Y.D., Park, J.H., Hwang, J., 2011. Effects of condensational growth on culturability of airborne bacteria: implications for sampling and control of bioaerosols. J. Aerosol Sci. 42, 213–223.

Parker, E.P., Trahan, M.W., Wagner, J.S., Rosenthal, S.E., Whitten, W.B., Gieray, R.A., et al., 2000. Detection and classification of individual airborne microparticles using laser ablation mass spectroscopy and multivariate analysis. Field Anal. Chem. Technol. 4 (1), 31–42.
Pasquarella, C., Pitzurra, O., Savino, A., 2000. The index of microbial air contamination. J. Hosp. Infect. 46, 241–256.
Pasquarella, C., Albertini, R., Dall'aglio, P., Saccani, E., Sansebastiano, G.E., Signorelli, C., 2008. Air microbial sampling: the state of the art. Ig. Sanita Pubbl. 64, 79–120.
Pasquarella, C., Vitali, P., Saccani, E., 2012. Microbial air monitoring in operating theatres: experience at the University Hospital of Parma. J. Hosp. Infect. 81, 50–57.
Paudyal, P., Semple, S., Niven, R., Tavernier, G., Ayres, J.G., 2011. Exposure to dust and endotoxin in textile processing workers. Ann. Occup. Hyg. 55, 403–409.
Pauvert, C., Buee, M., Laval, V., Edel-Hermann, V., Fauchery, L., Gautier, A., Lesur, I., Vallance, J., Vacher, C., 2019. Bioinformatics matters: the accuracy of plant and soil fungal community data is highly dependent on the metabarcoding pipeline. Fungal Ecol. 41, 23–33.
Pearson, C., Littlewood, E., Douglas, P., Robertson, S., Gant, T.W., Hansell, A.L., 2015. Exposures and health outcomes in relation to bioaerosol emissions from composting facilities: A systematic review of occupational and community studies. J. Toxicol. Environ. l Health Part B 18 (1), 43–69.
Peccia, J., Hernandez, M., 2006. Incorporating polymerase chain reaction-based identification, population characterization, and quantification of microorganisms into aerosol science: a review. Atmos. Environ. 40 (21), 3941–3961.
Peden, D., Reed, C.E., 2010. Environmental and occupational allergies. J. Allergy Clin. Immunol. 125, 150–160.
Pedroso, A.A., et al., 2006. Intestinal bacterial community and growth performance of chickens fed diets containing antibiotics. Poult. Sci. 85, 747–752.
Pegas, P.N., Evtyugina, M.G., Alves, C.A., 2010. Outdoor/indoor air quality in primary schools in Lisbon: a preliminary study. Quim Nova. 33, 1145–1149.
Persoons, R., Parat, S., Stoklov, M., Perdrix, A., Maitre, A., 2010. Critical working tasks and determinants of exposure to bioaerosols and MVOC at composting facilities. Int. J. Hyg. Environ. Health 213, 338–347.
Pillai, S.D., Ricke, S.C., 2002. Bioaerosols from municipal and animal wastes: background and contemporary issues. Can. J. Microbiol. 48, 681–696.
Polizzi, V., Adams, A., Malysheva, S.V., De Saeger, S., Van Peteghem, C., Moretti, A., et al., 2012. Identification of volatile markers for indoor fungal growth and chemotaxonomic classification of Aspergillus species. Fungal Biol. UK 116 (9), 941–953.
Poole, C.J.M., Wong, M., 2013. Allergic bronchopulmonary aspergillosis in garden waste (compost) collectors—occupational implications. Occup. Med. 63 (7), 517–519.
Posada-Cespedes, S., Seifert, D., Beerenwinkel, N., 2017. Recent advances in inferring viral diversity from high-throughput sequencing data. Virus Res. 239, 17–32.
Prussin, A.J., Marr, L.C., Bibby, K.J., 2014. Challenges of studying viral aerosol metagenomics and communities in comparison with bacterial and fungal aerosols. FEMS Microbiol. Lett. 357 (1), 1–9.
Qian, Y., Willeke, K., Ulevicius, V., Grinshpun, S.A., Donnelly, J., 1995. Dynamic size spectrometry of airborne microorganisms: laboratory evaluation and calibration. Atmos. Environ. 29, 1123–1129.
Radon, K., et al., 2001. Exposure assessment and lung function in pig and poultry farmers. Occup. Environ. Med. 58, 405–410.
Radon, K., Danuser, B., Iversen, M., Monso, E., Weber, C., Hartung, J., et al., 2002. Air contaminants in different European farming environments. Ann. Agric. Environ. Med. 9 (1), 41–48.

Rahav, E., Belkin, N., Paytan, A., Herut, B., 2019. The relationship between air-mass trajectories and the abundance of dust-borne prokaryotes at the SE Mediterranean sea. Atmosphere 10 (5), 280.

Rahkio, T., Korkeala, H., 1997. Airborne bacteria and carcass contamination in slaughterhouses. J. Food Prot. 60, 38–42.

Rahmani, A.R., Leili, M., Azarian, G., Poormohammadi, A., 2020. Sampling and detection of corona viruses in air: a mini review. Sci. Total Environ. 740, 140207.

Rao, C.Y., Burge, H.A., Chang, J.C., 1996. Review of quantitative standards and guidelines for fungi in indoor air. J. Air Waste Manage. Assoc. 46, 899–908.

Recer, G.M., Browne, M.L., Horn, E.G., Hill, K.M., Boehler, W.F., 2001. Ambient air levels of *Aspergillus fumigatus* and thermophilic actinomycetes in a residential neighborhood near a yard-waste composting facility. Aerobiologia 17, 99–108.

Redente, E.F., Massengale, R.D., 2006. A systematic analysis of the effect of corn, wheat, and poultry dusts on interleukin-8 production by human respiratory epithelial cells. J. Immunotoxicol. 3, 31–37.

Reponen, T., Seo, S.C., Grimsley, F., Lee, T., Crawford, C., Grinshpun, S.A., 2007. Fungal fragments in moldy houses: a field study in homes in New Orleans and Southern Ohio. Atmos. Environ. 41 (37), 8140–8149.

Reponen, T., Willeke, K., Grinshpun, S., 2011. Biological particle sampling. In: Kulkarni, P., Baron, P.A., Willeke, K. (Eds.), Aerosol Measurement: Principles, Techniques, and Applications. Wiley, New York.

Richter, J., Svozil, V., Král, V., Dobiášová, L.R., Vetvicka, V., 2015. β-Glucan affects mucosal immunity in children with chronic respiratory problems under physical stress: clinical trials. Ann. Transl. Med. 3 (4), 52–59.

Robertson, S., Douglas, P., Jarvis, D., Marczylo, E., 2019. Bioaerosol exposure from composting facilities and health outcomes in workers and in the community: a systematic review update. Int. J. Hyg. Environ. Health 222 (3), 364–386.

Rosario, K., Fierer, N., Miller, S., Luongo, J., Breitbart, M., 2018. Diversity of DNA and RNA viruses in indoor air as assessed via metagenomic sequencing. Environ. Sci. Technol. 52 (3), 1014–1027.

Roux, J.M., Sarda-Estève, R., Delapierre, G., Nadal, M.H., Bossuet, C., Olmedo, L., 2016. Development of a new portable air sampler based on electrostatic precipitation. Environ. Sci. Pollut. Res. 23 (9), 8175–8183.

Rule, A.M., Kesavan, J., Schwab, K.J., Buckley, T.J., 2007. Application of flow cytometry for the assessment of preservation and recovery efficiency of bioaerosol samplers spiked with *Pantoea agglomerans*. Environ. Sci. Technol. 41, 2467–2472.

Rule, A.M., Evans, S.L., Silbergeld, E.K., 2008. Food animal transport: a potential source of community exposures to health hazards from industrial farming (CAFOs). J. Infect. Public Health 1, 33–39.

Rylander, R., 1999. Health effects among workers in sewage treatment plants. Occup. Environ. Med. 56, 354–357.

Ryskevic, N., et al., 2010. Concept design of a UV light-emitting diode based fluorescence sensor for real-time bioparticle detection. Sens. Actuators B 148 (2), 371–378.

Samadi, S., Heederik, D.J., Krop, E.J., Jamshidifard, A.R., Willemse, T., Wouters, I.M., 2010. Allergen and endotoxin exposure in a companion animal hospital. Occup. Environ. Med. 67 (7), 486–492.

Samadi, S., Wouters, I.M., Heederik, D.J., 2013. A review of bio-aerosol exposures and associated health effects in veterinary practice. Ann. Agric. Environ. Med. 20, 206–221.

Sánchez-Monedero, M.A., Aguilar, M.I., Fenoll, R., Roig, A., 2008. Effect of the aeration system on the levels of airborne microorganisms generated at wastewater treatment plants. Water Res. 42 (14), 3739–3744.

Sattar, S.A., 2016. Indoor air as a vehicle for human pathogens: Introduction, objectives, and expectation of outcome. Am. J. Infect. Control 44 (9), S95–S101.

Sauvageat, E., Zeder, Y., Auderset, K., Calpini, B., Clot, B., Crouzy, B., Konzelmann, T., Lieberherr, G., Tummon, F., Vasilatou, K., 2020. Real-time pollen monitoring using digital holography. Atmos. Meas. Tech. 13 (3), 1539–1550. https://doi.org/10.5194/amt-13-1539-2020.

Scaltriti, S., Cencetti, S., Rovesti, S., Marchesi, I., Bargellini, A., Borella, P., 2007. Risk factors for particulate and microbial contamination of air in operating theatres. J. Hosp. Infect. 66, 320–326.

Schafer, M.P., Martinez, K.F., Mathews, E.S., 2003. Rapid detection and determination of the aerodynamic size range of airborne mycobacteria associated with whirlpools. Appl. Occup. Environ. Hyg. 18 (1), 41–50.

Scheff, P.A., Paulius, V.K., Curtis, L., Conroy, L.M., 2000. Indoor air quality in a Middle school, part II: development of emission factors for particulate matter and bioaerosols. Appl. Occup. Environ. Hyg. 15 (11), 835–842.

Schlosser, O., 2019. Bioaerosols and health: current knowledge and gaps in the field of waste management. Detritus 5. https://doi.org/10.31025/2611-4135/2019.13786.

Schlosser, O., Huyard, A., Cartnick, K., Yanez, A., Catalan, V., Do Quang, Z., 2009. Bioaerosol in composting facilities: occupational health risk assessment. Water Environ. Res. 81, 866–877.

Seifert, S.A., Susanna Von Essen, M.D., Jacobitz, K., Randall, R.N., Crouch, R.N., Lintner, C.P., 2003. Organic dust toxic syndrome: a review. J. Toxicol. Clin. Toxicol. 41 (2), 185–193.

Sengupta, A., Laucks, M., Dildine, N., Drapala, E., Davis, E., 2005. Bioaerosol characterization by surface-enhanced Raman spectroscopy (SERS). J. Aerosol Sci. 36 (5–6), 651–664.

Serrano-Silva, N., Calderón-Ezquerro, M.C., 2018. Environmental Pollution 235, 20–29.

Shah, N., Tang, H., Doak, T.G., Ye, Y., 2011. Comparing bacterial communities inferred from 16S rRNA gene sequencing and shotgun metagenomics. Biocomputing 2011, 165–176.

Sharma, A., Gilbert, J.A., 2018. Microbial exposure and human health. Curr. Opin. Microbiol. 44, 79–87.

Shintani, H., Taniai, E., Miki, A., Kurosu, S., Hayashi, F., 2004. Comparison of the trapping efficiency of the airborne microorganisms among air samplers. J. Hosp. Infection 56, 42–48.

Šikoparija, B., 2020. Desert dust has a notable impact on aerobiological measurements in Europe. Aeolian Res. 47, 100636. https://doi.org/10.1016/j.aeolia.2020.100636.

Simon, X., Bau, S., Boivin, A., Duquenne, P., Witschger, O., Görner, P., 2016. Physical performances and kinetics of evaporation of the CIP 10-m personal sampler's rotating cup containing aqueous or viscous collection fluid. Aerosol Sci. Tech. 50 (5), 507–520.

Smith, J., Adams, C.E., King, M.F., Noakes, C.J., Robertson, C., Dancer, S.J., 2018. Is there an association between airborne and surface microbes in the critical care environment? J. Hosp. Infect. 100 (3), e123–e129.

Spaan, S., Doekes, G., Heederik, D., et al., 2008. Effect of extraction and assay media on analysis of airborne endotoxin. Appl. Environ. Microbiol. 74, 3804–3811.

Speight, S.E., Hallis, B.A., Bennett, A.M., Benbough, J.E., 1997. Enzyme-linked immunosorbent assay for the detection of airborne microorganisms used in biotechnology. J. Aerosol Sci. 28 (3), 483–492.

Springston, J.P., Yocavitch, L., 2017. Existence and control of legionella bacteria in building water systems: a review. J. Occup. Environ. Hyg. 14 (2), 124–134.

Stetzenbach, L.D., 2009. Airborne infectious microorganisms. Encycl. Microbiol., 175–182.

Su, W.C., Tolchinsky, A.D., Chen, B.T., Sigaev, V.I., Cheng, Y.S., 2012a. Evaluation of physical sampling efficiency for cyclone-based personal bioaerosol samplers in moving air environments. J. Environ. Monit. 14 (9), 2430–2437.
Su, W.C., Tolchinsky, A.D., Sigaev, V.I., Cheng, Y.S., 2012b. A wind tunnel test of newly developed personal bioaerosol samplers. J. Air Waste Manage. Assoc. 62, 828–837.
Surette, M.D., Wright, G.D., 2017. Lessons from the environmental antibiotic resistome. Annu. Rev. Microbiol. 71, 309–329. https://doi.org/10.1146/annurev-micro-090816-093420.
SUVA, 2007. Grenzwerte am Arbeitsplatz. Luzern, Switzerland.
Swan, J.R.M., Kelsey, A., Crook, B., Gilbert, E.J., 2003. Occupational and Environmental Exposure to Bioaerosols From Composting and Potential Health Effect—A Critical Review of Published Data. Executive, H.A.S. Research Report, p. 130.
Szponar, B., Larsson, L., 2001. Use of mass spectrometry for characterising microbial communities in bioaerosols. Ann. Agric. Environ. Med. 8 (2), 111–117.
Tan, M., Shen, F., Yao, M., Zhu, T., 2011. Development of an automated electrostatic sampler (AES) for bioaerosol detection. Aerosol Sci. Tech. 45 (9), 1154–1160.
Tang, J.W., Bialasiewicz, S., Dwyer, D.E., Dilcher, M., Tellier, R., Taylor, J., Hua, H., Jennings, L., Kok, J., Levy, A., Smith, D., 2021. Where have all the viruses gone? Disappearance of seasonal respiratory viruses during the COVID-19 pandemic. J. Med. Virol. 93 (7), 4099.
Tang, C.-S., Wan, G.-H., 2013. Air quality monitoring of the post-operative recovery room and locations surrounding operating theatres in a medical centre in Taiwan. PLoS One 8 (4), e61093.
Terzieva, S., Donnelly, J., Ulevicius, V., Grinshpun, S.A., Willeke, K., Stelma, G.N., Brenner, K.P., 1996. Comparison of methods for detection and enumeration of airborne microorganisms collected by liquid impingement. Appl. Environ. Microbiol. 62 (7), 2264–2272.
Therkorn, J., Mainelis, G., 2013. Effects of plate agar volume on bioaerosol impactor measurement accuracy. Aerosol Sci. Tech. 47 (12), 1353–1362.
Therkorn, J., Thomas, N., Calderon, L., Scheinbeim, J., Mainelis, G., 2017. Design and development of a passive bioaerosol sampler using polarized ferroelectric polymer film. J. Aerosol Sci. 105, 128–144.
Thilsing, T., Madsen, A.M., Basinas, I., Schlünssen, V., Tendal, K., Bælum, J., 2015. Dust, endotoxin, fungi, and bacteria exposure as determined by work task, season, and type of plant in a flower greenhouse. Ann. Occup. Hyg. 59 (2), 142–157.
Thorn, J., Beijer, L., Rylander, R., 2002. Work related symptoms among sewage workers: a nationwide survey in Sweden. Occup. Environ. Med. 59, 562–566.
Tian, J.H., Yan, C., Nasir, Z.A., Garcia Alcega, S., Tyrrel, S., Coulon, F., 2020. Real time detection and characterisation of bioaerosol emissions from wastewater treatment plants. Sci. Total Environ. 721, 137629.
Toivola, M., Alm, S., Reponen, T., Kolari, S., Nevalainen, A., 2002. Personal exposures and microenvironmental concentrations of particles and bioaerosols. J. Environ. Monit. 4 (1), 166–174.
Tong, X., Xu, H., Zou, L., Cai, M., Xu, X., Zhao, Z., Xiao, F., Li, Y., 2017. High diversity of airborne fungi in the hospital environment as revealed by meta-sequencing-based microbiome analysis. Sci. Rep. 7, 39606.
Toyinbo, O., Hägerhed, L., Dimitroulopoulou, S., Dudzinska, M., Emmerich, S., Hemming, D., Park, J.H., Haverinen-Shaughnessy, U., Committee, S.T., 2022. Open database for international and national indoor environmental quality guidelines. Indoor Air 32 (4), e13028.
Tseng, C., Hsiao, P.K., Chang, K.C., Chen, W.T., Yiin, L.M., Hsieh, C.J., 2014. Optimization of propidium monoazide quantitative PCR for evaluating performances of bioaerosol samplers for sampling airborne Staphylococcus aureus. Aerosol Sci. Technol. 48 (12), 1308–1319. https://doi.org/10.1080/02786826.2014.985780.

Tseng, C.-C., Li, C.-S., 2005. Collection efficiencies of aerosol samplers for virus-containing aerosols. J. Aerosol Sci. 36 (5-6), 593–607.
Uhrbrand, K., Schultz, A.C., Madsen, A.M., 2011. Exposure to airborne noroviruses and other bioaerosol components at a wastewater treatment plant in Denmark. Food Environ. Virol. 3, 130–137.
Unterwurzacher, V., et al., 2018. Validation of a quantitative PCR based detection system for indoor mold exposure assessment in bioaerosols. Environ. Sci. Processes Impacts 20, 1454–1468.
USEPA, 1994. A Standardized EPA Protocol for Characterizing Indoor Air Quality in Large Office Buildings: Indoor Air Division, Washington D.C. and Atmospheric Research and Exposure Assessment Laboratory. Research Triangle Park, NC.
Van Hoek, A.H.A.M., Mevius, D., Guerra, B., Mullany, P., Roberts, A.P., Aarts, H.J.M., 2011. Acquired antibiotic resistance genes: an overview. Front. Microbiol. 2, 203.
Van Leuken, J.P.G., Swart, A.N., Brandsma, J., Terink, W., Van de Kassteele, J., Droogers, P., Sauter, F., Havelaar, A.H., Van der Hoek, W., 2016. Human Q Fever Incidence is Associated to Spatiotemporal Environmental Conditions. 2 One Heal, Amsterdam, Netherlands, pp. 77–87.
Verreault, D., Moineau, S., Duchaine, C., 2008. Methods for sampling of airborne viruses. Microbiol. Mol. Biol. Rev. 72, 413–444.
Verreault, D., Gendron, L., Rousseau, G.M., Veillette, M., Masse, D., Lindsley, W.G., Moineau, S., Duchaine, C., 2011. Detection of airborne lactococcal bacteriophages in cheese manufacturing plants. Appl. Environ. Microbiol. 77 (2), 491–497.
Vesper, S., McKinstry, C., Haugland, R., Wymer, L., Bradham, K., Ashley, P., Cox, D., Dewalt, G., Friedman, W., 2007. Development of an environmental relative moldiness index for US homes. J. Occup. Environ. Med. 49 (8), 829–833.
Vestlund, A.T., et al., 2014. Morphological classification of bioaerosols from composting using scanning electron microscopy. Waste Manag. 34 (7), 1101–1108.
Viani, I., Colucci, M.E., Pergreffi, M., et al., 2020. Passive air sampling: the use of the index of microbial air contamination. Acta Biomed. 91 (3-S), 92–105.
Viegas, C., Monteiro, A., dos Santos, M., Faria, T., Caetano, L.A., Carolino, E., Quintal Gomes, A., Marchand, G., Lacombe, N., Viegas, S., 2018. Filters from taxis air conditioning system: a tool to characterize driver's occupational exposure to bioburden? Environ. Res. 164, 522–529.
Wagner, J., Macher, J., 2012. Automated spore measurements using microscopy, image analysis, and peak recognition of near-monodisperse aerosols. Aerosol Sci. Tech. 46 (8), 862–873.
Waldman, L., Schmitt, K.H., 1996. Thermophoresis and diffusiophoresis of aerosol. In: Aerosol Science. Academic Press, London, United Kingdom, pp. 163–194.
Wallace, R.J., Gropp, J., Dierick, N., Costa, L.G., Martelli, G., Brantom, P.G., et al., 2016. Risks associated with endotoxins in feed additives produced by fermentation. Environ. Health 15, 5–9.
Walls, H.J., Ensor, D.S., Harvey, L.A., Kim, J.H., Chartier, R.T., Hering, S.V., Spielman, S.R., Lewis, G.S., 2016. Generation and sampling of nanoscale infectious viral aerosols. Aerosol Sci. Tech. 50 (8), 802–811.
Walser, S.M., Gerstner, D.G., Brenner, B., Höller, C., Liebl, B., Herr, C.E.W., 2014. Assessing the environmental health relevance of cooling towers—a systematic review of legionellosis outbreaks. Int. J. Hyg. Environ. Health 217, 145–154.
Walser, S.M., Gerstner, D.G., Brenner, B., Bünger, J., Eikmann, T., Janssen, B., Kolb, S., Kolk, A., Nowak, D., Raulf, M., et al., 2015. Evaluation of exposure–response relationships for health effects of microbial bioaerosols–a systematic review. Int. J. Hyg. Environ. Health 218 (7), 577–589.
Wan, G.-H., Huang, C.-G., Huang, Y.-C., 2012. Surveillance of airborne adenovirus and *Mycoplasma pneumoniae* in a hospital paediatric department. PLoS One 7 (3), e33974.

Wang, C.-H., Chen, B.T., Han, B.-C., Liu, A.-C.-Y., Hung, P.-C., Chen, C.-Y., Chao, H.J., 2015. Field evaluation of personal sampling methods for multiple bioaerosols. PLoS One 10 (3), e0120308.

Watson, H.H., 1958. The sampling efficiency of the thermal precipitator. Br. J. Appl. Phys. 9, 78–79.

Welker, M., Moore, E.R., 2011. Applications of whole-cell matrix-assisted laser-desorption/ionization time-of-flight mass spectrometry in systematic microbiology. Syst. Appl. Microbiol. 34 (1), 2–11.

Wery, N., 2014. Bioaerosols from composting facilities—a review. Front. Cell. Infect. Microbiol. 4, 42.

Wheeler, P.A., Stewart, I., Dumitrean, P., Donovan, B., 2001. Health Effects of Composting. A Study of Three Compost Sites and Review of Past Data. R&D Technical Report P1-315/TR, Environment Agency, UK.

Whitesides, G.M., 2006. The origins and the future of microfluidics. Nature 442, 368–373.

WHO, 2009. In: Heseltine, E., Rosen, J. (Eds.), Guidelines for Indoor Air Quality: Dampness and Mould. www.euro.who.int/_data/assets/pdf_file/0017/43325/E92645.pdf.

Whyte, R.T., 1993. Aerial pollutants and the health of poultry farmers. Worlds Poult. Sci. J. 49, 139–153.

Willeke, K., Baron, P.A., 1993. Aerosol Measurement: Principle, Techniques and Applications. John Wiley & Sons, New York.

Willeke, K., Macher, J.M., 1999. Section 11. Air sampling. In: Macher, J. (Ed.), Bioaerosols Assessment and Control. American Conference of Governmental Industrial Hygienists, Cincinnati.

Willeke, K., Grinshpun, S.A., Ulevicius, V., Terzieva, S., Donnelly, J., Stewart, and A. Juozaitis., 1995. Microbial stress, bounce and re-aerosolization in bioaerosol samplers. J. Aerosol Sci. 26, S883–S884.

Willeke, K., Lin, X., Grinshpun, S.A., 1998. Improved aerosol collection by combined impaction and centrifugal motion. Aerosol Sci. Tech. 28 (5), 439–456.

Williams, M.R., Stedtfeld, R.D., Guo, X., Hashsham, S.A., 2016. Antimicrobial resistance in the environment. Water Environ. Res. 88, 1951–1967.

Williams, B., Douglas, P., Roca Barcelo, A., Hansell, A.L., Hayes, E., 2019. Estimating Aspergillus fumigatus exposure from outdoor composting activities in England between 2005 and 14. Waste Manag. 1 (84), 235–244.

Womack, A.M., Artaxo, P.E., Ishida, F.Y., Mueller, R.C., Saleska, S.R., Wiedemann, K.T., et al., 2015. Characterization of active and total fungal communities in the atmosphere over the Amazon rainforest. Biogeosciences 12 (21), 6337–6349.

Woodhead, K., Taylor, E.W., Bannister, G., Chesworth, T., Hoffman, P., Humphreys, H., 2002. Behaviours and rituals in the operating theatre. J. Hosp. Infect. 51, 241–255.

Xie, S., Gu, A.Z., Cen, T., Li, D., Chen, J., 2019. The effect and mechanism of urban fine particulate matter (PM2.5) on horizontal transfer of plasmid-mediated antimicrobial resistance genes. Sci. Total Environ. 683, 116–123.

Xiong, H., Ye, X., Li, Y., Qi, J., Fang, X., Kong, J., 2021. Efficient microfluidic-based air sampling/monitoring platform for detection of aerosol SARS-CoV-2 on-site. Anal. Chem. 93 (9), 4270–4276.

Yamamoto, N., Hikono, M., Koyama, H., Kumagai, K., Fujii, M., Yanagisawa, Y., 2006. A passive sampler for airborne coarse particles. J. Aerosol Sci. 37 (11), 1442–1454.

Yamamoto, N., Matsuki, H., Yanagisawa, Y., 2007. Application of the personal aeroallergen sampler to assess personal exposures to Japanese cedar and cypress pollens. J. Exp. Sci. Environ. Epidemiol. 17 (7), 637.

Yamamoto, N., Kimura, M., Matsuki, H., Yanagisawa, Y., 2010. Optimization of a real-time PCR assay to quantitate airborne fungi collected on a gelatin filter. J. Biosci. Bioeng. 109 (1), 83–88.

Yang, W., Elankumaran, S., Marr, L.C., 2011. Concentrations and size distributions of airborne influenza A viruses measured indoors at a health centre, a day-care centre and on aeroplanes. J. R. Soc. Interface 8 (61), 1176–1184.

Yang, X., Wang, X., Zhang, Y., Lee, J., Su, J., Gates, R.S., 2013. Monitoring total endotoxin and (1→3)-beta-D-glucan at the air exhaust of concentrated animal feeding operations. J. Air Waste Manag. Assoc. 63 (10), 1190–1198.

Yao, M., Mainelis, G., 2006. Investigation of cut-off sizes and collection efficiencies of portable microbial samplers. Aerosol Sci. Tech. 40 (8), 595–606.

Yoo, K., Lee, T.K., Choi, E.J., Yang, J., Shukla, S.K., Hwang, S.I., Park, J., 2017. Molecular approaches for the detection and monitoring of microbial communities in bioaerosols: a review. J. Environ. Sci. 51, 234–247.

Zare, R.N., 2012. My life with LIF: a personal account of developing laser-induced fluorescence. Annu. Rev. Anal. Chem. 5, 1–14.

Zemouri, C., de Soet, H., Crielaard, W., Laheij, A., 2017. A scoping review on bio-aerosols in healthcare and the dental environment. PLoS One 12 (5), e0178007.

Zhen, H., Han, T., Fennell, D., Mainelis, G., 2013. Release of free DNA molecules by the membrane-impaired bacteria during aerosolization and sampling. Appl. Environ. Microbiol. 77, 7780–7789.

Zhen, H., Krumins, V., Fennell, D.E., Mainelis, G., 2018. Analysis of airborne microbial communities using 16s ribosomal RNA: potential bias due to air sampling stress. Sci. Total Environ. 621, 939–947.

Zuo, Z., Kuehn, T.H., Verma, H., Kumar, S., Goyal, S.M., Appert, J., et al., 2013. Association of airborne virus infectivity and survivability with its carrier particle size. Aerosol Sci. Tech. 47 (4), 373–382.

CHAPTER FOUR

A microbial solution to oil sand pollution: Understanding the microbiomes, metabolic pathways and mechanisms involved in naphthenic acid (NA) biodegradation

Corinne Whitby*
School of Life Sciences, University of Essex, Colchester, Essex, United Kingdom
*Corresponding author: e-mail address: cwhitby@essex.ac.uk

Contents

1. Introduction 232
   1.1 The challenges with naphthenic acids (NAs) 234
2. Natural vs anthropogenic naphthenic acid (NA) exposed microbiomes 237
   2.1 Natural exposed microbiomes: Freshwater sediments 239
   2.2 Groundwaters 243
   2.3 Marine ecosystems 245
   2.4 Bitumen saturated outcrop deposits 245
   2.5 Deep oil sand deposits 247
3. Anthropogenic naphthenic acid (NA) contaminated microbiomes 247
   3.1 Oil sand tailings ponds (OSTPs) 249
   3.2 Oil sands process affected water (OSPW) 253
   3.3 Mature fine tailings (MFT) 254
   3.4 End pit lakes (EPLs) and reclaimed ponds 256
   3.5 Biofilms, bioreactors and biofilters 260
4. Biodegradation of naphthenic acids (NAs) 263
   4.1 Factors affecting naphthenic acid (NA) biodegradation rates 264
   4.2 Aerobic vs anaerobic biodegradation of naphthenic acids (NAs) 265
   4.3 Naphthenic acid (NA)-degrading microorganisms 267
   4.4 Metabolic pathways of naphthenic acid (NA) biodegradation 268
   4.5 Mechanisms involved in naphthenic acid (NA) biodegradation and detoxification 269
5. Conclusions 273
6. Future perspectives 274
References 275

Advances in Ecological Research, Volume 67
ISSN 0065-2504
https://doi.org/10.1016/bs.aecr.2022.10.001

Abstract

Bituminous oil sands arise from the microbial oxidation of petroleum hydrocarbons over geological time, with one of the largest deposits found in northern Alberta, Canada. Associated with bitumen are mixtures of toxic and persistent organic acids, known as naphthenic acids (NAs). NAs can be found in various natural environments that are in contact with the bitumen, such as river and wetland sediments. However, NAs may also accumulate through anthropogenic activities that follow oil sand mining operations. During bitumen extraction, vast quantities of toxic oil sands process-affected water (OSPW), are generated that are stored in large oil sands tailings ponds (OSTPs), often for decades and cause serious impacts to the local environment. Bioremediation of NAs using microorganisms, has clear cost and environmental advantages, yet there is a critical lack of fundamental knowledge of the functional microbiome in these NA associated ecosystems. This chapter seeks to unravel the complex microbiomes present in natural and anthropogenic bitumen-rich environments, including the genes, enzymes and mechanisms involved in NA biodegradation.

Abbreviations

AEOs	Acid extractable organics
AOSR	Athabasca Oil Sands Region
BML	Base mine lake
CT	Composite (consolidated) tailings
CWTSs	Constructed wetland treatment systems
DGGE	Denaturing gradient gel electrophoresis
EPL	End pit lakes
ETF	External tailings facilities
ETA	External tailings area
FFT	Fluid fine tailings, having low solids content (e.g. $\leq$ 8 wt%)
MLSB	Mildred Lake Settling Basin
MFT	Mature fine tailings, having solids content $\geq$ 30 wt%
NAs	Naphthenic acids
NAFCs	Naphthenic acid fraction compounds (NAFCs)
OSPW	Oil sands process-affected water
OSTPs	Oil sands tailings ponds
PAHs	Polycyclic aromatic hydrocarbons
SAGD	Steam-assisted gravity drainage
TAN	Total acid number
TT	Thickened tailings
WSO	Water-soluble organics

1. Introduction

With increasing global demand for energy and the rapid depletion of conventional light crude-oil resources, unconventional oil resources, such as heavy oils, oil sands and bitumen are receiving growing interest as future energy sources (Zhang et al., 2020b). Bituminous oil sands are formed as a result of the partial oxidation of petroleum hydrocarbons by

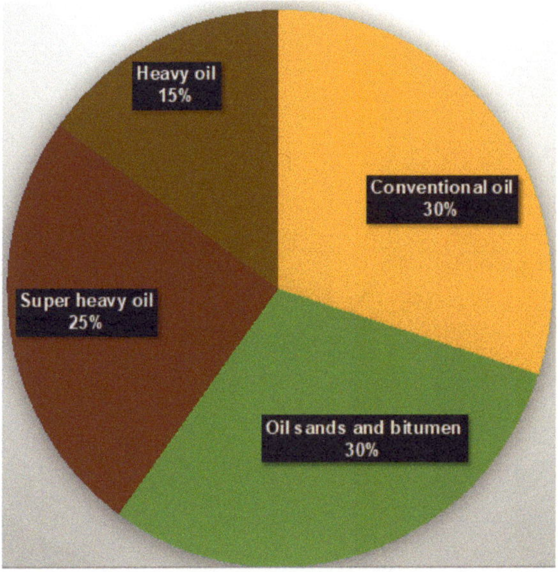

Fig. 1 Percentage of global oil reserves (conventional and unconventional). *Modified from Zhang et al. (2020b)*

microorganisms over geological time (Head et al., 2003). It is currently estimated that there are > 6 trillion barrels of heavy and super-heavy crude oils, bitumen and oil/tar sands worldwide, which together account for ~70% of total global oil reserves (Fig. 1) (Zhang et al., 2020b).

The largest known oil sand deposits are found in North and South America, with the largest in Venezuela, followed closely by the Athabasca Oil Sands Region (AOSR) in Northern Alberta, Canada (Fig. 2A) (Head et al., 2003). It has been estimated that > 169 billion barrels of recoverable bitumen is present in the AOSR (Brown and Ulrich, 2015; Chalaturnyk et al., 2002).

Compared to conventional crude oils, bitumen has fewer saturates, but contains more resins and asphaltenes (Woods et al., 2008). Bitumen-rich ores co-occur with clays, sand and silt (~10% bitumen, ~5% water and ~85% quartz sand, silt and clays) (Chalaturnyk et al., 2002). To extract the bitumen from surface-mined ores, the Clark hot water process is generally used, which involves mixing crushed ore with hot, alkaline water (historically ~ 70–80 °C; but now ~ 50 °C; pH ~8.5) (Foght et al., 2017). The process produces vast quantities of toxic wastewaters, known as oil sands process-affected water (OSPW), which have to be stored in large ponds, called oil sands tailings ponds (OSTPs) (Fig. 2B). These OSTPs are often in operation for decades, until approved strategies are available to effectively treat and safely release the OSPW back into the environment (Whitby, 2010).

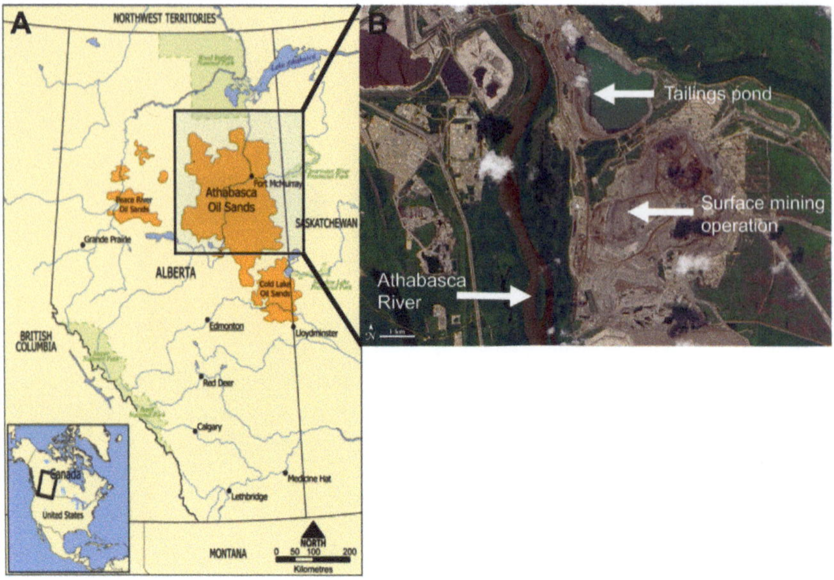

Fig. 2 (A) Map of the AOSR showing Peace River oil sands in western Alberta, Cold Lake oil sands and the larger Athabasca oil sands; (B) Athabasca oil sands mining operations. *Panel A: Modified from N. Einstein. Panel B: Modified from NASA Earth Observatory image 2009 (NASA, 2009).*

It is currently estimated that ~ 1 billion m^3 of tailings have accumulated in the AOSR, and this volume is set to increase as new operations are being developed, and until reclamation or their removal occurs (Foght et al., 2017).

Microorganisms play a fundamental role in both natural, and anthropogenic ecosystems that are associated with NAs. The *in situ* activities of microorganisms are not only important in the reclamation, removal and detoxification of NA-contaminated environments, but their metabolic processes also shape the very nature of the oil sand deposits themselves. Yet their importance and functional role in these environments is often overlooked, and there remains a lack of knowledge on the NA biodegradation mechanisms and the microorganisms involved. Recent advances in analytical and molecular techniques, however, has started to unravel the functional microbiomes found in these natural and anthropogenic ecosystems that are associated with NAs.

1.1 The challenges with naphthenic acids (NAs)

OSPW is known to contain sand, clay, unrecoverable bitumen, inorganic and a complex mixture of toxic, organic compounds termed "*Naphthenic*

Acids (NAs)" (Mikula et al., 1996; Whitby, 2010). NAs were traditionally defined as mixtures of acyclic and cyclic, saturated and (to a lesser extent) aromatic carboxylic acids with the empirical formula $C_nH_{2n+z}O_2$, where *n* represents the number of carbons and *Z* is either zero or a negative even integer representing "hydrogen deficiency" lost due to double-bond or ring formation (Fig. 3A and B) (Barros et al., 2022; Whitby, 2010). OSPW is also known to contain bi-, tri-, tetra-, penta-cyclic, mono-carboxylic and diamondoid carboxylic acids (Fig. 3C) (Rowland et al., 2011a, 2011b, 2011c; Wang et al., 2013; Wilde and Rowland, 2018; Wilde et al., 2015). Indeed, tricyclic (and bicyclic) diamondoid carboxylic acids are major components of OSPW, and unrefined oil sands (Rowland et al., 2011b; Bowman et al., 2014). In addition, oxy-NAs (Barrow et al., 2009; Wang et al., 2013), and structures with heteroatoms such as nitrogen and sulphur have been found (Bauer et al., 2015; Grewer et al., 2010; Hewitt et al., 2020). NAs that do not fit the "classical" formula are referred to as Naphthenic Acid Fraction Compounds (NAFCs) (Vander Meulen et al., 2021), which includes unsaturated and aromatic NA derivatives, acids with increased oxygen content, and compounds containing nitrogen and/or sulphur (Yang et al., 2019; Grewer et al., 2010). Current terminology also refer to the acid extractable organics (AEOs); water-soluble organic (WSO) and

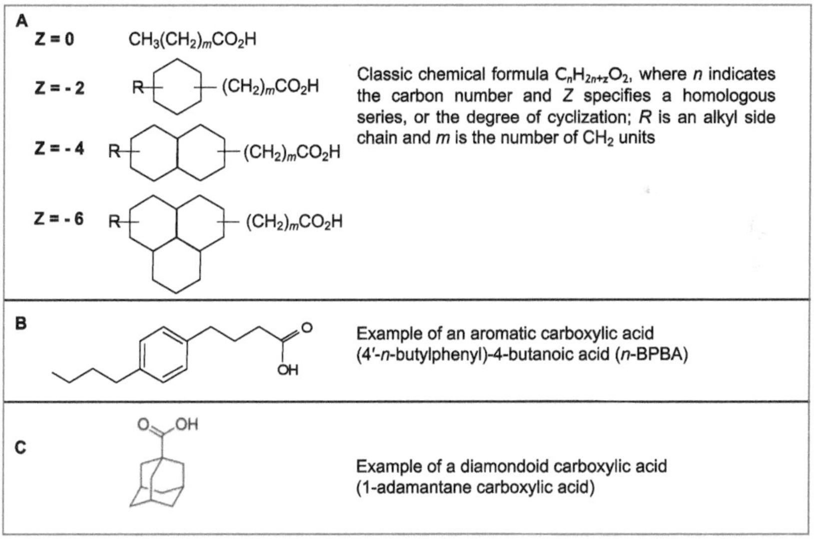

Fig. 3 Example structures of naphthenic acids (NAs) (A) NAs following the classical chemical formula; (B) example of an aromatic NA; (C) example of a diamondoid carboxylic acid. *Modified from Whitby (2010); Folwell et al. (2020); McKew et al. (2021).*

oil sands tailings water acid-extractable organics (OSTWAEO) (Yang et al., 2019; Grewer et al., 2010).

NAs pose several environmental challenges. For example, they demonstrate both acute and chronic toxicity to a variety of different organisms (Li et al., 2017; Morandi et al., 2015; Headley and McMartin, 2004; Frank et al. 2009; Beddow et al. 2016). NAs are highly toxic to fish at concentrations > 2.5–5 $mg L^{-1}$ (Dorn, 1992), and concentrations up to 2.8 $mg kg^{-1}$ have been detected in rainbow trout (*Oncorhynchus mykiss*) (Young et al., 2007). Toxicity effects in fish include reduced hatch success, a decrease in embryonic heart rates and cardiovascular abnormalities (Marentette et al., 2015). They are also known to act as endocrine disruptors (Rowland et al., 2011b). The degree of toxicity is generally related to molecular weight with higher molecular weight acids often demonstrating greater acute toxicity, as their increased acid content decreases hydrophobicity making them more bioavailable and facilitating bioaccumulation in cells (Holowenko et al., 2002). Acute toxicity may also be related to the surfactant properties of NAs, whereby NAs penetrate the cell wall, disrupting the membrane lipid bilayer, or change membrane properties (Frank et al., 2009; Schramm et al., 2000). Aromatic NAs, which comprise > 30% of the NAs in OSPW relative to the acyclic acids (Jones et al., 2012), contribute disproportionally to the overall toxicity of NAs (Johnson et al., 2011; Scarlett et al., 2013). Other toxic substances such as aromatic, olefinic, hydroxyl and dibasic acids (which by strict definition are not NAs) may also influence NA toxicity (Whitby, 2010).

NAs also act as oxidizing agents and are highly corrosive, leading to equipment failures, safety and reliability issues (Whitby, 2010). In particular, heterocyclic acids with multiple different heteroatoms are one of the main causes of corrosion during heavy oil processing (Tomczyk et al., 2001). The rate of corrosion depends on several factors, including temperature, the availability of the carboxylic acid group to form metal complexes, molecular size and chemical structure (Clemente and Fedorak, 2005). NAs can also form metal naphthenate precipitates (e.g. calcium naphthenates), which block pipelines (Whitby, 2010). The main fraction of calcium naphthenate deposits are a family of C_{80} isoprenoid tetracids (known as ARN acids) (Baugh et al., 2004) with structures closely resembling glycerol diphytanylglycerol tetraethers (GDGTs), which are core lipids found in some hyperthermophilic methanogens and Crenarchaeota (Albaiges et al., 1985; Chappé et al., 1982; Lutnaes et al., 2007). NAs also increase the total acid number (TAN), which may devalue the oil (Whitby, 2010).

A wide variety of physical and chemical methods have been applied to remove and detoxify NAs, including ozonation and UV irradiation treatments (Scott et al., 2005). However, such methods are expensive, often difficult to implement on a large scale, and may yield little results (MacKinnon and Boerger, 1986). Biologically based treatments, which exploit the degradative ability of microorganisms, have clear cost and environmental advantages to remove and detoxify the NAs (Folwell et al., 2020; Mahdavi et al., 2015; Yue et al., 2016). However, in order to better understand the mechanisms involved in NA biodegradation and the microorganisms involved, we need to further understand the functional role of the complex microbiomes found in the various natural and anthropogenic bitumen-rich environments.

2. Natural vs anthropogenic naphthenic acid (NA) exposed microbiomes

In northern Alberta, Canada, the oil sands deposits underlay a vast stretch of boreal forest with many lakes, streams, rivers, wetlands and groundwaters (Brandt et al., 2013). These natural ecosystems may be in direct contact with the natural bitumen deposits and consequently their biota may become exposed to varying levels of NAs (Del Rio et al., 2006; Clemente and Fedorak, 2005). NAs may also enter aquatic ecosystems via natural processes such as surface run-off, erosion of the riverbank oil deposits and precipitation (Richardson and Dacks, 2019; Ross et al., 2012). However, NAs may also enter natural systems (e.g. groundwaters) as a result of seepage of OSPW from nearby tailings ponds and this is an ongoing environmental threat in the AOSR (Ahad et al., 2013; Fennell and Arciszewski, 2019). Whilst infiltration of OSPW from tailings ponds into groundwaters does occur, seepage into surface waters is less common (Fennell and Arciszewski, 2019) (Fig. 4). However, any groundwater seepage reaching the rivers is rapidly diluted (Ferguson et al., 2009). Currently, the release of contaminants from anthropogenic activities to aquatic ecosystems is principally via atmospheric release such as dust from open areas (e.g. open pits) (Culp et al., 2021; Brook et al., 2019). Specifically, emissions of oxidized sulphur and nitrogen associated with mining activities have been reported, which acidify the surrounding boreal lake ecosystems via acid precipitation (Hazewinkel et al., 2008). This is exacerbated due to increased run-off as a result of land clearing destabilizing the soil matrix and reducing water retention (Connor et al., 2018; Culp et al., 2021). This higher run-off, in-turn, increases the sediment and nutrient loading of

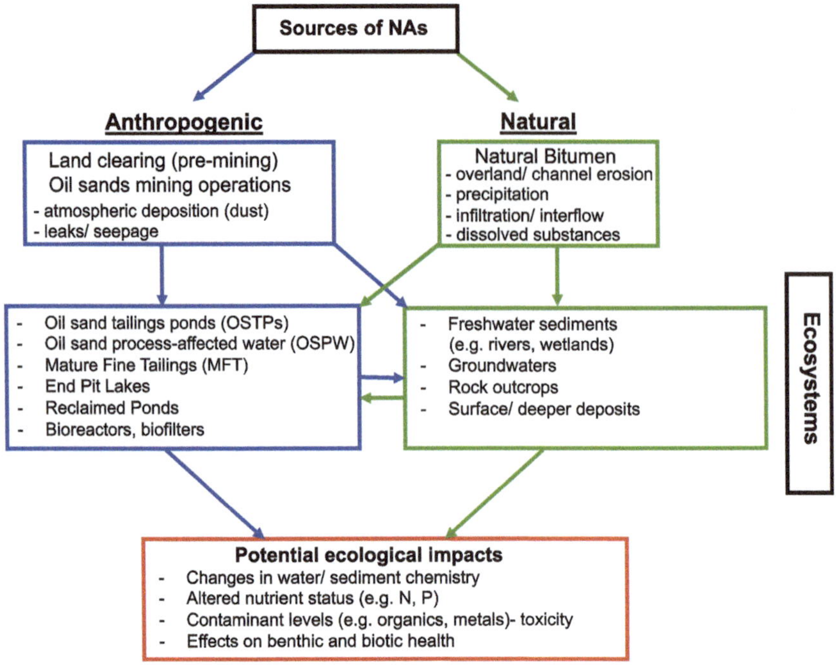

Fig. 4 Anthropogenic vs natural sources of naphthenic acids (NAs) and potential ecological impacts on ecosystems.

receiving waters leading to an increase in total suspended solids and the potential for eutrophication (Connor et al., 2018). Extensive upstream water extraction and river engineering in the oil sands region also reduces the volume of receiving waters in rivers and downstream wetlands, further exacerbating eutrophication as nutrient concentrations increase in the water column (Connor et al., 2018). In addition, the impacts from oil sands development are intensified by reduced water levels as a result of warmer and drier conditions associated with climate change in Canada's northern region (Rouse et al., 1997). Thus, the physicochemical conditions of aquatic ecosystems in the AOSR are driven by both natural processes (e.g. atmospheric deposition, runoff, erosion, groundwater flow) and human activities (e.g. deposition of contaminants, alterations to natural landscapes and water flow) (Culp et al., 2021).

Research into cause-and-effect pathways related to contaminant, sediment and nutrient inputs from oil sands development suggested that the impacts identified on the surrounding ecology appear to be associated with contaminant exposure (Culp et al., 2021). However, more research is required, especially as the source of this exposure is confounded by co-location of, and inability to differentiate between, oil sands mining operations (e.g. atmospheric emissions, seepage), and inputs from the natural bitumen outcrops (e.g. erosional material transported by surface and groundwater) (Culp et al., 2021) (see Chapter "Kinetic experiments for pyrolytic recycling of solid plastic waste" by Jain and Vinu). Assessment of exposure source is further complicated by the complex interplay between the natural environmental conditions, oil sands operations and other anthropogenic activities (e.g. nutrients and contaminants from municipal sewage effluent) (Culp et al., 2021). Critically, there is an urgent need for improved environmental monitoring to assess the potential ecological impacts of current oil sands developments especially any cumulative effects on freshwater resources (Brook et al., 2019). This is even more critical if the direct release of treated OSPW into the surrounding environment occurs in the future (Culp et al., 2021).

2.1 Natural exposed microbiomes: Freshwater sediments

In the area surrounding the Athabasca River and its tributaries, there are natural bituminous sediments that are composed of oil (10–12%), sands (83–85%) and water (4–5%) (Wang et al., 2014). In 2011, the Governments of Canada and Alberta established a Joint Oil Sands Monitoring Plan (JOSM) (now rebranded as the Oil Sands Monitoring Program (OSM) (Arciszewski et al., 2021) for the lower Athabasca River between Fort McMurray and its confluence with Lake Athabasca to monitor several parameters including water quality in the main stem of the Athabasca River, its tributaries, and the deltaic and wetland ecosystems at the river mouth) (Chambers et al., 2018). Tributary waters of the lower Athabasca watershed were characterized as moderately alkaline and moderately hard to hard water, due to the presence of Mg^{2+}, Ca^{2+} and bicarbonate, arising from the underlying layers of shale, sandstone and limestone in the region (Chambers et al., 2018). In the Athabasca River mainstem and its surrounding tributaries variable concentrations of NAs are found (e.g. between 0.1 and $0.9\,mg\,L^{-1}$ in the Athabasca River (Schramm et al. 2000), whilst in other locations NA concentrations may be as high as that found in natural bitumen (i.e. $7994\,\mu g\,g^{-1}$ with C_6 to C_{21} NAs (α-group) making up ~44.2% of the total NAs in Alberta oil sands bitumen) (Sun et al., 2017; Yang et al., 2019). High concentrations of other contaminants have also been found to exceed guidelines downstream of oil sands operations such as metals (e.g. aluminium, copper, iron, lead), and other organic compounds (e.g. pyrene) (Kelly et al., 2009, 2010). In addition, parts of the Athabasca River near the oil sands mining operations are considered nitrogen and phosphorus limited (Chambers et al., 2006; Herman et al., 1994; Lai et al., 1996). More research is required however, to determine whether it is the oil sands activities or whether it is the changing climatic conditions (or both), that are influencing the nutrient status of these aquatic ecosystems in the region (Laird et al., 2017).

Microorganisms such as those found in sediments overlying bitumen deposits are not only under additional nutrient stresses, but are continuously exposed to NAs, and must be able to thrive in these complex and toxic NA-rich environments (Brient et al., 1995). Indeed, water quality has been suggested as an important factor in shaping the microbial community (Bergsveinson et al., 2021). As a result, these ecosystems can harbour diverse microbiomes, and includes microorganisms with the ability to degrade NAs (Del Rio et al., 2006). Although, microorganisms of the Athabasca watershed are well-known to degrade NAs (Del Rio et al., 2006); microorganisms without any prior exposure to NAs may also metabolize NAs; including estuarine sediments (Smith et al., 2008) and soils (Johnson et al., 2011). Thus, more research is needed into the distribution, ecology, and mechanisms of NA-degrading microorganisms found in these various ecosystems.

2.1.1 Lotic systems (river sediments)

Freshwater (river) sediment microbiomes are crucial components of these ecosystems; yet there remains a lack understanding of the effects of NA exposure on the microbiomes present. Specifically, the microbiome of the Athabasca watershed is in continuous contact with NAs through oil sands erosion and oil-upgrading activities (Kelly et al., 2009, 2010; Kurek et al., 2013; Del Rio et al., 2006). Under the zero discharge policy in the region, only small releases have been approved by the regulator (i.e. $< 1\%$ of receiving river flows) (Culp et al., 2021). Although there is no deliberate discharge of contaminated tailings ponds into the nearby rivers and streams, leaching and seepage of tailings water may occur into nearby aquatic ecosystems, which influence their *in situ* microbiomes (Yergeau et al., 2012). Indeed, recent data suggests that OSPW infiltration may be present in the McLean Creek and Lower Beaver River, which are tributaries of the Athabasca River that flows through the AOSR and is adjacent to many tailings ponds (Fennell and Arciszewski, 2019).

Previous work has shown that proximity to tailings ponds is an important factor in shaping the bacterial and archaeal communities present in the Athabasca River sediments (Yergeau et al., 2012). Indeed, the river sediment microbiomes located close to the ponds were found to be more similar to each other (and to those from the tailings ponds), compared to the sediment microbiomes located further away (Yergeau et al., 2012). Generally, bacterial diversity was lower in sediments heavily disturbed by mining operations, and significant differences between bacterial communities of bitumen-associated and non-bitumen-associated streams from the Athabasca River and its tributaries (e.g. Firebag Creek, Steepbank Creek, and Ells River) have also been reported (Yergeau et al., 2012). Yet, diverse bacterial and eukaryotic communities have still been found in the Athabasca River, regardless of the degree of NA contamination (Bergsveinson et al., 2021).

Taxa found in sediments of the Athabasca River and its tributaries included *Pseudomonadota* (e.g. *Pseudomonas* sp., *Xanthomonas* sp.), with Alphaproteobacteria and Betaproteobacteria dominating, except in Steepbank Creek and the upper Athabasca tributary where Deltaproteobacteria dominated (Yergeau et al., 2012). *Bacteroidota, Bacillota,* and *Chloroflexota* have also been found in relatively high abundances (Yergeau et al., 2012). Moreover, all members of the *Actinomycetota,* Betaproteobacteria, and *Bacteroidota* correlated with the highest concentrations of PAHs and NAs in the river sediments, suggesting that these taxa could be used as putative bio-indicators of contamination in the region (Yergeau et al., 2012). *Mycobacterium* sp., *Nocardia* sp., and the yeast *Rhodotorula* sp. were also found in the Athabasca River region (Wyndham and Costerton, 1981a,b). High abundances of Chitinophagaceae, *Rhizobiales,* Rhodocyclaceae, Comamonadaceae, and Sphingomonadaceae were found in river and creek bed sediments from along the Horse River and Saline Creek (Wong et al., 2015).

Indeed significant differences in the abundance of specific genera have been found among the different habitats across the Athabasca River region (Bergsveinson et al., 2021). Moreover, analysis of NA concentration and profiles with water quality variables showed two tributaries to the Athabasca River- Beaver River and McLean Creek as possibly receiving OSPW seepage, demonstrating the importance of monitoring ambient water quality (Ross et al., 2012). There is also evidence to suggest that the river sediment community had adapted to the presence of hydrocarbons even before the influence of oil sands mining (Reid et al., 2018). Moreover, several processes have been found in these hydrocarbon-rich freshwater sediments; including sulphate- and nitrate-reduction, methanogenic and methanotrophic activities, suggesting an ecologically active microbial community is present in these river sediments (Reid et al., 2018).

2.1.2 Lentic systems (wetland sediments)

Natural wetlands provide many ecosystem services, such as water and carbon storage, climate regulation and ecological protection (Liu et al., 2022). In the AOSR, wetlands dominate the landscape, covering over 50,500 km^2 and equating to around 54% of total land cover (Volik et al., 2020). Generally, these wetlands have a low buffering capacity and are particularly vulnerable to acidification (Hazewinkel et al., 2008). Underlying these wetlands are poorly buffered soils where organic carbon is typically high (e.g. 25.05 ± 1.00 mg DOC L^{-1} and 28.74 ± 1.61 mg TOC L^{-1}) (Connor et al., 2018). The elevated levels of organic carbon may be attributed to the oil sands as a result of atmospheric deposition or surface water contributions from the upstream oil sands development (Connor et al., 2018). Thus, the rapid development of oil sands within the region is challenging ecosystem functionality. NA concentrations of < 45–$80\,mg\,L^{-1}$ in OSPW affected wetlands have been measured, compared to $< 2\,mg\,L^{-1}$ for non-OPSW affected wetlands (Del Rio et al., 2006; Pollet & Bendell-Young, 2000). Moreover, wetlands are generally established on land that has been significantly disturbed following mining operations, that consequently may include little or no vegetation (Foght et al., 2017). Yet despite their ecological importance, little is currently known about the microbiomes found in wetlands associated with NAs. Previous studies which were largely based on obtaining isolates from enrichment cultures derived from wetland sediments, have made some headway and identified microorganisms such as *Pseudomonas putida* and *Pseudomonas fluorescens* (Del Rio et al., 2006). Although both of these two microbes are very cosmopolitan species and are often found in culture-based studies (e.g. Johnson et al., 2013) and so may not be indicative of the composition of wetland microbiomes. However, there currently remains a lack of information on the *in situ* microbiomes of wetlands.

Many of the wetlands in the vicinity of the Athabasca oil sands are naturally exposed to low levels of bitumen and therefore have a history of long-term exposure to the associated NAs therein (Headley et al., 2000). Distinct microbial communities have been found in wetlands that receive process-affected water, compared to more heterogenous communities that are present in pristine wetlands with no previous exposure to NAs (Hadwin et al., 2006). Once wetlands are exposed to even low levels of OSPW (i.e. low-impact sites), the microbial communities may degrade certain NAs such as bicyclic NAs, (i.e. those belonging to the $Z = -4$ family) (Del Rio et al., 2006). Yet significant degradation of the bicyclic NA decahydronaphthoic acid (DHNA) was found to only occur when the microbial communities were exposed to process-affected water (Hadwin et al., 2006). With monocyclic NAs, such as cyclohexane carboxylic acid (CCA), wetland microbiomes were equally capable of their degradation, regardless of prior exposure to OSPW (Del Rio et al., 2006).

In general, the microorganisms found in NA exposed wetlands degrade NAs much faster than communities from non-exposed wetlands (Del Rio et al., 2006; Hadwin et al., 2006; Headley et al., 2000). Despite wetland communities being able to reduce the total NAs in OSPW, in some cases long residence times (i.e. 400 days) may be required before significant NA reduction occurs (Toor et al., 2013a). Furthermore, the more persistent NAs that are associated with residual chronic toxicity can often remain (Toor et al., 2013b). In some wetland microbiomes, nitrification may also inhibit NA mineralization as NA-degrading communities may be outcompeted by nitrifiers for available oxygen and nutrients, such as nitrogen and phosphorus (Lai et al., 1996). NA concentration has also been suggested as a major factor influencing wetland sediment microbiomes (Hadwin et al., 2006). For example, one study reported a shift towards *Actinomycetes* in wetlands associated with a decrease in NA metabolism (Hadwin et al., 2006). Although exposure to above ambient levels of NAs may shift the community structure to one that is capable of degrading NAs, the time required for this to alter the community is currently unknown and such information is crucial for wetland reclamation strategies. Indeed, bioaugmentation with NA-degrading sediment communities may be advantageous for wetland remediation (Hadwin et al., 2006). Currently there is limited information on the diversity of communities in these wetland ecosystems and more research is required to better understand the community dynamics of wetland microbiomes in relation to NA exposure and biodegradation.

2.2 Groundwaters

Bitumen reservoirs are often naturally associated with groundwater and/or aquifers which may be freshwater or brackish (Foght et al., 2017). However, a major environmental concern associated with the Athabasca oil sands mining operations is seepage of NAs from OSPW and tailings ponds into surface and/or groundwaters (Ahad et al., 2018; Sun et al., 2017; Fennell and Arciszewski, 2019). Ferguson et al., (2009) showed that leakage from the tailings pond and dkye system into underlying aquifers was possible. Indeed, infiltration of OSPW into groundwaters near some ponds has been reported (Fennell and Arciszewski, 2019) and high concentrations of NAs have been found in some groundwater aquifers (> $55\,mg\,L^{-1}$; Kilgour et al., 2019). Specifically, the tricyclic diamondoid acids have been found to persist in groundwaters following seepage from a low lying wetland site (Ahad et al., 2018). Recently however, it has been suggested that the clay and sand in the sediments underlying ponds may provide a protective mechanism reducing the infiltration of NAs into the sand-sediment layer, where groundwater channels occur (Lv et al., 2020). Leakage may also be restricted via the development of fine sediment layers at the bottom of these ponds as the tailings settle (Ferguson et al., 2009). Lv et al., (2020) analysed the microbial community of the clay sediment layer beneath an OSTP and showed that different electron acceptors, sediment types, and NAs sources were associated with specific microbial taxa and these factors explained the variation of microbial community structures observed. In addition, the importance of groundwater labile carbons in NA biodegradation was also reported (Lv et al., 2020).

To assess whether infiltration by OSPW into groundwaters has occurred is challenging, as groundwaters unaffected by OSPW can contain similar mixtures of potentially toxic compounds (i.e. metals, salts, organics) to those found in OSPW (Roy et al., 2016). These pollutants may enter groundwater ecosystems from atmospheric inputs (e.g. upgrader emissions, open-pit mine erosion) or acquired from the groundwater flow path from the geological formations in the area, such as the McMurray formation which contains natural oil sands (Roy et al., 2016). Biological-based strategies, such as using bioindicator species, may offer a potential approach for source differentiation. However, there is currently a paucity of information regarding the *in situ* biodegradation of NAs in groundwater systems and the microorganisms involved, making baseline comparisons impossible.

One study which used ^{13}C-labelled stable isotopes of specific NAs, investigated a shallow glacio-fluvial aquifer and followed NA biodegradation that was occurring along the groundwater flow-path from the OSTPs to the Athabasca River (Ahad et al., 2018). Biodegradation of cyclohexanecarboxylic acid (CHCA) and 1,2-cyclohexanedicarboxylic acid

(CHDCA) was reported and it was suggested that the indigenous microbial communities in the shallow subsurface groundwaters near to the tailings ponds could readily break down these compounds prior to surface water discharge (Ahad et al., 2018). There is currently a lack of information on the microbial diversity found in groundwater ecosystems contaminated with NAs. One study that tackled this knowledge gap by analysing water samples from a basal aquifer (up to 20 m in thickness underlying a 50–80 m thick layer of oil-saturated sands reservoir) found a dominance of *Epsilonproteobacteria* affiliated with *Sulfuricurvum*, *Arcobacter* and *Sulfurospirillum* sp. with lesser proportions of *Methanomicrobiales* (Hubert et al., 2012). Methane-oxidizing bacteria have also been found in contaminated groundwaters from a low lying wetland (Ahad et al., 2018). Although some headway has been made into unravelling the microbiomes of NA-contaminated groundwaters, more research is still required in this area, especially in relation to their functional role in these ecosystems.

2.3 Marine ecosystems

NAs may enter the marine environment either directly or indirectly from multiple sources, including surface runoff, natural leakages from oil sands tailing ponds, and accidental crude oil spills (Kannel and Gan, 2012). For example, NAs were detected in the surrounding waters and sediments following the Deepwater Horizon (Gulf of Mexico) and Hebei Spirit (Taean, South Korea) oil spills (Reddy et al., 2012; Wan et al., 2014). Concentrations of NAs have been found to be 50-100 times greater than total polycyclic aromatic hydrocarbons (PAHs) found in the sediments from the Taean spill (Wan et al., 2014). Furthermore, whilst acyclic $NAs_{n=5-20}$ were found to be easily degraded, the cyclic $NAs_{n=21-41}$ persisted (Wan et al., 2014). NAs were also found to be 10-30 times higher than that of PAHs in the sediment of Dalian bay, China after an oil spill accident (Zan et al., 2019). These findings highlight the potential ecological impact of NAs on marine environments. Yet the transportation, NA biodegradation and risk of NAs to the marine microbiome is largely unknown. More research is needed on NA biodegradation and the microbial communities involved in marine ecosystems.

2.4 Bitumen saturated outcrop deposits

The AOSR contain shallow bitumen deposits found at depths ranging from $\leq$ 50 TO 75 m below the surface that can be surface mined (Foght et al., 2017).

In addition, along many of the riverbanks in the area, natural bitumen-containing surface outcrops of the McMurray Formation are found (Wyndham and Costerton, 1981a; Wang et al., 2014; Wong et al., 2015). This includes exposed surface outcrops such as those located along the Athabasca-Clearwater River network (Wong et al., 2015). These bitumen-rich ores cooccur with sand, fine silt, and clay minerals ("fines") with an average of ~10 wt% bitumen, ~5% water and ~85% quartz sand, silt and clays (Chalaturnyk et al., 2002; Foght et al., 2017), and glycocalyx-mediated adhesion of bacteria to the surface of the bitumen may occur (Wyndham and Costerton, 1981b).

The *in situ* microbiota found in rocky outcrops must be able to survive the extreme temperature variations that occur in the region, which range from summer highs of +30 °C to winter lows between −40 and −50 °C (Richardson and Dacks, 2019; Wong et al., 2015). In these bitumen outcrops, there is some evidence for aerobic rather than anaerobic bitumen biotransformation, despite some known anaerobic phylotypes (e.g. methanogens) being detected (Wong et al., 2015). Metagenome analysis has revealed an abundance of genes encoding for known bacterial aerobic hydrocarbon biodegradation enzymes (such as mono- and di-oxygenases and fungal cytochrome P_{450} oxidases) but, not those involved in anaerobic hydrocarbon-degradation (Wong et al., 2015). Thus, in these exposed bitumen outcrops, bitumen degradation occurs and is primarily via aerobic processes (Wong et al., 2015), although these findings are only based on two metagenome samples, and that with further sampling contrasting results may be found.

Examination of the bitumen-saturated sandstone outcrops sampled from northeastern Alberta found diverse taxa including fungi, hydrocarbon-, methane-, and acetate-oxidizing heterotrophic bacteria (Wong et al., 2015). Specifically, two distinct networks were found in the outcrops with Network I comprising largely of aerobic putative hydrocarbon degraders (e.g. *Burkholderiales, Flavobacteriales, Pseudomonadales, Sphingomonadales,* anaerobic methanogens (e.g. *Methanomicrobiales, Methanosarcinales*, and *Methanobacteriales*), *Clostridiales*, and syntrophs (e.g. *Syntrophobacterales* and *Anaerolineales*), and Network II comprising aerobic bacteria and fungi (e.g. Protomyces and Coniosporium) as well as the putative hydrocarbon-, C_2-, or C_1-oxidizing bacteria (Wong et al., 2015). It was suggested that hydrocarbon degradation in these outcrops was initiated by fungi and the products of which were subsequently metabolized predominantly by the aerobic bacterial community (Wong et al., 2015). In addition, thermophilic microorganisms (e.g. *Methanothermobacter, Thermotaga, Thermomonas,*

Caldanaerobacter, Caldimicrobium, and *Thermanaeromonas*) were also found, and likely due to the higher summer temperatures (i.e. 55–60 °C) experienced on the outcrop slopes although the air temperatures immediately above the outcrop surface were between 33 °C and 38 °C, indicating that the bitumen-containing outcrop surface has effective heat-absorbing properties not unlike those of asphalt roads (Wong et al., 2015). Thus, these bitumen-containing outcrops are a unique ecosystem that harbors a diverse microbiome that may be important for NA biodegradation (Wong et al., 2015).

2.5 Deep oil sand deposits

Whilst surface-mining is generally used to extract the shallow bitumen deposits that are found tens of meters below the surface (i.e. $\leq$ 50–75 m depth), much deeper deposits (> 75 m depth) are also found that require alternative methods of extraction, such as Steam-Assisted Gravity Drainage (SAGD), which is an enhanced oil recovery technology involving an advanced form of steam stimulation (Foght et al., 2017; Gates and Larter, 2014). There is currently limited information on the microbiome of these deeper bitumen deposits, which is largely due to the difficulty in obtaining intact samples. However, one metagenomics/metabarcoding study, which analysed the microbiome of deep oil sand deposits from SAGD core drillings, found that both aerobic and anaerobic communities were in close proximity (An et al., 2013). Specifically, a network of aerobes were found which included the genera (orders): *Rhizobium* (*Rhizobiales*), *Cupriavidus* (*Burkholderiales*), *Brevundimonas* (*Caulobacterales*), *Delftia* (*Burkholderiales*), and *Methylobacterium* (*Rhizobiales*) (An et al., 2013). In addition, metagenome analysis identified several aerobic related oxygenase enzymes in the deep oil sand samples (An et al., 2013). These communities also contained hydrocarbon-degrading bacteria and methanogens such as *Methanosarcina*, which suggested limited oxygen ingress in these cores (An et al., 2013). Despite these findings, the microbiomes of deep oil sand deposits and their functional role in NA biodegradation remains understudied and further research is required in this area.

3. Anthropogenic naphthenic acid (NA) contaminated microbiomes

The oil sands mining operations in northern Alberta, Canada, started in 1967 and undertook a rapid expansion between 2000 and 2010 (Fennell and Arciszewski, 2019). Such a rapid development in the Athabasca watershed has raised considerable concerns regarding their effects on the surrounding ecosystems. Bitumen extraction disturbs the land, causing multiple environmental issues, including habitat fragmentation; freshwater abstraction; emission of biogenic greenhouse gases, and emissions of other pollutants, such as mercury and volatile hydrocarbons (Foght et al., 2017; Siddique et al., 2008; Simpson et al., 2010; Small et al., 2015). As a result of bitumen extraction, upgrading, storage and export operations, large volumes of OSPW are stored in vast tailings ponds, where NAs accumulate. However, the long-term storage of OSPW poses potential environmental risks to local and downstream ecosystems, where organic (e.g., residual hydrocarbons, NAs) and inorganic (e.g., heavy metals) contaminants from tailings ponds may enter surface and groundwaters as a result of effluent discharge, pond breaches, seepage and spills (Ahad et al., 2020; Fennell and Arciszewski, 2019). Indeed, seepage of OSPW (especially NAs) into surface and groundwaters from tailings ponds is a major environmental concern (Fennell and Arciszewski, 2019).

Differentiation of the source of NAs in the local environment (i.e. identifying whether from natural or anthropogenic origin) would help to determine whether any accidental NA incursions had occurred. Yet the differentiation of NA source is complicated by the diverse and complex nature of the mixtures of organic compounds found within all bitumen ecosystems (Hewitt et al., 2020). Moreover, the structural composition, concentration, toxicity, and chemistry of NAs may differ depending on the source of NAs, which in turn influences the *in situ* microbiomes found (Whitby, 2010). Several methods have been reported to differentiate between anthropogenic water soluble organics (WSOs) and naturally occurring bitumen impacted WSOs, including high resolution mass spectrometry techniques, ultraperformance liquid chromatography time-of-flight (ToF) mass spectrometry, and multidimensional gas chromatography with ToF mass spectrometry, among others (Hewitt et al., 2020). Recently, the use of orbitrap mass spectrometry with carbon isotopes (e.g. $\delta^{13}C_{pyr}$ and $\delta^{34}S$) has made some headway into distinguishing between naturally occurring and anthropogenic NA sources (Ahad et al., 2020). In addition to these analytical techniques, biologically based approaches to track specific NAs in aquatic environments could also prove useful. For example, the marine coccolithophore *Emiliania huxleyi* as well as other organisms such as cyanobacteria have been found to be highly sensitive to NAs, and could be used as potential bioindicators of NAs in the environment (Beddow et al., 2016; Yergeau et al., 2013). However, more research is required into their sensitivities towards the range of concentrations and types of NA compounds found in the various bitumen-associated ecosystems. Thus, having a better understanding of the source and transport of NAs in the environment will allow us to better understand the factors shaping the *in situ* microbiomes in these impacted ecosystems.

3.1 Oil sand tailings ponds (OSTPs)

In the AOSR, various different OSTP structures may be found. These include in-pit tailings ponds, which are modified from former mine pits, external tailings facilities (ETF) or external tailings area (ETA), which are dyked areas built aboveground and external to the active mining area (Fennell and Arciszewski, 2019; Jeeravipoolvarn et al., 2017). In general, OSTPs are constructed of coarse tailings (largely sand) that form a containment dyke (Fig. 5) (Roy et al., 2016). Within the dyke, there are collection systems to capture internal drainage, which may contain OSPW or any pond leakage (Roy et al., 2016). Any water that is captured is subsequently returned to the tailings ponds (Roy et al., 2016). In general, the ponds are underlain with a glacial till or clay, which has a low permeability, but where permeable materials do occur, the ponds are overlain with other lower permeability materials (Fennell and Arciszewski, 2019). Despite these measures,

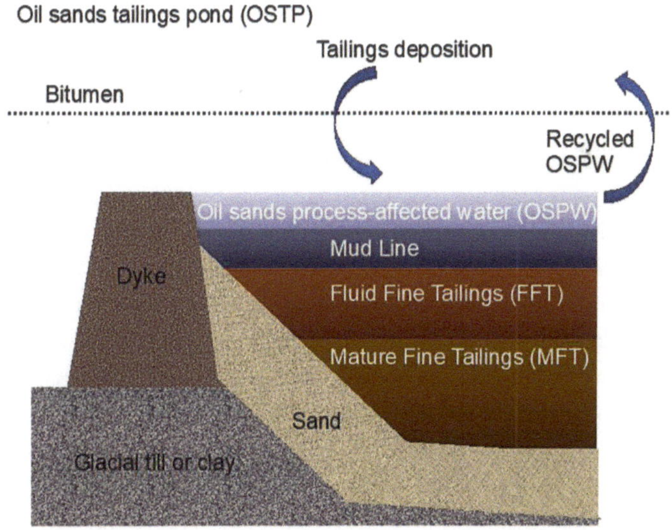

Fig. 5 Simplified schematic of the structure of an oil sands tailing pond (OSTP).

there remains a concern surrounding OSTPs and potential leakage of OSPW into the surrounding environments.

Tailings ponds are heterogenous ecosystems where unique microbiomes are found (Foght et al., 2017). For each pond, different environmental parameters and management practices exist which result in various physico-chemical gradients, and in turn influence the microbiomes present (An et al., 2013; Foght et al., 2017; Penner and Foght, 2010). For example, the addition of fresh tailings to ponds generate temperature gradients as they are still warm from processing when added (historically this was up to 60 °C) (MacKinnon, 1989). Generally, ponds are also nitrogen limited, as low concentrations of ammonium, nitrate and nitrite occur (Foght et al., 2017). Ponds are also typically stratified comprising of both oxic and anoxic zones. In general, oxygen ingress is limited to the upper surface layers (up to 1 m depth), and below this a sharp oxycline occurs, as the OSPW rapidly becomes anoxic (Ramos-Padrón et al., 2011). This anoxic zone is largely a result of aerobic and/or facultative microbial metabolism and the oxidation of microbially produced sulphides at depth (Foght et al., 2017). In some ponds, sulphate becomes depleted as a result of sulphate reduction generating hydrogen sulphide which is subsequently removed and removing the inhibitory effect on methanogens, facilitating a concomitant production of methane (Clemente and Fedorak, 2005; Fedorak et al., 2002). Consequently, deposited tailings can emit significant quantities of methane (CH_4) and carbon dioxide (CO_2), where residual hydrocarbons are being used in methanogenesis (Siddique et al., 2008). Indeed, (one of the largest and oldest tailings ponds, Mildred Lake Settling Basin (MLSB), has been estimated to emit ~43 million L CH_4 day^{-1} to the atmosphere (Siddique et al., 2008).

Within tailings ponds, the solid content increases with depth, and density increases from < 0.1 wt% in the upper layers to $> 10\%$ solids content in the lower layers (Lv et al., 2020). At the pond surface, there is an upper water layer of OSPW, which may range in depth from 2 to 10 m (Foght et al., 2017). In general, OSPW makes up only a small volumetric proportion of the ponds, largely due to re-use and recycling of the OSPW (which varies depending on the operator and/or seasonal changes) (Foght et al., 2017). Below the water layer is the solids enriched layers of fine silt, sand and clay minerals (known as "fines"). These "fines" include fluid fine tailings (FFT), which has a low solids content (Foght et al., 2017). Below this layer is the mature fine tailings (MFT) layer, which has a higher solids content (≥ 30 wt %), and is anoxic (Lv et al., 2020; Foght et al., 2017). The MFT layer can often remain undisturbed for decades and may reach depths of $\geq$ 40 m, depending on natural consolidation, dewatering, depth of the basin, age of the pond and management practice (Foght et al., 2017).

Some operators mix gypsum (calcium sulphate dihydrate) with fine tailings and coarse sand to produce composite tailings (CT) as a type of dry reclamation strategy. The idea is to neutralize the net negativity of clay minerals to act as a densifying agent and recover interstitial pore water, resulting in an increase in the solids content and flocculation (Ramos-Padrón et al., 2011; Mikula et al., 1996; Warren et al., 2016). It has been estimated that around $9.6 \times 10^8 m^3$ of CT are either in, or awaiting impoundment in surface pits within the AOSR (Warren et al., 2016). These chemically engineered CT tailings are slightly alkaline (pH 8), moderately saline and consist of a mixture of FFT (i.e. saline water, suspended Fe^{3+} rich clay minerals, and residual bitumen) and post-processed sand amended with gypsum (Matthews et al., 2002). However, the process causes an increase in the levels of Ca^{2+}, sulphate and NAs compared to MFT (Fedorak et al., 2002; Foght et al., 2017). The microbial diversity of CT has been reported to be relatively low compared with MFT, with common phyla found including Gamma- and Beta-proteobacteria, *Bacillota*, *Actinomycetota*, and microbes capable of sulphur metabolism (Warren et al., 2016). Thickened tailings (TT) which have been augmented with organic polymers to flocculate the solids to recover more water may also occur (Foght et al., 2017).

Tailings ponds can harbor a diverse range of microbiota. Facultative and/or aerobic microorganisms are reported to occur in high numbers (10^6–10^8 cells mL^{-1}) (Foght et al., 2017). Within the bacteria, members of the Beta-, Delta- and Epsilon-proteobacteria have been found (An et al., 2013; Yergeau et al., 2012). For example, *Thauera, Acidovorax, Thiobacillus, Brachymonas, Rhodoferax, Thiobacillus, Smithella, Schumannella, Hydrogenophaga, Azonexus, Salinimicrobium, Achromobacter, Gillisia, Alcaligenes* and *Acinetobacter* (Yergeau et al., 2012; Del Rio et al., 2006; Ramos-Padrón et al., 2011). In addition, enrichment cultures established from OSPW and oil sands have recovered *Pseudomonas* sp., *Alcaligenes* sp., *Acinetobacter* sp., and *Kurthia* sp. among others (Herman et al., 1994; Wyndham and Costerton, 1981a,b). However in some ponds, increased NA concentrations and/or limited carbon sources may select for a less diverse bacterial community compared to natural NA-contaminated ecosystems, such as Athabasca River sediments (Yergeau et al., 2012). Within the Archaea, Euryarchaeota (primarily *Methanomicrobia*) have been found, with the Crenarchaeota almost absent (Yergeau et al., 2012). Phototrophic and heterotrophic Eukarya including Fungi, Metazoa, Chlorophyta, Alveolates, Stramenopiles and Excavata occur (Aguilar et al., 2016). Specifically, 18S rRNA gene sequencing has revealed that the majority of sequences were from heterotrophs related to Rhizaria, Ciliata and Fungi which is consistent with other low-light and low-oxygen environments (Aguilar et al., 2016). In addition, photosynthetic taxa such as *Euglena* and Chlorophyta have also been reported (Aguilar et al., 2016).

The diverse microbial communities found in tailings ponds are capable of various metabolic functions including hydrocarbon biodegradation, element cycling (S and Fe) and gas production (CH_4, CO_2, H_2S) (Wilson et al., 2016; Warren et al., 2016). Anaerobic metabolism dominates OSTPs, particularly methanogenesis, due to the anoxic conditions during biodegradation of hydrocarbons such as residual bitumen (Siddique et al., 2008, 2015; Fedorak et al., 2003). Indeed, many of the microbiota found in tailings are involved in anaerobic hydrocarbon degradation, including fermentative and syntrophic bacteria, nitrate reducers, iron reducers, sulphate-reducing bacteria (SRB), hydrogenotrophic and acetoclastic methanogens (An et al., 2013; Golby et al., 2012; Ramos-Padrón et al., 2011). However, the presence of sulphate from the addition of gypsum by some operators may promote sulfidogenic activity and sulphate-reducing bacteria (e.g. *Desulfocapsa*, *Desulfurivibrio*, *Desulfobacterium*, and *Desulfuromonas*), may out-compete methanogens for available substrates inhibiting methanogenesis when sulphate is abundant (Fedorak et al., 2003; Ramos-Padrón et al., 2011). Yet, in the deep anaerobic zones (where sulphide and sulphate concentrations increases), methanogens (e.g. *Methanosaeta* (acetoclastic) and *Methanolinea* and *Methanoregula* (hydrogenotrophic) have been found (Ramos-Padrón et al., 2011).

Shifts in the microbial communities may also occur depending on the NA source, the hydrocarbon substrates present and the *in situ* conditions (Abu Laban et al., 2015c,d; Tan et al., 2013). For example, when aromatics are present, *Desulfosporosinus* (under methanogenic conditions) or *Desulfobulbaceae* (under sulfidogenic conditions) were found to dominate, whereas when alkanes were present, other taxa such as *Peptococcaceae* dominated (Abu Laban et al., 2015c,d; Tan et al., 2013). Furthermore, under nitrate-, and sulphate-reducing conditions, the relative abundance of *Rhizobium* increased when enriched with AEOs, whilst *Steroidobacter* and *Methanosarcina* increased under nitrate-reducing conditions in the presence of a commercial NA mixture (Lv et al., 2020). Shifts in the microbiome may also occur as a result of the various bioremediation treatments for tailings ponds, which can include UV irradiation, filtration, carbon adsorption, coagulation and ozonation (Quinlan and Tam, 2015). For example, following ozonation treatment, a decrease in the abundance of planktonic microbes and *Pseudomonadota* was reported (Islam et al., 2015). Specifically, *Pseudomonadales* which are generally dominant, were sensitive to ozonation and their abundance decreased, whilst others such as *Burkholderiales* increased in abundance, as ozonation reduced the toxicity of the OSPW and amount of labile hydrocarbons (Islam et al., 2015). In another study, treatment of OSPW by ozonation, also decreased the abundance of *Pseudomonadota* compared to the untreated community (Zhang et al., 2020a). Cyanobacteria and *Chloroflexota* have also been found to increase following ozonation of OSPW biofiltrates (Zhang et al., 2020a). Furthermore, *Bacteroidota* increased only in ozonated biofiltrated OSPW treatment, whereas *Actinomycetota* increased only after raw biofiltrated OPSW treatment (Zhang et al., 2020a). Thus, the functionally diverse microbiomes may alter through the various bioremediation treatment strategies and this is an important consideration for successful pond reclamation.

3.2 Oil sands process affected water (OSPW)

OSPW is composed of a mixture of various source waters including river water, formation water, recycled water and surface run-off (Mahaffey and Dubé, 2017), and is stored in ponds of various sizes and ages (Fennell and Arciszewski, 2019). It has been reported previously that the NA concentrations found in OSPW may be as high as $120\,\mathrm{mg\,L^{-1}}$ (Clemente and Fedorak, 2005; Holowenko et al., 2002). The constant recycling of OSPW may select for an enriched and adapted microbiome. Indeed, the surface waters of OSPW may harbor more diverse taxa (*Methyloversatilis*, *Azospirillum* and *Gemmata*) than the deeper layers, where strict anaerobes such as methanogens, sulphate reducers, nitrate reducers and iron reducers occur (An et al., 2013; Ramos-Padrón et al., 2011; Foght et al., 2017; Stasik et al., 2014). However, it has been reported that both methanogens and sulphate reducers are only minor members of OSPW communities (An et al., 2013). Additionally, thiosulphate oxidizers (Stasik et al., 2014), sulphur oxidizers (e.g. *Chromatiales* (*Thiocapsa*) and *Desulfuromonadales* (*Geobacter*)) and facultative species of *Rhodocyclales* (*Thauera*), *Burkholderiales* (*Acidovorax*, *Hydrogenophaga*, Alcaligenaceae) and *Flavobacteriales* (*Flavobacterium*) have been found in OSPW (An et al., 2013). In another study, enrichments of anoxic OSPW in the presence of model NAs were dominated by *Terrabacter*,

Derxia, Limnobacter, Lutibacter and *Pseudomonadota* under nitrate-reducing conditions, whilst *Anaerobacter, Clostridium, Syntrophus, Desulfobulbus, Desulfobaceterium* and *Desulfomicrobium* dominated under sulphate-reducing conditions (Clothier and Gieg, 2016). Also found in OSPW enriched with diamondoid NAs were *Pseudomonadota, Bacteroidota, Actinomycetota* and *Bacillota* (Folwell et al., 2020); and algae, such as *Chlorellales* and *Acutodesmus* (*Scenedesmus*) (Paulssen and Gieg, 2019). Methane oxidation to CO_2 by aerobic methanotrophic bacteria (predominantly *Methylocaldum*) has also been reported in OSPW, with ~17% of biogenic CH_4 at the pond surface being microbially oxidized, and potentially mitigating against some of the greenhouse gas emissions from OSTPs (Saidi-Mehrabad et al., 2013).

The microbiota found within OSPW also has the potential to aerobically biodegrade hydrocarbons (An et al., 2013). In the uppermost OSPW layer where it is aerated, aerobic hydrocarbon degradation has been reported, but is slower in the winter when the ponds are covered with ice (Foght et al., 2017). Data from metagenome analysis has found several putative mono-, di- oxygenases and O_2-dependent ring cleavage enzymes in the genomes of pond surface waters, suggesting that the microbial communities within OSPW have the potential for aerobic biodegradation of hydrocarbons (An et al., 2013). The yeast *Rhodotorula mucilaginosa* and fungus *Trichoderma harzianum* have also been enriched from OSPW amended with commercial NAs, demonstrating the capacity for mycoremediation of NAs including the diamondoid, 1-adamantane carboxylic acid (Miles et al., 2020). It is well known that microbial biodegradation decreases OSPW toxicity over time (Frank et al., 2009; Johnson et al., 2011) and therefore bioremediation using microorganisms is a potential strategy for both the detoxification and removal of OSPW.

3.3 Mature fine tailings (MFT)

Within the bacterial community, the MFT microbiome is dominated by *Pseudomonadota* (Penner and Foght, 2010). Notably, *Betaproteobacteria*, particularly *Albidoferax* (formerly *Rhodoferax*), *Acidovorax* sp., *Thauera, Desulfatibacillum, Burkholderiales* and *Hydrogenophilales* (Foght et al., 2017; Penner and Foght, 2010). In some MFT samples, a dominance of bacteria rather than archaea was found, with an increase in abundance of fermenters within the *Chloroflexota* (*Anaerolineales*) and *Actinomycetota* (*Coriobacteriales*) (Siddique et al., 2015; Collins et al., 2016; Mohamad Shahimin et al., 2016). However, it has also been shown that in MFT the relative abundances of archaea and bacteria change with increasing depth, and the proportions of archaeal sequences having a broad range of hydrocarbons and exhibiting active methanogenesis, decreased with depth where labile hydrocarbons were depleted, whilst the bacteria (notably, Alpha-, Beta- and Gammaproteobacteria) increased, at the expense of *Bacillota* and Deltaproteobacteria (Foght et al., 2017, Stasik et al., 2014). This supports the hypothesis that "dormant" MFT tend to be dominated by bacteria, whilst in "active" MFT archaea are dominant (Foght et al., 2017). It has also been reported that the deepest zones of the tailings pond harbor bacteria not found elsewhere (e.g. *Brachymonas, Thiobacillus*, and *Cellulomonas*), and that these microorganisms may help in some way with tailings densification or consolidation (Ramos-Padrón et al., 2011). Eukaryotes have also been found in MFT including Amoebozoa and Rhizaria (Aguilar et al., 2016).

Using mesocosms and laboratory incubations of MFT, it has been shown that bacterial and archaeal community succession patterns occur, whereby the original MFT is dominated by bacteria but shifts towards sulphidogenesis and a dominance of archaea following bioreactor manipulation (Chi Fru et al., 2013). In terms of the functional microbiome of MFT, an abundance of methanogens and sulphate reducers have been found with both methanogenesis and sulphidogenesis occurring in tailings ponds with their overall dominance depending on the operational practice (Ramos-Padrón et al., 2011; Foght et al., 2017). For example, where gypsum has been added, sulphate reduction may dominate (Foght et al., 2017). However, the methanogenic community is generally considered the most active in MFT (Lv et al., 2020; Fedorak et al., 2002, 2003). Indeed MFT enrichments have the potential for methanogenic degradation of hydrocarbons such as naphtha and short chain *n*-alkanes among other compounds (Foght et al., 2017). Within the methanogens, *Methanomicrobiales, Methanosaeta, Methanoregula* and *Methanosarcinales* were found to dominate (Penner and Foght, 2010; Foght et al., 2017). In the deeper anaerobic zones where sulphide and sulphate concentrations increase, the microbial community was dominated by syntrophs (*Pelotomaculum*, *Syntrophus*, and *Smithella* sp.) and sulphate- and sulphur-reducing bacteria (*Desulfocapsa, Desulfurivibrio* sp., *Desulfomicrobium, Desilfomonile, Desulfobulbus* and *Acetobacterium*) (An et al., 2013; Lv et al., 2020; Fedorak et al., 2002, 2003). A range of other anaerobic microbiota have also been found including: Comamonadaceae, Hydrogenophilaceae, Anaerolineaceae, iron reducers, fermenters, and nitrate reducers (e.g. Rhizobium, Steroidobacter and Ignavibacterium) (Lv et al., 2020; Fedorak et al., 2002, 2003). Although nitrate reduction occurs, it is generally less dominant as nitrate and nitrite concentrations are typically low (Fedorak et al., 2002; Penner and Foght, 2010). Despite this, facultative nitrate reducers are abundant and widely distributed throughout tailings depth profiles (Fedorak et al., 2002). Other anaerobic hydrocarbon-degrading taxa have also been recovered from MFT enrichments, including members of the *Clostridia* (*Peptococcaceae*, *Desulfosporosinus*), *Deltaproteobacteria* (*Desulfobulbaceae*), *Rhodocyclales* (An et al., 2013; Tan et al., 2014a,b; 2015a,b; Abu Laban et al., 2015a,b). Metagenome analysis has revealed genes for aerobic aromatic hydrocarbon biodegradation and aerobic methane oxidation (An et al., 2013). Strictly aerobic genera in this environment included the methanotrophic *Methylococcales*/Methylocaldum (An et al., 2013). Indeed, it is estimated that 20% of the methane formed in the deeper layers of tailings ponds is oxidized by methanotrophs in surface waters, which is promising for mitigation against the greenhouse gas emissions from ponds (Saidi-Mehrabad et al., 2013).

3.4 End pit lakes (EPLs) and reclaimed ponds

One of the major environmental challenges arising from the oil sands mining operations is the vast amount of tailings and OSPW that require reclamation (Foght et al., 2017). Under government regulations, it is mandatory for tailings ponds be reclaimed within 10 years of the end of a mining operation, either as terrestrial landscapes, wetlands or end-pit lakes (EPLs) (Paulssen and Gieg, 2019). Reclamation of tailings ponds however, remains a challenge due to slow rates of fine tailings densification and the presence of toxic, recalcitrant components associated with OSPW (Foght et al., 2017). Moreover biogenic greenhouse gas emissions remain a challenge with OSTPs accounting for 45% of total CH_4 emissions measured from the major surface mining facilities in the AOSR, whilst emissions from operations in the open pit mines accounted for $\sim$50% (Baray et al., 2018). Adopting nature-based solutions for OSPW and tailings bioremediation is attracting significant interest, but requires a thorough understanding of the *in situ* communities and their processes.

The microbiomes of tailings ponds are critical for pond reclamation and a diverse community is crucial for effective tailings bioremediation (Headley and McMartin, 2004; Yergeau et al., 2012; Ajaero et al., 2020). The flexibility and ability of the microbial community to adapt is also an important consideration for pond reclamation, especially as certain substrates are being selectively removed through the various microbial activities. Furthermore, pond reclamation and remediation must also consider issues such as H_2S production from constructed wetlands containing sulphate-rich tailings (Reid and Warren, 2016), acid mine drainage from surface deposition of tailings and the wider implications of EPLs (Foght et al., 2017). Under the Albertan Government's zero effluent discharge policy, operators undertake both "wet" and "dry" reclamation approaches (Foght et al., 2017). One strategy is to place de-watered tailings in exhausted mine pits and cover them with sand, soil and vegetation to regenerate the landscape. For dry landscaping, the MFT sediments may require chemical and/or physical manipulation to remove excess water and increase solids content, whereas MFT intended for wet reclamation (freshwater ecosystems) does not require additional dewatering (Foght et al., 2017). However, for the success of both wet and dry reclamation approaches, suitable microbial communities are required.

Wet capping, is one reclamation technique used with EPLs, whereby soft tailings or fluid fine tailings (FFT) are placed within decommissioned open pits and capped with freshwater (ca. 3–10 m depth) (Dompierre and Barbour, 2016). The layers of freshwater are continually replenished until the tailings are settled into a thick layer and the freshwater has diluted all potential contaminants to ecologically acceptable levels and oxygen levels are comparable to that found in natural lakes (Dompierre and Barbour, 2016; White and Liber, 2018). The design and configuration of EPLs and the type of tailings stored varies with operators, but their use allows oil sands operators to decrease the volumes of FFT stored in tailings impoundments. One of the major challenges associated with reclaiming oil sands tailings in EPLs is the slow consolidation rates due to the fine-grained nature of FFT (Cossey et al., 2021). To assess the efficacy of pond reclamation, heterotrophic protists have been suggested as bioindicators, as they are highly resilient, and represent a trophic level intermediate between bacteria and macrofauna (Richardson and Dacks, 2019). Another major issue surrounding EPLs is that high concentrations of dissolved constituents, NAs, unrecovered bitumen and petroleum hydrocarbons in the FFT pore water may be transported into the lake water (Dompierre and Barbour, 2016). It is proposed that this movement occurs via two key processes: (1) advective-diffusive mass transport with upward pore water flow caused by dewatering of the soft tailings; and (2) mixing by wind events or unstable density profiles through the lake water and upper portion of the FFT (Dompierre and Barbour, 2016). Heat transfer between the water cap and the FFT may also occur (Dompierre and Barbour, 2016). The fluid movements in the lake may cause mixing within the tailings influencing the *in situ* biogeochemistry, which in turn shape the indigenous microbiomes (Dompierre and Barbour, 2016). However, establishing a sustainable biological community in EPLs is yet to be fully assessed and more research is required in this area (Dompierre and Barbour, 2016).

One of the largest reclaimed sites for FFT in Canada is called Base Mine Lake (BML), which was established in 2012, and is approximately 780 hectares in size, comprising ca. 40 m depth of FFT overlain with 10 m of a freshwater cap (Risacher et al., 2018; Albakistani et al., 2022). Between 2014 and 2019, methane emissions averaged 610 kg ha^{-1} y^{-1} with the highest levels during spring thaw, but have since gradually declined over time (Albakistani et al., 2022). Methane/O_2 counter gradients occur in the water column. Specifically, dissolved methane concentration in the sediment is very high, and increases rapidly with depth (i.e. < 0.1 mM at the sediment-water interface, increasing to $>$ 2 mM between 1.0 and 1.5 m depth, and $\sim$ 4 mM between 1.5 and 3.5 m depth in the FFT layer) (Albakistani et al., 2022). During summer stratification, a methane gradient occurs in the water column (i.e. between 20 and 40 μM in the hypolimnion; decreasing to 0.38–1.2 μM in the epilimnion (Albakistani et al., 2022). However, dissolved O_2 concentrations decrease with increased depth in the water column, (i.e. between 70–85% saturation at the metalimnion, decreasing to 1–5% saturation at the sediment interface (Albakistani et al., 2022).

This dimictic pit lake exhibits methane cycling in the metalimnion, hypolimnion, and sediment-water interface, and methanotroph population dynamics are similar to that found in natural boreal lakes (Albakistani et al., 2022). Specifically, aerobic methanotrophic bacteria comprise up to 58% of the total bacterial communities (based on 16S rRNA gene sequencing reads) (Albakistani et al., 2022). Moreover, seasonal changes in methanotrophic activity and population abundances were also found in the lake water, with autumn and winter highs, which declined during summer stratification, especially in the epilimnion (Albakistani et al., 2022). Of the methanotroph genera found were: *Methylobacter*, *Methylovulum*, and *Methyloparacoccus* (Albakistani et al., 2022). Specifically, *Methylobacter* and *Methylovulum* populations peaked in the winter/spring, when methane oxidation activity was psychrophilic, and *Methyloparacoccus* populations increased through summer and autumn, when methane oxidation was mesophilic, which suggested that there was a temporal niche differentiation based on temperature, oxygen and methane concentration (Albakistani et al., 2022).

Seasonal changes in turbidity have also been reported in BML, which are thought to be due to the settling of fine particles suspended in the surface water and the resuspension of fine particles from the FFT-water interface (Lawrence et al., 2016). However, even small changes in turbidity may negatively impact on the microbial metabolic processes occurring (Shen et al., 2019). Biofilms which are formed on the FFT-water interface have a biostabilization effect, reducing the resuspension of sediments, and thus reduce turbidity in EPLs (in some cases by up to 99% depending on the biofilm age and mixing speed) (Cossey et al., 2021). The microbiomes of biofilms formed on the FFT-water interface contain diverse communities including photoautotrophs, such as cyanobacteria and Chlorophyta (green algae), and several heterotrophs including members of the Gammaproteobacteria, Desulfobulbia, and Anaerolineae (Cossey et al., 2021). In another study, it has been shown that when water from BML was used in a co-culture with *Chlorella kessleri*, biodegradation of model NAs was enhanced and the presence of *C. kessleri* concomitantly increased the diversity of the microbial community found in the BML water, with many known hydrocarbon and NA degraders (Yu et al., 2019).

Diverse microbial communities that contribute to nutrient cycling have also been found in other pond reclamation sites (Risacher et al., 2018). For example, methanotrophy and nitrification are important oxygen consuming process in these ecosystems (Risacher et al., 2018). However, during the early stages of pit lake establishment at low oxygen concentrations, nitrifiers appear to be outcompeted by methanotrophs (Risacher et al., 2018). Yet within four years of commissioning, active nitrification has been reported, affecting oxygen concentrations in the water column of pit lakes (Risacher et al., 2018). This is in contrast to OSTPs, where nitrification is often low, even when ammonium levels may be high (Risacher et al., 2018).

A promising low-energy method for OSPW treatment are constructed wetland treatment systems (CWTSs), which are designed to simulate natural wetlands (Foote, 2012; McQueen et al., 2017; Hendrikse et al., 2018; Ajaero et al., 2020; Simair et al., 2021). One pilot-scale CWTS was found to decrease the concentration of NAs, hydrocarbons, metals, and toxicity (McQueen et al., 2017). Indeed, several mature wetlands in the AOSR have been reported to promote NA degradation and detoxification (Vander Meulen et al., 2022). However, one laboratory scale constructed wetland, showed that the more complex NAs (i.e. those with increased carbon numbers such as 17–20) were more recalcitrant to biodegradation (Toor et al., 2013a,b). In addition, wetland reclamation involving CT deposits may increase sulphur levels, resulting in microbial sulphur cycling and H_2S generation (Reid and Warren, 2016). Indeed, high concentrations of H_2S (> 500 μM or 18 mg L$-^1$) have been found in the sand cap of a constructed wetland due to active dewatering causing an upward migration of sulphur rich water, and downwelling of labile organic carbon from the developing wetland which stimulated microbial H_2S production (Reid and Warren, 2016). In contrast, to H_2S levels, another study showed that methane emissions were lower at a constructed fen north of Fort McMurray compared to a nearby reference fen (Murray et al., 2017). Wetland reclamation after mining may also generate marshes with elevated salinity and residual hydrocarbons (Mollard et al., 2015). Bacterial diversity and abundance has been found to be relatively low in constructed wetlands with phyla including Gamma- and Beta-proteobacteria, *Bacillota*, *Actinomycetota*, and *Chloroflexota*) and some taxa distinct from OSTPs (Warren et al., 2016). Despite potential issues surrounding the long-term stability of constructed wetlands (Reid and Warren, 2016), they are currently being tested by the oil sands industry to treat OSPW, as operators face the challenge of reclamation under the Albertan Government's water-use restrictions.

3.5 Biofilms, bioreactors and biofilters

Often mixed cultures perform better at NA biodegradation than single cultures. For example, *Pseudomonas putida* and *Pseudomonas fluorescens* were better at NA degradation than as individual cultures (Del Rio et al., 2006). Moreover, mixed microbial consortia grown as biofilms may be effective in the removal of NAs, particularly the more recalcitrant NA compounds found in OSPW and tailings ponds (McKenzie et al., 2014; Golby et al., 2012; Demeter et al., 2015; Folwell et al., 2016). Indeed, differences in the NA-degrading microbiota has been reported between planktonic and biofilm microbiomes, with increased NA degradation rates associated with biofilm communities (Demeter et al., 2015; Folwell et al., 2016). Furthermore, overexpression of the proteins involved in biofilm formation may facilitate NA removal in OSPW and tailings (McKew et al., 2021). Indeed, a poly-beta-1,6-*N*-acetyl-D-glucosamine *N*-deacetylase protein, which is known to be involved in biofilm formation (Wang et al., 2004) was found to be significantly upregulated during NA biodegradation (McKew et al., 2021).

One approach to enhance biofilm formation is the use of bioreactors. Indeed it has been suggested that biofiltration of OSPW in fixed-bed biofilm reactors would allow the development of NA-degrading microbial communities within the biofilter, allowing for more successful removal (Zhang et al., 2020a). Bioreactors are engineered systems that can be used to enhance biodegradation rates of many recalcitrant organics including NAs (Islam et al., 2014a,b). These technologies may use materials such as clays to facilitate biofilm formation of NA-degrading microorganisms as well as to concentrate the NAs (Choi et al., 2014; Golby et al., 2012; Islam et al., 2014a; Paslawski et al., 2009). The use of petroleum coke (PC) adsorption to biofilters has also been shown to be effective in reducing the more structurally complex NAs (i.e. $12 \geq n \geq 18$ and $z = -10, -12$) and their toxicity in OSPWs, depending upon the PC content, pH and temperature (Kannel and Gan, 2012). Many different types of bioreactors have been reported with differing degrees of success. For example, a modified Ludzack-Ettinger membrane bioreactor (MLE-MBR) removed ~25% of NAs after 361 days of operation (Xue et al., 2016). Whilst sequencing batch reactors (SBR) inoculated with MFT achieved ~16% removal of NAs in OSPW (Choi et al., 2014). Granular activated carbon (GAC), which has a high adsorptive capacity for organics, removed > 86% and 99.5% of NAs from raw and ozonated OSPW, respectively (Islam et al., 2014b).

Characterization of the GAC biofilm community found *Pseudomonadota* dominated the biofilm community compared to planktonic samples (Islam et al., 2014b). Specifically, Alphaproteobacteria (e.g. *Rhizobiales, Rhodospirillales*, and *Rhodobacterales*) and Gammaproteobacteria (e.g. *Pseudomonadales, Alteromonadales, Chromatiales, Xanthomonadales, Oceanospirillales, Legionellales*, and *Methylococcales*) were abundant (Islam et al., 2014b). In addition, *Burkholderiales, Pseudomonadales, Sphingomonadales, Acidobacteriota, Verrucomicrobia, Bacteroidota*, and *Chloroflexota* were also recovered (Islam et al., 2014b). In another study which treated both raw and ozonated OSPW using biofilters, a shift in the indigenous microbial community was reported, with an increase in alpha diversity which corresponded with enhanced NA degradation (Zhang et al., 2020a). These findings suggested that continuous operation of OSPW in the bioreactors was in favour of shaping the overall microbiome towards better NA degradation (Zhang et al., 2020a). Therefore, to improve the design and efficiency of bioreactors, a better understanding of the functional microbiomes is required.

NA toxicity is reported to be one of the many factors that shape the functional microbiomes of biofilms used to treat NAs. Indeed, reduced photosynthetic activity has been observed in biofilms exposed to

NA-contaminated sediments (Yergeau et al., 2013). In the GAC biofilm community, an accumulation of toxic NAs on the GAC surface resulted in only those members of the community with high tolerances to NAs surviving (Islam et al., 2014b). Interestingly, Deltaproteobacteria were not able to grow on the GAC biofilm, which may be due to an increased sensitivity to higher concentrations of NAs found on the GAC surface (Islam et al., 2014b). In aerobic biofilms, a positive correlation between members of the *Pseudomonadota* and NA concentration has also been reported (Yergeau et al., 2013). Nitrospirae, which are potentially involved in denitrification, sulphur oxidation, and sulphate reduction, have been reported and may have been metabolizing NA compounds containing nitrogen and sulphur (Islam et al., 2014a). Immobilized soil/sediment bioreactors (ISBRs) were also dominated by ammonium- and nitrite-oxidizing bacteria, suggesting that nitrification was occurring during the treatment of NAs in OSPW (McKenzie et al., 2014). The dominant genera included *Nitrosomonas, Candidatus Nitrotoga, Arenimonas, Mesorhizobium, Bradyrhizobium, Nitrosospira, Mycobacterium, Limnobacter*, and *Commamonas*, with *Truepera, Flexibacter*, and *Saprospiraceae* and considered to be involved in the biodegradation of the more recalcitrant NAs (McKenzie et al., 2014). Another approach using the Calgary Biofilm Device (CBD) (a technology to facilitate the development of biofilms), demonstrated a dominance of *Pseudomonadota* (particularly Alphaproteobacteria) in the biofilms (Demeter et al., 2015; Golby et al., 2012). *Pseudomonadota*, particularly members of the Beta- and Gamma-proteobacteria were also dominant in a bioreactor inoculated with MFT (Choi et al., 2014).

Different aerobic and anerobic communities may occur, which affect NA removal rates. For example, anoxic biofilm reactors were reported to have at least twofold higher rates of NA removal (coupled with the reduction of nitrate) compared to aerobic biofilm reactors where bioreactor performance deteriorated (Gunawan et al., 2014). Within aerobic biofilms, *Rhodoferax, Acidovorax, Acinetobacter, Pseudomonadota*, and *Thioalkalispira* have been found, whilst *Hydrogenophaga, Rhodoferax, Methyloversatilis, Magnetospirillum*, and *Acidovorax* were present in anaerobic biofilms, but Archaea were absent (Golby et al., 2012). One study which used fixed-bed biofilters to biodegrade NAs in OSPW, found a high abundance of aerobic bacteria including (*Porphyrobacter, Legionella, Pseudomonadota*, and *Planctomycetota*) (Arslan and El-Din, 2021). However, redox conditions within the biofilters were anoxic and selected for anaerobic bacteria such as *Ruminicoccus, Eubacterium, Faecalibacterium, Dorea* and hydrogenotrophic methanogens (e.g. *Methanobrevibacter, Methanomassiliicoccus*), suggesting that methane production was likely occurring by syntrophic processes during OSPW remediation (Arslan and El-Din, 2021).

Bioaugmentation of fixed bed biofilters with NA-degrading bacterial strains showed the presence of potential hydrocarbons degraders (e.g. *Pseudomonadota, Pseudoxanthomonas*) and a high abundance of methylotrophs, notably *Methylobacillus* sp. (Arslan et al., 2022). Bioaugmentation of the biofilters not only rapidly improved OSPW remediation but aeration likely contributed to methane consumption in the top layer, minimizing its release into the environment (Arslan et al., 2022). In another study, a modified Ludzack-Ettinger membrane bioreactor (MLE-MBR) was dominated with hydrocarbon-degrading Betaproteobacteria; notably *Rhodocyclales* and *Sphingobacteriales* (Xue et al., 2016). In addition to increasing NA biodegradation rates, bioreactors may also provide a route for obtaining functional inoculants of NA degraders for application to further enhance NA bioremediation (Lemire et al., 2015). For example, two biofilm-associated biochar samples were shown to achieve 87% removal of NAs (Frankel et al., 2016). Thus, the use of bioreactors for enhancing mixed microbial communities grown as biofilms, and/or their use in obtaining functional isolates, may prove useful in future NA bioremediation strategies.

4. Biodegradation of naphthenic acids (NAs)

Both model and commercial NAs have been widely used in NA biodegradation studies to better understand the microorganisms involved in NA biodegradation (Whitby, 2010). Model NAs are chemically synthesized in the laboratory, whilst commercial NA mixtures are commercially-available mixtures of NA fractions that are obtained during petroleum processing and vary in composition and purity, with brands such as Kodak, Fluka and Merichem available (Whitby, 2010). Our previous understanding of the microorganisms driving NA biodegradation was largely through enriching consortia or obtaining isolates grown on either model or commercial NAs (Whitby, 2010). Such studies have been important, for example in providing a better understanding of how NA structure relates to rate of biodegradation, levels of NA persistence and the microorganisms involved in NA biodegradation (Whitby, 2010; Johnson et al., 2011, 2012). Yet model and commercial NAs, are not representative of those NAs found in the environment (Rowland et al., 2011a). In general, model and commercial NAs are degraded at a faster rate than those acid-extractable NAs that are recovered from environmental samples (Del Rio et al., 2006; Johnson et al., 2011; Scott et al., 2005). Since environmental NAs are often more recalcitrant than model or commercial NAs to biodegradation, more research is needed in this area (Bartlett et al., 2017; Lv et al., 2020; Johnson et al., 2011; Johnson et al., 2012; Johnson et al., 2013; Folwell et al., 2015; Scott et al., 2005; Bataineh et al., 2006; Headley et al., 2010). The real challenge however, lies in identifying the individual components of environmental NA mixtures, and the *in situ* microorganisms and processes involved in their respective metabolism.

4.1 Factors affecting naphthenic acid (NA) biodegradation rates

Several biotic and abiotic factors have been shown to affect NA biodegradation rates. For example, chemical structure, where linear and lower molecular weight compounds (< 22 carbons) are degraded preferentially, resulting in an increase in the relative amounts of recalcitrant hydrocarbons such as PAHs, NAs, branched and cyclic carboxylic acids, resins, and asphaltenes (Skeels and Whitby, 2018; Folwell et al., 2016; Clemente and Fedorak, 2005; Shuqing et al., 2008). Generally, lower molecular weight acids are more readily degraded leaving higher molecular weight compounds to persist in the environment (Biryukova et al., 2007; Whitby, 2010). For example, Biryukova et al. (2007) demonstrated that lower molecular mass NAs were preferentially degraded whilst the proportion of high molecular mass acids increased.

Generally, the most persistent NAs found in the environment contain multiple branched alkyl chains and methyl substitution of the cycloalkane rings (Han et al., 2008; Smith et al., 2008; Johnson et al., 2012) as well as the highly branched and multi-ringed diamondoid NAs found in tailings ponds (Demeter et al., 2015; Ahad et al., 2018; Paulssen and Gieg, 2019; Folwell et al., 2020; Rowland et al., 2011a). For example, the microbial biotransformation of branched and diamondoid NAs was found to be more difficult to achieve (Demeter et al., 2015; Johnson et al., 2011; Scott et al., 2005) than of those NAs that have single rings (Skeels and Whitby, 2018; Whitby, 2010). Recently, degradation of diamondoid NAs has been demonstrated by bacteria (e.g. *Pseudomonas* sp.) (Folwell et al., 2020) and algae (e.g. members of the genus *Acutodesmus* (*Scenedesmus*) and order *Chlorealles*) (Paulssen and Gieg, 2019).

In addition to the number of cycloalkane rings present, *cis*-isomerism in alicyclic acids also influences biodegradation rates and different degradation rates have been reported for different geometric isomers (Whitby, 2010). Specifically, heterotrophic microorganisms from the Athabasca River were reported to degrade trans-isomers more rapidly than the cis-isomers (Whitby, 2010). The surfactant properties of NAs and their associated toxicity may also impede microbial biodegradation (Zou et al., 1997). Abiotic factors such as nutrient availability (e.g. phosphorus, nitrogen), temperature, oxygen concentration, pH, salinity, redox potential, and light may also reduce NA degradation rates (Herman et al., 1994; Lai et al., 1996). Furthermore, addition of other carbon sources, such as glucose, may be important for maintaining the populations of NA degraders (Demeter et al., 2015). Indeed, the addition of organic compounds as co-metabolites may enhance NA degradation by some microorganisms (Dutta and Harayama, 2001). However, the added compounds or their metabolites may be preferentially degraded by the indigenous microbiota and/or may be toxic to other NA-degraders (Dutta and Harayama, 2001). The biodegradation of NAs by microorganisms is therefore crucial for the removal of recalcitrant NAs in the environment (Lai et al., 1996; Whitby, 2010). Yet, indigenous NAdegrading microbial populations must be able to thrive in the complex and toxic environments associated with NAs (Brient et al., 1995).

4.2 Aerobic vs anaerobic biodegradation of naphthenic acids (NAs)

NA biodegradation may occur both aerobically and anaerobically (Whitby, 2010). Generally, aerobic NA biodegradation occurs very rapidly (i.e. within days) (Smith et al., 2008; Biryukova et al., 2007), with fused ring structures notably susceptible to aerobic microbial degradation (Herman et al., 1993; Lai et al., 1996). One study using a commercial NA mixture found significant degradation (~85%) under aerobic conditions, but only a small fraction of the NA mixture was completely mineralized to CO_2 (Misiti et al., 2013). The use of phototrophic organisms therefore has great bioremediation potential (Kobayashi and Rittman, 1982). For example, phototrophic algae, not only may biotransform or bioaccumulate organic compounds but they have the added advantage of providing a constant supply of oxygen for the synergistic NA-degrading microbial communities (Kobayashi and Rittman, 1982). For example, one study showed that the addition of *Chlorella kessleri* increased the *in situ* microbial diversity, resulting in faster and more complete degradation of model NAs (Yu et al., 2019). Another study demonstrated that both *cis-* and *trans-*isomers of a model NA (4-methylcyclohexaneacetic acid) were completely degraded at high concentrations ($5.5\,mg\,L^{-1}$) by the diatom *Naviculla* sp. within two weeks, although the strain was unable to degrade NAs found in oil sands mixtures (Headley et al., 2008). Degradation of NAs in oil sands mixtures however, was demonstrated by the green alga, *Selenastrum* sp. (Headley et al., 2008). Furthermore, Quesnel et al. (2011) showed a removal of tailings associated NAs (i.e. 11–17 carbon compounds of the *Z* family -2) with the unicellular alga *Dunaliella tertiolecta.* These observed differences in degradation may be linked to different transport mechanisms in algae and diatoms, or the NAs inducing osmotic stress, or as a result of differences in NA concentration and structure (Headley et al., 2008).

In contrast to aerobic NA biodegradation, anaerobic NA biodegradation has received less attention. Transcriptome and enzyme activity data reported that the degradation pathway for the model NA, cyclohexylacetic acid (CHAA) induced under aerobic conditions could still work in anaerobic conditions with *Pseudoalteromonas* sp. (Zan et al., 2022). Another study which enriched a microbiome from an OSTP under anaerobic conditions (either sulphate-reducing or methanogenic) found no significant degradation of the NAs tested (i.e. adamantane-1-carboxylic acid and the acid-extractable NAs from OSPW) (Folwell et al., 2015). Folwell et al. (2016) further reported the slow anaerobic degradation of 2-methylnaphthalene (but not pyrene) by OSPW microbes and suggested that high molecular weight NAs and PAHs may persist in OSPW under anaerobic conditions (Folwell et al., 2016). In general, the mechanism for anaerobic NA biodegradation remains unclear (Whitby, 2010). However, it is known that under anaerobic conditions, the type and availability of electron acceptors is important in shaping the microbial NA-degrading community. One study using model NAs, showed that degradation depended on the anaerobic electron-accepting conditions (Clothier and Gieg, 2016). Specifically, under sulphate-reducing conditions *Desulfobulbus* and *Desulfomicrobium* were dominant, whilst under nitrate-reducing conditions, *Pseudomonadota, Terrabacter, Limnobacter, Lutibacter,* and *Derxia* were abundant (Clothier and Gieg, 2016). Under methanogenic conditions, *Clostridium* and *Methanosaeta* were dominant, whilst under iron-reducing conditions, methanogenic Archaea, such as *Methanosarcina* and *Methanoculleus* were found (Clothier and Gieg, 2016). The study also showed that most abundant bacterium under iron-reducing conditions was *Trichococcus*, a known citrate-fermenting microorganism, with a shift in the microbiome in NA-amended Fe (III)-reducing cultures from initially biodegrading NAs to utilizing citrate (Clothier and Gieg, 2016). Previously, Siddique et al. (2014) postulated that methanogens using H_2 to produce CH_4 may transfer some electrons to Fe(III), reducing it to Fe(II), thus suggesting a link between iron reduction and methanogenesis. Clothier and Gieg (2016) therefore suggested that the microbial communities in the NA-amended Fe(III)-reducing cultures most likely shifted over time, from initially biodegrading NAs to utilizing citrate.

4.3 Naphthenic acid (NA)-degrading microorganisms

If *in situ* microbial processes are to be exploited for the bioremediation of NA-contaminated environments, then a better understanding of NA-degrading microorganisms and their processes is required. Using a range of model NAs, several NA-degrading microorganisms have been identified, including *Mycobacterium* sp., *Brevibacterium* sp., *Achromobacter* sp., *Corynebacterium* sp., *Rhodococcus* sp., *Acinetobacter* sp., *Alcaligenes* sp., *Pseudomonas* sp., *Flavobacterium* sp., *Moraxella* sp., *Micrococcus* sp., and *Bacillus* sp. (Whitby, 2010). One study found a dominance of *Pseudomonas* sp., *Burkholderia* sp., and *Sphingomonas* sp. during NA biodegradation (Johnson et al. 2011). Whilst other NA-degraders, such as *Ochrobactrum* sp., *Brevundimonas* sp., and *Bacillus* sp. have also been shown to have higher metabolic activity on polycyclic aromatic NAs compared to other classes of NAs (Skeels and Whitby, 2018). Using model NAs, several different microorganisms, including *Pseudomonas citronellolis* and *Mycobacterium austroafricanum*, have demonstrated specialized metabolic capabilities to degrade carboxylic acids with alkyl-substituted aliphatic chains (Smith et al., 2008; Johnson et al., 2011). Studies have also shown that diverse bacteria utilize different commercial NAs as sole sources of carbon and energy, including *Corynebactrium*, *Arthrobacter*, *Acinetobacter*, *Alcaligenes*, *Pseudomonadota*, and *Bacillota* (Scott et al., 2005). One study using a commercial NA mixture showed that 80% of the 16S rRNA gene sequences belonged to the Gammaproteobacteria, including *P. putida* and *P. fluorescens*, as well as known hydrocarbon-degrading bacteria found in OSPW, such as *Microbulbifer* and *Xanthomonas* (Misiti et al., 2013). *Methylophilus* and *Methylobacillus*, which are known methanol-utilizing bacteria, were also present in a large proportion (63–64%) of the community, although their role in NA biodegradation is not yet fully understood, but likely as a result of methanol used in the synthetic wastewater (Misiti et al., 2013).

Many previous studies have demonstrated the ability of *Pseudomonadota* to degrade a range of different NAs (Whitby, 2010). For example, *P. fluorescens* and *P. putida* can degrade various model NAs (Blakley and

Papish, 1982; Del Rio et al., 2006; Johnson et al., 2013). *Pseudomonadota* were also reported to increase in abundance during the degradation of the highly branched aromatic NAs by mixed enrichment cultures (Johnson et al., 2011). Furthermore, *Pseudomonadota*, especially *P. stutzeri*, dominated enrichment cultures degrading the diamondoid adamantane-1-carboxylic acid (A1CA) (Folwell et al., 2020). Whereas other organisms, such as *Rhodococcus* sp. were not able to degrade the diamondoid tricyclic adamantane carboxylic acid, despite being able to degrade aliphatic and alicyclic carboxylic acid compounds (Presentato et al., 2018). In another study, a 17% reduction in the 10–18 carbon compounds of the Z family -2 to -14 was reported by *P. fluorescens* Pf-5 (McKew et al., 2021). It is therefore possible that under certain conditions, *Pseudomonadota* have a competitive advantage, not only by withstanding NA toxicity, but as likely NA-degrading genera. Thus, targeting pseudomonads in OSPW may be a bioremediation strategy for oil sands operators for enhanced NA removal. Indeed the use of *Pseudomonadota* has provided a further insight into the NA biodegradation mechanisms and the cellular stress responses following exposure to NAs (see Section 4.5).

4.4 Metabolic pathways of naphthenic acid (NA) biodegradation

In general, the pathways for aerobic NA biodegradation includes alpha-oxidation, beta-oxidation, and aromatization (Clemente and Fedorak, 2005; Whitby, 2010). Based on the metabolic intermediates detected, one study suggested that biodegradation of a model NA (CHCA) followed three possible metabolic pathways, namely: (1) via beta-oxidation, (2) via a pathway similar to benzoate degradation, and (3) via the aromatization of the cyclohexane ring (Clothier and Gieg, 2016). Previous work on aerobic NA biodegradation using model NAs showed that it proceeds via beta-oxidation of the carboxyl side chain, and depends on the degree of alkyl side chain branching (Johnson et al., 2011). Several microorganisms (e.g. *Pseudomonas putida, Acinetobacter anitratum, Alcaligenes faecalis*) among others can metabolise NAs by the beta-oxidation pathway (Blakley and Papish, 1982; Smith et al., 2008; Johnson et al., 2013; Clothier and Gieg, 2016). Biodegradation of the aromatic alkanoic NA; butyl phenyl butanoic acid (BPBA) by *Pseudomonas putida* KT2440 also followed the beta-oxidation pathway (Johnson et al., 2011). During the biodegradation of the model NA *n*-BPBA, a metabolite (*4′-n-butylphenyl)ethanoic acid* (*n-BPEA*) was produced, which was also found previously with mixed enrichment cultures (Johnson et al., 2011); pure cultures of *Mycobacterium* sp. (Johnson et al., 2012); and *Pseudomonas putida* KT2440 (Johnson et al., 2013). It was proposed that the biodegradation of the *n*-BPBA by *P. fluorescens* Pf-5 involved the removal of two carbons from the carboxyl side chain, which is indicative of beta-oxidation (Johnson et al., 2011, 2012, 2013). In addition, a *Mycobacterium* sp. isolate was shown to degrade aromatic NAs by switching between the beta- and omega-oxidation pathways (Johnson et al., 2012).

4.5 Mechanisms involved in naphthenic acid (NA) biodegradation and detoxification

Despite their toxicity and ubiquitous nature, very little is known about the microbial mechanisms involved during NA biodegradation. Until recently, the genes, proteins and mechanisms involved in NA biodegradation were largely unknown. Yet, such information is crucial to facilitate the rapid removal of these toxic, persistent pollutants from contaminated environments. Given that NAs are highly toxic to several organisms, including microorganisms (Frank et al., 2009; Whitby, 2010), the microbiota must first be able to protect itself from the toxic effects of NAs during NA metabolism. Recently, new insights have been reported into the general cellular responses to oxidative stress and cell detoxification mechanisms during NA degradation.

One study which used LC-MS/MS shotgun proteomics with the model organism *Pseudomonas fluorescens* Pf-5 reported multiple putative membrane porins and membrane transporters that were upregulated during the biodegradation of a model NA and a commercial NA mixture (McKew et al., 2021). Specifically, ATP (energy-dependent efflux pumps)-binding cassette (ABC) transporters were upregulated (McKew et al., 2021). Such findings are not surprising given that ABC transporters actively transport chemicals and their metabolites out of cells, protecting the cell from its toxicity effects (Bard, 2000; Klaassen and Lauren, 2010; Hessel et al., 2013; Alharbi et al., 2016). However, the ABC superfamily of transporter proteins may be inhibited by OSPW (and any NAs therein) (Alharbi et al., 2016). Other cellular responses to NA toxicity include the upregulation of putative outer membrane proteins (McKew et al., 2021), such as OmpA that allows slow membrane penetration by small compounds (Van der Heijden et al., 2016) and OprG (part of the OmpW family) involved in Fe transport and facilitates the repair of redox stress-induced damage (Andrews et al., 2003; McPhee et al., 2009). Several methyl-accepting chemotaxis proteins which are involved in cell motility also may play a protective role following NA exposure, such as in a chemotactic response to NA toxicity (McKew et al., 2021). Indeed, a positive correlation between chemotaxis and aerobic biodegradation genes has also been found in the marine environment with *Pseudoalteromonas* sp., with chemotaxis enhanced bacterium movement and NA biodegradation (Zan et al., 2022). Molybdenum and molybdoenzymes which are also involved in chemotaxis (Schwartz and Mendel, 2006; Baraquet et al., 2009; Leimkuhler and Lobbi-Nivol, 2016), and pollutant detoxification (Islam et al., 2004; Chovanec et al., 2012; Slyemi and Bonnefoy, 2012; Kruger et al., 2013) were also found as an environmental stress response (McKew et al., 2021). Moreover, molybdoenzymes require high affinity ABC transporters (Hagen, 2011), and supports the upregulation of ABC transporters and overexpression of a molybdenum (Moco) cofactor biosynthesis protein reported by McKew et al. (2021). Molybdenum metabolism is tightly connected to Fe-S cluster synthesis (Mendel, 2013) and bacterioferritin (Bfr2) which provides resistance to hydrogen peroxidase (Ma et al., 1999; Rivera, 2017), and was also significantly expressed during NA biodegradation (McKew et al., 2021). In *P. aeruginosa*, it was suggested that *Bfr2* (along with molybdoenzyme) upregulation was an oxidative stress response to NA toxicity (McKew et al., 2021).

Several other proteins relating to inorganic ion metabolism were also overexpressed as a cellular response to redox stress including a copper (Cu)-containing nitrite reductase (CuNiR) and a Cu-containing cupredoxin protein capable of electron transfer reactions, including being an electron donor to CuNiR in the denitrification pathway (Zumft, 1997; Vijgenboom et al., 1997). However, the role of CuNiR and MoCo biosynthesis proteins in NA degradation remains unknown (McKew et al., 2021). In another study, genome sequence analysis of a *Cupriavidus gilardii* strain CR3, isolated from a natural asphalt deposit, and utilized NAs, identified genes associated with xenobiotic biodegradation and metal resistance (Wang et al., 2015). Indeed, *C. gilardii* strain CR3 demonstrated heavy metal tolerance, achieved by self-detoxification through ion efflux, metal-complexation and metal-reduction and DNA self-repair mechanisms, suggesting that *C. gilardii* strain CR3 is well adapted to survive the harsh natural asphalt environments that contain high concentrations of NAs and heavy metals (Wang et al., 2015).

Recently, advances have been made into understanding the genetic and metabolic mechanisms involved in NA degradation. McKew et al. (2021) identified several proteins associated with fatty acid, lipid and amino acid metabolism during NA biodegradation. Notably, the formation of oleic, linoleic, palmitic, and steric acids increases during NA biodegradation and these fatty acids are common constituents of prokaryotic and some eukaryotic membranes and predominate in NA-degrading microorganisms (Biryukova et al., 2007; Clemente et al., 2004). Hopanoic and steroidal acids also increase during degradation, and are thought to be derived from the microorganisms responsible for oil biodegradation (Meredith et al., 2000). Specifically, acyl-CoA dehydrogenase, acyl-CoA thioesterase II, and an enoyl-CoA hydratase which metabolise various fatty acids by alpha- and beta-oxidation (Hunt et al., 2012) were identified during NA biodegradation (McKew et al., 2021). It is not surprising that proteins associated with alpha- and beta-oxidation of lipids were overexpressed, given that the aerobic NA biodegradation occurs via alpha-/beta-oxidation pathways (Blakley and Papish, 1982; Johnson et al., 2011) (see Section 4.4). Several putative dehydrogenases, oxidoreductases, carboxylases and transferases were also proposed to be involved in NA biodegradation, especially multiple proteins involved in sequential reactions in fatty acid degradation (McKew et al., 2021). It is not surprising that proteins involved in lipid metabolism were overexpressed during NA biodegradation, given the structural similarity between alkyl-carboxylic acid side chains in some model NAs and fatty acids (McKew et al., 2021). Moreover, it was suggested that *P. fluorescens Pf-5* was using its existing fatty acid catabolic pathways for NA biodegradation (McKew et al., 2021). Interestingly, since these fatty acid pathways are conserved, a wide variety of species are likely to have the enzymatic potential to biodegrade NAs (Whitby, 2010; Skeels and Whitby, 2018).

Recently, a marine *Pseudoalteromonas* strain was reported to degrade the model NA (cyclohexylacetic acid (CHAA) directly under aerobic conditions, but also anaerobically when the bacterium was induced with CHAA in aerobic conditions (Zan et al., 2022). Degradation was activated by acetyl-CoA transferase to form the corresponding cyclohexene, alcohol, and ketone and finally cleaved by hydroxymethylglutarate-CoA lyase (Zan et al., 2022). The biodegradation pathway involved the activation of the hydroxyl group by *atoB* and the anaerobic degradation pathway was consistent with the aerobic pathway after CHAA was activated by *atoB* (Zan et al., 2022). Finally, the aliphatic pimeloyl-CoA was converted into short-chain fatty acid by multistep beta-oxidation, and the product eventually entered into the TCA cycle (Zan et al., 2022). Furthermore, genome sequence analysis of *Cupriavidus gilardii* CR3 revealed that degradation of cyclohexane carboxylic acid (CHCA) undergoes an initial ring-cleavage, and that the products of which are further oxidised by beta-oxidation via several pathways including a mechanism similar to that used for fatty acid oxidation (Wang et al., 2015). The peripheral ring-cleavage process in *C. gilardii* CR3 observed by Wang et al. (2015), was similar to the proposed process for aerobic oxidation of CHCA by aerobic oxidation pathways in *Pseudomonas putida* (Blakley and Papish, 1982). In another study, the gene cluster *chcpca* was transcriptionally induced during the biodegradation of the model NAs CHCA and cyclopentanecarboxylic acid (CPCA) by *Rhodococcus aetherivorans* BCP1 and has been proposed to be involved in the beta-oxidation pathway (Presentato et al., 2018). In *C. gilardii*, homologs to *bad/ali* genes were suggested to be involved in the peripheral ring-cleavage pathway for NA degradation (Wang et al., 2015). The *badH* gene, whose predicted product is a member of the short-chain dehydrogenase/reductase family of enzymes, was reported to be involved in anaerobic benzoate degradation in *Rhodopseudomonas palustris* (Pelletier and Harwood, 2000). Other genes which play a role in CHCA metabolism in the *Arthrobacter* strain ATCC 51369, (previously *Corynebacterium cyclohexanicum*) include *pobA*, a regulator (*pobR*) and transporter (*pobK*) (Iwaki et al., 2005). Using RT (reverse Transcription)-qPCR, a gene cluster (*chpca*) was identified in *Rhodococcus aetherivorans* BCP1 that was transcriptionally induced during the growth on two model NAs, (cyclohexane carboxylic acid (CHCA) and cyclopentane carboxylic acid (CPCA)) (Presentato et al., 2018). The predicted products of the *chcpca* gene cluster are proposed to be involved in aerobic NA degradation in *R. aetherivorans* BCP1. Whole-genome analysis also revealed the presence of *pobA* and *chcpca* gene clusters putatively involved in NA degradation in *Rhodococcus opacus* R7 (Zampolli et al., 2020). Gene expression analysis demonstrated the specific induction of R7 *aliA1* gene, encoding for a long-chain-fatty-acid-CoA ligase, in the presence of CHCA and hexanoic acid and that *aliA1* could be targeted as a biomarker for NA degradation (Zampolli et al., 2020). Another study which used metatranscriptomics to investigate the treatment of both raw and ozonated OSPW using biofilters, identified genes involved in the degradation of organic acids and petroleum-related compounds, specifically, those involved in the metabolism of aromatic compounds and benzoate transport and degradation pathway (Zhang et al., 2020a). All of these studies have enhanced our knowledge of the genes, proteins and mechanisms involved in NA biodegradation. Yet further work is still needed to apply this information for improved bioremediation strategies in the future.

5. Conclusions

NAs are an important group of toxic pollutants that are found in naturally-occurring bitumen-rich environments (e.g. oil sands ores, river/wetland sediments, surface and groundwaters) or may accumulate as a result of anthropogenic activities from oil sand mining operations (e.g. process streams and tailings). These bitumen-rich ecosystems harbor dynamic microbiomes that not only have a profound effect on the composition and behavior of the NAs present, but are also major players in influencing their management and reclamation. It is therefore important to better understand the NA-degrading microbiomes across the different bitumen-rich ecosystems. Yet there is a lack of knowledge on their ecology, metabolic diversity and the mechanisms involved during NA degradation.

Both natural and anthropogenic bitumen-rich ecosystems have a rich taxonomic diversity (often regardless of the degree of NA contamination). Some microorganisms such as *Pseudomonas* sp. dominate across many NA-exposed ecosystems, whilst others, including certain methanogenic Archaea, occur in a few limited NA-contaminated environments. Many microbial processes occur in ecosystems associated with NAs, including sulphate and nitrate reduction, methanogenic, methanotrophic, sulfidogenic and hydrocarbon degradation. There are many drivers that shape these bitumen-rich microbiomes and their processes. For example, the structure and composition of NAs may select for various microbial groups. Nutrient limitations, water quality, co-occurrence of trace metals, and competition for substrates from other non-NA-degrading microbes may also play a role.

Until recently, little was known about the NA biodegradation mechanisms, and the enzymes and genes involved. Yet such information is crucial for the rapid removal of these toxic, persistent pollutants from NA-contaminated environments. However, to thrive in such bitumen-rich environments, the microbiota must first protect itself from the toxic effects of the NAs. Recently, proteomics and transcriptomics approaches have provided new insights into the general cellular responses to oxidative stress and the cell's detoxification mechanisms following exposure to NAs. Specifically, several putative membrane porins and membrane transporters, alongside methyl-accepting chemotaxis proteins, molybdenum and molybdoenzymes are known to be involved as a general stress response to NA toxicity.

During NA degradation, several proteins associated with fatty acid, lipid and amino acid metabolism have also been identified. Specifically, multiple proteins involved in sequential reactions in fatty acid metabolism were found. Given the structural similarity between some NAs and fatty acids, it is not surprising that proteins involved in lipid metabolism were over expressed during NA biodegradation. It has therefore been suggested that microorganisms use their existing fatty acid catabolic pathways for NA biodegradation. More importantly, since these fatty acid pathways are conserved, a wide variety of microbial species are likely to have the enzymatic potential to biodegrade NAs which is important for future bioremediation strategies.

6. Future perspectives

Biologically based treatments, which exploit the degradative ability of microorganisms, have both clear cost and environmental advantages to remove and detoxify NAs. Yet the real challenge is the removal of those NAs that are more persistent in the environment. In order to better understand NA degradation, it is important to elucidate the microbial ecology of NA degraders across different environments. Typically, microbial communities from environments that have a history of NA contamination, such as those found in oil sands and OSPW, may degrade NAs, especially the more recalcitrant NAs, more readily. However, prior exposure to NAs may not necessarily select for NA-degrading communities and NA-degraders have been identified from non-NA contaminated environments. Thus, more research is needed into the distribution, ecology, and mechanisms of NA-degrading microorganisms found in natural systems, especially more detailed spatial and temporal studies that focus on the interactions between natural processes and anthropogenic stressors across the different bitumen-associated habitats.

Significant advances in analytical and molecular methods, including high-throughput sequencing and "omics" approaches, have made some headway into elucidating the NA compounds and the microbiota present across natural and anthropogenic ecosystems. Despite this, we are still unable to unequivocally differentiate between contaminants supplied from natural and anthropogenic contaminant sources, and more research into using biologically-based markers (e.g. genes, proteins, indicator species) alongside new modelling tools may enable the sector to better predict and assess ecosystem health in the AOSR and elsewhere. The rich taxonomic and metabolic diversity found across both natural and anthropogenic ecosystems has provided fundamental knowledge applicable to other environments, particularly regarding anaerobic hydrocarbon biodegradation, and possible biotechnological applications therein. Although, microorganisms found in bitumen-rich environments are well known to degrade NAs, microorganisms without any prior exposure to NAs may also metabolize NAs. Therefore, for future remediation strategies, it may be possible to manipulate environmental conditions to select for only those microbial communities that show higher rates of NA degradation, as well as degrade the more persistent NAs. Indeed, identification of factors that affect the growth of NA-degrading microorganisms (either enhancing or inhibiting growth), such as the addition of nutrients, aeration, and ozonation, has important implications for the development of more effective NA bioremediation strategies. Although there are still significant challenges for developing effective NA bioremediation approaches, information on the distribution, ecology, and *in situ* microbial processes involved in NA-biodegradation will improve the effectiveness of new bioremediation technologies in the future. In conclusion, unlocking the metabolic potential of the microbiomes within natural and anthropogenic bitumen-rich environments will be fundamental to developing new energy technologies and managing our natural resources in the future.

References

Abu Laban, N., Tan, B.F., Dao, A., et al., 2015a. Draft genome sequence of un-cultivated toluene-degrading *Desulfobulbaceae* bacterium Tol-SR, obtained by stable isotope probing using [$^{13}C_6$] toluene. Genome Announc. 3. e01423–14.

Abu Laban, N., Tan, B.F., Dao, A., et al., 2015b. Draft genome sequence of uncultivated *Desulfosporosinus* sp. strain Tol-M, obtained by stable isotope probing using [$^{13}C_6$] toluene. Genome Announc. 3. e01422–14.

Abu Laban, N., Dao, A., Foght, J., 2015c. DNA stable-isotope probing of oil sands tailings pond enrichment cultures reveals different key players for toluene degradation under methanogenic and sulfidogenic conditions. FEMS Microbiol. Ecol. 91 (5).

Abu Laban, N., Dao, A., Semple, K., Foght, J., 2015d. Biodegradation of C_7 and C_8 iso-alkanes under methanogenic conditions. Environ. Microbiol. 17 (12), 4898–4915.

Aguilar, M., Richardson, E., Tan, B., et al., 2016. Next-generation sequencing assessment of eukaryotic diversity in oil sands tailings ponds sediments and surface water. J. Eukaryot. Microbiol. 263, 732–743.

Ahad, J.M.E., Pakdel, H., Savard, M.M., Calderhead, A.I., Gammon, P.R., Rivera, A., Peru, K.M., Headley, J.V., 2013. Characterization and quantification of mining-related "naphthenic acids" in groundwater near a major oil sands tailings pond. Environ. Sci. Technol. 4 (10), 5023–5030.

Ahad, J.M.E., Pakdel, H., Gammon, P.R., Siddique, T., Kuznetsova, A., Savard, M.M., 2018. Evaluating in situ biodegradation of ^{13}C-labelled naphthenic acids in groundwater near oil sands tailings ponds. Sci. Total Environ. 643, 392–399.

Ahad, J.M.E., Pakdel, H., Gammon, P.R., Mayer, B., Savard, M.M., Peru, K.M., Headley, J.V., 2020. Distinguishing natural from anthropogenic sources of acid extractable organics in groundwater near oil sands tailings ponds. Environ. Sci. Technol. 54, 2790–2799.

Ajaero, C., Meulen, I.V., Simair, M.C., le Roux, M., Parrott, J., Peru, K.M., McMartin, D.W., Headley, J.V., 2020. Developments in molecular level characterization of naphthenic acid fraction compounds degradation in a constructed wetland treatment system. Environments 7, 89.

Albaiges, J., Borbon, J., Walker, W., 1985. Petroleum isoprenoid hydrocarbons derived from catagenetic degradation of archaebacterial lipids. Org. Geochem. 8, 293.

Albakistani, E.A., Nwosu, F.C., Furgason, C., Haupt, E.S., Smirnova, A.V., et al., 2022. Seasonal dynamics of methanotrophic bacteria in a boreal oil sands end pit lake. Appl. Ind. Microbiol. 88 (3), e0145521.

Alharbi, H., Saunders, D.M., Al-Mousa, A., Alcorn, J., Pereira, A.S., Martin, J.W., Giesey, J.P., Wiseman, S.B., 2016. Inhibition of ABC transport proteins by oil sands process affected water. Aquat. Toxicol. 170, 81–88.

An, D., Brown, D., Chatterjee, I., Dong, X., Ramos-Padron, E., Wilson, S., et al., 2013. Microbial community and potential functional gene diversity involved in anaerobic hydrocarbon degradation and methanogenesis in an oil sands tailings pond 1. Genome 56, 612–618.

Andrews, S.C., Robinson, A.K., Rodríguez-Quiñones, F., 2003. Bacterial iron homeostasis. FEMS Microbiol. Rev. 27, 215–237.

Arciszewski, T.J., Roberts, D.R., Munkittrick, K.R., Scrimgeour, G.J., 2021. Challenges and benefits of models used to integrate monitoring programs at regional scales. Front. Environ. Sci. 9, 66698.

Arslan, M., El-Din, M.G., 2021. Bacterial diversity in petroleum coke based biofilters treating oil sands process water. Sci. Total Environ. 782, 146742.

Arslan, M., Müller, J.A., El-Din, M.G., 2022. Aerobic naphthenic acid-degrading bacteria in petroleum-coke improve oil sands process water remediation in biofilters: DNA-stable isotope probing reveals methylotrophy in Schmutzdecke. Sci. Total Environ. 815, 151961.

Baraquet, C., Theraulaz, L., Lobbi-Nivol, C., Mejean, V., Jourlin-Castelli, C., 2009. Unexpected chemoreceptors mediate energy taxis towards electron acceptors in *Shewanella oneidensis*. Mol. Microbiol. 73, 278–290.

Baray, S., Darlington, A., Gordon, M., Hayden, K.L., Leithead, A., Li, S.M., Liu, P.S., Mittermeier, R.L., Moussa, S.G., O'Brien, J., Staebler, R., 2018. Quantification of methane sources in the Athabasca Oil Sands Region of Alberta by aircraft mass balance. Atmos. Chem. Phys. 18 (10), 7361–7378.

Bard, S., 2000. Multi-xenobiotic resistance as a cellular defence mechanism in aquatic organisms. Aquat. Toxicol. 48, 357–389.

Barros, E.V., Filgueiras, P.R., Lacerda, V., Rodgers, R.P., Romão, W., 2022. Characterization of naphthenic acids in crude oil samples—a literature review. Fuel 319, 123775.

Barrow, M.P., Headley, J.V., Peru, K.M., Derrick, P.J., 2009. Data visualization for the characterization of naphthenic acids within petroleum samples. Energy Fuel 23, 2592–2599.

Bartlett, A.J., Frank, R.A., Gillis, P.L., Parrott, J.L., Marentette, J.R., Brown, L.R., Hooey, T., Vanderveen, R., McInnis, R., Brunswick, P., Shang, D., Headley, J.V., Peru, K.M., Hewitt, L.M., 2017. Toxicity of naphthenic acids to invertebrates: extracts from oil sands process-affected water versus commercial mixtures. Environ. Pollut. 227, 271–279.

Bataineh, M., Scott, A.C., Fedorak, P.M., Martin, J.W., 2006. Capillary HPLC/QTOFMS for characterizing complex naphthenic acid mixtures and their microbial transformation. Anal. Chem. 78, 8354–8361.

Bauer, A.E., Frank, R.A., Headley, J.V., Peru, K.M., Hewitt, L.M., Dixon, D.G., 2015. Enhanced characterization of oil sands acid-extractable organics fractions using electrospray ionization-high-resolution mass spectrometry and synchronous fluorescence spectroscopy. Environ. Toxicol. Chem. 34, 1001–1008.

Baugh, T.D., Wolf, N.O., Mediaas, H., Vindstad, J.E., Grande, K., 2004. Characterization of a calcium naphthenate deposit—The ARN acid discovery. Am. Chem. Soc. Div. Pet. Chem. 49, 274–276.

Beddow, J., Johnson, R.J., Lawson, T., Breckels, M.N., Webster, R.J., Smith, B.E., Rowland, S.J., Whitby, C., 2016. The effect of oil sands process-affected water and model naphthenic acids on photosynthesis and growth in *Emiliania huxleyi* and *Chlorella vulgaris*. Chemosphere 145, 416–423.

Bergsveinson, J., Lawrence, J.R., Schebel, A., Wasserscheid, J., Roy, J., Malcolm Conly, F., Sanschagrin, S., Korber, D.R., Tremblay, J., Greer, C.W., Droppo, I.G., 2021. Impact of sample collection on prokaryotic and eukaryotic diversity of niche environments of the oil-sand mining impacted Athabasca River. Can. J. Microbiol. 67 (11), 813–826.

Biryukova, O.V., Fedorak, P.M., Quideau, S.A., 2007. Biodegradation of naphthenic acids by rhizosphere microorganisms. Chemosphere 67, 2058–2064.

Blakley, E.R., Papish, B., 1982. The metabolism of cyclohexanecarboxylic acid and 3-cyclohexenecarboxylic acid by *Pseudomonas putida*. Can. J. Microbiol. 28, 1324–1329.

Bowman, D.T., Slater, G.F., Warren, L.A., McCarry, B.E., 2014. Identification of individual thiophene-, indane-, tetralin-, cyclohexane-, and adamantane-type carboxylic acids in composite tailings pore water from Alberta oil sands. Rapid Commun. Mass Spectrom. 28, 2075–2083.

Brandt, J.P., Flannigan, M.D., Maynard, D.G., Thompson, I.D., Volney, W.J.A., 2013. An introduction to Canada's boreal zone: ecosystem processes, health, sustainability, and environmental issues. Environ. Rev. 21, 207–226.

Brient, J.A., Wessner, P.J., Doly, M.N., 1995. Naphthenic acids. In: Kroschwitz, J.I. (Ed.), Encyclopedia of chemical technology, fourth ed. vol. 16. Wiley, New York, pp. 1017–1029.

Brook, J.R., Cober, S.G., Freemark, M., Harner, T., Li, S.M., Liggio, J., et al., 2019. Advances in science and applications or air pollution monitoring: a case study on oil sands monitoring targeting ecosystem protection. J. Air Waste Manag. Assoc. 69, 661–709.

Brown, L.D., Ulrich, A.C., 2015. Oil sands naphthenic acids: a review of properties, measurement, and treatment. Chemosphere 127, 276–290.

Chalaturnyk, R.J., Scott, J.D., Özüm, B., 2002. Management of oil sands tailings. Petrol. Sci. Technol. 20, 1025–1046.

Chambers, P.A., Culp, J.M., Glozier, N.E., Cash, K.J., Wrona, F.J., Noton, L., 2006. Northern rivers ecosystem initiative: nutrients and dissolved oxygen-issues and impacts. Environ. Monit. Assess. 113, 117–141.

Chambers, P.A., Alexander-Trusiak, A., Kirk, J., Manzano, C., Muir, D., Cooke, C., Hazewinkel, R., 2018. Surface Water Quality of Lower Athabasca River Tributaries. Oil Sands Monitoring Program Technical Report Series No. 1.3.

Chappé, B., Albrecht, P., Michaelis, W., 1982. Polar lipids of archaebacteria in sediments and petroleums. Science 217, 65–66.

Chi Fru, E., Chen, M., Walshe, G., et al., 2013. Bioreactor studies predict whole microbial population dynamics in oil sands tailings ponds. Appl. Microbiol. Biotechnol. 97, 3215–3224.

Choi, J., Hwang, G., El-Din, M.G., Liu, Y., 2014. Effect of reactor configuration and microbial characteristics on biofilm reactors for oil sands process-affected water treatment. Int. J. Biodeterior. Biodegrad. 89, 74–81.

Chovanec, P., Sparacino-Watkins, C., Zhang, N., Barsu, P., Stolz, J.F., 2012. Microbial reduction of chromate in the presence of nitrate by three nitrate respiring organisms. Front. Microbiol. 3, 416.

Clemente, J.S., Fedorak, P.M., 2005. A review of the occurrence, analyses, toxicity, and biodegradation of naphthenic acids. Chemosphere 60, 585–600.

Clemente, J.S., MacKinnon, M.D., Fedorak, P.M., 2004. Aerobic biodegradation of two commercial naphthenic acid preparations. Environ. Sci. Technol. 38, 1009–1016.

Clothier, L.N., Gieg, L.M., 2016. Anaerobic biodegradation of surrogate naphthenic acids. Water Res. 1 (90), 156–166.

Collins, V.C.E., Foght, J.M., Siddique, T., 2016. Co-occurrence of methanogenesis and N_2 fixation in oil sands tailings. Sci. Total Environ. 565, 306–312.

Connor, S.J., Alexander-Trusiak, A.C., Baird, D.J., 2018. Vulnerability of diatom communities in the peace-Athabasca Delta to environmental change. PeerJ 6, e5447. https://doi.org/10.7717/peerj.5447.

Cossey, H.L., Anwar, M.N., Kuznetsov, P.V., Ulrich, A.C., 2021. Biofilms for turbidity mitigation in oil sands end pit lakes. Microorganisms 9, 1443.

Culp, J.M., Droppo, I.G., di Cenzo, P.D., Alexander, A.C., Baird, D.J., Beltaos, S., Bickerton, G., Bonsal, B., Brua, R.B., Chambers, P.A., Dibike, Y., Glozier, N.E., Kirk, J.L., Levesque, L., McMaster, M., Muir, D., Parrott, J.L., Peters, D.L., Pippy, K., Roy, J.W., 2021. Ecological effects and causal synthesis of oil sands activity impacts on river ecosystems: water synthesis review. Environ. Rev. 29 (2), 315–327.

Del Rio, L.F., Hadwin, A.K.M., Pinto, L.J., MacKinnon, M.D., Moore, M.M., 2006. Degradation of naphthenic acids by sediment microorganisms. J. Appl. Microbiol. 101, 1049–1061.

Demeter, M.A., Lemire, J.A., Yue, G., Ceri, H., Turner, R.J., 2015. Culturing oil sands microbes as mixed species communities enhances ex situ model naphthenic acid degradation. Front. Microbiol. 6, 936.

Dompierre, K.A., Barbour, S.L., 2016. Characterization of physical mass transport through oil sands fluid fine tailings in an end pit lake: a multi-tracer study. J. Contam. Hydrol. 189, 12–26.

Dorn, P.B., 1992. Case histories-the petroleum industry. In: Ford, D.L. (Ed.), Toxicity Reduction: Evaluation and Control. Water Quality Management Library. vol. 3. Technomic Publishing Company, Lancaster, PA, pp. 183–223.

Dutta, T.K., Harayama, S., 2001. Biodegradation of n-alkylcycloalkanes and n-alkyl-benzens via new pathways in Alcanivorax sp. strain MBIC 4326. Appl. Environ. Microbiol. 67, 1970–1974.

Fedorak, P.M., Coy, D.L., Salloum, M.J., et al., 2002. Methanogenic potential of tailings samples from oil sands extraction plants. Can. J. Microbiol. 48, 21–33.

Fedorak, P.M., Coy, D.L., Dudas, M.J., et al., 2003. Microbially-mediated fugitive gas production from oil sands tailings and increased tailings densification rates. J. Environ. Eng. Sci. 2, 199–211.

Fennell, J., Arciszewski, T.J., 2019. Current knowledge of seepage from oil sands tailings ponds and its environmental influence in northeastern Alberta. Sci. Total Environ. 686, 968–985.

Ferguson, G.P., Rudolph, D.L., Barker, J.F., 2009. Hydro-dynamics of a large oil sand tailings impoundment and related environmental implications. Can. Geotechnol. J. 46, 1446–1460.

Foght, J.M., Gieg, L.M., Siddique, T., 2017. The microbiology of oil sands tailings: past, present, future. FEMS Microbiol. Ecol. 93, fix034.

Folwell, B.D., McGenity, T.J., Price, A., Johnson, R.J., Whitby, C., 2015. Exploring the capacity for anaerobic biodegradation of polycyclic aromatic hydrocarbons and naphthenic acids by microbes from oil sands process affected waters. Int. J. Biodeterior. Biodegrad. 108, 214–221.

Folwell, B.D., McGenity, T.J., Whitby, C., 2016. Biofilm and planktonic bacterial and fungal communities transforming high molecular weight polyaromatic hydrocarbons. Appl. Environ. Microbiol. 82, 2288–2299.

Folwell, B.D., McGenity, T.J., Whitby, C., 2020. Diamondoids are not forever: Microbial biotransformation of diamondoid carboxylic acids. J. Microbial. Biotechnol. 13, 495–508.
Foote, L., 2012. Threshold considerations and wetland reclamation in Alberta's mineable oil sands. Ecol. Soc. 17, 35.
Frank, R.A., Fischer, K., Kavanagh, R., Burnison, B.K., Arsenault, G., Headley, J.V., Peru, K.M., Van Der Kraak, G., Solomon, K.R., 2009. Effect of carboxylic acid content on the acute toxicity of oil sands naphthenic acids. Environ. Sci. Technol. 43, 266–271.
Frankel, M.L., Bhuiyan, T.I., Veksha, A., Demeter, M.A., Layzell, D.B., Helleur, R.J., Hill, J.M., Turner, R.J., 2016. Removal and biodegradation of naphthenic acids by biochar and attached environmental biofilms in the presence of co-contaminating metals. Bioresour. Technol. 216, 352–361.
Gates, I.D., Larter, S.R., 2014. Energy efficiency and emissions intensity of SAGD. Fuel 115, 706–713.
Golby, S., Ceri, H., Gieg, L., Chatterjee, I., Marques, L.L.R., Turner, R.J., 2012. Evaluation of microbial biofilm communities from an Alberta oil sands tailings pond. FEMS Microbiol. Ecol. 79, 240–250.
Grewer, D.M., Young, R.F., Whittal, R.M., Fedorak, P.M., 2010. Naphthenic acids and other acid-extractables in water samples from Alberta: what is being measured? Sci. Total Environ. 408, 5997–6010.
Gunawan, Y., Nemati, M., Dalai, A., 2014. Biodegradation of a surrogate naphthenic acid under denitrifying conditions. Water Res. 15 (51), 11–24.
Hadwin, A.K.M., Del Rio, L.F., Pinto, L.J., Painter, M., Routledge, R., Moore, M., 2006. Microbial communities in wetlands of the Athabasca oil sands: genetic and metabolic characterization. FEMS Microbiol. Ecol. 55, 68–78.
Hagen, W.R., 2011. Cellular uptake of molybdenum and tungsten. Coord. Chem. Rev. 255, 1117–1128.
Han, X., Scott, A.C., Fedorak, P.M., Batianeh, M., Martin, J.W., 2008. Influence of molecular structure on the biodegradability of naphthenic acids. Environ. Sci. Technol. 42, 1290–1295.
Hazewinkel, R.R.O., Wolfe, A.P., Pla, S., Curtis, C., Hadley, K., 2008. Have atmospheric emissions from the Athabasca oil sands impacted lakes in northeastern Alberta, Canada? Can. J. Fish. Aquat. Sci. 65 (8), 1554–1567.
Head, I.M., Jones, M., Larter, S., 2003. Biological activity in the deep subsurface and the origin of heavy oil. Nature 426, 344–352.
Headley, J.V., McMartin, D.W., 2004. A review of the occurrence and fate of naphthenic acids in aquatic environments. J. Environ. Sci. Health A 39, 1989–2010.
Headley, J.V., Conly, M., Dickson, L.C., Akve, C., Peru, K.M., 2000. Characterization of natural hydro- carbon release from oil sands deposits in tributaries of the Athabasca River Basin, Canada. Can. Tech. Rep. Fish Aquat. Sci. 2331, 89–90.
Headley, J.V., Du, J.L., Peru, K.M., Gurprasa, N., McMartin, D.W., 2008. Evaluation of algal phytodegradation of petroleum naphthenic acids. J. Environ. Sci. Health 43, 227–232.
Headley, J., Peru, K., Adenugba, A., Du, J.L., McMartin, D.W., 2010. Dissipation of naphthenic acids mixtures by lake biofilms. J. Environ. Sci. Health Part A 45, 1027–1036.
Hendrikse, M., Gaspari, D.P., McQueen, A.D., Kinley, C.M., Calomeni, A.J., Geer, T.D., Simair, M.C., Peru, K.M., Headley, J.V., Rodgers, J.H., et al., 2018. Treatment of oil sands process-affected waters using a pilot-scale hybrid constructed wetland. Ecol. Eng. 115, 45–57.
Herman, D.C., Fedorak, P.M., Costerton, J.W., 1993. Biodegradation of cycloalkane carboxylic acids in oil sand tailings. Can. J. Microbiol. 39, 576–580.
Herman, D.C., Fedorak, P.M., MacKinnon, M.D., Costerton, J.W., 1994. Biodegradation of naphthenic acids by microbial populations indigenous to oil sands tailings. Can. J. Microbiol. 40, 467–477.

Hessel, S., Seidel, J.A., Lampen, A., 2013. Multidrug resistance-associated proteins are involved in the transport of the glutathione conjugates of the ultimate carcinogen of benzo[a]pyrene in human Caco-2 cells. Arch. Toxicol. 87, 269–280.

Hewitt, L.M., Roy, J.W., Rowland, S.J., Bickerton, G., DeSilva, A., Headley, J.V., Milestone, C.B., Scarlett, A.G., Brown, S., Spencer, C., West, C.E., Peru, K.M., Grapentine, L., Ahad, J.M.E., Pakdel, H., Frank, R.A., 2020. Advances in distinguishing groundwater influenced by oil sands process-affected water (OSPW) from natural bitumen-influenced groundwaters. Environ. Sci. Technol. 54, 1522–1532.

Holowenko, F.M., MacKinnon, M.D., Fedorak, P.M., 2002. Characterization of naphthenic acids in oil sands wastewaters by gas chromatography-mass spectrometry. Water Res. 36, 2843–2855.

Hubert, C.R.J., Oldenburg, T.B.P., Fustic, M., et al., 2012. Massive dominance of *Epsilonproteo bacteria* in formation waters from a Canadian oil sands reservoir containing severely biodegraded oil. Environ. Microbiol. 14, 387–404.

Hunt, M.C., Siponen, M.I., Alexson, S.H.E., 2012. The emerging role of acyl-CoA thioesterases and acyltransferases in regulating peroxisomal lipid metabolism. Biochim. Biophys. Acta 1822, 1397–1410.

Islam, F.S., Gault, A.G., Boothman, C., Polya, D.A., Charnock, J.M., Chatterjee, D., Lloyd, J.R., 2004. Role of metal-reducing bacteria in arsenic release from Bengal delta sediments. Nature 430, 68–71.

Islam, M.S., Dong, T., Sheng, Z., Zhang, Y., Liu, Y., Gamal El-Din, M., 2014a. Microbial community structure and operational performance of a fluidized bed biofilm reactor treating oil sands process-affected water. Int. J. Biodeterior. Biodegrad. 91, 111–118.

Islam, M.S., Dong, T., McPhedran, K.N., Sheng, Z., Zhang, Y., Liu, Y., Gamal El-Din, M., 2014b. Impact of ozonation pre-treatment of oil sands process affected water on the operational performance of a GAC-fluidized bed biofilm reactor. Biodegradation 25, 811–823.

Islam, M.S., Zhang, Y., McPhedran, K.N., Liu, Y., Gamal, E.-D.M., 2015. Next-generation pyrosequencing analysis of microbial biofilm communities on granular activated carbon in treatment of oil sands process-affected water. Appl. Environ. Microbiol. 81 (12), 4037–4048.

Iwaki, H., Saji, H., Abe, K., Hasegawa, Y., 2005. Cloning and sequence analysis of the 4-hydroxybenzoate 3-hydroxylase gene from a cyclohexanecarboxylate-degrading Gram-positive bacterium, "*Corynebacterium cyclohexanicum*" strain ATCC 51369. Microb. Environ. 20, 144–150.

Jeeravipoolvarn, S., Miller, W., Scott, D., Kabwe, L., Wilson, W., 2017. Modeling effect of bitumen extraction process on oil sand tailings ponds. J. Civil Eng. Architec. 11, 48–59.

Johnson, R., Smith, B.E., Sutton, P., McGenity, T.J., Rowland, S.J., Whitby, C., 2011. Microbial biodegradation of aromatic alkanoic naphthenic acids is affected by the degree of alkyl side chain branching. ISME J. 5, 486–496.

Johnson, R.J., West, C., Swaih, A.M., Folwell, B.D., Smith, B.E., Rowland, S.J., Whitby, C., 2012. Aerobic biotransformation of alkyl branched aromatic alkanoic naphthenic acids via two different pathways by a new isolate of *Mycobacterium*. Environ. Microbiol. 14, 872–882.

Johnson, R.J., Smith, B.E., Rowland, S.J., Whitby, C., 2013. Biodegradation of alkyl branched aromatic alkanoic naphthenic acids by *Pseudomonas putida* KT2440. Int. J. Biodeterior. Biodegrad. 81, 3–8.

Jones, D., West, C.E., Scarlett, A.G., Frank, R., Rowland, S.J., 2012. Isolation and estimation of the 'aromatic' naphthenic acid content of an oil sands process-affected water extract. J. Chromatogr. A 1247, 171–175.

Kannel, P.R., Gan, T.Y., 2012. Naphthenic acids degradation and toxicity mitigation in tailings wastewater systems and aquatic environments: a review. J. Environ. Sci. Health A 47 (1), 1–21.

Kelly, E.N., Short, J.W., Schindler, D.W., Hodson, P.V., Ma, M., Kwan, A.K., Fortin, B.L., 2009. Oil sands development contributes polycyclic aromatic compounds to the Athabasca River and its tributaries. Proc. Natl. Acad. Sci. U. S. A. 106, 22346–22351.

Kelly, E.N., Schindler, D.W., Hodson, P.V., Short, J.W., Radmanovich, R., Nielsen, C.C., 2010. Oil sands development contributes elements toxic at low concentrations to the Athabasca River and its tributaries. Proc. Natl. Acad. Sci. U. S. A. 107, 16178–16183.

Kilgour, B., Mahaffey, A., Brown, C., Hughes, S., Hatry, C., Hamilton, L., 2019. Variation in toxicity and ecological risks associated with some oil sands groundwaters. Sci. Total Environ. 659, 1224–1233.

Klaassen, C., Lauren, M., 2010. Xenobiotic, bile acid, and cholesterol transporters: function and regulation. Pharmacol. Rev. 62, 1–96.

Kobayashi, H., Rittman, B.E., 1982. Microbial removal of hazardous organic compounds. Environ. Sci. Technol. 16, 170A–183A.

Kruger, M.C., Bertin, P.N., Heipieper, H.J., Arsène-Ploetze, F., 2013. Bacterial metabolism of environmental arsenic—mechanisms and biotechnological applications. Appl. Microbiol. Biotechnol. 97, 3827–3841.

Kurek, J., Kirk, J.L., Muir, D.C.G., Wang, X., Evans, M.S., Smol, J.P., 2013. Legacy of half a century of Athabasca oil sands development recorded by lake ecosystems. Proc. Natl. Acad. Sci. U. S. A. 110, 1761–1766.

Lai, J.L., Pinto, W.S., Kiehlmann, E., Bendell-Young, L.I., Moore, M.M., 1996. Factors that affect the degradation of naphthenic acids in oil sands wastewater by indigenous microbial communities. Environ. Toxicol. Chem. 15, 1482–1491.

Laird, K.R., Das, B., Hesjedal, B., Leavitt, P.R., Mushet, G.R., Scott, K.A., Simpson, G.L., Wissel, B., Wolfe, J., Cumming, B.F., 2017. Paleolimnological assessment of nutrient enrichment on Diatom assemblages in a priori defined nitrogen- and phosphorus-limited lakes downwind of the Athabasca oil sands. J. Limnol. 76 (3), 488–502.

Lawrence, G.A., Tedford, E.W., Pieters, R., 2016. Suspended solids in an end pit lake: potential mixing mechanisms. Can. J. Civ. Eng. 43, 211–217.

Leimkuhler, S., Lobbi-Nivol, C., 2016. Bacterial molybdoenzymes: old enzymes for new purposes. FEMS Microbiol. Rev. 40, 1–18.

Lemire, J.A., Demeter, M.A., George, I., Ceri, H., Turner, R.J., 2015. A novel approach for harnessing biofilm communities in moving bed biofilm reactors for industrial wastewater treatment. AIMS Bioeng. 2 (4), 387–403.

Li, C., Fu, L., Stafford, J., Belosevic, M., Gamal El-Din, M., 2017. The toxicity of oil sands process-affected water (OSPW): a critical review. Sci. Total Environ. 601–602, 1785–1802 (2).

Liu, H., Yang, G., Jia, H., Sun, B., 2022. Crude oil degradation by a novel strain *Pseudomonas aeruginosa* AQNU-1 isolated from an oil-contaminated lake Wetland. Processes 10 (2), 307.

Lutnaes, B.F., Krane, J., Smith, B.E., Rowland, S.J., 2007. Structure elucidation of C80, C81 and C82 isoprenoid tetracids responsible for naphthenate deposition in crude oil production. Org. Biomol. Chem. 5, 1873–1877.

Lv, X., Ma, B., Cologgi, D., Lee, K., Ulrich, A., 2020. Naphthenic acid anaerobic biodegrading consortia enriched from pristine sediments underlying oil sands tailings ponds. J. Hazard. Mater. 394, 122546.

Ma, J.-F., Ochsner, U.A., Klotz, M.G., Nanayakkara, V.K., Howell, M.L., Johnson, Z., Posey, J.E., Vasil, M.L., Monaco, J.J., Hassett, D.J., 1999. Bacterioferritin a modulates catalase a (KatA) activity and resistance to hydrogen peroxide in *Pseudomonas aeruginosa*. J. Bacteriol. 181, 3730–3742.

MacKinnon, M.D., 1989. Development of the tailings pond at Syn-crude's oil sands plant: 1978–1987. AOSTRA J. Res. 5, 109–133.

MacKinnon, M.D., Boerger, H., 1986. Desription of two treatment methods for detoxifying oil sands tailings pond water. Water Pollut. Res. J. Can. 21, 496–512.
Mahaffey, A., Dubé, M., 2017. Review of the composition and toxicity of oil sands process-affected water. Environ. Rev. 25 (1), 97–114.
Mahdavi, H., Prasad, V., Liu, Y., Ulrich, A.C., 2015. In situ biodegradation of naphthenic acids in oil sands tailings pond water using indigenous algae-bacteria consortium. Bioresour. Technol. 187, 97–105.
Marentette, J.R., Frank, R.A., Bartlett, A.J., Gillis, P.L., Hewitt, L.M., Peru, K.M., Headley, J.V., Brunswick, P., Shang, D., Parrott, J.L., 2015. Toxicity of naphthenic acid fraction components extracted from fresh and aged oil sands process-affected waters, and commercial naphthenic acid mixtures, to fathead minnow (*Pimephales promelas*) embryos. Aquat. Toxicol. 164, 108–117.
Matthews, J., Shaw, W., Mackinnon, M., Cuddy, R., 2002. Development of composite tailings technology at Syncrude. Int. J. Surf. Min. Reclam. Environ. 16, 24–29.
McKenzie, N., Yue, S., Liu, X., Ramsay, B.A., Ramsay, J.A., 2014. Biodegradation of naphthenic acids in oils sands process waters in an immobilized soil/sediment bioreactor. Chemosphere 109, 164–172.
McKew, B., Johnson, R., Clothier, L., Skeels, K., Ross, M., Frenzel, M., Gieg, L., Martin, J., Hough, M., Whitby, C., 2021. Differential protein expression during growth on model and commercial mixtures of naphthenic acids in *Pseudomonas fluorescens* Pf-5. Microbiol. Open 10 (4), e1196.
McPhee, J.B., Tamber, S., Bains, M., Maier, E., Gellatly, S., Lo, A., Benz, R., Hancock, R.E.W., 2009. The major outer membrane protein OprG of *Pseudomonas aeruginosa* contributes to cytotoxicity and forms an anaerobically regulated, cation-selective channel. FEMS Microbiol. Lett. 296, 241–247.
McQueen, A.D., Hendrikse, M., Gaspari, D.P., Kinley, C.M., Rodgers, J.H., Castle, J.W., 2017. Performance of a hybrid pilot-scale constructed wetland system for treating oil sands process-affected water from the Athabasca oil sands. Ecol. Eng. 102, 152–165.
Mendel, R.R., 2013. The molybdenum cofactor. J. Biol. Chem. 288, 13165–13172.
Meredith, W., Kelland, S.-J., Jones, D.M., 2000. Influence of biodegradation on crude oil acidity and carboxylic acid composition. Org. Geochem. 31, 1059–1073.
Mikula, R.J., Kasperski, K.L., Burns, R.D., et al., 1996. Nature and fate of oil sands tailings. In: Schramm, L.L. (Ed.), Suspensions: Fundamentals and Applications in the Petroleum Industry. American Chemical Society, Washington, DC, pp. 677–723.
Miles, S.M., Asiedu, E., Balaberda, A.L., Ulrich, A.C., 2020. Oil sands process affected water sourced Trichoderma harzianum demonstrates capacity for mycoremediation of naphthenic acid fraction compounds. Chemosphere 258, 127281.
Misiti, T.M., Tezel, U., Tandukar, M., 2013. Aerobic biotransformation potential of a commercial mixture of naphthenic acids. Water Res. 47, 5520–5534.
Mohamad Shahimin, M.F., Foght, J.M., Siddique, T., 2016. Preferential methanogenic biodegradation of short-chain *n*-alkanes by microbial communities from two different oil sands tailings ponds. Sci. Total Environ. 553, 250–257.
Mollard, F.P., Roy, M.C., Foote, A.L., 2015. Performance of wetland forbs transplanted into marshes amended with oil sands processed water. Environ. Monit. Assess. 187 (3), 1–8.
Morandi, G.D., Wiseman, S.B., Pereira, A., Mankidy, R., Gault, I.G.M., Martin, J.W., Giesy, J.P., 2015. Effects-directed analysis of dissolved organic compounds in oil sands process-affected water. Environ. Sci. Technol. 49, 12395–12404.
Murray, K.R., Barlow, N., Strack, M., 2017. Methane emissions dynamics from a constructed fen and reference sites in the Athabasca Oil Sands Region, Alberta. Sci. Total Environ. 1 (583), 369–381.
NASA, 2009. Earth Observatory Image. http://earthobservatory.nasa.gov/IOTD/view.php?id=40997.

Paslawski, J., Nemati, M., Hill, G., Headley, J.V., 2009. Biodegradation kinetics of trans-4-methyl-1-cyclohexane carboxylic acid in continuously stirred tank and immobilized cell bioreactors. J. Chem. Technol. Biotechnol. 84, 992–1000.

Paulssen, J.M., Gieg, L.M., 2019. Biodegradation of 1 adamantane carboxylic acid by algal-bacterial microbial communities derived from oil sands tailings ponds. Algal Res. 41, 1–7.

Pelletier, D.A., Harwood, C.S., 2000. 2-Hydroxycyclohexanecarboxyl coenzyme A dehydrogenase, an enzyme characteristic of the anaerobic benzoate degradation pathway used by *Rhodopseudomonas palustris*. J. Bacteriol. 182, 2753–2760.

Penner, T.J., Foght, J.M., 2010. Mature fine tailings from oil sands processing harbour diverse methanogenic communities. Can. J. Microbiol. 56 (6), 459–470.

Pollet, I., Bendell-Young, L.I., 2000. Amphibians as indicators of wetland quality in wetlands formed from oil sands effluent. Environ. Toxicol. Chem. 19, 2589–2597.

Presentato, A., Cappelletti, M., Sansone, A., Ferreri, C., Piacenza, E., Demeter, M.A., Crognale, S., Petruccioli, M., Milazzo, G., Fedi, S., Steinbüchel, A., Turner, R.J., Zannoni, D., 2018. Aerobic growth of *Rhodococcus aetherivorans* BCP1 using selected naphthenic acids as the sole carbon and energy sources. Front. Microbiol. 9, 672.

Quesnel, D.M., Bhaskar, I.M., Gieg, L.M., Chua, G., 2011. Naphthenic acid biodegradation by the unicellular alga Dunaliella tertiolecta. Chemosphere 84 (4), 504–511.

Quinlan, P., Tam, K., 2015. Water treatment technologies for the remediation of naphthenic acids in oil sands process-affected water. Chem. Eng. J. 279, 696–714.

Ramos-Padrón, E., Bordenave, S., Lin, S., et al., 2011. Carbon and sulfur cycling by microbial communities in a gypsum-treated oil sands tailings pond. Environ. Sci. Technol. 45, 439–446.

Reddy, C.M., Arey, J.S., Seewald, J.S., Sylva, S.P., Lemkau, K.L., Nelson, R.K., Carmichael, C.A., McIntyre, C.P., Fenwick, J., Ventura, G.T., Van Mooy, B.A., 2012. Composition and fate of gas and oil released to the water column during the deepwater horizon oil spill. Proc. Natl. Acad. Sci. U. S. A. 109 (50), 20229–20234.

Reid, M.L., Warren, L.A., 2016. S reactivity of an oil sands composite tailings deposit undergoing reclamation wetland construction. J. Environ. Manage. 166, 321–329.

Reid, T., Chaganti, S.R., Droppo, I.G., Weisener, C.G., 2018. Novel insights into freshwater hydrocarbon-rich sediments using metatranscriptomics: opening the black box. Water Res. 136, 1–11.

Richardson, E., Dacks, J.B., 2019. Microbial eukaryotes in oil sands environments: heterotrophs in the spotlight. Microorganisms 7, 178.

Risacher, F.F., Morris, P.K., Arriaga, D., Goad, C., Nelson, T.C., Slater, G.F., Warren, L.A., 2018. The interplay of methane and ammonia as key oxygen consuming constituents in early stage development of Base Mine Lake, the first demonstration oil sands pit lake. Appl. Geochem. 93, 49–59.

Rivera, M., 2017. Bacterioferritin: structure, dynamics, and protein–protein interactions at play in iron storage and mobilization. Acc. Chem. Res. 50, 331–340.

Ross, M.S., Pereira, A.D.S., Fennell, J., Davies, M., Johnson, J., Sliva, L., Martin, J.W., 2012. Quantitative and qualitative analysis of naphthenic acids in natural waters surrounding the Canadian oil sands industry. Environ. Sci. Technol. 46, 12796–12805.

Rouse, W.R., Douglas, M.S.V., Hecky, R.E., Hershey, A.E., Kling, G.W., Lesack, L., Marsh, P., McDonald, M., Nicholson, B.J., Roulet, N.T., Smol, J.P., 1997. Effects of climate change on the freshwaters of arctic and subarctic North America. Hydrol. Process. 11 (8), 873–902.

Rowland, S.J., Scarlett, A.G., Jones, D., West, C.E., Frank, R.A., 2011a. Diamonds in the rough: identification of individual naphthenic acids in oil sands process water. Environ. Sci. Technol. 45, 3154–3159.

Rowland, S.J., West, C.E., Jones, D., Scarlett, A.G., Frank, R.A., Hewitt, L.M., 2011b. Steroidal aromatic naphthenic acids in oil sands process-affected water: Structural comparisons with environmental estrogens. Environ. Sci. Technol. 45, 9806–9815.

Rowland, S.J., Jones, D., Scarlett, A.G., West, C.E., Hin, L.P., Boberek, M., Tonkin, A., Smith, B.E., Whitby, C., 2011c. Synthesis and toxicity of some metabolites of the microbial degradation of synthetic naphthenic acids. Sci. Total Environ. 409, 2936–2941.

Roy, J.W., Bickerton, G., Frank, R.A., Grapentine, L., Hewitt, L.M., 2016. Assessing risks of shallow riparian groundwater quality near an oil sands tailings pond. Groundwater 54, 545–558.

Saidi-Mehrabad, A., He, Z., Tamas, I., et al., 2013. Methanotrophic bacteria in oilsands tailings ponds of northern Alberta. ISME J. 7, 908–921.

Scarlett, A.G., Reinhardy, H.C., Henry, T.B., West, C.E., Frank, R.A., Hewitt, L.M., Rowland, S.J., 2013. Acute toxicity of aromatic and non-aromatic fractions of naphthenic acids extracted from oil sands process-affected water to larval zebrafish. Chemosphere 93, 415–420.

Schramm, L.L., Stasiuk, E.N., MacKinnon, M., 2000. Surfactants in Athabasca oil sands slurry conditioning, flotation recovery, and tailings processes. In: Schramm, L.L. (Ed.), Surfactants: Fundamentals and Applications in the Petroleum Industry. Cambridge University Press, Cambridge, pp. 365–430.

Schwartz, G., Mendel, R.R., 2006. Molybdenum cofactor biosynthesis and molybdenum enzymes. Annu. Rev. Plant Biol. 57, 623–647.

Scott, A.C., MacKinnon, M.D., Fedorak, P.M., 2005. Naphthenic acids in Athabasca oil sands tailings waters are less biodegradable than commercial naphthenic acids. Environ. Sci. Technol. 39, 8388–8394.

Shen, X., Sun, T., Su, M., Dang, Z., Yang, Z., 2019. Short-term response of aquatic ecosystem metabolism to turbidity disturbance in experimental estuarine wetlands. Ecol. Eng. 136, 55–61.

Shuqing, Z., Haiping, H., Yuming, L., 2008. Biodegradation and origin of oil sands in the Western Canadian sedimentary basin. Pet. Sci. 5, 87–94.

Siddique, T., Gupta, R., Fedorak, P.M., MacKinnon, M.D., Foght, J.M., 2008. A first approximation kinetic model to predict methane generation from an oil sands tailings settling basin. Chemosphere 72, 1573–1580.

Siddique, T., Kuznetsov, P., Kuznetsova, A., Li, C., Young, R., Arocena, J.M., Foght, J.M., 2014. Microbially-accelerated consolidation of oil sands tailings. Pathway II: solid phase biogeochemistry. Front. Microbiol. 5, 107.

Siddique, T., Shahimin, M.F., Zamir, S., et al., 2015. Long-term incubation reveals methanogenic biodegradation of C_5 and C_6 *iso*-alkanes in oil sands tailings. Environ. Sci. Technol. 49, 14732–14739.

Simair, M.C., Parrott, J.L., Varun Gupta, M., Frank, R.A., Peru, K.M., Ajaero, C., McMartin, D.W., Headley, J.V., 2021. Treatment of oil sands process affected waters by constructed wetlands: evaluation of designs and plant types. Sci. Total Environ. 772, 145508.

Simpson, I.J., Blake, N.J., Barletta, B., et al., 2010. Characterization of trace gases measured over Alberta oil sands mining operations: 76 speciated C_2-C_{10} volatile organic compounds (VOCs), CO_2, CH_4, CO, NO, NO_2, NO_y, O_3 and SO_2. Atmos. Chem. Phys. 10, 11931–11954.

Skeels, K., Whitby, C., 2018. Microbial ecology of naphthenic acid degradation. In: McGenity, T.J. (Ed.), Microbial Communities Utilizing Hydrocarbons and Lipids: Members, Metagenomics and Ecophysiology, Handbook of Hydrocarbon and Lipid Microbiology. Springer, Cham, https://doi.org/10.1007/978-3-319-60063-5_5-1.

Slyemi, D., Bonnefoy, V., 2012. How prokaryotes deal with arsenic. Environ. Microbiol. Rep. 4, 571–586.

Small, C.C., Cho, S., Hashisho, Z., et al., 2015. Emissions from oil sands tailings ponds: Review of tailings pond parameters and emission estimates. J. Petrol. Sci. Eng. 127, 490–501.

Smith, B.E., Lewis, A., Belt, S., Whitby, C., Rowland, S.J., 2008. Effects of alkyl chain branching on the biotransformation of naphthenic acids. Environ. Sci. Technol. 42, 9323–9328.

Stasik, S., Loick, N., Knöller, K., Weisener, C., Wendt-Potthoff, K., 2014. Understanding biogeochemical gradients of sulfur, iron and carbon in an oil sands tailings pond. Chem. Geol. 382, 44–53.

Sun, C., Shotyk, W., Cuss, C.W., Donner, M.W., Fennell, J., Javed, M., et al., 2017. Characterization of naphthenic acids and other dissolved organics in natural water from the Athabasca Oil Sands Region. Environ. Sci. Technol. 51 (17), 9524–9532.

Tan, B., Dong, X., Sensen, C.W., Foght, J., 2013. Metagenomic analysis of an anaerobic alkane-degrading microbial culture: potential hydrocarbon-activating pathways and inferred roles of community members. Genome 56 (10), 599–611.

Tan, B.F., Charchuk, R., Li, C., et al., 2014a. Draft genome sequence of uncultivated Firmicutes (*Peptococcaceae* SCADC) single cells sorted from methanogenic alkane-degrading cultures. Genome Announc. 2, e00909–e00914.

Tan, B.F., de Araujo Silva, R., Rozyki, T., et al., 2014b. Draft genome sequences of three *Smithella* sp. obtained from a methanogenic alkane-degrading culture and oil field produced water. Genome Announc. 2. e01085–14.

Tan, B., Fowler, S.J., Laban, N.A., et al., 2015a. Comparative analysis of metagenomes from three methanogenic hydrocarbon- degrading enrichment cultures with 41 environmental samples. ISME J. 9, 2028–2045.

Tan, B.F., Semple, K., Foght, J., 2015b. Anaerobic alkane biodegradation by cultures enriched from oil sands tailings ponds involves multiple species capable of fumarate addition. FEMS Microbiol. Ecol. 91, fiv042.

Tomczyk, N.A., Winans, R.E., Shinn, J.H., Robinson, R.C., 2001. On the nature and origin of acidic species in petroleum. 1. Detailed acid type distribution in a California crude oil. Energy Fuel 15, 1498–1504.

Toor, N.S., Franz, E.D., Fedorak, P.M., MacKinnon, M.D., Liber, K., 2013a. Degradation and aquatic toxicity of naphthenic acids in oil sands process-affected waters using simulated wetlands. Chemosphere 90, 449–458.

Toor, N.S., Han, X., Franz, E., Mackinnon, M.D., Martin, J.W., Liber, K., 2013b. Selective biodegradation of naphthenic acids and a probable link between mixture profiles and aquatic toxicity. Environ. Toxicol. Chem. 32, 2207–2216.

Van der Heijden, J., Reynolds, L., Deng, W., Scholz, R., Imani, K., Foster, L.J., Duong, F., Finlay, B.B., 2016. Salmonella rapidly regulates membrane permeability to survive oxidative stress. MBio 7 (4). e01238-16.

Vander Meulen, I.J., Klemish, J.L., Peru, K.M., Chen, D.D.Y., Pyle, G.G., Headley, J.V., 2021. Molecular profiles of naphthenic acid fraction compounds from mine lease wetlands in the Athabasca Oil Sands Region. Chemosphere 272, 129892.

Vander Meulen, I.J., Schock, D.M., Parrott, J.L., Simair, M.C., Mundy, L.J., Ajaero, C., Pauli, B.D., Peru, K.M., McMartin, D.W., Headley, J.V., 2022. Transformation of bitumen-derived naphthenic acid fraction compounds across surface waters of wetlands in the Athabasca Oil Sands Region. Sci. Total Environ. 806, 150619.

Vijgenboom, E., Busch, J.E., Canters, G.W., 1997. *In vivo* studies disprove an obligatory role of azurin in denitrification in *Pseudomonas aeruginosa* and show that *azu* expression is under control of RpoS and ANR. Microbiology 143, 2853–2863.

Volik, O., Elmes, M., Petrone, R., Kessel, E., Green, A., Cobbaert, D., Price, J., 2020. Wetlands in the Athabasca Oil Sands Region: the nexus between wetland hydrological function and resource extraction. Environ. Rev. 28 (3), 246–261.

Wan, Y., Wang, B., Khim, J.S., Hong, S., Shim, W.J., Hu, J., 2014. Naphthenic acids in coastal sediments after the Hebei Spirit oil spill: a potential indicator for oil contamination. Environ. Sci. Technol. 48 (7), 4153–4162.

Wang, X., Preston 3rd, J.F., Romeo, T., 2004. The *pgaABCD* locus of *Escherichia coli* promotes the synthesis of a polysaccharide adhesin required for biofilm formation. J. Bacteriol. 186, 2724–2734.

Wang, B., Wan, Y., Gao, Y., Yang, M., Hu, J., 2013. Determination and characterization of oxy-naphthenic acids in oilfield wastewater. Environ. Sci. Technol. 47, 9545–9554.

Wang, Z., Yang, C., Parrott, J.L., Frank, R.A., Yang, Z., Brown, C.E., Hollebone, B.P., Landriault, M., Fieldhouse, B., Liu, Y., Zhang, G., 2014. Forensic source differentiation of petrogenic, pyrogenic, and biogenic hydrocarbons in Canadian oil sands environmental samples. J. Hazard. Mater. 271, 166–177.

Wang, X., Chen, M., Xiao, J., Hao, L., Crowley, D., Zhang, Z., Yu, J., Huang, N., Huo, M., Wu, J., 2015. Genome sequence analysis of the naphthenic acid degrading and metal resistant bacterium *Cupriavidus gilardii* CR3. PLoS One 10 (8), e0132881.

Warren, L.A., Kendra, K.E., Brady, A.L., Slater, G.F., 2016. Sulfur biogeochemistry of an oil sands composite tailings deposit. Front. Microbiol. 6, 1533.

Whitby, C., 2010. Naphthenic acid biodegradation. Adv. Appl. Microbiol. 70, 93–125.

White, K.B., Liber, K., 2018. Early chemical and toxicological risk characterization of inorganic constituents in surface water from the Canadian oil sands first large-scale end pit lake. Chemosphere 211, 745–757.

Wilde, M.J., Rowland, S.J., 2018. Naphthenic acids in oil sands process waters: identification by conversion of the acids or esters to hydrocarbons. Org. Geochem. 115, 188–196.

Wilde, M.J., West, C.E., Scarlett, A.G., Jones, D., Frank, R.A., Hewitt, L.M., Rowland, S.J., 2015. Bicyclic naphthenic acids in oil sands process water: identification by comprehensive multidimensional gas chromatography–mass spectrometry. J. Chromatogr. A 1378, 74–87.

Wilson, S.L., Li, C., Ramos-Padrón, E., et al., 2016. Oil sands tailings ponds harbour a small core prokaryotic microbiome and diverse accessory communities. J. Biotechnol. 235, 187–196.

Wong, M.L., An, D., Caffrey, S.M., Soh, J., Dong, X., Sensen, C.W., Oldenburg, T.B.P., Larter, S.R., Voordouw, G., 2015. Roles of thermophiles and fungi in bitumen degradation in mostly cold oil sands outcrops. Appl. Environ. Microbiol. 81, 6825–6838.

Woods, J., Kung, J., Kingston, D., Kotlyar, L., Sparks, B., McCracken, T., 2008. Canadian crudes: a comparative study of SARA fractions from a modified HPLC separation technique. Oil Gas Sci. Technol. Rev. IFP 63, 151–163.

Wyndham, R.C., Costerton, J.W., 1981a. Heterotrophic potentials and hydrocarbon biodegradation potentials of sediment microorganisms within the Athabasca oil sands deposit. Appl. Environ. Microbiol. 41, 783–790.

Wyndham, R.C., Costerton, J.W., 1981b. In vitro microbial degradation of bituminous hydrocarbons and in situ colonization of bitumen surfaces within Athabasca oil sands deposit. Appl. Environ. Microbiol. 41, 791–800.

Xue, J., Zhang, Y., Liu, Y., El-Din, M.G., 2016. Treatment of oil sands process-affected water (OSPW) using a membrane bioreactor with a submerged flat-sheet ceramic microfiltration membrane. Water Res. 1 (88), 1.

Yang, C., Zhang, G., Serhan, M., Koivu, G., Yang, Z., Hollebone, B., Lambert, P., Brown, C.E., 2019. Characterization of naphthenic acids in crude oils and refined petroleum products. Fuel 255, 115849.

Yergeau, E., Lawrence, J.R., Sanschagrin, S., Waiser, M.J., Korber, D.R., Greer, C.W., 2012. Next generation sequencing of microbial communities in the Athabasca River and its tributaries in relation to oil sands mining activities. Appl. Environ. Microbiol. 78, 7626–7637.

Yergeau, E., Lawrence, J.R., Sanschagrin, S., Roy, J.L., Swerhone, G.D.W., Korber, D.R., 2013. Aerobic biofilms grown from Athabasca watershed sediments are inhibited by increasing concentrations of bituminous compounds. Appl. Environ. Microbiol. 79, 7398–7412.

Young, R.F., Orr, E.A., Goss, G.G., Fedorak, P.M., 2007. Detection of naphthenic acids in fish exposed to commercial naphthenic acids or oil-sands process-affected water. Chemosphere 68, 518–527.

Yu, X., Lee, K., Ulrich, A.C., 2019. Model naphthenic acids removal by microalgae and base mine lake cap water microbial inoculum. Chemosphere 234, 796–805.

Yue, S., Ramsay, B.A., Wang, J., Ramsay, J.A., 2016. Biodegradation and detoxification of naphthenic acids in oil sands process affected waters. Sci. Total Environ. 572, 273–279.

Zampolli, J., Di Canito, A., Cappelletti, M., Collina, E., Lasagni, M., Di Gennaro, P., 2020. Biodegradation of naphthenic acids: identification of *Rhodococcus opacus* R7 genes as molecular markers for environmental monitoring and their application in slurry microcosms. Appl. Microbiol. Biotechnol. 104, 2675–2689.

Zan, S., Wang, J., Wang, F., Han, Y., Du, M., Fan, J., 2019. Variation and distribution of naphthenic acids in Dalian Bay sediment. Mar. Pollut. Bull. 140, 597–602.

Zan, S., Wang, J., Wang, F., Li, Z., Du, M., Cai, Y., 2022. A novel degradation mechanism of naphthenic acids by marine Pseudoalteromonas sp. J. Hazard. Mater. 424, 127534.

Zhang, L., Zhang, Y., Patterson, J., Arslan, M., Zhang, Y., Gamal El-Din, M., 2020a. Biofiltration of oil sands process water in fixed-bed biofilm reactors shapes microbial community structure for enhanced degradation of naphthenic acids. Sci. Total Environ. 718, 137028.

Zhang, J., Gao, H., Xue, Q., 2020b. Potential applications of microbial enhanced oil recovery to heavy oil. Crit. Rev. Biotechnol. 40 (4), 459–474.

Zou, L., Han, B., Yan, H., Kasperski, K.L., Xu, Y., Hepler, L.G., 1997. Enthalpy of adsorption and isotherm of adsorption of naphthenic acid onto clays. J. Colloid Interface Sci. 190, 472–475.

Zumft, W.G., 1997. Cell biology and molecular basis of denitrification. Microbiol. Mol. Biol. Rev. 61, 533–616.

CHAPTER FIVE

The gut microbiome in health and disease: Inflammatory bowel diseases

Salma El-Sahhar[a] and Patrick Varga-Weisz[a,b,*]
[a]School of Life Sciences, University of Essex, Colchester, United Kingdom
[b]São Paulo Excellence Chair, Department of Genetics and Evolution, Microbiology and Immunology, Institute of Biology, University of Campinas, Campinas, SP, Brazil
*Corresponding author: e-mail address: patrick.varga-weisz@essex.ac.uk

Contents

1. Introduction 290
2. Genetic analysis highlights the role of the gut microbiota in IBD 295
3. How do changes in the microbiome influence IBD? 303
4. How does the environment contribute to gut microbiome changes? 309
5. Can microbial-based treatments be the answer? 314
6. Conclusion and outlook 318
References 320

Abstract

In the era of high throughput DNA sequencing technologies, there has been an explosion in the last 10 years of our knowledge of human microbiomes, especially the gut microbiome and its contribution to health and disease. This understanding has confirmed that the gut is an incredible complex ecosystem with critical health implications. Inflammatory bowel diseases, primarily Crohn's disease and ulcerative colitis, are diseases where a direct link to the gut microbiome is best understood. These pathologies are difficult to treat and dramatically on the rise globally, possibly linked to lifestyle changes, like diet, urbanisation, and antibiotics exposure. Here we canvas our current knowledge of the link between the gut microbiome and these diseases. Large-scale and longitudinal population-wide studies of gut microbiomes in combination with appropriate animal models will be instrumental to propel the field from cataloguing correlations to understanding causation.

Abbreviations

CD Crohn's disease
DSS dextran sodium sulphate
FDA Food and Drug Administration (USA)
FISH fluorescence in situ hybridisation
FMT faecal matter transplantation

Advances in Ecological Research, Volume 67
ISSN 0065-2504
https://doi.org/10.1016/bs.aecr.2022.09.005

GIT	gastrointestinal tract
GWAS	genome-wide association studies
HMP	human microbiome project
IBD	inflammatory bowel disease
IECs	intestinal epithelial cells
IL	interleukin
MDP	muramyl dipeptide
MUC	mucin
SCFAs	short-chain fatty acids
SPF	specific-pathogen free
UC	ulcerative colitis

1. Introduction

When Japan's long-standing prime minister Shinzo Abe resigned in 2020 because of his ulcerative colitis, it put the debilitating nature of inflammatory bowel diseases in the international press. What is largely not appreciated is that these diseases seem to reflect an aberrant reaction to what is normally an innocuous gut microbiome. Despite their bad reputation in relation to health, microbes are important for maintaining a state of mental and physical health and overall wellness. In fact, there are at least as many microbes inhabiting our bodies, primarily in the colon, as there are human cells (Gilbert and Stephens, 2018; Sender et al., 2016).

Bacteria, viruses, fungi and archaea reside in communities throughout the body, and are collectively known as the human microbiome. Yet, when most studies mention the microbiome, they are usually referring to bacterial communities rather than the whole assembly of microbial organisms. The term "microbiome" was coined by the Nobel laureate, Joshua Lederberg, in 2001 (Prescott, 2017). Research into the topic dates back to the 19th century (Farré-Maduell and Casals-Pascual, 2019). However, technical limitations in culturing (often anaerobe) constituents of the human microbiomes were the main reason why this field was not broadly explored. Recent advancements in high throughput DNA sequencing technologies and decreasing sequencing costs have been followed by a surge of investigations into the microbiome (Fig. 1).

Over the last decade, there has been an influx of new findings about the microbiome expedited by the Human Microbiome Project (HMP) (Fig. 1). The HMP is a group of interdisciplinary projects launched throughout the US, Europe and Asia with an aim to discover new biomarkers associated

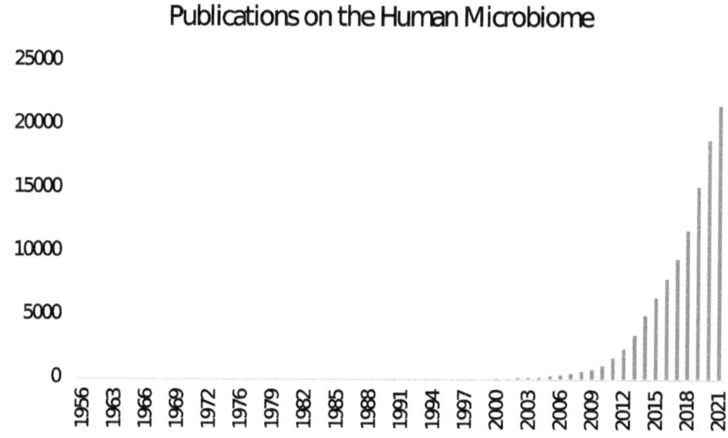

Fig. 1 The number of studies on the human microbiome is on the rise since the boom in next generation sequencing in the early 2000s, generated by a basic search on PubMed.

with health and disease (Turnbaugh et al., 2007). The first phase of the HMP laid the groundwork for successive studies on the interplay between the human host and microbes. It focused on characterising the structural and functional properties of microbial populations inhabiting five key body areas from a cohort of healthy, Western adults (Huttenhower et al., 2012). 5177 Microbial taxonomic profiles were identified using 16S ribosomal RNA sequencing while whole genome shotgun/metagenomics sequencing elucidated microbial functional properties (Methé et al., 2012). The results showed that the make-up and diversity of the microbial populations differed markedly at each body site. Oral and gut communities were the most diverse compared to the populations present at other sites. However, microbial populations found in saliva among subjects were more or less similar (Huttenhower et al., 2012). Vaginal sites accommodated the simplest of communities, low in diversity. In fact, high microbial diversity in vaginal tissues has been correlated with vaginosis (Gilbert et al., 2018). The skin microbiome was shown to be unique for each person. Overall, an individual's microbiome is stable over time and has a relatively unique makeup that contributes to the metabolic pathways essential for human health (Caporaso et al., 2011). Fluctuations in the biodiversity and abundance of certain taxa in the microbiome are constantly occurring with environmental changes at different temporal rates depending on the individual (Flores et al., 2014). There are many factors that influence microbial composition, particularly in the gut, which is the most studied community and is the focus of this chapter. Genetics, sex, age, the immune system, birth mode, diet, lifestyle,

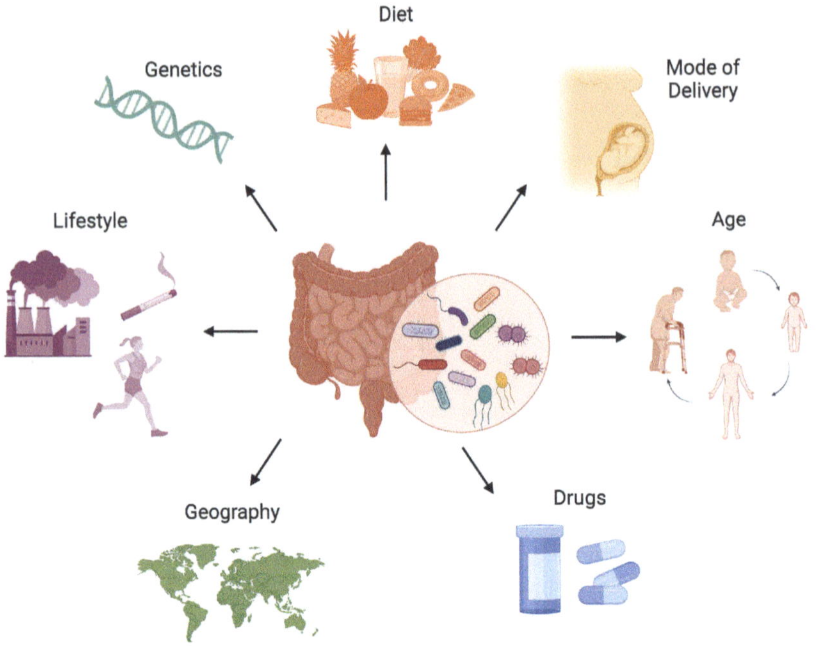

Fig. 2 A summary graphic illustrating factors that influence the gut microbiome. The figure was created with BioRender.com.

geography and exposure to antibiotics and other drugs have all been shown to play a significant role in shaping the host's gut microbial communities (Fig. 2).

The second phase of the HMP explored the interplay between the host and their microbiota over time in representative microbiome-associated conditions, one of which is inflammatory bowel disease (IBD) (Proctor et al., 2019). IBD are a group of disorders that are thought to be both affected by and in turn affect the gut microbiome (Frank et al., 2007; Manichanh et al., 2006). These disorders involve an altered immune response towards gut microbiota in genetically predisposed individuals, under the influence of certain environmental factors, such as diet, antibiotic exposure, lifestyle (Piovani et al., 2019). Whether **dysbiosis** is a causative or risk factor in developing IBD is yet to be clearly understood. The multidimensionality of these conditions makes it very challenging to draw causal links between the gut microbiota and IBD, despite it being the most widely studied disease group in relation to the microbiome. Extensive research examining the associations between the microbiome and IBD will be discussed thoroughly throughout this chapter. We include a Glossary (Table 1) for the explanation of some of the concepts and technical terms in bold.

Table 1 Glossary.

Term	Definition
Dysbiosis	An imbalance of the microbiome causing compositional and functional changes in its constituents
Intestinal Mucosa	The inner lining of the intestinal tract composed of mucus, intestinal epithelium, lamina propria and microbiota
Mucin	Glycoproteins that are the major components of the mucus, which lubricate and protect the gastrointestinal tract
Lamina propria	Thin layer of loose connective tissue that forms part of the intestinal mucosa
Crypts	Intestinal glands
Granuloma	Aggregation of macrophages forming in response to chronic inflammation
Paneth cells	Secretory cells residing in the base of intestinal crypts
Colectomy	A surgical procedure that involves a segmental resection of the colon
Genome-wide association studies (GWAS)	Studies based on genetic variation, especially single nucleotide polymorphisms (SNPs), exploring the link between genotype and phenotype using many DNA samples from large populations. These are compared, e.g., using microarrays build around SNPs.
Pattern recognition receptors (PRR)	A class of receptors that detect specific molecular structures on the surface of microbes.
Toll-like receptor (TLR)	A class of receptors present on sentinel cells that play a role in the innate immune system by detecting microbial derived molecules
Nod-like receptor (NLR)	PRRs that detect pathogens and regulate innate immune responses
RIG-I-like receptors	Cytosolic PRRs that detect viruses
Autophagy	A cellular process that involves the engulfment of organelles and macromolecules by vesicles for degradation
Nuclear factor-κB (NF-κB)	A transcription factor involved in cell survival and cytokine production
Immunoglobulin A (IgA)	An antibody that plays a role in mucosal immunity
T-helper 17 cells (Th17)	A subset of pro-inflammatory T-helper cells that produce interleukin 17 (IL-17)
Transforming growth factor beta (TGF-β)	A pleiotropic cytokine that regulates immune responses

Continued

Table 1 Glossary.—cont'd

Term	**Definition**
Cytokine	A small signalling protein involved in regulating immune responses
16S rRNA gene	A gene encoding for the prokaryotic small ribosomal subunit 16S rRNA commonly used in phylogenetic studies
18S rRNA gene	A gene encoding for the eukaryotic small ribosomal subunit 18S rRNA commonly used in phylogenetic studies
Internal Transcribed Spacer (ITS)	A region of non-coding DNA situated between the small-subunit ribosomal RNA (rRNA) and the large-subunit rRNA coding genes
Fluorescence in situ hybridization (FISH)	A cytogenetic technique that targets chromosomal locations using fluorescent probes detected by a fluorescent microscope
Taxonomy Nomenclature	As this chapter predominantly refers to published literature that uses the old nomenclature, we use it here throughout, noting that the new nomenclature is not universally accepted. The new nomenclature can be found here https://www.microbiologyresearch.org/content/journal/ijsem/10.1099/ijsem.0.005056
Bacteriocin	Toxic proteins produced by specific species of bacteria to inhibit the growth of other related species
Regulatory T cells (Treg cells)	A specialised subpopulation of T cells that maintain homeostasis by suppressing immune responses
Histone deacetylase (HDAC)	An enzyme that plays a role in modulating chromatin architecture by removing acetyl groups from lysine residues in histone tails
Hypoxia-inducible factor 1-alpha (HIF1-α)	A subunit of transcription factor HIF1
Adherent-Invasive *E. coli* (AIEC)	Strains of *E. coli* that adhere to IECs and colonise the mucosa
Dendritic cells	Antigen presenting cells of the immune system
Occludin	A plasma membrane protein present in tight junctions
Ileal pouch anastomosis (IPAA)	A surgical procedure done post colectomies to restore gastrointestinal continuity
Perianal disease	A type of disease that occurs in CD patients that causes inflammation around the anus
Luminal disease	A type of disease that occurs in CD patients that causes inflammation of the intestinal lining and/or wall

2. Genetic analysis highlights the role of the gut microbiota in IBD

Digestive diseases, particularly Inflammatory Bowel Diseases (IBD), are one of the most explored areas within microbiome research (Integrative HMP iHMP Research Network Consortium, 2014). IBD are one of the most prevalent gastrointestinal diseases globally, with rising incidence rates in newly industrialised countries in Asia, Africa and South America (Guan, 2019; Freeman et al., 2021; Ng et al., 2017). Europe and North America have reported the highest prevalence of IBD (Ng et al., 2017). However, the mortality rates associated with the disease are low (Freeman et al., 2021). It is usually diagnosed in patients within the 20–40 years age bracket (Guan, 2019). The common symptoms include diarrhoea, bloody stools, abdominal pain, weight loss, obstruction of the gastrointestinal tract (GIT) and, in some cases, extra-gastrointestinal manifestations like rashes, arthritis and eye complications (Frank et al., 2007; Ghosh and Premchand, 2015). IBD is characterised by chronic inflammation of the **intestinal mucosa** mediated by the immune system and driven by genetics as well as environmental factors, including diet, smoking and socio-economic status (Ni et al., 2017). Crohn's disease (CD) and Ulcerative Colitis (UC) are the two main forms of IBD, about 5% of IBD cases remain unclassified due to unclear symptoms. UC is distinguished by mucosal inflammation starting in the rectum and extending proximally into the colon, while CD involves areas of transmural inflammation surrounded by normal mucosa that can span the entire gastrointestinal tract (Ni et al., 2017). These traits are clearly observed in histopathology procedures, where biopsies from patients with UC show inflammation restricted to the mucosa and submucosa with **mucin** depletion and increased neutrophil migration to the **lamina propria** and **crypts** forming micro-abscesses (Guan, 2019; Xavier and Podolsky, 2007). CD, on the other hand, is characterised by different histological features, including **granulomas** infiltrated by aggregations of macrophages and other inflammatory cells, as well as Paneth cell deficiency (Xavier and Podolsky, 2007). These histopathological hallmarks together with the symptoms presented by patients are used in determining the severity of the disease which impacts the course of treatment. Topical medication is prescribed to mild cases of UC, while mild-to-moderate cases require oral anti-inflammatories progressing to immunomodulatory drugs for severe cases (Ghosh and Premchand, 2015). Steroids are given to subdue flare-ups post remission periods as long and frequent UC flares pose a risk of bowel cancer. About 25% of UC patients require **colectomy** surgery due the risk of cancer development or poor response to medical treatment (Ghosh and Premchand, 2015). Meanwhile, 80% of CD cases undergo some sort of surgery be it as minor as an abscess drainage or as complex as a bowel resection. However, the standard treatment for CD is immunomodulatory drugs and biological therapy for more severe cases (Ghosh and Premchand, 2015).

There is an underlying genetic component to the pathogenesis of IBD. **Genome-wide association studies (GWAS)** have been conducted to pinpoint IBD genetic risk loci (Huang et al., 2017; Jostins et al., 2012; Liu et al., 2015). Over 200 IBD-associated loci have been identified (Graham and Xavier, 2020; Jostins et al., 2012; Liu et al., 2015). Only a fraction of those markers has been validated and shown to have a causal role on development of IBD. A majority of the genes implicated in the pathology of the disease are involved in pathways that regulate mucosal integrity (barrier function), immunity and host-microbe interactions, thus highlighting the role of the microbiota in the disease (Graham and Xavier, 2020). Below we discuss some of the key genes that have been linked to IBD.

The intestinal mucosal barrier plays a critical role in maintaining homeostasis and disease progression (Turner, 2009) (Fig. 3) by segregating intestinal epithelial cells (IECs) from luminal contents, like macromolecules, digestive enzymes and microbes – both commensals and pathogens (Okumura and Takeda, 2018). It acts as a physical barrier as well as a chemical one that signals to the immune system when needed. The mucosal lining is comprised of an undulating monolayer of specialised epithelial cells such as enterocytes, goblet cells and enteroendocrine cells and their precursors, organised into crypts and villi (Peterson and Artis, 2014). The crypts are comprised of intestinal stem cells (ISCs), transit amplifying cells as well as **Paneth cells** (in the small intestine), while villi consist of terminally differentiated cells, like absorptive enterocytes and goblet cells. This epithelial layer is physically protected via a mucus layer and the glycocalyx (Maloy and Powrie, 2011). The latter is a network of glycolipids and glycoproteins covering the villi, particularly absorptive IECs, preventing the passing of bacteria or unwanted molecules. A viscous gel-like fluid enriched in Mucin2 (MUC2), a glycoprotein secreted by goblet cells, forms the mucus layer. The large intestine has a double mucus layer for added protection, since it holds the largest microbial community in the body (Peterson and Artis, 2014) (Fig. 3, right side). The inner mucus layer is anchored to the intestinal epithelium and is usually impermeable, due to polymerisation and stratification of MUC2,

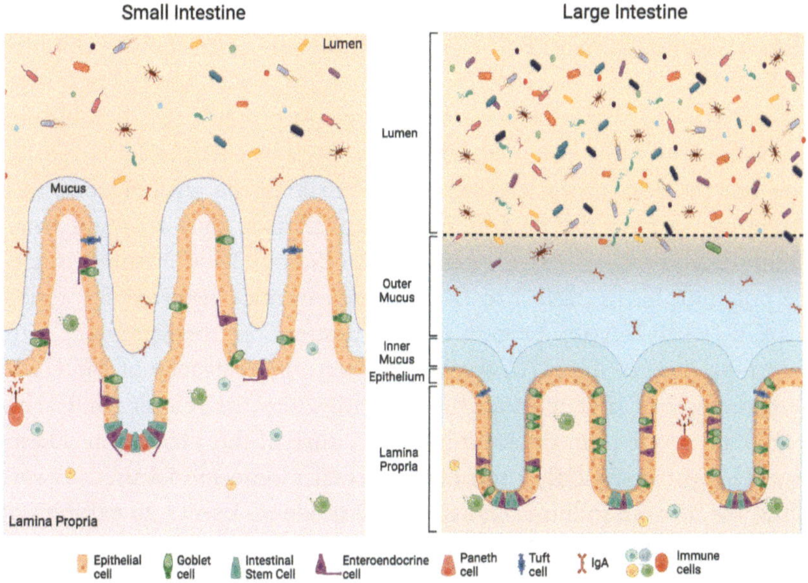

Fig. 3 The intestinal epithelial barrier. The intestinal mucosa acts both as a chemical and physical barrier that protects IECs from harmful molecules and microbes present in the intestinal lumen. The physical barrier is comprised of a single mucus layer in the small intestine while a double layer of a firm inner mucus layer and a looser outer mucus layer. The porous outer mucus layer allows for some commensal habitation. AMPs, immunoglobulins, cytokines and other signalling molecules secreted by IECs and immune cells of the mucosa make up the chemical barrier. The figure was created with BioRender.com.

blocking the entry of bacteria. Whereas the looser outer mucus layer is more porous thereby harbouring some bacterial species (Okumura and Takeda, 2018).

Patients with UC tend to have a disjointed and diminished mucus layer, especially during flare ups, allowing for direct contact between microbes and the epithelium (Kang et al., 2022). A decrease in MUC2 secretion accompanied by reduced numbers of goblet cells account for this phenotype which is reversed upon remission (Sun et al., 2016). In contrast, CD patients tend to have a thicker mucus layer compared to healthy individuals, however, the properties of the mucus seem to be compromised (Sun et al., 2016). GWA studies have linked IBD pathogenesis to mutations in several *MUC* genes, including *MUC2*, *MUC3* and *MUC19* (Kyo et al., 1999; McCole, 2014; Rivas et al., 2011; Visschedijk et al., 2016). A study done on two Dutch and one German cohorts has drawn a link between *MUC2* and

UC susceptibility, although this association was lacking in the German cohort (Visschedijk et al., 2016). A correlation between rare *MUC3A* alleles and both UC and CD was examined by Kyo et al. (2001). They hypothesised that conformational changes in mucin proteins encoded by these rare alleles could lead to a decrease in glycosylation—a mechanism which gives mucins their protective properties—making the glycoproteins more susceptible to bacterial proteases and degradation, thus affecting barrier stability (Kyo et al., 1999). Despite GWAS linking *MUC19* single nucleotide polymorphisms (SNPs) to IBD, a study in mice reported the absence of *MUC19* mRNA and protein in the intestines but their presence in the oral cavity (Das et al., 2010) which questions the relevance of *MUC19* in the pathology of IBD. Further investigations into the role of *MUC* genes in IBD are required in determining the nature of the association. Despite this, the most commonly used experimental mouse model used to study colitis, the dextran sodium sulphate (DSS) model, is known to exhibit swift changes in the colon mucus layer, rendering it permeable to bacteria, upon DSS administration, bringing about mucosal inflammation (Johansson et al., 2010). Therefore, it can be concluded that mucins play a part in IBD, given their integral role in maintaining the physical mucosal barrier, but their exact role in the pathogenesis of the disease should be investigated further.

The *Cadherin 1* (*CDH1*) gene encodes for transmembrane epithelial protein E-cadherin, essential for cell adhesion and maintaining tight junctions. It is present in the *IBD1* locus, along with a few other IBD susceptibility genes, and is known to be associated with CD and UC (Barrett et al., 2009; Muise et al., 2009). Mutations in *CDH1* can lead to increased mucosal permeability, allowing for the infiltration of luminal microbes through the intestinal wall, with the exact mechanisms still unknown (Younis et al., 2020). SNPs in *CDH1* in in vitro models result in a cytoplasmic accumulation of the protein rather than localisation to the plasma membrane—its expected location (Muise et al., 2009). This observation is shared in CD patients carrying the same SNPs, with increased cytoplasmic E-cadherin staining. Deficient E-cadherin compromises the integrity of the epithelial barrier increasing the risk of inflammation.

Hepatocyte nuclear factor 4 α (*HNF4A*) is another gene involved in the maintenance of the mucosal barrier and associated with IBD. As a DNA binding transcription factor, it regulates the expression of genes involved in epithelial cell proliferation and differentiation as well as tight junction formation. GWA studies in the Netherlands and the UK have shown a strong link between variations in *HNF4A* and UC (Barrett et al., 2009;

Van Sommeren et al., 2011). Moreover, a study investigating the role of intestinal *HNF4A* in mice revealed that selective deletion of the gene in IECs increased mucosal permeability and caused severe DSS-induced colitis in mice (Ahn et al., 2008). *HNF4A* seems to play a protective role against UC by maintaining mucosal barrier function and architecture.

Aside from acting as a physical barrier to foreign bodies and toxic materials, the mucosa and its constituents also provide a chemical barrier. IECs have pathogen recognition and processing mechanisms in addition to secretory defences that keep infiltrators at bay. They express an array of **pattern recognition receptors (PRRs)**, such as **toll-like receptors (TLRs)**, **Nod-like receptors (NLRs)** and **RIG-I-like receptors,** that screen the environment for foreign microbes and, in turn, orchestrate immune cell responses in the mucosa (Peterson and Artis, 2014). Mucins are one of the secretory defences produced by goblet cells, forming the first line of defence in the intestine. Goblet cells also regulate the physical mucosal barrier via secretion of trefoil factor 3 (TFF3) and resistin-like molecule-β (RELMβ). TTF3 is a protein involved in epithelial repair, IEC migration, anti-apoptotic pathways and mucin crosslinking—adding structural integrity to the mucus layer (Peterson and Artis, 2014). In addition to promoting MUC2 secretion, RELMβ plays a role in regulating immune responses during inflammation particularly of macrophages and T-cells (Artis et al., 2004; Nair et al., 2009). Intestinal homeostasis is fortified further by the secretion of antimicrobial proteins (AMPs) by several IECs. For instance, Paneth cells release several AMPs, including defensins, cathelicidins, and lysozymes, that disrupt cell walls and membrane surfaces in both Gram-positive and Gram-negative bacteria via peptidoglycan breakage and pore-formation, respectively (Peterson and Artis, 2014). REGIIIγ is another enterocyte and Paneth cell-derived AMP which is essential in maintaining mucosal homeostasis by spatially segregating members of the microbiota from the small intestinal lining (Vaishnava et al., 2011). IECs are not only sentinels but also mediators, transporting immunoglobulins released by secretory B-cells from the lamina propria across the epithelial barrier in response to pathogens (Peterson and Artis, 2014).

Autophagy and the unfolded protein response (UPR) are processes that regulate immune responses and integral in regulating secretory IECs. The invasion of microbes into the epithelium is limited by autophagy and interruptions in this process can lead to spontaneous inflammation (Peterson and Artis, 2014). UPRs promotes protein folding and translocation in the endoplasmic reticulum (ER) as well as regulation of protein synthesis, arresting of the cell cycle and induction of apoptosis. Regulation of these mechanisms prevent ER stress in secretory cells averting inflammation (Kaser et al., 2008). The range of pathogen recognition and processing mechanisms described reinforces the intestinal barrier. Any defects in these operations leaves the mucosa exposed to inflammation and can result in health complications, like in IBD.

Nucleotide-binding oligomerization domain 2 (*NOD2*) is the most studied susceptibility gene in IBD and is commonly mutated in CD patients of European descent with a prevalence of ~33% (Guan, 2019). Certain polymorphisms in *NOD2* are linked to poor prognosis for CD patients, however, not all carriers of these develop CD. The *NOD2* gene encodes for a cytosolic receptor expressed in IECs, particularly Paneth cells, and antigen presenting cells (APCs), like macrophages and **dendritic cells** (Guan, 2019). The function of NOD2 may vary depending on cell-type, but antibacterial immunity seems to be the primary function of NOD2 (Xavier and Podolsky, 2007). Broadly, the leucine rich repeat (LRR) domains in NOD2 recognises the muramyl dipeptide (MDP) component present in bacterial peptidoglycans, which leads to the activation of **nuclear factor-κB (NF-κB)** through a signalling cascade (Xavier and Podolsky, 2007). This, in turn, induces the secretion of either defensins or pro-inflammatory **cytokines** (like IL-12), depending on the cell type. NOD2 is also known to initiate autophagy by interacting with the protein ATG16L1 through MDP (Guan, 2019). Mutations in *NOD2* can impair pathogen recognition and processing leading to an attenuated immune response, resulting in inflammation. CD patients carrying loss-of-function *NOD2* mutations tend to have lower expression of the AMP α-defensin (Younis et al., 2020). Studies have shown that there is a 10-fold increase in risk for CD with bi-allelic risk carriers versus individuals heterozygous for *NOD2* variants (Hugot et al., 2001, 2007; Ogura et al., 2001).

Genetic variations in the *autophagy-related 16-like 1 protein* (*ATG16L1*) are also strongly associated with CD (Fowler et al., 2008; Liu et al., 2015; McGovern et al., 2015). The product of this gene is an autophagy protein that is widely expressed throughout the intestinal epithelium in several cell types, including APCs, T cells, and B cells (Xavier and Podolsky, 2007). Histology samples from CD patients homozygous for *ATG16L1* T300A substitution show aberrant numbers of Paneth cells with irregular morphologies (Graham and Xavier, 2020). Mutated *ATG16L1* impairs the ability of Paneth cells to properly identify and eliminate pathogens partly due to aberrant TLR signalling (Guan, 2019). This interrupted signalling hinders autophagy in Paneth cells, affecting AMP production, thus allowing bacteria to breach the epithelial barrier which then elicits an immune response bringing about inflammation (Younis et al., 2020). Mice with selective deletion of *ATG16L1* in T cells exhibit spontaneous intestinal inflammation, highlighting that ATG16L1 is essential in maintaining stability in the mucosa (Kabat et al., 2016). The *T300A* genotype is also implicated in ER stress pathways (Graham and Xavier, 2020). Carriers of this variant, both healthy individuals and CD patients, experience polyreactive **immunoglobulin A (IgA)** responses following elevated epithelial ER stress which target both pathogenic and commensal bacteria, potentially leading to dysbiosis (Grootjans et al., 2019; Guan, 2019). IgA plays an important role in gut mucosal immunity by targeting pathogens and promoting tolerance to commensal microbes (Jackson et al., 2021). Thus, ATG16L1 plays a pivotal role in maintaining homeostasis in the gut.

Deficient immune responses are one of the main causes of inflammation. Interleukin 23 (IL-23) is a pro-inflammatory cytokine involved in adaptive immunity and an orchestrator in IBD inflammation, along with its receptor (IL23R). IL-23 is upregulated in the mucosa of both CD and UC patients and it can be protective or a risk factor (Xavier and Podolsky, 2007; Younis et al., 2020). Following bacterial recognition, the cytokine is released by dendritic cells and macrophages promoting the differentiation of pro-inflammatory **T-helper 17 cells (Th17)** from naïve T cells via **TGF-β** and IL-6 signalling. In addition to inducing differentiation, TGF-β and IL-6 initiate the expression of IL23R on Th17 cell surfaces (McGovern and Powrie, 2007). IL23R is expressed in dendritic cells and macrophages, thus in *IL23R* and *IL-23* variants, this cytokine cascade can trigger a state of unrestrained inflammation through a continuous supply of pro-inflammation molecules like IL-6, IL-12, IL-17, INF-γ, TNF-α and IL-23 (Younis et al., 2020). IL-23 serves as an effective drug target against defective immune responses; risankizumab -rzaa, a monoclonal antibody targeting the p19 subunit of the cytokine has been approved by the FDA to treat cases of moderate to severe CD (D'Haens et al., 2022).

Interleukin 10 (IL-10) is another cytokine with anti-inflammatory properties associated with very early-onset IBD in ages 6 and under (Younis et al., 2020). It is secreted by APCs, lymphocytes, and macrophages to dampen rogue immune responses avoiding development of autoimmune diseases. Genetic variants of the encoding gene cause the downregulation of IL-10 eliciting an aggravated immune response through pro-inflammatory cytokines, like TNF-α and IL-12. Both humans and mice with defects in

IL-10 signalling experience severe colitis (Kotlarz et al., 2012; Kühn et al., 1993). However, *IL10* aberrance is not enough to cause IBD, and its function is tightly linked with the microbiota. Studies show that the microbiota is essential in spontaneous colitis development and immune system activation in the *IL10* mouse model (Sellon et al., 1998; Wirtz and Neurath, 2007). Members of the microbiota induce IL-10 secretion via TLR signalling, downregulating immune response and preventing inflammation (Levast et al., 2015).

Genetic predisposition to IBD has been mapped by GWA studies. However, these studies provide correlations to IBD and do not identify causal genes. What the genetics of IBD reveals is that defects in bacterial recognition in innate immunity responses is at the core of IBD, highlighting the importance of host-microbiota interactions in the disease. More functional studies need to be carried out to better understand the associations between genetic polymorphisms and IBD. Interestingly, only a small percentage of IBD-associated risk loci carriers develop IBD which would imply that genetics is not the main driver of IBD. The development of IBD is caused by a convoluted mix of genetic risk factors, and alterations to the microbiome and its interactions with the mucosal immune system as well as environmental factors. The development of IBD, particularly UC, seems to be highly dependent on and influenced by gut microbes and their products (Roy et al., 2017; Sellon et al., 1998). Studies have linked imbalances in the composition of the gut microbiota, dysbiosis, to IBD.

The interdependency between the microflora and occurrence of colitis has been Observed in several IBD mouse models in the studies discussed throughout this section. The DSS model (briefly described in text above) chemically induces colitis through IEC damage and exposure of the lamina propria and submucosal environment to luminal contents (Low et al., 2013). It is one of the most commonly used mouse models to study UC, and its effectiveness depends on several factors, only one of which is microbiota composition. Germ-free (GF) mice were shown to have reduced DSS colonic inflammation when compared to partial microbial depletion using antibiotics (Hernández-Chirlaque et al., 2016). Roy et al. (2017) and Kazakevych et al. (2020) reported that a DSS-mediated colitis response is determined by the composition of an otherwise Specific-Pathogen Free (SPF) microbiome. 2, 4, 6 Trinitrobenzene (TNBS) is a chemical agent that causes mucosal inflammation in mice that resembles CD and is known to reduce gut microbial diversity in mice (Kozik et al., 2019; Low et al., 2013). Its colitic effect can be alleviated with administration of butyrate producing *Butyricicoccus* in rats (Eeckhaut et al., 2013). Efficacy of immunologically induced colitis models using bacterial agents, like Adhesion invasive *E. coli* (AIEC) and *Salmonella typhimurium*, is dependent on commensal bacteria (Low et al., 2013). Antibiotics are generally administered in these models to deplete biofilm-forming bacteria, like *Lactobacillus casei*, and other commensals, allowing for better colonisation for the inflammation causing pathological agents. Gene knockout animal models of IBD are generally used to study the immune responses and inflammatory pathways involved in mucosal inflammation (Wirtz and Neurath, 2007). Gut microbes are necessary for development of spontaneous colitis in both $IL2^{-/-}$ and $IL10^{-/-}$ mice (Rakoff-Nahoum et al., 2006). Mice deficient in IL-10 or IL-2 do not develop colitis in germ-free conditions unless transferred to a specific pathogen-free environment (Hoshi et al., 2012; Sellon et al., 1998). However, intestinal inflammation mediated by the microbiota in both animal models are regulated by different signalling pathways (Kiesler et al., 2015; Rakoff-Nahoum et al., 2006). Another method of inducing colitis in mice is through adoptive T cell transfer in immunodeficient mouse models, like $Rag^{-/-}$ or SCID mice (Low et al., 2013). Susceptibility of T cell induced colitis is reliant on gut microbial composition in this model, too (Webb et al., 2018). In conclusion, animal models used to study IBD indicate the significance of the microbiome in the aetiology of the disease. Studies have linked dysbiosis to IBD, recent findings on this topic will be discussed in detail throughout the next section of this chapter.

3. How do changes in the microbiome influence IBD?

Identifying and quantifying the taxa present within the gut microbiota is crucial in understanding their role in the human body. There are a several established methods that work on the classification of gut microbes like amplicon sequencing, metagenomic sequencing and shotgun sequencing (Young et al., 2021). Amplicon sequencing, particularly of the **16S rRNA gene**, is the most widely used technique in characterising the microbiome despite it lacking taxonomic resolution at the species/strain level (Johnson et al., 2019). 16S sequencing is paired with other phylogenetic markers most commonly the **18S rRNA gene** and **ITS** region, these are used to identify eukaryotic and fungal members of the microbiome. On the other hand, metagenomic sequencing is better suited for understanding the functional properties of microbes and microbial communities, since the technique samples all genes instead of the single 16S rRNA gene; this allows for both microbial classification as well as highlighting putative metabolic functions (Iyer, 2016). Functional properties of the microbiome can be revealed further by transcriptomic, proteomic and metabolomic analyses. Microbiota community structure is an area much less explored and can be visualised using **fluorescence in situ hybridization (FISH)**, which labels bacterial communities in host tissues by utilising 16S rRNA fluorescent probes (Iyer, 2016). Most sampling of the gut microbiome is non-invasive and is done through stool analysis, which is considered representative of the colonic luminal community. Microbes inhabiting the mucosal layer are sampled via endoscopic tissue biopsies or luminal brushings and other less frequently used techniques, like intestinal fluid aspiration (Tang et al., 2020). When analysing microbiome data potential issues of each sampling approach should be taken into consideration, like the bias introduced in faecal specimens as well as the effect of bowel preparation prior endoscopies on the microbiota and the invasiveness of the procedure. Despite this, using a combination of the currently available techniques, we have gained great insight into the composition, structure, and dynamics of the gut microflora in health and disease.

Studies over the past two decades have shown that, the most prevalent bacteria in GIT are broadly classified into four phyla: Firmicutes, Bacteriodetes, Actinobacteria and Proteobacteria (see glossary for **taxonomy nomenclature**) (Cresci and Izzo, 2019; Hillman et al., 2017) while bacteria belonging to the Fusobacteria, Tenericutes, Spirochaetes, Cyanobacteria and TM7 phyla are rarer (Hillman et al., 2017). The diversity and richness of the microbes varies axially and longitudinally along the digestive tract, changing dramatically throughout the first few years of life, then stabilising and eventually declining with old age (Wilmanski et al., 2021). As summarised in Fig. 2, the architecture of gut microflora communities and their dynamic stability is shaped by both intrinsic and extrinsic factors including, the birthing process, diet, gender, geographical location, pharmaceutical drugs and age (Cresci and Izzo, 2019; Hillman et al., 2017; Schmidt et al., 2018). Seeding of the gut microbiome is believed to start during the birthing process, where the newborn is inoculated with microbes from either the vaginal-anal area or the skin, depending on the mode of delivery (Browne et al., 2022). Feeding style post birth affects early-life colonisation, where breast-fed infants have higher numbers of *Bifidobacterium* species selectively enhanced by oligosaccharides present in breast milk (McBurney et al., 2019). In early years, members of the microbiota are mostly aerobic shifting with time to facultatively anaerobic bacteria and finally to largely anaerobic microbes (Cresci and Bawden, 2015).

Various patterns of microbial colonisation have been linked to numerous health states, whether steady or disease conditions. Despite the extensive research carried out in the field over the last two decades, there is a lack of a clear definition of a "healthy microbiome". Essentially, health is a dynamic state; therefore it is proposed that the associated microbiota should be diverse and resilient in order to overcome perturbations and return to equilibrium (McBurney et al., 2019). Microbial community composition varies greatly among individuals, it is almost as unique as a thumbprint. About 70% of interindividual microbiota variation is unexplained, while the remaining deviation is due to both genetic and environmental factors (McBurney et al., 2019; Rothschild et al., 2018; Wang et al., 2018). This evidence is a plausible justification to why a state of equilibrium cannot be defined by a distinct microbial composition. However, there are certain traits and phenotypes that correlate with a diverse gut microbiota, including longer colonic transit times and stool with soft to firm consistency (Roager et al., 2016; Vandeputte et al., 2016).

Scientists have developed a method of classifying microbiomes of individuals based on their composition, into what they termed as enterotypes in attempt to classify microbial markers with diagnostic potential (Arumugam et al., 2011). Arumugam and colleagues identified three main enterotypes that are recognisable by variations in the levels of three main genera. Two of the clusters are always denoted by *Bacteroides* and *Prevotella*, while the third cluster is driven by different groups, *Ruminococcus*, *Blautia*, and unclassified Lachnospiraceae depending on the studied dataset. Further analysis revealed that enterotypes are influenced by long-term dietary habits (Wu et al., 2011). A study has shown that infants have two enterotypes driven by Firmicutes and *Bifidobacterium* dominance which shifts to *Bacteroides* and *Prevotella* as they get older (Xiao et al., 2021). However, the concept of enterotypes has been challenged since most microbiome data exhibit a continuous gradient of dominant taxa and not microbial populations with discrete taxa (Knights et al., 2014). Despite all efforts, there is also a lack of valid microbial biomarkers which adds further challenges to the present complexities of defining a "healthy microbiome". Patterns of colonisation associated with steady and disease states are very difficult to identify due to the lack of longitudinal studies and conflicting observations. For instance, *Akkermansia* is often linked to healthier metabolic features in obese individuals but negatively correlated with incidence of multiple sclerosis – suggesting that microbial behaviour is context dependent just as found in the natural environment (Alzarhani et al., 2019; Dao et al., 2016; McBurney et al., 2019). However, a recent multi-generational study on a large Dutch cohort investigating environmental factors causing microbiome variations highlighted over 1000 associations between disease and bacterial taxa (Gacesa et al., 2022). *Alistipes senegalensi* was one of the microbes identified by Gacesa et al. (2022) that is associated with 43 disease phenotypes, including IBD. Overall, it might be more pertinent to identify a set of "healthy" microbial metabolic functions rather than a "healthy" microbial composition (Bäckhed et al., 2012).

Human metabolic processes are highly reliant on members of the microbiota. Intestinal bacteria are involved in the synthesis of amino acids, enzymes, and vitamins, in addition to nutrient and mineral absorption, and fermentation. For instance, *Bifidobacterium* spp. are known to synthesise vitamin K and B (Nishida et al., 2017). Many plant-derived dietary fibres are indigestible by the human body and can only be fermented by commensal bacteria. Short-chain fatty acids (SCFAs, primarily acetate, propionate and butyrate) are by-products of this process and are important in maintaining homeostasis in the gut. Butyrate is the main energy source of IECs and prevents dysbiosis by enhancing IEC uptake of oxygen through β oxidation, maintaining a hypoxic environment for commensal anaerobes, reviewed in Valdes et al. (2018). On the other hand, acetate is vital for the growth of other gut flora and is involved in central appetite regulation, cholesterol metabolism and lipogenesis (Frost et al., 2014). Firmicutes such as, *Faecalibacterium prausnitzii*, *Clostridium leptum* as well as *Eubacterium rectale* and *Roseburia* spp. of the family Lachnospiraceae, are mainly butyrate producers while members of Bacteroidetes mostly produce acetate and propionate (reviewed in Venegas et al., 2019). *Eubacterium hallii* and *Anaerostipes* spp. are secondary producers of butyrate from lactate and acetate. Members of the genus *Bifidobacterium* produce acetate and lactate, while *Akkermansia* spp. generate propionate and acetate (Venegas et al., 2019). A shift in abundance of these SCFA producers, among other species, alters the metabolic profile of an individual thus disrupting the functional properties of the intestinal ecosystem (Cummings and Macfarlane, 1991; Venegas et al., 2019).

Research shows that a less diverse microbiome renders an individual more susceptible to disease (Bäckhed et al., 2012). IBD is one of many diseases associated with dysbiosis and specific commensal microbes without an established causal relationship. Changes in metabolic pathways linked to inflammation and the flourishing of pathobionts could potentially act as microbial stressors, supporting the outgrowth of species associated with dysbiosis. Anaerobic SCFA producers are depleted in the intestinal mucosa and faeces of IBD patients, leaving space for the expansion of populations of pathobionts, like *E. coli* and *Salmonella* (Venegas et al., 2019). Studies have shown that IBD patients have low numbers of gut microbes belonging to the phyla Bacteroidetes and Firmicutes alongside an increased abundance of Proteobacteria, Actinobacteria and Bacilli (Frank et al., 2007; Ni et al., 2017; Venegas et al., 2019). The SCFA producers, *F. prausnitzii* and *Roseburia intestinalis* are particularly reduced in CD cases (Franzosa et al., 2019; Ni et al., 2017; Venegas et al., 2019). *F. prausnitzii* elicits the production of the anti-inflammatory IL-10 and prevents the release of inflammatory cytokines (IL-12 and INF-γ) thus maintaining mucosal equilibrium (Nishida et al., 2017). Therefore, altering the presence of *F. prausnitzii* can disrupt mucosal stability, bringing about inflammation. Inflammation is strongly tied to the suppression of genes involved in SCFA utilisation and metabolism. Genes encoding enzymes involved with butyrate uptake and oxidation are downregulated in dysbiotic UC patients and this reduction can be reversed with anti-TNF-α antibody treatment (Venegas et al., 2019).

Commensal bacteria maintain gut homeostasis also by mediating intestinal mucosal immunity while others suppress pathobionts directly through nutrient competition and secretion of antimicrobial substances. *Bacillus thuringiensis*, for instance, releases a **bacteriocin** targeting spore-forming bacteria such as other *Bacillus* species or Clostridia (Nishida et al., 2017). *Candidatus Arthomitus*, also known as segmented filamentous bacteria (SFB), promote the maturation of the mucosal immune system, IgA secretion, as well as Th17 development (Nishida et al., 2017). SCFA producers can indirectly modulate immune system interactions, such as AMP (REGIIIγ and β-defensins) production, induction of the anti-inflammatory IL-10, and T-cell differentiation (Nishida et al., 2017; Vaga et al., 2020). Butyrate produced by Clostridium clusters IV, XIVa and XVIII enhances the function of colonic **regulatory T cells (Treg cells)** (Ni et al., 2017; Nishida et al., 2017). Moreover, butyrate has the ability of preventing inflammation by suppressing NF-κB activation, inhibiting proinflammatory cytokines, like TNF-α, INF-γ, and IL-6, through **HDAC** inhibition and stabilising **HIF-α** (Ni et al., 2017).

Adherent-Invasive *E. coli* (AIEC) is another bacterial species linked to IBD that affects the permeability of the intestinal mucosa and is increased in CD patients (Ni et al., 2017; Nishida et al., 2017). IBD has also been associated with an increase in the sulphate-reducing bacteria *Desulfovibrio*, where the hydrogen sulphate produced from that process causes damage to IECs and inflammation (Nishida et al., 2017). Elevated levels of tryptophan metabolites in IBD patients indicates an increase in tryptophan

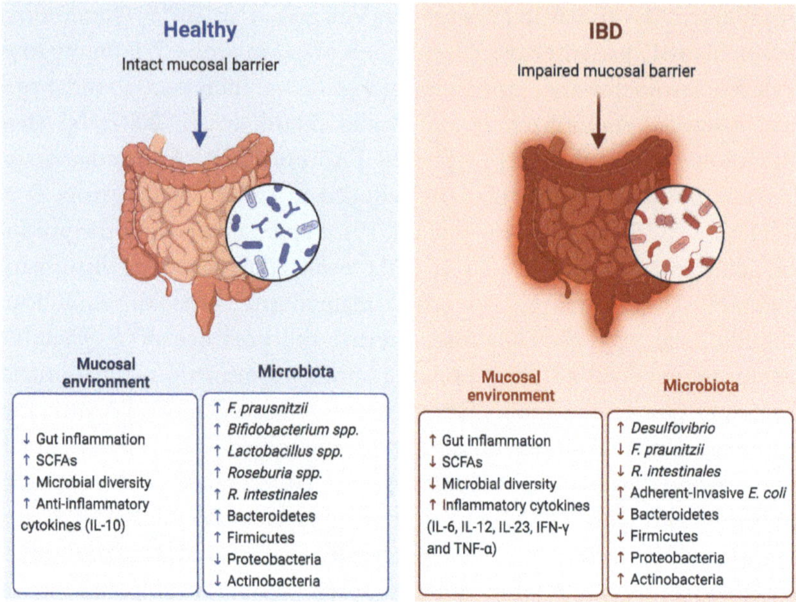

Fig. 4 Summary of some of the key features linked to a healthy microbiome or a microbiome linked to IBD (right). The figure was created with BioRender.com.

degradation which, can be linked to the abundance of *Bacteroides* species involved in tryptophan metabolism in IBD (Lee and Chang, 2021; Nikolaus et al., 2017; Vaga et al., 2020). Secondary bile acids generated by microbiota has anti-inflammatory properties and plays a role in maintaining mucosal homeostasis. IBD patients exhibit imbalances in bile acid absorption and deficiencies in secondary bile acids which can induce the growth of bile-tolerant pathobionts, like the sulphite-reducing *Bilophila wadsworthia*, as shown in IL10-deficient mice (Devkota et al., 2012; Lee and Chang, 2021; Natividad et al., 2018). A visual summary of some of the key findings related to the gut microbiome and IBD is presented in Fig. 4.

All the findings discussed earlier are correlations with IBD and causality has not been shown. Most microbial associations have not been replicated, some have even been contradicted, implying that they may be context-dependent or heavily influenced by environmental factors. Only 2% of taxa at the genus level are found to be different in IBD when compared to healthy individuals as opposed to 12% of change in metabolic pathways (Ni et al., 2017). A study has shown over 2700 differentially abundant metabolites in faecal samples from IBD patients, of which 244 metabolites are significantly increased (Franzosa et al., 2019). This suggests that perhaps identifying metabolic pathways and metabolite profiles associated with IBD rather than a microbial signature is a better way to understand the role of the microbiota in disease progression.

4. How does the environment contribute to gut microbiome changes?

Diet, lifestyle, physical activity, smoking and exposure to pollution are some of the environmental factors that shape the gut microbiome and, in some cases, are associated with IBD prevalence. Incidence rates of IBD are higher in urban and western environments, particularly in areas that have undergone a period of industrialization over the past few decades (Benchimol et al., 2017; Ng et al., 2013). A large-scale population-based study showed that, within 50 years, there has been a 10-fold increase in incidence of IBD in Japan, South Korea, China and India (Ng et al., 2013). Incidence rates are still higher in western populations, however, IBD is an emerging threat in developing countries. The link between urbanisation and IBD development could be mediated by multiple lifestyle factors, like diet, exposure to air pollution, hygiene practices, and antimicrobials (Ananthakrishnan et al., 2017). There have been changes in microbial patterns associated with urbanisation, like an increase in Bacteroidetes, particularly *Alistipes*, the Firmicutes *Blautia*, *Faecalibacterium* and *Ruminococcus*, as well as *Bifidobacterium* and *Bilophila* of the phyla Actinobacteria and Proteobacteria, respectively (Zuo et al., 2018). Generally, there is a shift in microbial composition from gram positive to gram negative species with pathogenic tendencies (Hills et al., 2019). Studies investigating variation within microbiota structure in relation to geography have pointed out that there are deviations across countries as well as between ethnicities within the same country (Cresci and Bawden, 2015; Prideaux et al., 2013). Gut microbial communities of Malawian and Amerindian populations (of the Venezuelan Amazonia) are divergent from those of the US, with less diversity in the latter (Yatsunenko et al., 2012). These differences can be attributed to diet since they have been mapped to metabolic pathways involved in carbohydrate metabolism and vitamin B biosynthesis (Yatsunenko et al., 2012). Malawian and Amerindian diets are carbohydrate rich, high in corn and cassava, while a US diet is rich in protein (Cresci and Bawden, 2015). Another study comparing microbiotas of African and Western children showed that microbial communities in children from Burkina Faso coevolved with their diets, maximising energy intake from fibrous foods (Filippo et al., 2010). There was a higher prevalence of Bacteroidetes, notably Prevotella species involved in cellulose and xylan hydrolysis, and less of Firmicutes in children from Burkina Faso, compared to a total absence of these microbes in an Italian cohort (Filippo et al., 2010). The difference in gut microbial composition can be linked back to the African diet, which is high in plant fibre (Zuo et al., 2018).

Diet is one of the environmental factors that has a huge impact on the microbiome. Metagenomic studies on twins have shown that the influence of dietary habits and cohabitation on the microbiome outweighs the effect of genetics on the structure and functionality of the gut microflora (Gacesa et al., *2022;* Hills et al., *2019;* Rothschild et al., 2018). This starts at infancy where breastfeeding selects for a microbiome dominated by Bifidobacterium due to the oligosaccharides present in breastmilk and can reverse infancy dysbiosis from intrapartum antibiotic exposure (Azad et al., 2016; Cresci and Bawden, 2015). Development of IBD is negatively correlated with breastfeeding, however, the causal links are not clear (Ananthakrishnan et al., 2017). Dietary habits continue to shape the microbiome throughout life with various ramifications. A western diet—one typically high in simple carbohydrates, fats, additives and low in fibre—has a negative effect on microbiome diversity as well as host mucosal barrier and immunity (Hills et al., 2019; Levine et al., 2018). An abundance of *Bifidobacteria* and a reduction in *Lactobacillus*, *Streptococcus* and *Roseburia* are commonly associated with diets high in simple carbohydrates (Hills et al., 2019). Reduced mucosal barrier function favouring AIEC colonisation and mucin-degrading *Ruminococcus* accompanied by dysbiosis is induced by a western diet in both CD patients and mouse models (Agus et al., 2016; Martinez-Medina et al., 2014). Histological assays on mice fed a western diet showed depletion of goblet cells followed by a reduction in MUC2 expression and increased intestinal permeability (Martinez-Medina et al., 2014). Moreover, the dysbiosis caused by a western diet decreases production of SCFAs and expression of the SCFA receptor GPR43 modulating the immune response via Treg cell reduction (Agus et al., 2016). Interestingly, biopsies from CD patients also showed reduced expression of the butyrate receptor GPR43 highlighting the role of butyrate in modulating inflammation (Agus et al., 2016). A few studies have linked diets high in fat with intestinal inflammation. Mice put on high fat diets exhibited an increase in pro-inflammatory markers, including TNF-α and INF-γ, decreased Treg cells, triggered recruitment of dendritic cells and Th17 cells along with downregulation of tight junction proteins collectively exacerbating intestinal inflammation (Gruber et al., 2013; Ma et al., 2008; Suzuki and Hara, 2010). Furthermore, structure and function of the microbiota were highly impacted by a diet high in fat, presenting reduced numbers of Ruminococcaceae, altered antimicrobial pathways, defective bile metabolism and shifts in nutrient metabolism highlighted by a large set of differentially abundant bacterial proteins and metabolites (Daniel et al., 2014). Risk of UC is significantly reduced with high polyunsaturated dietary fat intake as opposed to unfavourable trans-unsaturated fats (Levine et al., 2018).

High-fibre diets, on the other hand, are linked to a highly diverse microbiome. Increased dietary intake of fibre in IL10$^{-/-}$ mice suppresses gut inflammation which protects against acute colitis through several mechanisms, including increased expression and proper localisation of tight junction proteins which maintain the mucosal barrier as well as a decrease in production of the pro-inflammatory IFN-γ accompanied by an increase in SCFA and Treg production (Bassaganya-Riera et al., 2011; Silveira et al., 2017; Wang et al., 2016). Supplementing the diet of mice with fermentable fibres also increased SCFA production and decreased mucosal permeability through an increase in expression of **occludin** and other tight junction proteins, reducing inflammation (Hung and Suzuki, 2016). However, increased fibre intake did not resolve the low levels of SCFAs in UC patients despite an established link between low fibre and the condition, suggesting that not all findings in mouse models are translatable in humans (James et al., 2015; Levine et al., 2018). Regardless, a study revealed a bloom of bile tolerant bacteria, like *Alistipes* and *Bilophila* and Bacteroides (a pattern previously linked to IBD), in the microbiota of individuals following an animal-based diet compared to those on a plant-based diet, which supports the connection between animal protein and IBD (David et al., 2014; Levine et al., 2018). There have been no dietary patterns tied to CD. However, a diet low in fermentable oligosaccharides, disaccharides, monosaccharides, and polyols, known as the FODMAP diet, seems to provide symptom relief for IBD patients with no attenuation of inflammation (Cox et al., 2020). This diet is only recommended for the short-term due to its low levels of fibres and effect on the microbiome, with lower abundances of *Bifidobacterium* species and *F. prausnitzii* (Cox et al., 2020; Hills et al., 2019).

Processed foods are high in food additives which proved to have harmful effects on both mucosal stability and the gut microflora (Hills et al., 2019; Levine et al., 2018). Carboxymethylcellulose is an emulsifier commonly used in food products and was found to increase intestinal permeability through depletion of the mucus layer and distension of inter-villi spaces, allowing for bacterial invasion in IL10$^{-/-}$ mice (Chassaing et al., 2015; Swidsinski et al., 2009). Another emulsifier, polysorbate 80, was shown to cause similar deleterious effects on the intestinal mucosa as well as the microbiome (Chassaing et al., 2017). Both emulsifiers alter the microbial composition to a less diverse one, with an increase in mucin-degrading bacteria (like *Ruminococcus gnavus*), Verrucomicrobia and inflammatory Proteobacteria and Enterobacteriaceae species along with a decrease in key Bacteriodaceae (Chassaing et al., 2017). Maltodextrin is an artificial sweetener linked to gut inflammation through elevated levels of IgA (Miyazato et al., 2016). The sweetener has been shown to cause dysbiosis promoting the expansion of AIEC and biofilm formation (Nickerson and Mcdonald, 2012). It is evident that additive-rich processed foods have adverse effects on mucosal immunity and the microbiome. However, not all foods have a solely harmful effect on health and the microbiome, as some may provide context dependent beneficial effects. Administering dietary alcohol in mice promoted colonic hyperpermeability and decreased butyrate and other SCFA levels in stool (Forsyth et al., 2017). Yet, red wine in moderation exhibits prebiotic benefits—owed to high levels of polyphenols—with an increase in Proteobacteria, Fusobacteria, Firmicutes and Bacteroidetes species, especially *F. prausnitzii* and *Bifidobacterium* (Hills et al., 2019; Queipo-Ortuño et al., 2012). These findings have not been replicated with other kinds of alcoholic beverages.

With the significant effect of diet on both the gut microflora and IBD, investigations into IBD and obesity comorbidities has been an area of interest. Generally, obesity is reflected in gut microbiota shifts in abundance of Bacteroidetes and Firmicutes, but no connection can be drawn due to the minimal research done on the co-morbidity of IBD and obesity within the context of the microbiome.

Another major player in microbiome shifts is drug exposure. Intake of drugs, like antibiotics and non-steroidal anti-inflammatory drugs (NSAIDs), result in significant changes in gut microbial composition and are considered a major predisposition for IBD (Ananthakrishnan et al., 2017; Dethlefsen et al., 2008). A 5-day course of the antibiotic ciprofloxacin in humans is enough to reduce microbial diversity and richness of the microbiome within 3–4 days of antibiotic administration (Dethlefsen and Relman, 2011). This effect is mostly reversible within 4 weeks to 6 months with the exception of a few taxa (Dethlefsen et al., 2008). Studies investigating the use of antibiotics in early life and IBD predisposition found a strong link in Western populations, but only with CD and not UC (Ungaro et al., 2014). Another study showed in a cohort of 36 IBD subjects, 21 (58%) had at least 1 dispensation of antibiotics in their first year of life compared to 39% of the controls (Shaw et al., 2010). Furthermore, children receiving antibiotics during their first year of life were almost three times as likely to develop IBD compared to those that did not (Shaw et al., 2010). An increase in use of NSAIDs poses a higher risk of developing IBD (Ananthakrishnan et al., 2017). The group of drugs is known to cause intestinal mucosal permeability and oxidative stress (Bjarnason et al., 2018). Changes in gut microbial composition have been reported upon NSAID intake, these alterations vary depending on the type of NSAID and tend to favour taxa able to metabolise these drugs (Maseda and Ricciotti, 2020). *Prevotella* spp., *Bacteroides* spp. Ruminococaceae and *Barnesiella* spp. were increased with the use of aspirin, whereas an enrichment of Acidaminococcaceae and Enterobacteriacceae were associated with ibuprofen and celecoxib users (Rogers and Aronoff, 2016). Another study reported no changes in microbiota upon celecoxib use but a decrease in butyrate production as well as inflammatory markers in vitro (Hernandez-Sanabria et al., 2020).

Other lifestyle factors, like supplement intake, smoking, and exposure to pollution, have been connected to both the microbiome and IBD, although, how these factors influence IBD are poorly understood (Ananthakrishnan et al., 2012; Piovani et al., 2019). There has been a lot of speculation on the role of vitamin D in IBD pathogenesis. However, a population study on 169 participants showed that vitamin D deficiency is not associated with development of the disease (Opstelten et al., 2018). Vitamin D and its receptor play a role in modulating the immune system through the butyrate producing bacteria but there is no causative link between that and pathogenesis of IBD (Battistini et al., 2021). Smoking is a lifestyle habit with controversial effects on IBD. It is considered a risk factor in CD development and somehow protective against UC, however, this protective effect was only observed in some populations (Piovani et al., 2019). No associations were found between smoking and IBD in Asian and Jewish populations. On the other hand, smoking affects the human microbiome and causes dysbiosis in mice with reduced numbers of Firmicutes, Actinobacteria, *Bifidobacterium*, and *Lactococcus* and an enrichment in Bacteroidetes, *Prevotella*, *Bacteroides*, Proteobacteria and Clostridium, reviewed in (Savin et al., 2018). Functional studies need to be carried out to assess the nature of the association between smoking, IBD and the microbiome. Pollution is another environmental factor connected to socioeconomic status and urbanisation that affects the microbiome and has been linked to IBD. Early exposure to air pollution was associated with an increased risk of IBD in children and young adults (Kaplan et al., 2010). However, this link has been questioned by Opstelten et al. (2016). Particulate matter (PM) exposure in mice was shown to increase intestinal permeability via rearrangement of tight junction proteins of the epithelium and promoted inflammation through IL-6 release and NF-κB activation (Mutlu et al., 2011; Salim et al., 2014). A similar pro-inflammatory cytokine response in $IL10^{-/-}$ mice was observed along with increased gut permeability with PM supplemented chow (Kish et al., 2013). The mice experience shifts in Bacteroidetes, Firmicutes and Verrucobacteria leading to altered SCFA production and reduced butyrate levels. Again, these observations are insufficient for assigning solid associations between smoking and IBD, let alone suspect causative effects. The lifestyle factors discussed are clearly connected to microbiome perturbations but have a less causal effect on IBD.

5. Can microbial-based treatments be the answer?

As briefly mentioned in the first section of this contribution, IBD is managed through different treatment protocols depending on the condition, severity of the disease and response to treatment. A large number of IBD patients fail to respond to initial treatment and a fraction of those who respond to medication initially become secondary non-responders, especially with biological therapy (Chudy-Onwugaje et al., 2019; Lamb et al., 2019). Current regimens target the aberrant immune responses involved in IBD and can change during active and quiescent phases of the disease. The first line of treatment for mild to moderate pancolitis in UC patients is administration of oral and/or enema 5-aminosalicylic acid (5-ASA), which is an anti-inflammatory drug that captures damaging free radicals in the bowel (Kuehbacher et al., 2008; Lamb et al., 2019). 5-ASA is also used for maintenance therapy in UC patients. Patients experiencing major side effects or are refractory to 5-ASA are prescribed corticosteroids. Due to their severe side effects, corticosteroids are only used in moderate to severe cases as well as refractory cases of both UC and CD (Chudy-Onwugaje et al., 2019; Lamb et al., 2019). Immunosuppressive or immunomodulator therapies in the form of biological agents, like human anti-TNF-α monoclonal antibodies, are used in the event of failure in response to second line drugs (Kuehbacher et al., 2008; Lamb et al., 2019). In the case of complete resistance to medication and persistence of health complications, a colectomy followed by **ileal pouch anastomosis (IPAA)** is the remaining option (Lamb et al., 2019).

Primary therapy in CD patients with moderately to severely active disease includes systemic corticosteroids (Lamb et al., 2019). However, nutritional intervention through exclusive enteral nutrition (EEN) has proven to be an effective first line of treatment, inducing remission with a success rate of 73% but only in children (Eindor-Abarbanel et al., 2021; Lamb et al., 2019). EEN involves the administration of a nutritionally-whole liquified diet orally or through nasogastric tubes for 4–8 weeks (Borrelli et al., 2006; Lamb et al., 2019). The mechanism of action for this nutritional therapy remains poorly understood, yet it has been shown to lower gut microbial diversity with reduction in *Bacteroides* species and *Clostridium coccoides* (Maclellan et al., 2017). When compared to children treated with corticosteroids, those on EEN exhibited significant mucosal healing but lower numbers of butyrate-producing bacteria, which seems very contradictory since lower levels of butyrate are generally associated with active disease (Pigneur et al., 2019). Nutritional intervention is not as effective in adults, early use of biological agents is considered with adult CD patients who have poor prognoses (Lamb et al., 2019). Failure of conventional therapy and recurrence of the disease can be addressed by ileocaecal resection to induce and maintain remission.

Antibiotics are also part of the IBD treatment regimens discussed above and have been largely used in the incidence of pouchitis (infection post IPAA), abdominal abscesses, enteric infections, IBD associated *C. difficile* infection, **perianal disease** and **luminal disease** (Eindor-Abarbanel et al., 2021; Ledder et al., 2014). They are also used to modulate gut microflora by targeting bacteria associated with inflammation in CD patients and selecting for the beneficial ones, since patients with CD tend to harbour mucosal infiltrating pathobionts (Eindor-Abarbanel et al., 2021). However, exposure to antibiotics could have a negative effect on the microbiome on its already unstable state. A recent study has shown reduced microbial diversity and enrichment of *Escherichia* species along with a reduced UC activity index in young UC patients post antibiotic treatment (Turner et al., 2019). Two other studies reported similar conflicting responses to antibiotics in CD patients, where the microbiota community in both study groups was less diverse but faecal calprotectin (an inflammation marker) was reduced in patients from one of the studies (Levine et al., 2018; Sprockett et al., 2019). Interestingly, increased abundance of *Bifidobacterium* and

F. prausnitzii along with an increase in SCFA were observed post antibiotic treatment in 4 CD patients (Maccaferri et al., 2010). However, the findings of all the studies combined with a small remainder of studies are not enough to draw a strong conclusion on the effect of antibiotics on IBD and microbiota of IBD patients.

The standard treatment protocol is not always effective and usually comes with serious side effects, thus the search remains for a more robust alternative treatment. Since the microbiome is heavily associated with IBD, there has been an interest in microbe-based or microbe-related interventions. Probiotics are live bacteria and fungi that are present in the form of food products or supplements that provide health benefit to the host by restoring balance and function in the gut microbial communities when administered in the right dosage (Martyniak et al., 2021). Probiotic strains have been shown to maintain the mucosal barrier function and modulate immune responses (Ng and Hart, 2014). There are countless probiotic preparations on the market, and they are designed in a way so that the microbes can survive transit until they reach the larger bowel (Eindor-Abarbanel et al., 2021). The main microbial strains used in most probiotic supplements include *E. coli* Nissle 1917, *E. coli* (serotype 06:K5:H1), *Lactobacillus reuteri*, *L. casei* ATCC PTA-3945, *Bifidobacterium* and yeast species *Saccharomyces boulardii* (Eindor-Abarbanel et al., 2021). VSL#3 is one of the most commonly administered and investigated probiotic in IBD, and it is comprised of *Streptococcus thermophilus*, three bifidobacteria strains (*Bifidobacterium longum*, *B. breve*, and *B. infantis*) as well as 4 *Lactobacillus* strains (*L. casei*, *Lactobacillus acidophilus*, *L. bulgaricus*) (Cheng et al., 2020). There have been contradictory observations from studies evaluating the efficacy of probiotics in inducing and maintaining remission in IBD patients; a few patients had positive outcomes (Palumbo et al., 2016; Shadnoush et al., 2015; Tamaki et al., 2016; Yoshimatsu et al., 2015) while others showed no significant effects (Bjarnason et al., 2018; Fedorak et al., 2015; Matsuoka et al., 2018). Meta-analysis studies and reviews also present opposing findings, where two groups found probiotics to be beneficial in UC but not CD patients (Ganji-Arjenaki and Rafieian-Kopaei, 2018; Zhang et al., 2021) while Astó et al. (2019) reported a significant effect of probiotics inducing remission but not maintaining it.

The effects of probiotics are short-lived and cease once treatment is stopped, which is probably due to the lack of nutrients or substrates required for the microorganisms to flourish (Tannock et al., 2000). Instead, patients can be prescribed prebiotics, which are compounds (commonly found in foods) that are selectively utilised by members of the microbiota and promote their growth and function, thus providing health benefit to the host (Eindor-Abarbanel et al., 2021). Glucooligosaccharides, fructooligosaccharides, lactulose and inulin are the most widely used prebiotics and are indigestible by humans (Martyniak et al., 2021). Generally, they are known to enhance the growth of lactic acid-producing bacteria, like *Lactobacillus* as well as *Bifidobacterium* and *Bacteroides* species, and promote SCFA production (Cummings and Macfarlane, 2002; Martyniak et al., 2021). Clinical studies assessing the effect of prebiotics on IBD patients have been contradictory; Benjamin et al. (2011) reported deterioration of CD patients on prebiotics while another group reported no significant improvement or worsening in both UC and CD on prebiotics (Hafer et al., 2007). A few studies have reported improvement in clinical activity of the disease and remission states with different prebiotics (Fernandez-Banares et al., 1999; Hallert et al., 1991; Kanauchi et al., 2003). These findings counter each other, and presently no significant positive effect can be linked to prebiotics intake and IBD. However, the combination of prebiotics and probiotics (collectively referred to as synbiotics) has been shown to have a beneficial effect (Zhang et al., 2021). Usually, synbiotic supplementation comes in the form of *Lactobacillus* GG and/or *Bifidobacteria* coupled with fructooligosaccharides and/or inulin (Martyniak et al., 2021). Compared to a placebo group, 13 CD patients put on *B. longum* and a mix of inulin and fructoligosaccharides showed significant clinical and histopathological improvement, in addition to enrichment of *Bifidobacterium* species and reduction in inflammatory markers. However, this was only observed in the first 3 months of treatment and not the total 6 months (Steed et al., 2010). Chermesh et al. (2007) reported no significant improvement in CD patients on synbiotics. Again, results are conflicting with no significant effect highlighted and investigations in UC and synbiotics were equally paradoxical (Martyniak et al., 2021).

Just like probiotics, faecal matter transplantation (FMT) works through modulation of the recipient's immune response via microbiota compositional changes. FMT involves the delivery of stool derived microbes from a healthy donor to the gastrointestinal tract of a patient (Ni et al., 2017). It is argued that it is a more robust method to treat dysbiosis as it has been shown to increase microbial diversity due to a larger and more varied microbial load during transfer (Bojanova and Bordenstein, 2016; Hamilton et al., 2013; Khoruts et al., 2010). However, the screening process for donors is very complex to avoid infections and potential allergic reactions that can be passed on through the faecal matter (Bibbò et al., 2020; Weingarden and Vaughn, 2017). This procedure has been widely explored to restore microbial composition and successfully used to treat *C. difficile* infections. The success rate of over 90% in recurrent *C. difficile* infection has made FMT a therapeutic of interest (Drekonja et al., 2015). However, the clinical response to FMT is not equally positive in IBD patients. Success rates varied from 25% to 79% in UC patients, while response to FMT in CD and placebo groups was mostly not significant due to small cohort size and differences in procedure protocols (Costello et al., 2019; Crothers et al., 2021; Paramsothy et al., 2017; Sokol et al., 2020; Sood et al., 2019; Yang et al., 2020).

In general, more comprehensive investigations into the mentioned treatments should be done as current work is inconsistent and is based on small cohorts. Furthermore, the efficacy of current therapeutics should be linked to its functional effects on the microbiome. A study has shown that two thirds of commonly prescribed drugs are metabolised by at least 1 bacterial strain of the microflora and that each strain can process from 11 to 95 different drugs (Zimmermann et al., 2019). The authors identified and validated 30 microbiome-associated enzymes able to convert 20 known drugs into 59 candidate metabolites. This could be a reason why many patients are resistant to conventional treatment. Identifying microbes associated with current IBD medication metabolism could help alleviate refractory therapy. With the steady drop of sequencing expenses, personalised treatment for IBD patients could be implemented and potentially solve inefficiencies of current IBD treatment regimens.

6. Conclusion and outlook

The gut microbiota plays a role in maintaining homeostasis and disease pathogenesis. Despite the lack of clear causal relationships between the human microbiome and many health conditions, the evidence for links between commensals and health are intriguing. The gut microbiome is a very complex network built on intricate interactions among microbes and the crosstalk between host and microbes. Current research into the field is largely based on the bacterial component of the microbiome and fails to address the importance of the other members, like fungi, viruses, and archaea. There is very little work done on investigating fungal and viral communities in the gut and even less on archaea. Different bacterial strains can contribute to the same metabolic pathways conferring a health or diseased state. Perhaps there could be an unknown co-dependency between other members of the gut microflora influencing major functional properties involved in immunity or metabolism. Exploring the neglected part of the human microbiome would give us better, more holistic insight into its role in health and disease. The added complexity of environmental variables causing microflora perturbations makes it very challenging to assign causal relationships. This could explain why many of the findings in mice are not replicated in humans. For instance, the mouse models used to study IBD focus on isolated events or a number of variables in a tightly controlled environment; however, it is very difficult to do the same in a clinical setting due to great variability in environmental and genetic factors.

These difficulties highlight the importance of animal models in identifying microbiome-host association with disease. To correctly identify microbiome links to disease the appropriate model must be chosen. In the case of IBD, the spontaneous CD model SJL/J is less favoured, especially with oxazolone induced inflammation, due to the development of other autoimmune diseases along with CD via Th1 mediated inflammation (Baydi et al., 2021). Furthermore, the sex of the mice used can be as influential as the model used especially when looking into gut microbial composition. It is critical to consider such technical matters when designing a study to avoid bias. Similar considerations should be taken in clinical studies as well, most of the clinical studies investigating the effect environmental factors in IBD have very low statistical power due to the small number of patients in their study groups. More effort should be put into designing larger cohort studies in order to get a more accurate representation of the disease and its risk factors.

In the future, it will be pivotal to perform longitudinal studies in large patient- and control-cohorts, to capture associations and changes in the gut microbiome and linking these to disease or disease progression, paving the way to identify causations rather than just correlations. As the power of DNA sequencing technologies and bioinformatic analysis increase, these will be combined with proteomic, transcriptomic and metabolomic analysis to capture the dynamic nature and function of the human microbiome in greater detail, allowing us to understand the ecology of this system with respect to health and disease.

Genetics are less influential within the context of IBD and the microbiome, whereas environmental factors, especially diet, play a leading role in gut microbial composition and can affect disease susceptibility.

Further investigations are required for the identification of optimal manipulations of gut microbes whether prophylactically or in treatment form. As our understanding of the human microbiome increases, we are beginning to unlock its secrets in order to promote health and fight disease.

References

Agus, A., et al., 2016. Western diet induces a shift in microbiota composition enhancing susceptibility to adherent-Invasive E. coli infection and intestinal inflammation. Sci. Rep. https://doi.org/10.1038/srep19032.

Ahn, S., et al., 2008. Hepatocyte nuclear factor 4α in the intestinal epithelial cells protects against inflammatory bowel disease. Inflamm. Bowel Dis. 14 (7), 908–920. https://doi.org/10.1002/ibd.20413.

Alzarhani, A.K., et al., 2019. Are drivers of root-associated fungal community structure context specific? ISME J. 13 (5), 1330–1344. https://doi.org/10.1038/s41396-019-0350-y.

Ananthakrishnan, A.N., et al., 2012. Higher predicted vitamin D status is associated with reduced risk of Crohn's disease. Gastroenterology 142 (3), 482–489. https://doi.org/10.1053/j.gastro.2011.11.040.

Ananthakrishnan, A.N., et al., 2017. Environmental triggers in IBD: a review of progress and evidence. Nat. Rev. Gastroenterol. Hepatol. 15 (1), 39–49. https://doi.org/10.1038/nrgastro.2017.136.

Artis, D., et al., 2004. RELMβ/FIZZ2 is a goblet cell-specific immune-effector molecule in the gastrointestinal tract. Proc. Natl. Acad. Sci. U. S. A. 101 (37), 13596–13600. https://doi.org/10.1073/pnas.0404034101.

Arumugam, M., et al., 2011. Enterotypes of the human gut microbiome. Nature 473 (7346), 174–180. https://doi.org/10.1038/nature09944.

Astó, E., et al., 2019. The efficacy of probiotics, prebiotic inulin-type fructans, and synbiotics in human ulcerative colitis: a systematic review and meta-analysis. Nutrients 11 (2). https://doi.org/10.3390/nu11020293.

Azad, M.B., et al., 2016. Impact of maternal intrapartum antibiotics, method of birth and breastfeeding on gut microbiota during the first year of life: a prospective cohort study. BJOG 123 (6), 983–993. https://doi.org/10.1111/1471-0528.13601.

Bäckhed, F., et al., 2012. Defining a healthy human gut microbiome: current concepts, future directions, and clinical applications. Cell Host Microbe 12 (5), 611–622. https://doi.org/10.1016/j.chom.2012.10.012.

Barrett, J.C., et al., 2009. Genome-wide association study of ulcerative colitis identifies three new susceptibility loci, including the HNF4A region. Nat. Genet. 41 (12), 1330–1334. https://doi.org/10.1038/ng.483.

Bassaganya-Riera, J., et al., 2011. Soluble fibers and resistant starch ameliorate disease activity in interleukin-10-deficient mice with inflammatory bowel disease. J. Nutr. 141 (7), 1318–1325. https://doi.org/10.3945/jn.111.139022.

Battistini, C., et al., 2021. Vitamin D modulates intestinal microbiota in inflammatory bowel diseases. Int. J. of Mol. Sci. 22 (1). https://doi.org/10.3390/ijms22010362.

Baydi, Z., et al., 2021. An update of research animal models of inflammatory bowel disease. ScientificWorldJournal. https://doi.org/10.1155/2021/7479540.

Benchimol, E.I., et al., 2017. Rural and urban residence during early life is associated with a lower risk of inflammatory bowel disease: a population-based inception and birth cohort study. Am. J. Gastroenterol. 112 (9), 1412–1422. https://doi.org/10.1038/ajg.2017.208.

Benjamin, J., et al., 2011. Randomised, double-blind, placebo-controlled trial of fructo-oligosaccharides in active Crohn's disease. Gut 60 (7), 923–929. https://doi.org/10.1136/gut.2010.232025.

Bibbò, S., et al., 2020. Fecal microbiota transplantation: screening and selection to choose the optimal donor. J. Clin. Med. 9 (6). https://doi.org/10.3390/jcm9061757.
Bjarnason, I., et al., 2018. Mechanisms of damage to the gastrointestinal tract from non-steroidal anti-inflammatory drugs. Gastroenterology 154 (3), 500–514. https://doi.org/10.1053/j.gastro.2017.10.049.
Bojanova, D.P., Bordenstein, S.R., 2016. Fecal transplants: what is being transferred ? PLoS Biol. 14 (7). https://doi.org/10.1371/journal.pbio.1002503.
Borrelli, O., et al., 2006. Polymeric diet alone versus corticosteroids in the treatment of active pediatric crohn's disease: a randomized controlled open-label trial. Clin. Gastroenterol. Hepatol. 4 (6), 744–753. https://doi.org/10.1016/j.cgh.2006.03.010.
Browne, H.P., Shao, Y., Lawley, T.D., 2022. Mother – infant transmission of human microbiota. Curr. Opin. Microbiol. 69. https://doi.org/10.1016/j.mib.2022.102173.
Caporaso, J.G., et al., 2011. Moving pictures of the human microbiome. Genome Biol. 12 (5). https://doi.org/10.1186/gb-2011-12-5-r50.
Chassaing, B., et al., 2015. Dietary emulsifiers impact the mouse gut microbiota promoting colitis and metabolic syndrome. Nature 519 (7541), 92–96. https://doi.org/10.1038/nature14232.
Chassaing, B., et al., 2017. Dietary emulsifiers directly alter human microbiota composition and gene expression ex vivo potentiating intestinal inflammation. Gut 66 (8), 1414–1427. https://doi.org/10.1136/gutjnl-2016-313099.
Cheng, F., et al., 2020. Probiotic mixture VSL#3: an overview of basic and clinical studies in chronic diseases. World J. Clin. Cases 8 (8), 1361–1385. https://doi.org/10.12998/wjcc.v8.i8.1361.
Chermesh, I., et al., 2007. Failure of Synbiotic 2000 to prevent postoperative recurrence of Crohn's disease. Dig. Dis. Sci. 52 (2), 385–389. https://doi.org/10.1007/s10620-006-9549-7.
Chudy-Onwugaje, K.O., et al., 2019. A state-of-the-art review of new and emerging therapies for the treatment of IBD. Inflamm. Bowel Dis. 25 (5), 820–830. https://doi.org/10.1093/ibd/izy327.
Costello, S.P., et al., 2019. Effect of fecal microbiota transplantation on 8-week remission in patients with ulcerative colitis a randomized clinical trial. JAMA 321 (2), 156–164. https://doi.org/10.1001/jama.2018.20046.
Cox, S.R., et al., 2020. Effects of low fodmap diet on symptoms, fecal microbiome, and markers of inflammation in patients with quiescent inflammatory bowel disease in a randomized trial. Gastroenterology 158 (1), 176–188.e7. https://doi.org/10.1053/j.gastro.2019.09.024.
Cresci, G.A., Bawden, E., 2015. Gut microbiome: what we do and don't know. Nutr. Clin. Pract. 30 (6), 734–746. https://doi.org/10.1177/0884533615609899.
Cresci, G.A.M., Izzo, K., 2019. Gut Microbiome. In: Corrigan, L., Roberts, K., Steiger, E. (Eds.), Adult Short Bowel Syndrome, pp. 45–54, https://doi.org/10.1016/b978-0-12-814330-8.00004-4.
Crothers, J.W., et al., 2021. Daily, oral FMT for long-term maintenance therapy in ulcerative colitis: results of a single-center, prospective, randomized pilot study. BMC Gastroenterol. 21 (1). https://doi.org/10.1186/s12876-021-01856-9.
Cummings, J.H., Macfarlane, G.T., 1991. The control and consequences of bacterial fermentation. J. Appl. Bacteriol. 70 (6), 443–459. https://doi.org/10.1111/j.1365-2672.1991.tb02739.x.
Cummings, J.H., Macfarlane, G.T., 2002. Gastrointestinal effects of prebiotics. Br. J. Nutr. 87 (2), S145–S151. https://doi.org/10.1079/BJN/2002530.
D'Haens, G., et al., 2022. Risankizumab as induction therapy for Crohn's disease: results from the phase 3 ADVANCE and MOTIVATE induction trials. The Lancet 339 (10340), 2015–2030.

Daniel, H., et al., 2014. High-fat diet alters gut microbiota physiology in mice. ISME J. 8 (2), 295–308. https://doi.org/10.1038/ismej.2013.155.
Dao, M.C., et al., 2016. Akkermansia muciniphila and improved metabolic health during a dietary intervention in obesity: relationship with gut microbiome richness and ecology. Gut 65 (3), 426–436. https://doi.org/10.1136/gutjnl-2014-308778.
Das, B., et al., 2010. Tissue distibution of murine Muc19/smgc gene products. J. Histochem. Cytochem. 58 (2), 141–156. https://doi.org/10.1369/jhc.2009.954891.
David, L.A., et al., 2014. Diet rapidly and reproducibly alters the human gut microbiome. Nature 505 (7484), 559–563. https://doi.org/10.1038/nature12820.
Dethlefsen, L., Relman, D.A., 2011. Incomplete recovery and individualized responses of the human distal gut microbiota to repeated antibiotic perturbation. Proc. Natl. Acad. Sci. U. S. A. 108 (Suppl1), S4554–S4561. https://doi.org/10.1073/pnas.1000087107.
Dethlefsen, L., et al., 2008. The pervasive effects of an antibiotic on the human gut microbiota, as revealed by deep 16S rRNA sequencing. PLoS Biol. 6 (11). https://doi.org/10.1371/journal.pgen.1000255.
Devkota, S., et al., 2012. Dietary-fat-induced taurocholic acid promotes pathobiont expansion and colitis in Il10−/− mice. Nature 487 (7405), 104–108. https://doi.org/10.1038/nature11225.
Drekonja, D., et al., 2015. Fecal microbiota transplantation for clostridium difficile infection: a systematic review. Ann. Intern. Med. 162 (9), 630–638. https://doi.org/10.7326/M14-2693.
Eeckhaut, V., et al., 2013. Butyricicoccus pullicaecorum in inflammatory bowel disease. Gut 62 (12), 1745–1752. https://doi.org/10.1136/gutjnl-2012-303611.
Eindor-Abarbanel, A., Healey, G.R., Jacobson, K., 2021. Therapeutic advances in gut microbiome modulation in patients with inflammatory bowel disease from pediatrics to adulthood. Int. J. Mol. Sci. 22 (22). https://doi.org/10.3390/ijms222212506.
Farré-Maduell, E., Casals-Pascual, C., 2019. The origins of gut microbiome research in Europe: from escherich to Nissle. Hum. Microbiome J. 14. https://doi.org/10.1016/j.humic.2019.100065.
Fedorak, R.N., et al., 2015. The probiotic VSL#3 has anti-inflammatory effects and could reduce endoscopic recurrence after surgery for Crohn's disease. Clin. Gastroenterol. Hepatol. 13 (5), 928–935.e2. https://doi.org/10.1016/j.cgh.2014.10.031.
Fernandez-Banares, F., et al., 1999. Randomized clinical trial of Plantago ovata seeds (dietary Fiber) as compared with mesalamine in maintaining remission in ulcerative colitis. Spanish Group for the Study of Crohn's disease and ulcerative colitis (GETECCU). Am. J. Gastroenterol. 94 (2), 427–433. https://doi.org/10.1111/j.1572-0241.1999.872_a.x.
Filippo, C.D., et al., 2010. Impact of diet in shaping gut microbiota revealed by a comparative study in children from Europe and rural Africa. Proc. Natl. Acad. Sci. U. S. A. 107 (33), 14691–14696. https://doi.org/10.1073/pnas.1005963107.
Flores, G.E., et al., 2014. Temporal variability is a personalized feature of the human microbiome. Genome Biol. 15 (12), 531. https://doi.org/10.1186/s13059-014-0531-y.
Forsyth, C.B., et al., 2017. Alcohol feeding in mice promotes colonic hyperpermeability and changes in colonic organoid stem cell fate. Alcohol. Clin. Exp. Res. 41 (12), 2100–2113. https://doi.org/10.1111/acer.13519.
Fowler, E.V., et al., 2008. ATG16L1 T300A shows strong associations with disease subgroups in a large Australian IBD population: further support for significant disease heterogeneity. Am. J. Gastroenterol. 103 (10), 2519–2526. https://doi.org/10.1111/j.1572-0241.2008.02023.x.
Frank, D.N., et al., 2007. Molecular-phylogenetic characterization of microbial community imbalances in human inflammatory bowel diseases. Proc. Natl. Acad. Sci. U. S. A. 104 (34), 13780–13785. https://doi.org/10.1073/pnas.0706625104.

Franzosa, E.A., et al., 2019. Gut microbiome structure and metabolic activity in inflammatory bowel disease. Nat. Microbiol. 4 (2), 293–305. https://doi.org/10.1038/s41564-018-0306-4.

Freeman, K., et al., 2021. The incidence and prevalence of inflammatory bowel disease in UK primary care: a retrospective cohort study of the IQVIA medical research database. BMC Gastroenterol. 21 (1). https://doi.org/10.1186/s12876-021-01716-6.

Frost, G., et al., 2014. The short-chain fatty acid acetate reduces appetite via a central homeostatic mechanism. Nat. Commun. https://doi.org/10.1038/ncomms4611.

Gacesa, R., et al., 2022. Environmental factors shaping the gut microbiome in a Dutch population. Nature 604 (7907), 732–739. https://doi.org/10.1038/s41586-022-04567-7.

Ganji-Arjenaki, M., Rafieian-Kopaei, M., 2018. Probiotics are a good choice in remission of inflammatory bowel diseases: a meta analysis and systematic review. J. Cell. Physiol. 233 (3), 2091–2103. https://doi.org/10.1002/jcp.25911.

Ghosh, N., Premchand, P., 2015. A UK cost of care model for inflammatory bowel disease. Front. Gastroenterol. 6 (3), 169–174. https://doi.org/10.1136/flgastro-2014-100514.

Gilbert, J.A., Stephens, B., 2018. Microbiology of the built environment. Nat. Rev. Microbiol. 16 (11), 661–670. https://doi.org/10.1038/s41579-018-0065-5.

Gilbert, J.A., et al., 2018. Current understanding of the human microbiome. Nat. Med. 24 (4), 392–400.

Graham, D.B., Xavier, R.J., 2020. Pathway paradigms revealed from the genetics of inflammatory bowel disease. Nature 578 (7796), 527–539. https://doi.org/10.1038/s41586-020-2025-2.

Grootjans, J., et al., 2019. Epithelial endoplasmic reticulum stress orchestrates a protective IgA response. Science 363 (6430), 993–998. https://doi.org/10.1126/science.aat7186.

Gruber, L., et al., 2013. High fat diet accelerates pathogenesis of murine Crohn's disease-like ileitis independently of obesity. PLoS One 8 (8). https://doi.org/10.1371/journal.pone.0071661.

Guan, Q., 2019. A comprehensive review and update on the pathogenesis of inflammatory bowel disease. J. Immunol. Res. https://doi.org/10.1155/2019/7247238.

Hafer, A., et al., 2007. Effect of oral lactulose on clinical and immunohistochemical parameters in patients with inflammatory bowel disease: a pilot study. BMC Gastroenterol. 7 (1). https://doi.org/10.1186/1471-230X-7-36.

Hallert, C., et al., 1991. Ispaghula husk may relieve gastrointestinal symptoms in ulcerative colitis in remission. Scand. J. Gastroenterol. 26 (7), 747–750. https://doi.org/10.3109/00365529108998594.

Hamilton, M.J., et al., 2013. High-throughput DNA sequence analysis reveals stable engraftment of gut microbiota following transplantation of previously frozen fecal bacteria. Gut Microbes 4 (2), 125–135. https://doi.org/10.4161/gmic.23571.

Hernández-Chirlaque, C., et al., 2016. Germ-free and antibiotic-treated mice are highly susceptible to epithelial injury in DSS colitis. J. Crohns Colitis 10 (11), 1324–1335. https://doi.org/10.1093/ecco-jcc/jjw096.

Hernandez-Sanabria, E., et al., 2020. Short-term supplementation of celecoxib-shifted butyrate production on a simulated model of the gut microbial ecosystem and ameliorated in vitro inflammation. NPJ Biofilms and Microbiomes. 6 (1). https://doi.org/10.1038/s41522-020-0119-0.

Hillman, E.T., et al., 2017. Microbial ecology along the gastrointestinal tract. Microbes Environ. 32 (4), 300–313. https://doi.org/10.1264/jsme2.ME17017.

Hills, R.D., et al., 2019. Gut microbiome: profound implications for diet and disease. Nutrients 11 (7). https://doi.org/10.3390/nu11071613.

Hoshi, N., et al., 2012. MyD88 signalling in colonic mononuclear phagocytes drives colitis in IL-10 deficient mice. Nat. Commun. 3. https://doi.org/10.1038/ncomms2113.

Huang, H., et al., 2017. Fine-mapping inflammatory bowel disease loci to single-variant resolution. Nature 547 (7662), 173–178. https://doi.org/10.1038/nature22969.

Hugot, J.P., et al., 2001. Association of NOD2 leucine-rich repeat variants with susceptibility to Crohn's disease. Nature 411 (6837), 599–603. https://doi.org/10.1038/35079107.

Hugot, J.P., et al., 2007. Prevalence of CARD15/NOD2 mutations in Caucasian healthy people. Am. J. Gastroenterol. 102 (6), 1259–1267. https://doi.org/10.1111/j.1572-0241.2007.01149.x.

Hung, T.V., Suzuki, T., 2016. Dietary fermentable fiber reduces intestinal barrier defects and inflammation in colitic mice. J. Nutr. 146 (10), 1970–1979. https://doi.org/10.3945/jn.116.232538.

Huttenhower, C., et al., 2012. Structure, function and diversity of the healthy human microbiome. Nature 486 (7402), 207–214. https://doi.org/10.1038/nature11234.

Integrative HMP (iHMP) Research Network Consortium, 2014. The integrative human microbiome project: dynamic analysis of microbiome-host omics profiles during periods of human health and disease. Cell Host Microbe 16 (3), 276–289. https://doi.org/10.1016/j.chom.2014.08.014.

Iyer, N., 2016. Methods in microbiome research. Lab. Anim. (NY) 45 (9), 323–326. https://doi.org/10.1038/laban.1093.

Jackson, M.A., et al., 2021. Accurate identification and quantification of commensal microbiota bound by host immunoglobulins. Microbiome 9 (1). https://doi.org/10.1186/s40168-020-00992-w.

James, S.L., et al., 2015. Abnormal fibre usage in UC in remission. Gut 64 (4), 562–570. https://doi.org/10.1136/gutjnl-2014-307198.

Johansson, J., et al., 2010. Bacteria penetrate the inner mucus layer before inflammation in the dextran sulfate colitis model. PLoS One 5 (8). https://doi.org/10.1371/journal.pone.0012238.

Johnson, J.S., et al., 2019. Evaluation of 16S rRNA gene sequencing for species and strain-level microbiome analysis. Nat. Commun. 10 (1). https://doi.org/10.1038/s41467-019-13036-1.

Jostins, L., et al., 2012. Host-microbe interactions have shaped the genetic architecture of inflammatory bowel disease. Nature 491 (7422), 119–124. https://doi.org/10.1038/nature11582.

Kabat, A.M., et al., 2016. The autophagy gene Atg16l1 differentially regulates Treg and TH2 cells to control intestinal inflammation. Elife 5. https://doi.org/10.7554/eLife.12444.

Kanauchi, O., et al., 2003. Treatment of ulcerative colitis patients by long-term administration of germinated barley foodstuff: multi-center open trial. Int. J. Mol. Med. 12 (5), 701–704.

Kang, Y., et al., 2022. The role and function of mucins and its relationship to inflammatory bowel disease. Front. Med. https://doi.org/10.3389/fmed.2022.848344.

Kaplan, G.G., et al., 2010. The inflammatory bowel diseases and ambient air pollution: a novel association. Am. J. Gastroenterol. 105 (11), 2412–2419. https://doi.org/10.1038/ajg.2010.252.

Kaser, A., et al., 2008. XBP1 links ER stress to intestinal inflammation and confers genetic risk for human inflammatory bowel disease. Cell 134 (5), 743–756. https://doi.org/10.1016/j.cell.2008.07.021.

Kazakevych, J., et al., 2020. Smarcad1 mediates microbiota-induced inflammation in mouse and coordinates gene expression in the intestinal epithelium. Genome Biol. 21 (1). https://doi.org/10.1186/s13059-020-01976-7.

Khoruts, A., et al., 2010. Changes in the composition of the human fecal microbiome after bacteriotherapy for recurrent Clostridium difficile-associated diarrhea. J. Clin. Gastroenterol. 44 (5), 354–360. https://doi.org/10.1097/MCG.0b013e3181c87e02.

Kiesler, P., Fuss, I.J., Strober, W., 2015. Experimental models of inflammatory bowel diseases. Cell. Mol. Gastroenterol. Hepatol. 1 (2), 154–170. https://doi.org/10.1016/j.jcmgh.2015.01.006.
Kish, L., et al., 2013. Environmental particulate matter induces murine intestinal inflammatory responses and alters the gut microbiome. PLoS One 8 (4). https://doi.org/10.1371/journal.pone.0062220.
Knights, D., et al., 2014. Rethinking "enterotypes". Cell Host Microbe 16 (4), 433–437. https://doi.org/10.1016/j.chom.2014.09.013.
Kotlarz, D., et al., 2012. Loss of interleukin-10 signaling and infantile inflammatory bowel disease: implications for diagnosis and therapy. Gastroenterology 143 (2), 347–355. https://doi.org/10.1053/j.gastro.2012.04.045.
Kozik, A.J., et al., 2019. Comparison of the fecal, cecal, and mucus microbiome in male and female mice after TNBS-induced colitis. PLoS One 14 (11). https://doi.org/10.1371/journal.pone.0225079.
Kuehbacher, T., et al., 2008. Intestinal TM7 bacterial phylogenies in active inflammatory bowel disease. J. Med. Microbiol. 57 (12), 1569–1576. https://doi.org/10.1099/jmm.0.47719-0.
Kühn, R.K., et al., 1993. Interleukin-10-deficient mice develop chronic enterocolitis. Cell 75 (2), 263–274. https://doi.org/10.1016/0092-8674(93)80068-p.
Kyo, K., et al., 1999. Association of ulcerative colitis with rare VNTR alleles of the human intestinal mucin gene, MUC3. Hum. Mol. Genet. 8 (2), 307–311. https://doi.org/10.1093/hmg/8.2.307.
Kyo, K., et al., 2001. Associations of distinct variants of the intestinal mucin gene MUC3A with ulcerative colitis and Crohn's disease. J. Hum. Genet. 46 (1), 5–20. https://doi.org/10.1007/s100380170118.
Lamb, C.A., et al., 2019. British Society of Gastroenterology consensus guidelines on the management of inflammatory bowel disease in adults. Gut 68 (Suppl 3), S1–S106. https://doi.org/10.1136/gutjnl-2019-318484.
Ledder, O., et al., 2014. Clinical patterns and outcome of early-onset inflammatory bowel disease. J. Pediatr. Gasteroenterol. Mutr. 59 (5), 562–564. https://doi.org/10.1097/MPG.0000000000000465.
Lee, M., Chang, E.B., 2021. Inflammatory bowel diseases (IBD) and the microbiome—searching the crime scene for clues. Gastroenterology 160 (2), 524–537. https://doi.org/10.1053/j.gastro.2020.09.056.
Levast, B., Li, Z., Madrenas, J., 2015. The role of IL-10 in microbiome-associated immune modulation and disease tolerance. Cytokine 75 (2), 291–301. https://doi.org/10.1016/j.cyto.2014.11.027.
Levine, A., Sigall Boneh, R., Wine, E., 2018. Evolving role of diet in the pathogenesis and treatment of inflammatory bowel diseases. Gut 67 (9), 1726–1738. https://doi.org/10.1136/gutjnl-2017-315866.
Liu, J.Z., et al., 2015. Association analyses identify 38 susceptibility loci for inflammatory bowel disease and highlight shared genetic risk across populations. Nat. Genet. 47 (9), 979–986. https://doi.org/10.1038/ng.3359.
Low, D., Nguyen, D.D., Mizoguchi, E., 2013. Animal models of ulcerative colitis and their application in drug research. Drug Des. Devel. Ther. 7, 1341–1356. https://doi.org/10.2147/DDDT.S40107.
Ma, X., Hua, J., Li, Z., 2008. Probiotics improve high fat diet-induced hepatic steatosis and insulin resistance by increasing hepatic NKT cells. J. Hepatol. 49 (5), 821–830. https://doi.org/10.1016/j.jhep.2008.05.025.
Maccaferri, S., et al., 2010. Rifaximin modulates the colonic microbiota of patients with Crohn's disease: an in vitro approach using a continuous culture colonic model system. J. Antimicrob. Chemother. 65 (12), 2556–2565. https://doi.org/10.1093/jac/dkq345.

Maclellan, A., et al., 2017. The impact of exclusive enteral nutrition (EEN) on the gut microbiome in Crohn's disease: a review. Nutrients 9 (5). https://doi.org/10.3390/nu9050447.

Maloy, K.J., Powrie, F., 2011. Intestinal homeostasis and its breakdown in inflammatory bowel disease. Nature 474 (7351), 298–306. https://doi.org/10.1038/nature10208.

Manichanh, C., et al., 2006. Reduced diversity of faecal microbiota in Crohn's disease revealed by a metagenomic approach. Gut 55 (2), 205–211. https://doi.org/10.1136/gut.2005.073817.

Martinez-Medina, M., et al., 2014. Western diet induces dysbiosis with increased E. coli in CEABAC10 mice, alters host barrier function favouring AIEC colonisation. Gut 63 (1), 116–124. https://doi.org/10.1136/gutjnl-2012-304119.

Martyniak, A., et al., 2021. Prebiotics, probiotics, synbiotics, paraprobiotics and postbiotic compounds in IBD. Biomolecules 11 (12). https://doi.org/10.3390/biom11121903.

Maseda, D., Ricciotti, E., 2020. NSAID–gut microbiota interactions. Front. Pharmacol. 11. https://doi.org/10.3389/fphar.2020.01153.

Matsuoka, K., et al., 2018. Evidence-based clinical practice guidelines for inflammatory bowel disease. J. Gastroenterol. 53 (3), 305–353. https://doi.org/10.1007/s00535-018-1439-1.

McBurney, M.I., et al., 2019. Establishing what constitutes a healthy human gut microbiome: state of the science, regulatory considerations, and future directions. J. Nutr. 149 (11), 1882–1895. https://doi.org/10.1093/jn/nxz154.

McCole, D.F., 2014. IBD candidate genes and intestinal barrier regulation. Inflamm. Bowel Dis. 20 (10), 1829–1849. https://doi.org/10.1097/MIB.0000000000000090.

McGovern, D., Powrie, F., 2007. The IL23 axis plays a key role in the pathogenesis of IBD. Gut 56 (10), 1333–1336. https://doi.org/10.1136/gut.2006.115402.

McGovern, D.P.B., Kugathasan, S., Cho, J.H., 2015. Genetics of inflammatory bowel diseases. Gastroenterology 149 (5), 1163–1176.e2. https://doi.org/10.1053/j.gastro.2015.08.001.

Methé, B.A., et al., 2012. A framework for human microbiome research. Nature 486 (7402), 215–221. https://doi.org/10.1038/nature11209.

Miyazato, S., et al., 2016. Continuous intake of resistant maltodextrin enhanced intestinal immune response through changes in the intestinal environment in mice. Biosci. Microbiota Food Health 35 (1), 1–7. https://doi.org/10.12938/bmfh.2015-009.

Muise, A.M., et al., 2009. Polymorphisms in E-cadherin (CDH1) result in a mislocalised cytoplasmic protein that is associated with Crohn's disease. Gut 58 (8), 1121–1127. https://doi.org/10.1136/gut.2008.175117.

Mutlu, E.A., et al., 2011. Particulate matter air pollution causes oxidant-mediated increase in gut permeability in mice. Part. Fibre Toxicol. 8. https://doi.org/10.1186/1743-8977-8-19.

Nair, M.G., et al., 2009. Alternatively activated macrophage-derived RELM-α is a negative regulator of type 2 inflammation in the lung. J. Exp. Med. 206 (4), 1200–1207. https://doi.org/10.1084/jem.20082048.

Natividad, J.M., et al., 2018. Bilophila wadsworthia aggravates high fat diet induced metabolic dysfunctions in mice. Nat. Commun. 9 (1). https://doi.org/10.1038/s41467-018-05249-7.

Ng, S.C., et al., 2017. Worldwide incidence and prevalence of inflammatory bowel disease in the 21st century: a systematic review of population-based studies. The Lancet 390 (10114), 2769–2778. https://doi.org/10.1016/S0140-6736(17)32448-0.

Ng, S.C., Hart, A.L., 2014. Probiotics in Inflammatory Bowel Disease: Modulation of the Pathological Immune or Inflammatory Activation. Their Potential Benefits in the Different Phases of the Disease.

Ng, S.C., et al., 2013. Incidence and phenotype of inflammatory bowel disease based on results from the Asia-pacific Crohn's and colitis epidemiology study. Gastroenterology 145 (1), 158–165.e2. https://doi.org/10.1053/j.gastro.2013.04.007.

Ni, J., et al., 2017. Gut microbiota and IBD: causation or correlation? Nat. Rev. Gastroenterol. Hepatol. 14 (10), 573–584. https://doi.org/10.1038/nrgastro.2017.88.

Nickerson, K.P., Mcdonald, C., 2012. Crohn's disease-associated adherent-invasive Escherichia coli adhesion is enhanced by exposure to the ubiquitous dietary polysaccharide maltodextrin. PLoS One 7 (12). https://doi.org/10.1371/journal.pone.0052132.

Nikolaus, S., et al., 2017. Increased tryptophan metabolism is associated with activity of inflammatory bowel diseases. Gastroenterology 153 (6), 1504–1516.e2. https://doi.org/10.1053/j.gastro.2017.08.028.

Nishida, A., et al., 2017. Gut microbiota in the pathogenesis of inflammatory bowel disease. Clin. J. Gastroenterol. 11 (1). https://doi.org/10.1007/s12328-017-0813-5.

Ogura, Y., et al., 2001. A frameshift mutation in NOD2 associated with susceptibility to Crohn's disease. Nature 411 (6837), 603–606. https://doi.org/10.1038/35079114.

Okumura, R., Takeda, K., 2018. Maintenance of intestinal homeostasis by mucosal barriers. Inflamm. Regen. 38 (1). https://doi.org/10.1186/s41232-018-0063-z.

Opstelten, J.L., et al., 2016. Exposure to ambient air pollution and the risk of inflammatory bowel disease: a european nested case–control study. Dig. Dis. Sci. 61 (10), 2963–2971. https://doi.org/10.1007/s10620-016-4249-4.

Opstelten, J.L., et al., 2018. Prediagnostic serum vitamin D levels and the risk of crohn's disease and ulcerative colitis in european populations: a nested case-control study. Inflamm. Bowel Dis. 24 (3), 633–640. https://doi.org/10.1093/ibd/izx050.

Palumbo, V.D., et al., 2016. The long-term effects of probiotics in the therapy of ulcerative colitis: a clinical study. Biomed. Pap. Med. Fac. Univ. Palacky Olomouc Czech Repub. 160 (3), 372–377.

Paramsothy, S., et al., 2017. Faecal microbiota transplantation for inflammatory bowel disease: a systematic review and meta-analysis. J. Crohns Colitis 11 (10), 1180–1199. https://doi.org/10.1093/ecco-jcc/jjx063.

Peterson, L.W., Artis, D., 2014. Intestinal epithelial cells: regulators of barrier function and immune homeostasis. Nat. Rev. Immunol. 14 (3), 141–153. https://doi.org/10.1038/nri3608.

Pigneur, B., et al., 2019. Mucosal healing and bacterial composition in response to enteral nutrition mucosal healing and bacterial composition in response to enteral nutrition vs steroid-based induction therapy — a randomised prospective clinical trial in children with Crohn's disease. J. Crohns Colitis 13 (7), 846–855. https://doi.org/10.1093/ecco-jcc/jjy207.

Piovani, D., et al., 2019. Environmental risk factors for inflammatory bowel diseases: an umbrella review of meta-analyses. Gastroenterology 157 (3), 647–659.e4. https://doi.org/10.1053/j.gastro.2019.04.016.

Prescott, S.L., 2017. History of medicine: origin of the term microbiome and why it matters. Hum. Microbiome J. 4 (1), 24–25. https://doi.org/10.1016/j.humic.2017.05.004.

Prideaux, L., et al., 2013. Impact of ethnicity, geography, and disease on the microbiota in health and inflammatory bowel disease. Inflamm. Bowel Dis. 19 (13), 2906–2918. https://doi.org/10.1097/01.MIB.0000435759.05577.12.

Proctor, L.M., et al., 2019. The Integrative human microbiome project. Nature 569 (7758), 641–648. https://doi.org/10.1038/s41586-019-1238-8.

Queipo-Ortuño, M.I., et al., 2012. Influence of red wine polyphenols and ethanol on the gut microbiota ecology and biochemical biomarkers. Am. J. Clin. Nutr. 95 (6), 1323–1334. https://doi.org/10.3945/ajcn.111.027847.

Rakoff-Nahoum, S., Hao, L., Medzhitov, R., 2006. Role of toll-like receptors in spontaneous commensal-dependent colitis. Immunity 25 (2), 319–329. https://doi.org/10.1016/j.immuni.2006.06.010.

Rivas, M.A., et al., 2011. Deep resequencing of GWAS loci identifies independent rare variants associated with inflammatory bowel disease. Nat. Genet. 43 (11), 1066–1073. https://doi.org/10.1038/ng.952.

Roager, H.M., et al., 2016. Colonic transit time is related to bacterial metabolism and mucosal turnover in the gut. Nat. Microbiol. 1 (9). https://doi.org/10.1038/nmicrobiol.2016.93.

Rogers, M.A.M., Aronoff, D.M., 2016. The influence of nonsteroidal anti-inflammatory drugs on the gut microbiome. Clin. Microbiol. Infect. 22 (2). https://doi.org/10.1016/j.cmi.2015.10.003.

Rothschild, D., et al., 2018. Environment dominates over host genetics in shaping human gut microbiota. Nature 555 (7695), 210–215. https://doi.org/10.1038/nature25973.

Roy, U., et al., 2017. Distinct microbial communities trigger colitis development upon intestinal barrier damage via innate or adaptive immune cells. Cell Rep. 21 (4), 994–1008. https://doi.org/10.1016/j.celrep.2017.09.097.

Salim, S.Y., Kaplan, G.G., Madsen, K.L., 2014. Air pollution effects on the gut microbiota: a link between exposure and inflammatory disease. Gut Microbes 5 (2), 215–219. https://doi.org/10.4161/gmic.27251.

Savin, Z., et al., 2018. Smoking and the intestinal microbiome. Arch. Microbiol. 200 (5), 677–684. https://doi.org/10.1007/s00203-018-1506-2.

Schmidt, T.S.B., Raes, J., Bork, P., 2018. The human gut microbiome: from association to modulation. Cell 172 (6), 1198–1215. https://doi.org/10.1016/j.cell.2018.02.044.

Sellon, R.K., et al., 1998. Resident enteric bacteria are necessary for development of spontaneous colitis and immune system activation in interleukin-10-deficient mice. Infect. Immun. 66 (11), 5224–5231. https://doi.org/10.1128/iai.66.11.5224-5231.1998.

Sender, R., Fuchs, S., Milo, R., 2016. Are we really vastly outnumbered? Revisiting the ratio of bacterial to host cells in humans. Cell 164 (3), 337–340. https://doi.org/10.1016/j.cell.2016.01.013.

Shadnoush, M., et al., 2015. Effects of probiotics on gut microbiota in patients with inflammatory bowel disease: a double-blind, placebo-controlled clinical trial. Korean J. Gastroenterol. 65 (4), 215–221. https://doi.org/10.4166/kjg.2015.65.4.215.

Shaw, S.Y., Blanchard, J.F., Bernstein, C.N., 2010. Association between the use of antibiotics in the first year of life and pediatric inflammatory bowel disease. Am. J. Gastroenterol. 105 (12), 2687–2692. https://doi.org/10.1038/ajg.2010.398.

Silveira, A.L.M., et al., 2017. Preventive rather than therapeutic treatment with high fiber diet attenuates clinical and inflammatory markers of acute and chronic DSS- induced colitis in mice. Eur. J. Nutr. 56 (1), 179–191. https://doi.org/10.1007/s00394-015-1068-x.

Sokol, H., et al., 2020. Fecal microbiota transplantation to maintain remission in Crohn's disease: a pilot randomized controlled study. Microbiome 8 (1). https://doi.org/10.1186/s40168-020-0792-5.

Sood, A., et al., 2019. Role of faecal microbiota transplantation for maintenance of remission in patients with ulcerative colitis: a pilot study. J. Crohns Colitis 13 (10), 1311–1317. https://doi.org/10.1093/ecco-jcc/jjz060.

Sprockett, D., et al., 2019. Treatment-specific composition of the gut microbiota is associated with disease remission in a pediatric crohn's disease cohort. Inflamm. Bowel Dis. 25 (12), 1927–1938. https://doi.org/10.1093/ibd/izz130.

Steed, H., et al., 2010. Clinical trial: the microbiological and immunological effects of synbiotic consumption - a randomized double-blind placebo-controlled study in active Crohn's disease. Ailment Pharmacol. Ther. 32 (7), 872–883. https://doi.org/10.1111/j.1365-2036.2010.04417.x.

Sun, J., et al., 2016. Therapeutic potential to modify the mucus barrier in inflammatory bowel disease. Nutrients 8 (1). https://doi.org/10.3390/nu8010044.

Suzuki, T., Hara, H., 2010. Dietary fat and bile juice, but not obesity, are responsible for the increase in small intestinal permeability induced through the suppression of tight junction protein expression in LETO and OLETF rats. Nutr. Metab. 7 (1). https://doi.org/10.1186/1743-7075-7-19.

Swidsinski, A., et al., 2009. Bacterial overgrowth and inflammation of small intestine after carboxymethylcellulose ingestion in genetically. Inflamm. Bowel Dis. 15 (3), 359–364. https://doi.org/10.1002/ibd.20763.

Tamaki, H., et al., 2016. Efficacy of probiotic treatment with Bifidobacterium longum 536 for induction of remission in active ulcerative colitis: a randomized, double-blinded, placebo-controlled multicenter trial. Dig. Endosc. 28 (1), 67–74. https://doi.org/10.1111/den.12553.

Tang, Q., et al., 2020. Current sampling methods for gut microbiota: a call for more precise devices. Front. Cell. Infect. Microbiol. 10 (1). https://doi.org/10.3389/fcimb.2020.00151.

Tannock, G.W., et al., 2000. Analysis of the fecal microflora of human subjects consuming a probiotic product containing Lactobacillus rhamnosus DR20. Appl. Environ. Microbiol. 66 (6), 2578–2588. https://doi.org/10.1128/AEM.66.6.2578-2588.2000.

Turnbaugh, P.J., et al., 2007. The human microbiome project. Nature 449 (7164), 804–810. https://doi.org/10.1038/nature06244.

Turner, J.R., 2009. Intestinal mucosal barrier function in health and disease. Nat. Rev. Immunol. 9 (11), 799–809. https://doi.org/10.1038/nri2653.

Turner, D., et al., 2019. Antibiotic cocktail for pediatric acute severe colitis and the microbiome: the PRASCO randomized controlled trial. Inflamm. Bowel Dis. 26 (11), 1733–1742. https://doi.org/10.1093/ibd/izz298.

Ungaro, R., et al., 2014. Antibiotics associated with increased risk of new-onset Crohn's disease but not ulcerative colitis: a meta-analysis. Am. J. Gastroenterol. 109 (11), 1728–1738. https://doi.org/10.1038/ajg.2014.246.

Vaga, S., et al., 2020. Compositional and functional differences of the mucosal microbiota along the intestine of healthy individuals. Sci. Rep. 10 (1). https://doi.org/10.1038/s41598-020-71939-2.

Vaishnava, S., et al., 2011. The antibacterial lectin RegIIIγ promotes the spatial segregation of microbiota and host in the intestine. Science 334 (6053), 255–258. https://doi.org/10.1126/science.1209791.

Valdes, A.M., et al., 2018. Role of the gut microbiota in nutrition and health. BMJ 361. https://doi.org/10.1136/bmj.k2179.

Van Sommeren, S., et al., 2011. HNF4α and CDH1 are associated with ulcerative colitis in a Dutch cohort. Inflamm. Bowel Dis. 17 (8), 1714–1718. https://doi.org/10.1002/ibd.21541.

Vandeputte, D., et al., 2016. Stool consistency is strongly associated with gut microbiota richness and composition, enterotypes and bacterial growth rates. Gut 65 (1), 57–62. https://doi.org/10.1136/gutjnl-2015-309618.

Venegas, D.P., et al., 2019. Short chain fatty acids (SCFAs)-mediated gut epithelial and immune regulation and its relevance for inflammatory bowel diseases. Front. Immunol. https://doi.org/10.3389/fimmu.2019.00277.

Visschedijk, M.C., et al., 2016. Pooled resequencing of 122 ulcerative colitis genes in a large Dutch cohort suggests population-specific associations of rare variants in MUC2. PLoS One 11 (8). https://doi.org/10.1371/journal.pone.0159609.

Wang, H., et al., 2016. Dietary non-digestible polysaccharides ameliorate intestinal epithelial barrier dysfunction in IL-10 knockout mice. J. Crohns Colitis 10 (9), 1076–1086. https://doi.org/10.1093/ecco-jcc/jjw065.

Wang, J., et al., 2018. Meta-analysis of human genome-microbiome association studies: the MiBioGen consortium initiative. Microbiome 6 (1). https://doi.org/10.1186/s40168-018-0479-3.

Webb, C.R., et al., 2018. Differential susceptibility to T cell-induced colitis in mice: role of the intestinal microbiota. Inflamm. Bowel Dis. 24 (2), 361–379. https://doi.org/10.1093/ibd/izx014.

Weingarden, A.R., Vaughn, B.P., 2017. Intestinal microbiota, fecal microbiota transplantation, and inflammatory bowel disease. Gut Microbes 8 (3), 238–252. https://doi.org/10.1080/19490976.2017.1290757.

Wilmanski, T., et al., 2021. Gut microbiome pattern reflects healthy ageing and predicts survival in humans. Nat. Metab. 3 (1), 274–286. https://doi.org/10.1038/s42255-021-00348-0.

Wirtz, S., Neurath, M.F., 2007. Mouse models of inflammatory bowel disease ☆. Adv. Drug Deliv. Rev. 59 (11), 1073–1083. https://doi.org/10.1016/j.addr.2007.07.003.

Wu, G.D., et al., 2011. Linking long-term dietary patterns with gut microbial enterotypes. Science 334 (6052), 105–108. https://doi.org/10.1126/science.1208344.

Xavier, R.J., Podolsky, D.K., 2007. Unravelling the pathogenesis of inflammatory bowel disease. Nature 448 (7152), 427–434. https://doi.org/10.1038/nature06005.

Xiao, L., et al., 2021. Deterministic transition of enterotypes shapes the infant gut microbiome at an early age. Genome Biol. 22 (1). https://doi.org/10.1186/s13059-021-02463-3.

Yang, Z., et al., 2020. Fecal microbiota transplant via endoscopic delivering through small intestine and colon: no difference for crohn's disease. Dig. Dis. Sci. 65 (1), 150–157. https://doi.org/10.1007/s10620-019-05751-y.

Yatsunenko, T., et al., 2012. Human gut microbiome viewed across age and geography. Nature 486 (7402), 222–227. https://doi.org/10.1038/nature11053.

Yoshimatsu, Y., et al., 2015. Effectiveness of probiotic therapy for the prevention of relapse in patients with inactive ulcerative colitis. World J. Gastroenterol. 21 (19), 5985–5994. https://doi.org/10.3748/wjg.v21.i19.5985.

Young, R.B., et al., 2021. Key technologies for progressing discovery of microbiome-based medicines. Front. Microbiol. 12 (1). https://doi.org/10.3389/fmicb.2021.685935.

Younis, N., Zarif, R., Mahfouz, R., 2020. Inflammatory bowel disease: between genetics and microbiota. Mol. Biol. Rep. 47 (4), 3053–3063. https://doi.org/10.1007/s11033-020-05318-5.

Zhang, X.-F., et al., 2021. Clinical effects and gut microbiota changes of using probiotics, prebiotics or synbiotics in inflammatory bowel disease: a systematic review and meta-analysis. Eur. J. Nutr. 60 (5), 2855–2875. https://doi.org/10.1007/s00394-021-02503-5.

Zimmermann, M., et al., 2019. Mapping human microbiome drug metabolism by gut bacteria and their genes. Nature 570 (7762), 462–467. https://doi.org/10.1038/s41586-019-1291-3.

Zuo, T., et al., 2018. Urbanization and the gut microbiota in health and inflammatory bowel disease. *Nat. Rev. Gastroenterol.* Hepatol. 15 (7), 440–452. https://doi.org/10.1038/s41575-018-0003-z.

CHAPTER SIX

Mind the gaps: What do we know about how multiple chemical stressors impact freshwater aquatic microbiomes?

Alessia Bani[a,†], Kate C. Randall[a,b,†], Dave R. Clark[a,c,†], Benjamin H. Gregson[a], Drew K. Henderson[a], Erin C. Losty[a], and Robert M.W. Ferguson[a,*]

[a]School of Life Sciences, University of Essex, Wivenhoe Park, Colchester, Essex, United Kingdom
[b]Applied Sciences, University of Northumbria, Newcastle Upon Tyne, United Kingdom
[c]Institute for Analytics and Data Science, University of Essex, Wivenhoe Park, Colchester, Essex, United Kingdom
*Corresponding author: e-mail address: rmwfer@essex.ac.uk

Contents

1. Introduction 332
2. The freshwater aquatic chemical exposome 335
3. Selection of chemicals and literature search 345
4. Responses of freshwater microbial communities to eight important chemicals: What do we know so far? 350
5. Embracing complexity: Multiple stressors in freshwaters 355
6. Everything is connected: Indirect effects of chemical stressors 357
7. Antimicrobial resistance (AMR): Are freshwaters the key link in the one health approach? 358
8. Nutrient pollution: Does nothing else matter? 360
9. Beyond taxonomy: A new paradigm for biomonitoring of freshwater ecosystems 362
10. Bridging the gap 366
11. Summary and future directions 367
Acknowledgements 368
References 368

† Equally contributing first authors.

Advances in Ecological Research, Volume 67
ISSN 0065-2504
https://doi.org/10.1016/bs.aecr.2022.09.003

Abstract

Freshwaters are critical to human survival and well-being. However, over-reliance on freshwaters has made them among the most threatened ecosystems in the world. Freshwaters also connect terrestrial and marine ecosystems, making them a melting pot for anthropogenic stressors, in-particular agricultural and urban runoff containing chemical and nutrient pollution. Many of the ecosystem services provided by freshwaters are directly or indirectly supported by microorganisms, which form the basal layer of aquatic food webs, drive nitrogen fixation, oxygen production, and nutrients cycling. Yet the effects of multiple interacting stressors on microbial populations and communities is poorly understood and rarely considered in policy development. Here, we evaluate the current state of knowledge around how key anthropogenic stressors, pesticide and antibiotic pollution, affect freshwater microbiomes. Based on their concentrations in EU rivers and prevalence in agriculture, we review freshwater microbial community responses to eight selected chemicals pollutants: glyphosate, diflufenican, oxytetracycline, amoxicillin, metaldehyde, imidacloprid, tebuconazole, and chlorothalonil. We focused on mesocosm-based studies as these balance environmental realism and experimental control, providing mechanistic insights into ecological effects. Our review revealed a systematic bias towards a limited selection of chemicals, studied in isolation, on a subset of microbial taxa. For example, studies focussing on glyphosate represent approximately ⅓ of all the relevant publications identified, while other prevalent chemical pollutants such as metaldehyde, chlorothalonil, and diflufenican are ignored. More research is urgently needed to fill this gap, especially studies investigating the interacting effects of chemical mixtures with other anthropogenic stressors, for example nutrient pollution and climate change. We propose a move away from the over-simplistic study of chemical pollutants towards a research agenda that incorporates more realistic scenarios and comprehensive biomonitoring combining taxonomic and trait-based information.

1. Introduction

Freshwater ecosystems are globally important repositories of biodiversity, essential resources for human civilisation, and significant drivers of biogeochemical cycles in which all major nutrients and greenhouse gases are consumed, cycled, or emitted. Each of these properties is underpinned by the functionality and activity of microbial communities which are key players in freshwater environments, from ecological and biogeochemical perspectives (Falkowski et al., 2008).

Rivers, streams, and other freshwater systems play key roles in the composition of Earth's atmosphere, especially with respect to greenhouse gases. Globally, rivers are estimated to transport 0.4 Pg of carbon to the sea each year (Cauwet, 2002), while more than half of the carbon transported by rivers is retained in the ecosystem or lost through gaseous emissions (Cole et al., 2007). Furthermore, recent meta-analyses implicate freshwater systems as the source of >92% of global methane emissions, almost exclusively produced by methanogenic microorganisms (Rosentreter et al., 2021). Similarly for the cycling of other macronutrients such as nitrogen, microbial activity in rivers can convert ~40% of terrestrial nitrogen (N) runoff to the atmospheric gas N_2 (Galloway et al., 2004), while microbial transformations of nitrogen are also estimated to account for >10% of all anthropogenic nitrous oxide emissions (Beaulieu et al., 2011). Predicting how freshwater environments may change in the future therefore relies on us understanding the environmental and anthropogenic variables that influence microbial community diversity, composition and functionality.

While natural environmental gradients are important in structuring freshwater microbiomes (e.g. Clark et al., 2022), the impacts of anthropogenic activities are also gaining more attention. These factors include warming, drought, nutrient addition, and chemical pollution. In particular, chemical pollution caused by terrestrial runoff is increasingly recognised as an important determinant of water quality and driver of ecological shifts (Tan et al., 2015). Rivers contain complex mixtures of chemicals, the most dominant sources for these are human pharmaceuticals (e.g. antibiotics), agriculture (pesticides and herbicides), and industry (hydrocarbons and heavy metals), with many classified as contaminants of emerging concern by the EU and named on the US Contaminant Candidate List (Richardson et al., 2022). Many of these chemicals enter rivers directly from terrestrial runoff, or indirectly via wastewater treatment systems. Chemical pollution from these routes will be exacerbated in the coming century by increased use of pesticides, increased human population and urbanisation, intensification of agriculture, and increased runoff and overflow of wastewater resulting from more frequent extreme precipitation events (Christensen and Christensen, 2003). For a review of tracking these complex mixtures of chemicals through the environment see Escher et al. (2020). Traditionally, metal pollution has received the widest attention as it is thought to represent the largest ecological threat for 'higher' taxa (animals and plants) (Johnson et al., 2020). However, any increases in runoff will likely also raise the concentrations of other chemicals (e.g. pesticides). Thus, a broader perspective is required to understand how chemical stressors will affect the traditionally overlooked microbial world.

Traditionally, the ecological responses of 'macroorganisms' have been used to indicate water quality, making them the subject of most biomonitoring strategies (Zolkefli et al., 2020). Fast generation times and sensitivity to environmental change would seemingly make microorganisms ideal for biomonitoring (Cavicchioli et al., 2019; Sagova-Mareckova et al., 2021), however microbial biomonitoring is generally limited to faecal indicators such as *Escherichia coli* and enterococci (e.g. ISO 9308-1:2014, n.d.; ISO 7899-2:2000, n.d.). Research into the impacts of chemical stressors on microbiomes is patchy, generally focussing on only a few taxa or the isolated impacts of a small subset of chemicals (Clarke et al., 2019; Lu et al., 2020; Pascault et al., 2014), and often in a laboratory setting (Deng et al., 2022; Peng et al., 2019; Zhang et al., 2022). Such studies inevitably lack realism as they cannot incorporate the enormous diversity of environmental microbiomes and the ecological interactions therein, and rarely consider the possibility of multiple interacting stressors (both chemical and other stressors). Together, these caveats limit the utility of this research as an evidence base for development of effective policy to manage freshwaters. Molecular methods (nucleic acid based) enable us to resolve entire aquatic communities across all trophic levels (including microorganisms) from the gene to the ecosystem levels of organisation both *in situ* or in controlled mesocosm experiments. Using these molecular tools it is possible to disentangle the complex interacting pressures that chemicals exert on microbial communities, shedding light on the impacts of chemicals on microbially-mediated processes, and potentially making microbial communities powerful targets for biomonitoring approaches. To date, the body of research shedding light on the impacts of chemical stressors on microbial communities has not been properly reviewed.

Towards this goal, we set out to review the body of research on the effects of four major groups of chemical stressors that target different key components of freshwater ecosystems (herbicides, insecticides/molluscicides, fungicides, and antibiotics) on aquatic microbial communities. Based on usage and reported aquatic concentrations we identify the two key chemicals with differing modes of action within each group, resulting in a total of eight focal chemicals. Here, we synthesise and aggregate existing research on the effects of these chemicals on freshwater microbial communities, examine knowledge gaps present in the literature, and describe opportunities to better integrate microbial communities in biomonitoring of chemical pollution in freshwater ecosystems

2. The freshwater aquatic chemical exposome

To get an overview of the chemical exposome in European rivers we used the aggregated Waterbase—Water Quality ICM data available from the European Environment Agency (EEA) under the broader Waterbase dataset (EEA, 2022). Waterbase—Water Quality ICM, herein referred to as the Waterbase dataset, contains time series concentrations of nutrients, organic matter, hazardous substances and other chemicals in rivers, lakes, groundwater, transitional, coastal and marine waters from 38 European Economic Area (EEA) member countries, plus the United Kingdom, from 1965 to 2021. We selected data from river water samples, and chemical determinants only (e.g. removed biological, and nutrient indicators such as BOD, and total nitrogen) leaving 2,124,791 measurements representing 529 chemicals.

These data reflect increasing effort across the EU to characterise the chemical exposome of rivers (and waterbodies in general), with both the number and diversity of chemicals reported steadily rising from five in 1972 (comprising heavy metals such as copper and lead) to 391 in 2020 (comprising a wide diversity of pesticides, hydrocarbons, personal care products, and pharmaceuticals) (Fig. 1). Pesticides and hydrocarbons dominate the data, with 176 determinants being either pesticides or metabolites of pesticides, and 127 being poly/aromatic hydrocarbons, chlorocarbons, or hydrocarbons. Pharmaceuticals are not well represented, with only

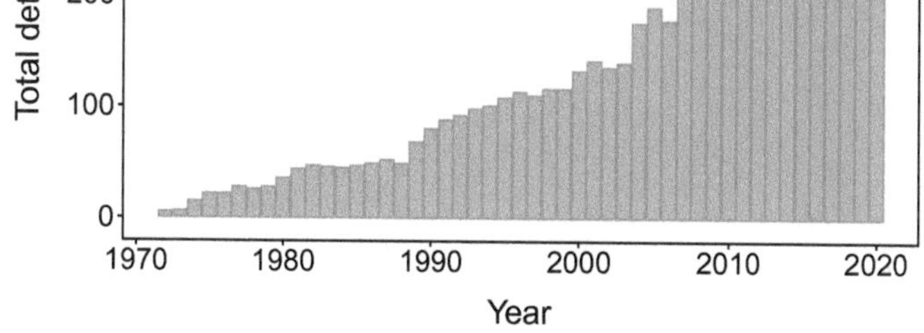

Fig. 1 Total chemical determinants reported in 'Waterbase' from 1970 to 2020 (EEA, 2022).

24 determinants, of which six are antibiotics. To focus on a list of prevalent chemical pollutants, we removed data points for which the recorded concentration was below the detection limit and anomalous data points (records three orders of magnitude greater than the median). Remaining pollutants are summarised below across four major classes (fungicides, insecticides, herbicides and pharmaceuticals).

Fungicides: Historically, fenpropimorph and the azole propiconazole were the most monitored fungicides, with records of their concentration in rivers from the early 1990s. Notably, concentrations of both have decreased over the last decade, now replaced by tebuconazole. It is unclear whether this is due to wider monitoring efforts which only began in 2018 (Fig. 2, Table 1), but does coincide with increased relative usage of propiconazole and tebuconazole on arable crops over the last 20 years (Fera, 2020), likely implicating arable farming as a major source of these pollutants. Fungicides are limited to a few classes and modes of action, most of which are widely used in both health care and agriculture including azoles, polyenes, echinocadins, flucytosines, and strobilurins (Wall and Lopez-Ribot, 2020). Chlorothalonil is the most used fungicide (in UK agriculture) in terms of area and mass applied to arable crops, with tebuconazole second (Fera, 2020). Despite limited detections of chlorothalonil in Waterbase, research suggests that it is an abundant pollutant in surface water and sediments, with concentrations as high as 46.5 μg/L (Callicott and Hooper-Bùi, 2019; Sakkas et al., 2002; Voulvoulis et al., 2000).

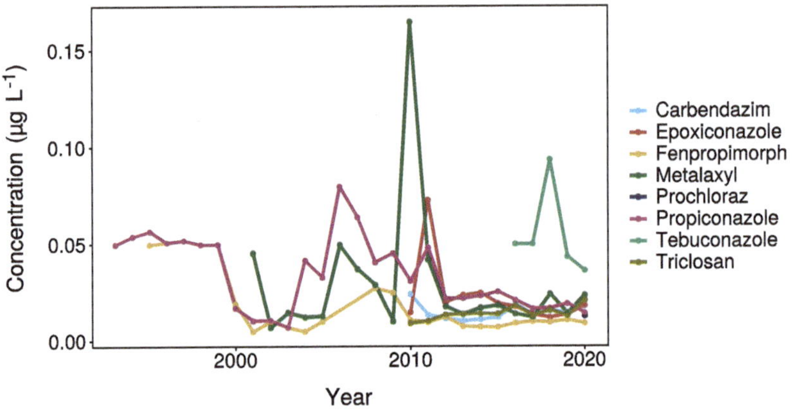

Fig. 2 Yearly mean concentrations of dominant fungicides in EU rivers from 1993 to 2021 (EEA, 2022).

Table 1 Number of detections greater than the limit of detection (LOD), minimum, mean, and maximum concentrations of fungicides, herbicides, insecticides, pharmaceuticals, and antimicrobials detected in river water as reported in Waterbase (EEA, 2022).

Class	Chemical	CAS	Concentration (µg/L)			
			>LOD	min	mean	max
Fungicide	Captan	133-06-2	59	0.010	0.010	0.010
	Carbendazim	10605-21-7	502	0.001	0.058	21.000
	Chlordecone	143-50-0	187	0.005	0.009	0.510
	Epoxiconazole	133855-98-8	250	0.001	0.010	1.400
	Fenpropimorph	67564-91-4	220	0.000	0.012	0.280
	Fluquinconazole	136426-54-5	113	0.005	0.008	0.050
	Imazalil	35554-44-0	108	0.001	0.071	4.300
	Metalaxyl	57837-19-1	767	0.001	0.031	14.000
	Propiconazole	60207-90-1	978	0.000	0.023	3.100
	Quinoxyfen	124495-18-7	55	0.001	0.015	0.270
	Quintozene	82-68-8	135	0.000	0.006	0.050
	Tebuconazole	107534-96-3	161	0.005	0.035	6.000
Herbicide	2,4,5-T	93-76-5	540	0.003	0.023	20.000
	2,4,5-trichlorophenol	95-95-4	147	0.025	0.049	2.200
	2,4,6-trichlorophenol	88-06-2	177	0.000	0.032	3.000
	2-4 D	94-75-7	1921	0.000	0.027	20.000
	2,6-dichlorobenzamide	2008-58-4	280	0.002	0.040	0.300
	Acetochlor	34256-82-1	800	0.001	0.027	7.700
	Acetochlor ESA	187022-11-3	744	0.010	0.049	1.400
	Acetochlor OA	194992-44-4	350	0.005	0.037	1.300
	Aclonifen	74070-46-5	170	0.000	0.016	1.569
	Alachlor	15972-60-8	4162	0.000	0.025	9.000
	Alachlor ESA	142363-53-9	921	0.010	0.110	2.000
	Alachlor OA	171262-17-2	100	0.010	0.039	0.460
	Ametryn	834-12-8	381	0.001	0.019	0.200

Continued

Table 1 Number of detections greater than the limit of detection (LOD), minimum, mean, and maximum concentrations of fungicides, herbicides, insecticides, pharmaceuticals, and antimicrobials detected in river water as reported in Waterbase (EEA, 2022).—cont'd

			Concentration (μg/L)			
Class	**Chemical**	**CAS**	**>LOD**	**min**	**mean**	**max**
	AMPA	1066-51-9	1789	0.025	0.652	20.200
	Atrazine	1912-24-9	8803	0.000	0.043	18.000
	Benfluralin	1861-40-1	334	0.003	0.012	0.125
	Bentazone	25057-89-0	2511	0.001	0.032	19.000
	Bifenox	42576-02-3	124	0.000	0.006	0.160
	Bromacil	314-40-9	245	0.001	0.027	2.000
	Bromoxynil	1689-84-5	198	0.001	0.019	1.400
	Chloridazon	1698-60-8	741	0.000	0.035	4.700
	Chlorsulfuron	64902-72-3	95	0.005	0.009	0.250
	Chlorthiamid	1918-13-4	108	0.005	0.005	0.005
	Clopyralid	1702-17-6	205	0.000	0.052	2.400
	Cyanazine	21725-46-2	303	0.001	0.021	0.100
	Desethylatrazine	6190-65-4	2495	0.001	0.066	8.000
	Desethylterbuthylazine	30125-63-4	1650	0.001	0.083	19.000
	Dicamba	1918-00-9	595	0.002	0.034	6.600
	Dichlobenil	1194-65-6	506	0.001	0.021	0.580
	Dichlorprop (2,4-DP)	120-36-5	331	0.001	0.018	1.450
	Diflufenican	83164-33-4	308	0.000	0.004	0.110
	Dimethachlor	50563-36-5	276	0.005	0.011	0.893
	Diuron	330-54-1	6950	0.000	0.057	21.000
	Ethofumesate	26225-79-6	687	0.001	0.056	10.000
	Flufenacet	142459-58-3	117	0.001	0.015	0.650
	Glyphosate	1071-83-6	3069	0.001	0.256	21.500
	Hexazinone	51235-04-2	821	0.001	0.018	0.640
	Ioxynil	1689-83-4	245	0.005	0.029	0.420

Table 1 Number of detections greater than the limit of detection (LOD), minimum, mean, and maximum concentrations of fungicides, herbicides, insecticides, pharmaceuticals, and antimicrobials detected in river water as reported in Waterbase (EEA, 2022).—cont'd

Class	Chemical	CAS	>LOD	Concentration (µg/L) min	mean	max
	Isoproturon	34123-59-6	6646	0.000	0.043	20.000
	Lenacil	01/08/2164	368	0.005	0.024	1.800
	Linuron	330-55-2	2915	0.000	0.032	12.500
	MCPA	94-74-6	3812	0.000	0.042	19.100
	MCPB	94-81-5	264	0.005	0.021	1.400
	Mecoprop	7085-19-0	2339	0.000	0.024	3.100
	Mecoprop-P (MCPP-P)	16484-77-8	98	0.005	0.017	2.900
	Metamitron	41394-05-2	205	0.001	0.035	2.900
	Metazachlor	67129-08-2	2196	0.000	0.025	15.000
	Metazachlor ESA	172960-62-2	1022	0.010	0.252	6.030
	Metazachlor OA	1231244-60-2	807	0.005	0.104	7.790
	Metolachlor	51218-45-2	4815	0.001	0.060	21.000
	Metolachlor ESA	171118-09-5	1172	0.010	0.082	5.160
	Metolachlor OA	152019-73-3	956	0.003	0.036	2.500
	Metribuzin	21087-64-9	911	0.001	0.030	9.000
	Nicosulfuron	111991-09-4	112	0.002	0.010	3.000
	Oxadiazon	19666-30-9	248	0.001	0.053	4.000
	Oxyfluorfen	42874-03-3	93	0.001	0.007	1.000
	Pendimethalin	40487-42-1	1271	0.001	0.018	4.300
	Prometon	1610-18-0	166	0.005	0.024	0.100
	Prometryn	7287-19-6	771	0.000	0.019	0.866
	Propazine	139-40-2	637	0.001	0.031	6.000
	Propyzamide	23950-58-5	544	0.002	0.028	4.700

Continued

Table 1 Number of detections greater than the limit of detection (LOD), minimum, mean, and maximum concentrations of fungicides, herbicides, insecticides, pharmaceuticals, and antimicrobials detected in river water as reported in Waterbase (EEA, 2022).—cont'd

			Concentration (μg/L)			
Class	**Chemical**	**CAS**	**>LOD**	**min**	**mean**	**max**
	Pyridate	55512-33-9	136	0.005	0.024	0.500
	Sebuthylazine	7286-69-3	110	0.005	0.010	0.025
	Secbumeton	26259-45-0	198	0.005	0.021	0.050
	Simazine	122-34-9	6282	0.000	0.037	17.000
	Terbumeton	33693-04-8	464	0.010	0.021	0.500
	Terbuthylazine	5915-41-3	4819	0.000	0.059	17.000
	Terbutryn	886-50-0	1732	0.001	0.026	21.000
	Trifluralin	1582-09-8	3149	0.000	0.022	19.000
Insecticide	Acetamiprid	135410-20-7	55	0.001	0.018	0.200
	Aldrin	309-00-2	4260	0.000	0.047	20.000
	Alpha-Endosulfan	959-98-8	2456	0.000	0.009	6.400
	Alpha-HCH	319-84-6	2695	0.000	0.019	13.000
	Beta-Endosulfan	33213-65-9	1311	0.000	0.007	1.250
	Beta-HCH	319-85-7	2854	0.000	0.015	18.200
	Carbofuran	1563-66-2	692	0.000	0.035	9.611
	Chlordane	57-74-9	339	0.000	0.015	0.050
	Chlorfenvinphos	470-90-6	2656	0.000	0.021	7.419
	Chlorpyrifos	2921-88-2	3889	0.000	0.019	18.000
	Chlorpyrifos-methyl	5598-13-0	457	0.000	0.014	0.370
	Clothianidin	210880-92-5	112	0.005	0.013	0.170
	Cypermethrin	52315-07-8	338	0.000	0.009	1.100
	DDT, o,p′	789-02-6	2265	0.000	0.014	2.500
	DDT, p,p′	50-29-3	4109	0.000	0.054	20.800
	Demeton-S-methyl	919-86-8	469	0.005	0.050	0.500
	Diazinon	333-41-5	396	0.000	0.015	2.000

Table 1 Number of detections greater than the limit of detection (LOD), minimum, mean, and maximum concentrations of fungicides, herbicides, insecticides, pharmaceuticals, and antimicrobials detected in river water as reported in Waterbase (EEA, 2022).—cont'd

			Concentration (μg/L)			
Class	**Chemical**	**CAS**	**>LOD**	**min**	**mean**	**max**
	Dichlorvos	62-73-7	1325	0.000	0.015	5.000
	Dicofol	115-32-2	429	0.000	0.010	0.620
	Dieldrin	60-57-1	4444	0.000	0.029	18.000
	Diethyltoluamide	134-62-3	774	0.010	0.052	11.600
	Dimethoate	60-51-5	1386	0.000	0.029	7.064
	Endosulfan	115-29-7	934	0.000	0.007	1.600
	Endrin	72-20-8	4289	0.000	0.031	10.000
	Fenitrothion	122-14-5	1137	0.001	0.014	1.250
	Gamma-HCH (Lindane)	58-89-9	5035	0.000	0.067	21.000
	Heptachlor	76-44-8	1450	0.000	0.019	13.390
	Heptachlor epoxide	1024-57-3	592	0.000	0.005	0.100
	Imidacloprid	138261-41-3	419	0.001	0.021	0.923
	Imidacloprid (alternative code)	105827-78-9	205	0.000	0.005	0.110
	Iodofenphos	18181-70-9	111	0.001	0.010	0.100
	Isodrin	465-73-6	3027	0.000	0.022	18.000
	Malathion	121-75-5	1342	0.001	0.022	17.100
	Methamidophos	10265-92-6	276	0.005	0.022	0.250
	Methomyl	16752-77-5	166	0.005	0.063	5.530
	Methoxychlor	72-43-5	933	0.000	0.009	1.130
	Omethoate	1113-02-6	301	0.000	0.029	0.250
	Parathion	56-38-2	899	0.000	0.018	5.000
	Parathion-methyl	298-00-0	817	0.000	0.016	10.000
	Permethrin	52645-53-1	486	0.001	0.022	1.400

Continued

Table 1 Number of detections greater than the limit of detection (LOD), minimum, mean, and maximum concentrations of fungicides, herbicides, insecticides, pharmaceuticals, and antimicrobials detected in river water as reported in Waterbase (EEA, 2022).—cont'd

				Concentration (μg/L)		
Class	**Chemical**	**CAS**	**>LOD**	**min**	**mean**	**max**
	Phosalone	2310-17-0	305	0.005	0.019	0.100
	Pirimicarb	23103-98-2	370	0.000	0.012	1.500
	Tebufenozide	112410-23-8	111	0.005	0.020	0.068
	Terbufos	13071-79-9	69	0.005	0.018	0.100
	Thiacloprid	111988-49-9	163	0.001	0.014	0.641
	Thiamethoxam	153719-23-4	176	0.001	0.014	1.000
Pharmaceutical	Atenolol	29122-68-7	305	0.005	0.014	0.220
	Caffeine	58-08-2	544	0.002	0.215	6.700
	Carbamazepin	298-46-4	1106	0.001	0.053	1.850
	Diclofenac	15307-86-5	1496	0.002	0.071	9.200
	Estrone (E1)	53-16-7	61	0.000	1.274	13.260
	Ibuprofen	15687-27-1	1035	0.005	0.071	10.000
	Metoprolol	37350-58-6	466	0.004	0.041	0.964
	Naproxen	22204-53-1	370	0.010	0.022	0.518
Antibacterial	Triclosan	3380-34-5	210	0.000	0.012	0.458
Antibiotic	Clarithromycin	81103-11-9	511	0.001	0.035	1.410
	Erythromycin	114-07-8	177	0.010	0.043	2.100
	Sulfamethoxazol	723-46-6	662	0.006	0.047	2.900
	Trimethoprim	738-70-5	292	0.001	0.016	0.335
	Azitromycin	83905-01-5	113	0.010	0.064	1.510
Biocide	Cybutryne	28159-98-0	148	0.000	0.004	0.320
	Pentachlorophenol	87-86-5	2675	0.000	0.083	19.200

Only chemicals with >50 detections above the reported limit of detection were included.

Insecticides: Endrin is the most monitored insecticide, first detected in the 1970s, but with concentrations generally low and stable over this period with two peaks in the early 80s and around 2010 (Fig. 3). From 1980 to 2005 chlorfenvinphos and chlorpyrifos were the most dominant insecticides (EEA, 2022). However, none of these pesticides are currently approved for use in the EU/UK and their concentrations have generally decreased. Over the last 10 years (2010–2020) neonicotinoids (e.g. imidacloprid and thiacloprid) emerged as the most detected insecticide groups, with imidacloprid being the most abundant (Fig. 3). Despite limits set in 2013 by the EU and the recent total EU ban (2018) on outdoor use of most neonicotinoids, a survey of UK Environment Agency data on insecticides concentrations in English rivers from 2016 to 2018 nevertheless found imidacloprid in 66% of samples, at an average concentration of 0.0317 μg/L (max 36 μg/L) (Perkins et al., 2021). Imidacloprid is widely used as a flea treatment for pets (e.g. spot-on solutions, sprays, and collars) with >30 licensed veterinary products containing imidacloprid in the UK (Tyler et al., 2019; VMD, 2022). Usage of the remaining approved neonicotinoids (thiacloprid and acetamiprid) has increased dramatically since the EU/UK bans. In particular, use of acetamiprid on arable crops increased >5 times (by kg) from 2019 to 2020 (Fera, 2020). It is therefore likely that neonicotinoids such as imidacloprid, along with thiacloprid and acetamiprid (currently still licenced for outdoor use in the UK) will remain part of the aquatic chemical exposome despite the EU/UK bans. Usage of Lambda-cyhalothrin on arable crops in the UK has approximately doubled in the last 10 years to 10,588 kg per year (Fera, 2020), possibly as farmers look to alternatives to neonicotinoids.

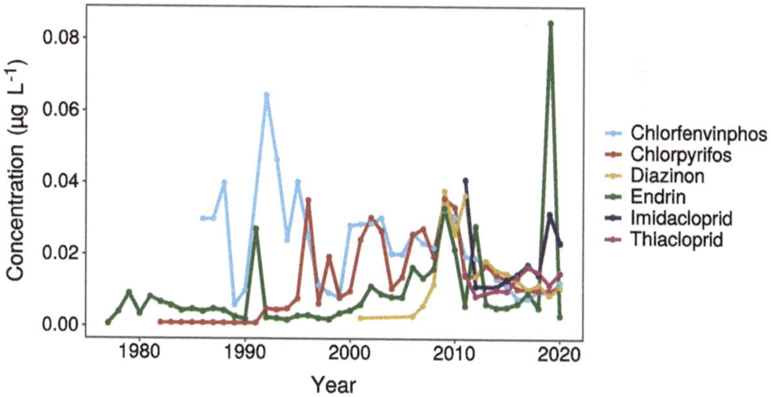

Fig. 3 Yearly mean concentrations of dominant insecticides in EU rivers from 1975 to 2021 (EEA, 2022).

Herbicides: Compared to insecticides and fungicides there is a far greater diversity of herbicides used and monitored. Historically, atrazine was found at high concentrations in rivers (mean 0.43 μg/L and max 18 μg/L, Table 1), but is no longer widely detected following its 2004 ban in the UK and EU. More recently glyphosate dominates both usage and detection in freshwaters, with glyphosate and its metabolite aminomethylphosphonic acid (AMPA), consistently detected over the last 15 years, with mean concentrations of 0.26–0.65 μg/L and maxima of 21.5 and 20.2 μg/L for glyphosate and AMPA, respectively (Fig. 4, Table 1). Glyphosate accounted for >20% of all herbicides applied to arable land (by area) in 2020—more than double the next two most widely applied herbicides combined, metsulfuron and diflufenican which accounted for 10% and 9% of total area applied respectively (Fera, 2020).

Pharmaceuticals: In general, pharmaceuticals are not well represented in the Waterbase dataset. Since 2008, caffeine (orginiting from both pharmaceutical and other sources), diclofenac, ibuprofen, azitromycin, estrone, ethanimidamide, carbamazepine and the antibiotics, amoxicillin, clarithromycin and erythromycin predominate, with caffeine the dominant determinant (Fig. 5, Table 1). High concentrations of clarithromycin and erythromycin are consistent with predicted concentrations of macrolides in the River Thames (UK), based on prescription data (Singer et al., 2019). However, given the small number of pharmaceuticals recorded in Waterbase, these data are not a robust reflection of pharmaceutical pollution in UK rivers. A recent global survey (1052 samples from 104 countries) of pharmaceutical pollution in rivers also found caffeine and

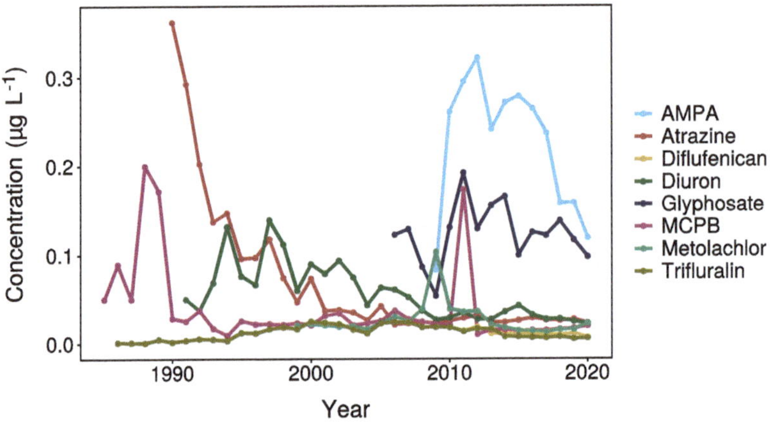

Fig. 4 Yearly mean concentrations of dominant herbicides in EU rivers from 1985 to 2021 (EEA, 2022).

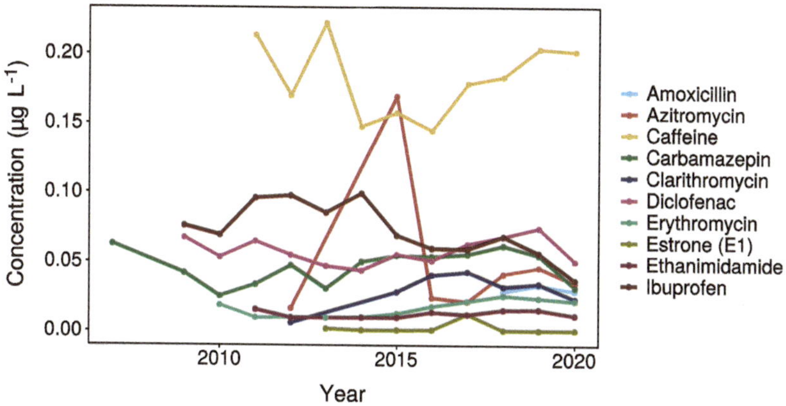

Fig. 5 Yearly mean concentrations of dominant Pharmaceuticals in EU rivers from 2008 to 2021 (EEA, 2022).

carbamazepine to be abundant (>50% of samples) (Wilkinson et al., 2022). Other abundant drugs detected by Wilkinson et al. (2022), like paracetamol, metformin, gabapentin, nicotine are not recorded in Waterbase. There is also a significant blind spot for illicit drugs, with cocaine and ketamine frequently detected in rivers, e.g. see Miller et al. (2019) and Miller et al. (2021). Other frequently detected compounds include psychoactive drugs such as the antidepressants Citalopram and Alprazolam. The effects of these compounds on freshwaters are not monitored, meaning the impact of any potential exposure is poorly understood.

3. Selection of chemicals and literature search

We selected four major classes (targeting different nodes of aquatic food webs) of pollutants: fungicides, insecticides/molluscicides, herbicides, and antibiotics to investigate the possible effects of aquatic chemical pollution on microbes, and how this might influence aquatic ecosystem functioning and the provision of critical ecosystem services. For each class we selected the two most common chemicals considering both UK usage and reported concentrations in Waterbase (eight chemicals total, Table 2), with differing modes of action to encompass a broader range of ecological responses. The fungicides were tebuconazole (dominant in EU/UK rivers and widely used on arable crops) which suppresses ergosterol biosynthesis, and chlorothalonil (not monitored in Waterbase but the most applied fungicide in the UK) which deactivates glutathione and induces oxidative stress in the cell. The insecticide imidacloprid was selected because, despite the 2018 EU/UK ban, it remains

Table 2 Results of Web of Science literature search.

Type of chemical	Target compound	Number of hits from Web of Science	Number of suitable publications
Fungicide	Tebuconazole	11	6
Fungicide	Chlorothalonil	5	0
Insecticide	Metaldehyde	1	0
Insecticide	Imidacloprid	7	2
Herbicide	Glyphosate	40	9
Herbicide	Diflufenican	0	0
Antibiotic	Amoxicillin	21	3
Antibiotic	Oxytetracycline	41	6

Table 3 Physico-chemical properties of the eight chemicals reviewed here.

Chemical	Class	CAS No	MW	WS	Log KOW	DT_{50}
Glyphosate	Organophosphate herbicide	1071-83-6	169.07	100000	−6.28	16.1
Diflufenican	Carboxamide herbicide	83164-33-4	394.29	0.05	4.20	94.5
Imidacloprid	Neonicotinoid insecticide	138261-41-3	255.66	610	0.57	187
Metaldehyde	Tetroxocane molluscicide	108-62-3	176.21	188	0.12	5.1
Chlorothalonil	Chloronitrile fungicide	1897-45-6	265.91	0.81	2.94	3.5
Tebuconazole	Triazole fungicide	107534-96-3	307.82	36	3.7	365
Amoxicillin	Antibacterial Antibiotic	26787-78-0	365.40	3430	0.87	35.6[b]
Oxytetracycline	Bactericide, Antimicrobial antibiotic	79-57-2	460.44	1000	−0.9[a]	148[b]

[a]https://comptox.epa.gov/dashboard/chemical/properties (no temperature specified).
[b]https://comptox.epa.gov/dashboard/chemical/properties (predicted average).
Compound properties from (http://sitem.herts.ac.uk/aeru/vsdb/index.htm) MW, Molecular weight (g/mol); SW, water solubility (20 °C, mg/L); LOG KOW, logarithm (base 10) of the octanol-water partition coefficient at pH 7, 20 °C; DT_{50} lab, 50% dissipation time under laboratory conditions (days), unless otherwise stated.

an abundant pollutant which blocks neurotransmission by acetylcholine receptors. The molluscicide was metaldehyde (banned in the UK in 2022), which after ingestion induces rapid hydrolysation of acetaldehyde and excessive production of mucus. The herbicides were glyphosate, which inhibits aromatic amino acid synthesis and is the most abundant herbicide currently detected (Fig. 4), and diflufenican which inhibits photosynthesis and is widely applied to arable crops. The antibiotics selected were the most commonly prescribed; amoxicillin, which inhibits cell wall production in bacteria and oxytetracycline which inhibits translation in bacterial cells and is widely used in farming (Grenni, 2022; Singh et al., 2019). For a summary of the predicted environmental hazard for each chemical see Box 1.

BOX 1 Environmental hazard summary for each chemical; persistence, mobility, toxicity and bioaccumulation

Persistence, mobility and toxicity (PMT) in the environment are used to identify chemical pollutants that are a concern to surface waters and drinking water resources (ECHA, 2017; Stockholm Convention, 2019; United Nations Environment Programme, 2019). Pollutants that are highly mobile and persistent, with slow degradation in the environment can travel great distances from their source and, over time, bioaccumulate to harmful levels within individuals and within food webs (Crookes and Fisk, 2018; Hale et al., 2020).

Several physico-chemical characteristics can be used to assess the potential PMT risk of chemicals in surface and drinking water sources. Bioaccumulation potential of chemicals can be measured via LogKow values and Bioconcentration factors. LogKow is the logarithm (base 10) partition coefficient of a substance between n-octanol and water phases, with a value below 2.7 indicating low bioaccumulation and above 3.0 a high bioaccumulation potential (Crookes and Fisk, 2018; McGinley et al., 2022; Rüdel et al., 2020). Bioconcentration factor (BCF) is the ratio between the concentration of a chemical taken up by an organism and its concentration in the environment (usually water) at equilibrium, during uptake through respiration and contact. A BCF of 100–5000 is considered the threshold for concern and above 5000 signifies a high potential for bioaccumulation (Crookes and Fisk, 2018; Garber, 2009). Adsorption of chemicals to soil particles is one factor that determines the mobility of chemicals in the environment. Mobility can be assessed using the LogKoc value which is the logarithm (base 10) partition coefficient of a substance between organic carbon and water phases. The higher the LogKoc value, the stronger the tendency to adhere to soil, and thus the lower the mobility whereas chemicals with a lower LogKoc have a higher affinity to water and are considered mobile (Crookes and Fisk, 2018; Rüdel et al., 2020). The persistence of a chemical is linked to the amount of time that it remains present in the environment and can be quantified by its

Continued

BOX 1 Environmental hazard summary for each chemical; persistence, mobility, toxicity and bioaccumulation—cont'd

DT_{50}, which is the halving-time of a chemical in soil or water (Crookes and Fisk, 2018; McGinley et al., 2022). Additionally, Henry's law constant is a gas law used to measure the volatility of a chemical, indicating the likelihood of its displacement to the air, relative to water. A value less than 0.1 is considered non-volatile and thus indicates greater persistence in the environment, while a value above 100 is considered volatile and suggests lower persistence (Crookes and Fisk, 2018). The criteria for persistent, bioaccumulative and toxic compounds are set out in the Annex XIII of the REACH Regulations (2006) and are applied to glyphosate, diflufenican, imidacloprid, metaldehyde, chlorothalonil, tebuconazole, amoxicillin and oxytetracycline in Table 3 (ECHA, 2017; United Nations Environment Programme, 2019).

Tebuconazole has a low water solubility (36 mg/L) but has high potential for bioaccumulation (logKow = 3.7) and is highly persistent in soil (DT_{50} = 365 days). Therefore, this fungicide is considered moderately persistent, stable, and slightly mobile and given a moderate alert for environmental fate, but in aquatic environments, tebuconazole has acute toxicity and is given a high alert for ecotoxicology and human health (ECHA, 2006;*University Hertfordshire*, 2007 ; Williams et al., 2017).

Chlorothalonil has low water solubility, is slightly mobile (logKow = 2.94) and is considered non persistent in soil (DT_{50} = 3.53 days) (*University Hertfordshire*, 2007). With a BCF of 100, chlorothalonil is on the lower end of the 100–5000 threshold for concern according to criteria set out by US EPA, but is still allocated a high alert for aquatic ecotoxicology and human health (ECHA, 2006; INCHEM, 2002).

Metaldehyde is moderately soluble, highly volatile, and non-persistent with a soil degradation DT_{50} of 5.1 days. Furthermore, it is moderately mobile in soil (logKoc = 2.38) and has low potential for bioaccumulation (BCF = 11 L/kg) and is therefore a moderate alert for both environmental fate and ecotoxicology (*University Hertfordshire*, 2007; Williams et al., 2017).

Imidacloprid has a low bioaccumulation potential (measured by logKow) and volatility, but is highly water-soluble (610 mg/L) and persistent in soil (DT_{50} = 187 days). Its degradation in water-sediment is slow and is considered moderately mobile and stable, but is considered highly leachable when accounting for its low adherence to soil and persistence (Gustafson, 1989). Therefore, imidacloprid has a high alert for environmental fate and is considered highly toxic in the environment, especially towards birds, crustacea, fish and bees (INCHEM, 2002; Williams et al., 2017).

Glyphosate is classified as a moderate alert for environmental fate, ecotoxicity and human health (Crookes and Fisk, 2018; United Nations Environment Programme, 2019). Glyphosate has slight mobility (logKoc = 3.15), relatively low bioaccumulation potential (logKow = −6.28), and is not considered persistent in

BOX 1 Environmental hazard summary for each chemical; persistence, mobility, toxicity and bioaccumulation—cont'd

the environment (DT_{50} = 16.11 days), although it is highly water-soluble and has low volatility (Henry's law constant = 2.10×10^{-08} Pa m^3 mol^{-1}).

Diflufenican herbicide has low solubility in water (0.05 mg/L), but a high logKow and low vapour pressure of 4.2 and 4.25×10^{-03} mPa, respectively, meaning there is a high potential for particle bound transport. Diflufenican is also stable and moderately persistent with a slow degradation (BCF = 1276 L/kg). The high potential for particle-bound transport (Wauchope et al., 1992) and bioaccumulation mean that Diflufenican has a high alert for environmental fate and toxicity to aquatic life (*University Hertfordshire*, 2007).

Amoxicillin is highly water-soluble (3430 mg/L), slightly mobile (logKoc = 2.94) and moderately persistent (DT_{50} of 35.5 days). These properties, combined with amoxicillin's low bioaccumulation potential (logKow = 0.87) mean that it is considered a moderate alert for environmental fate (Williams et al., 2017). Although more research is needed to fill in the gaps to make robust predictions for environmental fate, ecotoxicology and toxicology, it is considered a moderate environmental and health hazard (ECHA, 2007; Williams et al., 2017).

Oxytetracycline (usually used as the hydrochloride form) has high solubility in water (1000 mg/L), low bioaccumulation potential (logKow = -0.9), and low mobility in soil (logKoc = 2.43) and is therefore categorised as a moderate alert for environmental fate and ecotoxicity, despite acute toxicity to aquatic life after short term exposure (*University Hertfordshire*, 2007; Williams et al., 2017).

A literature review was conducted to collate the current understanding of the response of freshwater microbial communities to the eight chemical stressors identified above using the Web of Science online database, accessed through the University of Essex (June 2022) (summarised in Table 2). The following search terms were used for each chemical:

(Name of target chemical) AND freshwater AND microb* NOT marine

The initial focus was to compare, contrast and summarise the molecular microbial community responses (based on analyses of nucleic acids such as DNA or RNA) reported in these papers. The titles and abstracts of papers were manually screened to discard irrelevant studies. For most chemicals screened, and particularly for both herbicides and fungicides, one chemical did not produce any output. In general, there was a lack of studies utilising molecular techniques, with only 12 out of 26 publications using molecular techniques to analyse changes in the microbial community and/or that focus on freshwater environments. In addition, there was a lack of dose concentration gradients represented in the literature, with many studies conducting simpler experiments based on chemical presence/absence (which were also discarded). Antibiotic studies were dominated by culture-dependent methods, across a range of concentrations and antibiotics, with a focus on screening known pathogens (e.g. Ali et al., 2016; Henríquez-Núñez, 2012).

For chemicals where no relevant studies were found, a broader search criteria was applied. For metaldehyde, broader search terms ('metaldehyde AND bacteria AND freshwater NOT marine') produced no relevant articles. Similarly for diflufenican, this search attempt returned no results. An attempt to search for general 'bleaching herbicides', linking to the broader mode of action for this herbicide also failed. The substitution of diflufenican with its compound name and a broader chemical name of pyridinecarboxamide (National Centre for Biotechnology Information, 2022) was also unsuccessful. No relevant outputs were found for chlorothalonil (Table 2).

4. Responses of freshwater microbial communities to eight important chemicals: What do we know so far?

Fungicides: Of the two fungicides, chlorothalonil failed to produce any relevant results, while tebuconazole was shown to negatively impact fungal biomass and reproduction (Donnadieu et al., 2016; Pimentão et al., 2020). Molecular data showed that initial responses of fungal communities to tebuconazole are able to recover over time (Pesce et al., 2016). For other non-target microbial groups, such as bacteria, results are limited, but indicate a general robustness to tebuconazole exposure (Dimitrov et al., 2014; Pascault et al., 2014; Pesce et al., 2016). Additionally, in some instances, planktonic bacterial biomass (non-molecular data) increased post tebuconazole addition (Artigas et al., 2014; Donnadieu et al., 2016). This has been attributed to the death of benthic biofilm communities acting as a nutrient source for bacterioplankton communities when released into the water column (Artigas et al., 2014).

Insecticide/molluscicides: Screening for insecticide impacts on microbial communities lead to few relevant publications, supporting our earlier assertion that non-target groups of freshwater microbes are overlooked in ecotoxicology studies. Metaldehyde has not been investigated in association with freshwater microbial communities. However, other studies have investigated if bacterial communities in other environments, such as soils, are capable of metabolising it (Balashova et al., 2020; Castro-Gutiérrez et al., 2020). Several isolated bacterial strains were able to metabolise metaldehyde with the putative genes responsible having been acquired through horizontal gene transfer (Castro-Gutiérrez et al., 2020). Bacterial metaldehyde degradation has also been shown to be enhanced following acclimation to the chemical (Castro-Gutiérrez et al., 2020; Rolph et al., 2019). Repeated exposures to metaldehyde are likely due to its high solubility and leaching into water bodies, this could result in bacterial communities adapted to degrade it via horizontal gene transfer across multiple bacterial strains in the environment. Imidacloprid, which like metaldehyde, is also banned from outdoor use in EU and UK, is the subject of a single publication. In that study, the planktonic bacterial community was unresponsive to a dose gradient, with differences in alpha and beta diversity only observed in combination with glyphosate, although the study's authors suggest this can be attributed to the herbicide effect alone (da Costa Naíla et al., 2021). Again, more attention has been given to other environments, such as soil, and the ability of their bacterial communities to metabolise this compound. Repeated addition of imidacloprid led to the enrichment of bacteria such as *Klebsiella pneumonia* capable of its degradation (Diaz et al., 2017) and alterations to the nitrogen cycle, with nitrification decreasing and ammonification increasing, leading to a higher urea concentration (Diaz et al., 2017). Given that urea is a key input for freshwater nitrogen cycling and a nutrient for aquatic algae (Krausfeldt et al., 2019), chemical pollutants that alter its availability could have significant implications for freshwater biochemistry.

Herbicides: Diflufenican failed to return any relevant articles. In contrast, glyphosate returned the greatest number of relevant hits across all eight chemicals. Bacterioplankton community responses were governed by the dose concentration (y García et al., 2022). Concentrations lower than 15 mg/L had no effect on bacterial communities, whereas at 40 mg/L, irreversible compositional changes occurred (da Costa Naíla et al., 2021). Responses to glyphosate varied between bacterial taxonomic groups, with a significant increase in the relative abundance of the genus *Agrobacterium* with glyphosate, and a negative impacts on Bacteroidetes, Actinobacteria and Cyanobacteria at the same concentration.

At higher concentrations glyphosate serves as a phosphorus source for freshwater microbes (Drzyzga and Lipok, 2018; Lu, 2020; Vera et al., 2012) and has been shown to increase the abundance of antimicrobial resistant genes (ARGs) in the community (da Costa Naíla et al., 2022). Phosphorus availability is influenced by the addition of glyphosate (Drzyzga and Lipok, 2018; Pizarro et al., 2016) particularly for cyanobacteria and heterotrophic bacteria that contains the *phnJ* gene (encoding the C-P lyase responsible for glyphosate degradation) (Saxton et al., 2011). Picocyanobacteria capable of degrading glyphosate have been found in laboratory studies (Pérez et al., 2007; Pizarro et al., 2016) and also in natural systems (Berman et al., 2020). Additionally, *phnJ* genes attributed to *Pantoea*, *Pseudomonas*, and *Klebsiella* were detected in urban freshwaters in Ecuador (Hernández-Alomia et al., 2022).

Antibiotics: Culture-dependent techniques are still at the core of microbial research for antibiotics. Some studies have focused on the characterisation of microbial communities in freshwater bodies that have higher levels of these antibiotics due to effluent from WWTPs or hospitals in the surrounding area (Guan et al., 2018; Li et al., 2011; Lorenzo et al., 2018). Bacterial richness was found to be higher in aerobic WWTP effluent enriched in amoxicillin, while oxytetracycline had no clear effect on bacterial richness (Harnisz et al., 2015; Li et al., 2011). Complementary to biomonitoring studies, functional groups of bacteria including methane oxidisers and sulphate reducing bacteria have been studied by qPCR analysis of key functional genes, including the particulate methane monooxygenase gene (*pmoA*) and sulphite reductase gene (*dsrB*). After exposure to oxytetracycline, both functional groups were found to increase in abundance over 30 days, accompanied by a shift in bacterial community composition (Fernández et al., 2015; Pérez et al., 2007). This suggests unexplored impacts of antibiotics on freshwater microbiomes and implications for key ecosystems processes.

Our literature review (see Table 4 for summary) highlighted a clear knowledge gap in freshwater microbial ecotoxicological research. For three out of the eight key chemicals investigated (diflufenican, metaldehyde and chlorothalonil) no relevant matches were found in the literature. For the remaining five chemicals the number of appropriate studies were limited, ranging from a minimum of two for imidacloprid, to six for tebuconazole and oxytetracycline, reinforcing the idea that freshwater microbial communities have been overlooked in pollutant toxicity assessments. Chemical pollutants are heated objects of public debate, yet despite this (e.g. the recent UK ban enforced for metaldehyde in 2022) our literature screening found

Table 4 Summary of relevant molecular publications for our eight target chemicals and their effects on freshwater microbial communities.

Target compound	Type of chemical	Bacteria	Fungi	Others
Tebuconazole	Fungicide	• No effect (Dimitrov et al., 2014; Pascault et al., 2014; Pesce et al., 2016)	• Short-term responses of community composition to additions followed by a recovery (Donnadieu et al., 2016; Pimentão et al., 2020) • Seasonality (drought) can influence the effect (Pesce et al., 2016)	No published research available
Chlorothalonil	Fungicide	No published research available	No published research available	No published research available
Metaldehyde	Insecticide	No published research available	No published research available	No published research available
Imidacloprid	Insecticide	• No effect along a gradient from 0 to 60 μg/L (da Costa Naíla et al., 2021)	No published research available	No published research available
Glyphosate	Herbicide	• Low dosage <15 mg/L no effect, but 40 mg/L irreversible effect on composition, regardless of pre-exposure (da Costa Naíla et al., 2021, 2022) • Cyanobacteria can use glyphosate as a phosphorus source (Lu et al., 2020; Saxton et al., 2011) • Persistence to survive and thrive post addition is related to the number of ARGs within a genome (da Costa Naíla et al., 2022)	No published research available	• Change in community linked more to seasonality rather than glyphosate (Pesce et al., 2009)

Continued

Table 4 Summary of relevant molecular publications for our eight target chemicals and their effects on freshwater microbial communities.—cont'd

Target compound	Type of chemical	Bacteria	Fungi	Others
Diflufenican	Herbicide	No published research available	No published research available	No published research available
Amoxicillin	Antibiotic	• Slight increase in richness downstream from sites releasing the antibiotics vs upstream (Li et al., 2011) • Found increase in resistant genes post treatment release into freshwaters (downstream) (Lorenzo et al., 2018)	No published research available	No published research available
Oxytetracycline	Antibiotic	• Methanotrophs and sulphate reducing bacteria increased in abundance post exposure (Fernández et al., 2015; Tong and Xie, 2019) • Change in methanotroph and sulphate reducing bacterial community composition (Fernández et al., 2015; Tong and Xie, 2019) • No difference in richness and abundance of bacteria upstream vs downstream of sites releasing antibiotic into freshwaters (Harnisz et al., 2015; Li et al., 2011)	No published research available	No published research available

no studies for metaldehyde impacts on freshwater microbiomes. In contrast, glyphosate, which has received similar publicity, had nine relevant studies. A common pattern emerging from this review is that many studies are confined to single microbial species, or a small subset of species, rather than taking a community-level approach. This was especially striking for antibiotics; there is a rich body of literature that focuses on culturable bacteria and their ability to grow at different antibiotic concentrations, especially for strains connected with human health or economically important species, but relatively only four community-focussed studies. Papers by da Costa Naíla et al. (2022) and Lu et al. (2020) highlight the potential knowledge to be gained from the use of molecular tools to investigate microbial responses to chemical pollutants from the gene to the ecosystem levels of organisation, but only recently have robust experiments been conducted that incorporate dose concentration gradients, supported by in-depth molecular microbial analyses (da Costa Naíla et al., 2021, 2022; Lu et al., 2020).

5. Embracing complexity: Multiple stressors in freshwaters

Only one study has included two of the chemicals considered in this review (da Costa Naíla et al., 2021, 2022). The combined effect of glyphosate and imidacloprid on the planktonic bacterial community revealed that glyphosate has an effect at high concentrations, whereas imidacloprid was effective on its own. Environmental stressors are rarely present independently, and often combine in unpredictable ways (Crain et al., 2008; Jackson, 2016; Ormerod et al., 2010; Orr et al., 2020). This includes the effects of chemicals when they are combined in complex mixtures, in succession (Løkke et al., 2013), and the interaction with other abiotic stressors (e.g. warming, acidification, nutrients, hydrology) (Moe et al., 2013). This can also be confounded by biotic stressors, as the indirect effects of stressors propagate through food webs (Jochum et al., 2012). The simplest way for multiple stressors to combine is additively, i.e., the net effect is the sum of the two individual stressors (Fig. 6). If this was always the case then predicting the effects of combined stressors would be straightforward given the results of single stressor experiments. It could be argued that the regulatory framework in freshwaters still follows this model, for example pesticide compliance is still based on the sum of total chemical concentrations in the UK/EU. However, for more than a decade, research in marine and

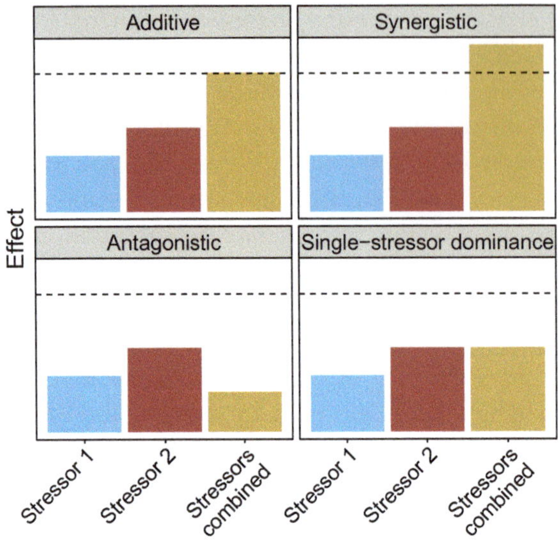

Fig. 6 Conceptual framework for the combined effects of two interacting stressors. Blue and red bars are the effects of a single stressor, while the yellow bars are the possible effects when these chemcials combined as a mixture.

freshwaters have focused on a multi stressor framework (Christensen et al., 2006; Jackson, 2016; Ormerod et al., 2010; Orr et al., 2020). Meta-analyses show that the interactions between stressors in aquatic systems (biotic, abiotic, or xenobiotic) are frequently interactive (Holmstrup et al., 2010; Jackson et al., 2016; Laskowski et al., 2010); the net effect is either synergistic (greater than the sum of the individual stressors), or antagonistic (less than the sum of the two stressors) (Fig. 6). For example, Ferguson et al. (2021) showed the positive effects of warming and the negative effects of nutrient enrichment on bacterial diversity in coastal mesocosms almost exactly cancelled out, meaning that when both stressors combined there was no net effect. An additional way that multiple stressors may interact is via single stressor dominance, where the effect of one stressor overwhelms the others (Fig. 6) as shown by da Costa Naíla et al. (2021) with a mixture of glyphosate and imidacloprid. This may happen due to overlapping modes of action, relative differences in the severity of effects, or environmental concentrations of the two stressors (particularly in the case of chemical contaminants). Similarly in lakes, an overriding stressor, such as nutrient pollution, overwhelmed the effects of other stressors (Birk et al., 2020). This could wrongly result in considering only the dominant stressor. From the articles available, we highlight a substantial lack of studies combining multiple stressor effects on community responses (additive, synergistic or negative effect). More multi stressor experiments are required, however, a purely empirical approach to disentangling the effects of multiple interacting stressors in freshwaters is not realistic. The resulting unexpected combination of multiple stressors could be translated into an impracticable number of testing possibilities, especially if we consider that some effects could be context-dependent and be modulated by temporal, spatial, or taxonomic scales. A predictive framework is therefore necessary to understand the effects of multiple stressors on microbial communities. This would require us to move towards a more mechanistic understanding of stressor interactions (see Orr et al., 2020) by, for example, generalising stressor interactions and accounting for chemical modes of action, functional traits, and ecological complexity in our predictions of how multiple stressors combine.

6. Everything is connected: Indirect effects of chemical stressors

As with chemicals, the microorganisms themselves are not present in isolation but are part of wider ecological networks. Consequently, we must also account for the impacts of chemicals propagating through food webs. Indirect effects can be mediated via top-down, bottom-up, and competitor effects. For example, after a catastrophic spill of the insecticide chlorpyrifos in the River Kennet (UK), the abundance of ammonia-oxidizing bacteria increased because of the ammonia released by the decay of dead invertebrates (Thompson et al., 2016). Other indirect effects on microbes included release from top-down control, with changes in diatom population size structure (larger diatoms at impacted sites following the elimination of their consumers), and competitive release (elevated microbial litter decomposition rates compensating for the reduced biomass of the invertebrate detritivore *Gammarus pulex*) (Thompson et al., 2016). Bacteria are also commonly subject to bottom-up effects of chemical contaminants as they may use them to fuel metabolism, as also observed by Thompson et al. (2015), and commonly observed for hydrocarbons (Ferguson et al., 2017; Thomas et al., 2020). This is not strictly an indirect effect, but could lead to indirect (bottom-up) effects higher in the food web, in addition to indirect competitive effects within the microbial community.

Indirect effects of anthropogenic stressors (Kidd et al., 2014) including trophic cascades (Jochum et al., 2012; McElroy et al., 2015) are well documented in aquatic systems for higher taxa, but how these propagate to microbial communities at the base of aquatic systems is often unclear (Ferguson et al., 2021). A comprehensive review of indirect effects of contaminants in aquatic systems by Fleeger et al. (2003) found only three (out of 150) studies considered effects on bacteria and/or diatoms, and none for fungi. The review found that top-down effects (trophic cascades) were common (60% of manipulative studies), with increases in primary producers after selective effects against grazers for example with insecticides (Fleeger et al., 2003). However, bottom-up effects caused by decreases in primary producers with herbicides were also common. As discussed, bacteria may use the chemical as a metabolic substrate. This stimulation of microbial populations at the base of the food web might invoke a bottom-up response in bacterivores or have lasting effects on nutrient cycling stimulating primary producers (Fleeger et al., 2003). The reality is that these effects are not present in isolation and an aquatic microbial community may be concurrently subject to any combination and likely all these indirect effects, with different units of the microbial community responding differently. The use of increasingly sophisticated statistical methods underpinned by increasingly big molecular datasets, and advances in and access to greater computing power (including via the cloud) will be key in further elucidating these complex indirect interactions (Clark et al., 2018). It is important that both data analysis and experimental design consider the bacterial community as integral to the wider food web. In the same way that we need experiments with realistic chemical mixtures, we need experimental systems with realistic ecological networks (e.g. mesocosms) to analyse the entire food web response.

7. Antimicrobial resistance (AMR): Are freshwaters the key link in the one health approach?

Antibiotics have led to vast gains in life expectancy, but the rising tide of AMR threatens these gains. AMR is predicted to cause the deaths of at least 10 million people annually by 2050 (O'Neil, 2014), overtaking cancer and heart disease as the leading cause of death (The World Health Organisation, 2020). In parallel, rising resistance in crop pathogens threatens our ability to feed an ever-growing human population (Fisher et al., 2018). As such, AMR presents a serious public health threat. But AMR is not an emerging future problem, it is estimated that 1.27 million deaths were directly attributable to AMR infection in 2019, significantly more than for HIV/AIDS or malaria (Murray et al., 2022). These figures are likely underestimated as they do not account for resistant fungi, the impact of which is often overlooked (Almeida et al., 2019).

Ultimately, overconsumption and industrial-scale production has led to the widespread dissemination of antibiotics in the environment. This has initiated a global selection pressure for antibiotic resistance (Larsson, 2014). Antimicrobial resistant genes (ARGs) observed in clinical isolates likely originated from environmental microbes (Munita and Arias, 2016). Therefore, the environment can be thought of as an ARG reservoir and a potential source of novel resistance mechanisms in human pathogens. The drivers for the development and dispersal of ARGs and AMR microorganisms in freshwaters is therefore an urgent research priority.

Most antibiotics are partially metabolised and excreted in a biologically active form (Kümmerer and Henninger, 2003). As a result, undegraded antibiotics are often detected in hospital and domestic wastewater (Dinh et al., 2017). Likewise, wastewater produced by the pharmaceutical industry may contain sublethal and lethal concentrations of antibiotics (Fick et al., 2009; Larsson et al., 2007). Antibiotics that enter WWTPs are either biodegraded, absorbed into sewage sludge, or discharged into freshwater systems. As well as antibiotics, treated wastewater often contains ARGs and antibiotic-resistant bacteria. These drivers of resistance can become enriched in freshwaters downstream of WWTP outputs (Lorenzo et al., 2018; Murphy et al., 2021). Freshwaters also receive agricultural and urban runoff. Consequently, they are vulnerable to heavy metals, antibiotics, and in particular pesticide pollution. Fungicide pollution (e.g. azoles) from agricultural runoff may select for fungal resistance (Cao et al., 2021; Gisi, 2014; Verweij, 2020). Bacteria and fungi also possess a variety of resistance genes that can be co-selected (or up-regulated) in the presence of multiple chemical stressors. A single gene may confer resistance to several chemicals (cross-resistance). For example, glyphosate stress reduces intracellular antibiotic concentrations by upregulating the activity of non-specific multidrug efflux pumps (Kurenbach et al., 2017). The net direction and magnitude of the interactions between co-pollutants determine selection pressure for AMR. Both the number and types of chemicals are important (Bordas et al., 1997; Kurenbach et al., 2017). Therefore, the diversity of chemical pollutants found in the freshwaters has important consequences for the development of AMR.

Many people rely on freshwaters for drinking, recreation, crop irrigation and animal husbandry. The humans and animals that use freshwaters are potential vectors of environmental ARGs. The One Health perspective highlights the importance of tackling the spread of AMR across this human-animal-environment interface. As of 2018, over 100 countries have developed National AMR Action Plans (Chua et al., 2021). AMR has been studied extensively in clinical settings, but the role of the environment is not as well-understood (Singer et al., 2016). Freshwaters are a critical link across the One Health interface and therefore constitute a major gap in our knowledge base for developing effective AMR action plans.

8. Nutrient pollution: Does nothing else matter?

Nitrogen and phosphorus have long been considered the primary pollutants in freshwaters. These two nutrients are key as they drive the trophic status of surface waters, e.g., eutrophication. There is considerable focus on eutrophication as the cause of algal blooms, depletion of dissolved oxygen/ hypoxia, fish kills, loss of biodiversity, and changes in ecological structure (in particular reduction in macrophytes) (Smith and Schindler, 2009; Wurtsbaugh et al., 2019). All of these impacts degrade crucial ecosystem services provided by freshwaters (e.g. food and drinking water supply, climate regulation, flood reduction, cultural services) and have therefore been a considerable focus of regulation and monitoring. Despite this, between 1970 and 2000 the number of rivers with significant nutrient pollution increased by one-third globally (Liu, 2012; Seitzinger et al., 2010). EEA data shows that, despite a decrease in reported total organic nitrogen since 1970, stable oxidised forms of nitrogen have increased or remained stable (Fig. 7). By 2050 it is predicted that nutrient pollution will continue to increase, and become more widespread globally especially in Asia, Africa, and South America (van der Struijk and Kroeze, 2010; Yasin et al., 2010). Nutrients inputs can come from agriculture (e.g. fertiliser), urban and industrial runoff,

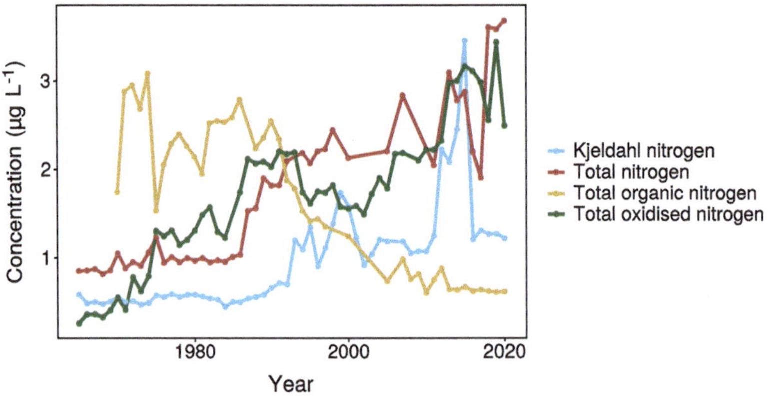

Fig. 7 Mean reported concentrations of nitrogen compounds in freshwater from 1972 to 2021 (EEA, 2022).

sewage discharge, and atmospheric deposition, all of which can be driven by increasing population and/or anthropogenic climate change. For example, to feed an ever growing population the use of agricultural fertiliser will increase or runoff will be aggravated by increased extreme precipitation. However, even in this context it may be possible to mitigate nutrient pollution into rivers with careful management strategies (Ulén and Fölster, 2007).

Microorganisms are key players in the biotransformations of nutrients entering freshwater ecosystems, particularly in nitrogen and phosphorus cycling. They are directly and indirectly affected by nutrient pollution resulting for example in a decreased microbial diversity and changes in nitrogen cycling populations (Ferguson et al., 2021; Graves et al., 2016; Kearns et al., 2016; Wang et al., 2019) and given their place at the base of food webs propagate these effects through the entire freshwater food web. There is also a potential for priming effects, whereby an input of nitrogen or phosphorus into freshwaters stimulate the microbial community resulting in greater microbial nutrient cycling and potentially release of CO_2, or other greenhouse gases (Bingeman et al., 1953), however little evidence are available to support or refute this in freshwaters (Bengtsson et al., 2014; Catalán et al., 2015; Danger, 2013; Kuehn et al., 2014; Wang et al., 2021). Arguably, given nutrient levels are generally stable or increasing globally (Fig. 7); nutrients and not chemicals may still be the major threat to freshwaters. We can detect thousands of chemicals in a drop of water, many of which are at diminishingly low concentrations, orders of magnitude below thresholds at which they might have a measurable effect. Under these realistic conditions do the effects of nutrient pollution simply overwhelm those of other chemical stressors? This is a critical question as nutrient and chemical pollution often enter freshwaters via the same routes and co-occur, as for example in sewage discharge and biosolid soil amendments both contain a complex mixture of nutrients and chemicals (Kinney et al., 2006). A meta-analysis of manipulative multi stressor experiments in freshwaters by Birk et al. (2020) found that nutrients were by far the most overriding stressor in freshwaters, particularly in lakes. In rivers the picture is more complex; while the effects of nutrient enrichment are still prominent, they are more strongly modulated by secondary stressors (Birk et al., 2020). da Costa Naíla et al. (2022) considered the effects of the pesticide glyphosate crossed with nutrients on ARGs in bacterioplankton communities. They found only a weak effect of nutrients, concluding that glyphosate increases the abundance of ARGs. However, the concentrations of glyphosate used (0.3 and 15 mg/L) are several orders of magnitude greater than typically observed in freshwaters (Fig. 4) with nutrient concentrations, high but within reasonable ranges for eutrophic waters (15–60 μg TP/L and 231–924 μg TN/L). Given more realistic environmental concentrations of the two stressors, would the relative importance of each stressor have changed? There is a critical lack of research that considers the effects of chemicals in the context of nutrient pollution, and this remains a critical research gap.

9. Beyond taxonomy: A new paradigm for biomonitoring of freshwater ecosystems

Biomonitoring is the use of indicator or sentinel species to assess the health or contamination of an environment. A classic example is the historical use of canaries (and small mammals) to detect carbon monoxide in mines, first proposed by Haldane (1895) as a more sensitive alternative to gas lamps. There is a long history of biomonitoring (often based on macroinvertebrates) to assess water quality in freshwaters. The 'Saprobien System' developed by Kolkwitz and Marsson (1909) is one of the earliest biomonitoring tools for freshwaters, and is based on indicator species associated with stages of deterioration in waters receiving waste. The foundational concept is that polluted waters contain different organisms than clean waters, and the basic underpinning of freshwater biomonitoring has endured over the last century (Patrick, 1949). Attempts to integrate information about abundance, diversity, and taxonomic traits of biomonitoring targets, rather than just their presence or absence, has spawned a number of indices relevant for freshwater monitoring. For example, for macroinvertebrates the Trent biotic index, BMWP score, Hilsenhoff's index, SIGNAL Index (Woodiwiss, 1964; Armitage et al., 1983; Hilsenhoff, 1987; Chessman, 1995, respectively); for diatoms the trophic diatom index (Kelly and Whitton, 1995); and for protozoa the species/community pollution value (Jiang, 2006). Indeed, current biomonitoring systems for freshwaters, including the River Invertebrate Prediction and Classification System (UK, Wright et al., 1993), the Australian River Assessment System (Simpson and Norris, 2000), Canadian Aquatic Biomonitoring Network (Government of Canada, 2012), and The MEDiterranean Prediction And Classification System (Spain, Poquet et al., 2009) are still based on the taxonomic identity of monitored species. This form of biomonitoring is relatively simple and inexpensive to implement, with minimal specialist equipment required, and sampling procedures can be easily disseminated to support broad-scale monitoring. However, a key consideration of these methods are they were primarily designed to detect nutrient pollution, limiting their applicability in detecting other chemical pollutants (Bo et al., 2016). Furthermore, both the sampling and sample analysis are labour intensive and may require specialist taxonomic expertise making standardisation challenging, while the logistical issues of working in or around flowing water constrain the range of sites that can be safely sampled. Some of these problems can be overcome with rigorous and regular training, but this adds further costs that are magnified at the national scale. Alternatively, molecular methods, including those developed to study microbial communities (Box 2) could provide a route forward.

BOX 2 Data in every drop of water: the value of culture independent methods

While culture-dependent techniques remain an important tool in environmental microbiology, difficulties in culturing the vast majority of microbial biodiversity (<1% of microbes thought to be 'straightforward' to culture—Chan et al., 2019; Hugenholtz et al., 1998) have resulted in wider uptake of culture-independent techniques, including molecular methods, as a way of biomonitoring microbial communities. The effects of chemical pollution on freshwater ecosystems cut across all taxonomic groups and encompass effects from the genome to the ecosystem level. Molecular methods allow us to address the breadth of these impacts and are therefore an essential tool in biomonitoring strategies.

Metabarcoding (also known as amplicon sequencing) is the most widely used molecular approach to characterise freshwater microbial community composition. Here, a region of a phylogenetic (e.g. 16S/18S rRNA gene or Internal Transcribed Spacer region (ITS)) or functional marker gene is amplified via PCR and sequenced in a high-throughput manner (Clark et al., 2018; Grossart et al., 2020), providing a snapshot of the microbial community and allowing changes in diversity or composition to be detected. However, metabarcoding does not provide functional information and can suffer from limited taxonomic resolution, due to both primer selection, and the relatively short sequences produced by the sequencing platforms commonly used in such studies (Johnson et al., 2019; Min and Hickey, 2007). Recent developments in long-read sequencing platforms (e.g. PacBio or Oxford Nanopore) enable sequencing of entire marker genes, achieving species or even strain-level resolution (Johnson et al., 2019; Tedersoo and Anslan, 2019), thereby attenuating the issue of limited taxonomic resolution. Even so, the issue of missing functional information remains.

A suite of bioinformatic tools now exist that allow prediction of microbiome functional profiles from metabarcoding data (see Djemiel et al., 2022 and

Continued

BOX 2 Data in every drop of water: the value of culture independent methods—cont'd

references therein), but all are constrained by taxonomic resolution and the phylogenetic relatedness of reference genomes used to make predictions (Poretsky et al., 2014). Metagenomic sequencing circumvents these issues by directly sequencing the entire genomic content of an environmental sample, resolving both the taxonomic composition and functionality of communities (Quince et al., 2017). Unlike metabarcoding, metagenomic sequencing does not rely on PCR-amplified material (although random PCR amplification may be required in low biomass environments), thereby removing issues related to the availability/suitability of primers and amplification bias, and offers strain-level taxonomic resolution through the assembly of individual genomes (Anyansi et al., 2020; Ranjan et al., 2016). While metagenomic sequencing may trade-off sequencing depth for breadth and requires more demanding computational/bioinformatic requirements for data analysis (Laudadio et al., 2018; Shakya et al., 2013), the ability to simultaneously determine both taxonomic and functional profiles offers great potential insight into how chemical stressors impact microbial communities.

The proliferation of molecular (nucleic acid) based methods has revolutionised ecology, with high-throughput methods enabling us to extract 'data from every drop of water' see Box 2 (Clark et al., 2018). Molecular methods have the potential to improve current monitoring by bringing them into the 'big data' era of ecology. Metabarcoding can be used to target taxonomic barcodes across all domains of life (Baird and Hajibabaei, 2012; Taberlet, 2012), reducing the need for specialist taxonomic expertise, and opening the possibility of automated sampling methods to support real-time biomonitoring via the cloud (Bohan et al., 2017), increasing the number and range of sites that can be monitored simultaneously. The main drawback of metabarcoding methods would be the lack of quantitative information on the number and or size of taxa. Compositional information can be inferred for unicellular organisms, but even this comes with caveats, and for multicellular organisms the basis for gleaning any quantitative information becomes increasingly precarious, although not impossible (Maruyama, 2014; Yates et al., 2019). The most immediate benefit of metabarcoding based biomonitoring is the ability to assess a broader range of taxa simultaneously, which could bring the overlooked microbial world into the focus of biomonitoring which, given their sensitivity to environmental change and importance in freshwater ecosystem functioning should be a priority (Cavicchioli et al., 2019; Clark et al., 2018). As we have shown here (Table 4), this initiative is currently limited by the lack of available data to support the development of appropriate microbiome-based pollution indices to do so.

Molecular methods may also yield more informative biomonitoring of chemical pollutants as they allow us to incorporate ecological information beyond taxonomy into our assessment. There have long been calls for a more holistic approach to biomonitoring that accounts for measures of ecological structure and non-taxonomic traits-based metrics alongside, or even instead of species composition (Karr, 1981). Such metrics could quantify traits (e.g. body size, morphology, growth rate, life history, functionality) and community structure (composition, diversity, species interactions) (see Culp, 2011; Dolédec and Statzner, 2008; Menezes et al., 2010). The SPEcies At Risk (SPEAR) method is one example, which links pesticide concentrations to the composition of invertebrate communities within a trait based framework (Liess and Von Der Ohe, 2005). Advantages of such an approach include, mechanistic linkage between biotic response and environment, consistent descriptors across broad spatial scales, improved temporal stability, easy integration into current biomonitoring programs, and greater utility for ecological risk assessment (Culp et al., 2011). Molecular biomonitoring tools could mesh well with such schemes for several reasons. First, these tools allow us to resolve ecological networks across multiple trophic levels, thereby enabling the identification of more appropriate sentinel species (or groups of species) or interactions to focus biomonitoring efforts (Derocles et al., 2018; Gray et al., 2014). For example, the identification and monitoring of microbial keystone species (as highly-connected nodes in a network) could provide an early warning system for chemical pollution events (Gray et al., 2014).

Additionally, molecular methods allow us to take a 'genes to ecosystems' based approach by exploiting the functional information contained within metagenomic data (Thompson et al., 2016). Monitoring for shifts in the identity and abundance of microbial functional groups has already proved effective in predicting shifts in freshwater biogeochemical cycles (e.g. Thompson et al., 2016; Zhu et al., 2020), and could be better exploited in biomonitoring to predict potential changes in ecosystem functionality due to chemical pollutants. In much the same way that taxonomic networks can be used to identify keystone taxa and nodes (Gray et al., 2014), metagenomic sequencing can be used to construct ecologically relevant community-scale functional networks under different environmental scenarios (chemical mixtures, nutrients) (Baumann et al., 2022; Buck et al., 2021; Linz et al., 2018; Tran et al., 2021; Zhou et al., 2022). In addition to providing taxa-independent biomonitoring targets (e.g. genes that can be monitored with quantitative methods like qPCR), it allows us to focus biomonitoring on relevant ecosystem services provided by fresh waters, e.g., changes in biogeochemical cycling, thus allowing the development of value-based biomonitoring. Rather than simply assessing if the system is perturbed from a 'pristine' state, we would be able to determine if the core functioning and services provided by the ecosystem are compromised. A key emergent question for developing value based monitoring would therefore be, what services do we value?

10. Bridging the gap

As discussed, there is a need for empirical research into the effects of chemical stressors to include realistic ecological networks under realistic environmental contexts. We highlight the value of mesocosms as providing the balance between environmental complexity and experimental control required to develop a mechanistic understanding of the effects of chemical stressors, and to develop hypotheses to test in larger-scale *in situ* experiments (Chará-Serna et al., 2019; da Costa Naíla et al., 2021, 2022; Pizarro et al., 2016; Vera et al., 2012). In a practical sense, mesocosm experiments control for independent variables of interest and treatment combinations, while allowing equal exposure to background abiotic fluctuations during the same timeframe. Experiments can therefore simultaneously incorporate a diversity of treatments and environmental or biotic conditions. This would otherwise be hard to replicate at field level due to logistical and economic constraints and the potential lack of suitable study sites. Depending on the size of mesocosms and budget, mesocosm experiments allow for adequate treatment replication and fully factorial designs to tease apart causality (Stewart et al., 2013). They provide the opportunity to manage food web complexity and combine this with abiotic factors (Ferguson et al., 2021) to build an understanding of trophic interactions in combination with scenarios that threaten the future of aquatic systems, namely climate change, agricultural intensification, and urbanisation. This can be obscured at greater spatial scales due to the increased potential for unmeasured confounding variables (Sharma et al., 2021). Mesocosm experiments are an invaluable tool to study the ecotoxicological impacts of chemicals in different scenarios that should not be released into the natural environment, due to an insufficient understanding of the consequences to aquatic ecosystems (Sharma et al., 2021). At the field level this is only possible under rare serendipitous circumstances whereby researchers can capitalise on real-world pollution events (i.e., Andújar et al., 2018; Thompson et al., 2016). Once causality has been identified, these types of mesocosm experimental systems pave the way for hypothesis driven, *in-situ* field research that captures more realism (Stewart et al., 2013), and together can provide strong evidence for future predictions (Fordham, 2015). As identified in this review article, mesocosm experiments that incorporate mixtures of chemical stressors, of different sources (pesticide, pharmaceutical, drug and industry) should be considered and assessed along concentration gradients. At present most of these experiments are performed in pond mesocosms, but there is a need to look at the impacts on running water bodies as stream channel mesocosms are far less frequent (for examples see Donnadieu et al., 2016; Pesce et al., 2016).

11. Summary and future directions

This review has highlighted an important research gap in freshwater ecology and ecotoxicology. We cannot continue to overlook the role of microbial communities in freshwaters as they are the foundation of many of the critical ecosystem services they provide. There is a lack of studies on how the freshwater microbiome responds to multiple chemical stressors, including at realistic concentrations or mixture complexity. It is also fundamental to understand how chemical stressors interact with other anthropogenic climate stressors, but critically the role of nutrients which remain possibly the key stressor in freshwaters. Given the seemingly innumerable potential chemical mixtures and stressor combinations, a mechanistic understanding of how different anthropogenic stressors combine to predict their combined effects will be required to do this. Future work should also consider anthropogenic impacts across whole aquatic communities, and not focus on single taxa or groups. By exploiting molecular tools to study the impact of chemical stress across realistically complex freshwater food webs, such as mesocosm studies, we can usher in a new era of freshwater biomonitoring programmes that provides a more holistic understanding of the impacts of chemical stressors on freshwater ecosystems. Equally, a more nuanced functional or trait based approach to biomonitoring, taking advantage of advances in molecular methods is required. However, it requires reflection on what criteria should be considered to assess the ecological condition of freshwater habitats. Should biomonitoring focus on an idealised 'natural' state, or on optimising key ecosystem services (for example pollutant removal)? There must also be greater consideration of the interaction between public and environmental health given the central role of freshwaters as a melting pot for pollutants, pathogens, and AMR microorganisms and their dispersal across the 'One Health' interface.

Acknowledgements

This work arises from an early career writing group that was supported by funds from two UK Natural Environment Research Council (NERC) projects (NE/S000291/1 and NE/S005560/1).

References

Ali, L., et al., 2016. Nutrient-induced antibiotic resistance in Enterococcus faecalis in the eutrophic environment. J. Glob. Antimicrob. Resist. 7, 78–83.

Almeida, F., Rodrigues, M.L., Coelho, C., 2019. The still underestimated problem of fungal diseases worldwide. Front. Microbiol. 10, 214.

Andújar, C., et al., 2018. Metabarcoding of freshwater invertebrates to detect the effects of a pesticide spill. Mol. Ecol. 27 (1), 146–166.

Anyansi, C., Straub, T.J., Manson, A.L., Earl, A.M., Abeel, T., 2020. Computational methods for strain-level microbial detection in colony and metagenome sequencing data. Front. Microbiol. 11, 1925.

Armitage, P.D., Moss, D., Wright, J.F., Furse, M.T., 1983. The performance of a new biological water quality system based on macroinvertebrates over a wide range of unpolluted running-water sites. Water Res. 17, 333–347.

Artigas, J., et al., 2014. Comparative sensitivity to the fungicide tebuconazole of biofilm and plankton microbial communities in freshwater ecosystems. Sci. Total Environ. 468, 326–336.

Baird, D.J., Hajibabaei, M., 2012. Biomonitoring 2.0: a new paradigm in ecosystem assessment made possible by next-generation DNA sequencing. Mol. Ecol., 2039–2044.

Balashova, N., et al., 2020. Ubiquity of microbial capacity to degrade metaldehyde in dissimilar agricultural, allotment and garden soils. Sci. Total Environ. 704, 135412.

Baumann, K.B.L., et al., 2022. Microbial nitrogen transformation potential in sediments of two contrasting lakes is spatially structured but seasonally stable. Msphere 7 (1), e01013–e01021.

Beaulieu, J.J., Tank, J.L., Hamilton, S.K., Wollheim, W.M., Hall Jr., R.O., Mulholland, P.J., Thomas, S.M., 2011. Nitrous oxide emission from denitrification in stream and river networks. Proc. Natl. Acad. Sci. U.S.A. 108 (1), 214–219.

Bengtsson, M.M., et al., 2014. No evidence of aquatic priming effects in hyporheic zone microcosms. Sci. Rep. 4, 5187.

Berman, M.C., et al., 2020. Field evidence supports former experimental claims on the stimulatory effect of glyphosate on picocyanobacteria communities. Sci. Total Environ. 701, 134601.

Bingeman, C.W., Varner, J.E., Martin, W.P., 1953. The effect of the addition of organic materials on the decomposition of an organic soil. Soil Sci. Soc. Am. J. 17 (1), 34–38.

Birk, S., et al., 2020. Impacts of multiple stressors on freshwater biota across spatial scales and ecosystems. Nat. Ecol. Evol. 4 (8), 1060–1068.

Bo, T., Doretto, A., Laini, A., Bona, F., Fenoglio, S., 2016. Biomonitoring with macroinvertebrate communities in Italy: what happened to our past and what is the future? J. Limnol. 76 (s1), 21–28. https://doi.org/10.4081/jlimnol.2016.1584.
Bohan, D.A., et al., 2017. Next-generation global biomonitoring: large-scale, automated reconstruction of ecological networks. Trends Ecol. Evol. 32 (7), 477–487.
Bordas, A.C., Brady, M.S., Siewierski, M., Katz, S.E., 1997. In vitro enhancement of antibiotic resistance development—interaction of residue levels of pesticides and antibiotics. J. Food Prot. 60 (5), 531–536. https://doi.org/10.4315/0362-028X-60.5.531.
Buck, M., et al., 2021. Comprehensive dataset of shotgun metagenomes from oxygen stratified freshwater lakes and ponds. Sci. Data 8 (1), 1–10.
Callicott, K., Hooper-Bùi, L., 2019. An investigation of environmental concentrations of atrazine, chlorothalonil, and fipronil in a Mississippi River-influenced marsh in southern Louisiana. Microchem. J. 146, 1241–1248.
Cao, D., et al., 2021. Prevalence of azole-resistant Aspergillus fumigatus is highly associated with azole fungicide residues in the fields. Environ. Sci. Technol. 55 (5), 3041–3049.
Castro-Gutiérrez, V., et al., 2020. Genomic basis for pesticide degradation revealed by selection, isolation and characterisation of a library of metaldehyde-degrading strains from soil. Soil Biol. Biochem. 142, 107702.
Catalán, N., et al., 2015. Absence of a priming effect on dissolved organic carbon degradation in lake water. Limnol. Oceanogr. 60 (1), 159–168.
Cauwet, G., 2002. DOM in the coastal zone. In: Hansell, D.A., Carlson, C.A. (Eds.), Biogeochemistry of Marine Dissolved Organic Matter. Academic Press, London, UK, pp. 579–609.
Cavicchioli, R., et al., 2019. Scientists' warning to humanity: microorganisms and climate change. Nat. Rev. Microbiol. 17 (9), 569–586.
Chan, A.W., Naphtali, J., Schellhorn, H.E., 2019. High-throughput DNA sequencing technologies for water and wastewater analysis. Sci. Prog. 102, 351–376.
Chará-Serna, A.M., et al., 2019. Nutrients and sediment modify the impacts of a neonicotinoid insecticide on freshwater community structure and ecosystem functioning. Sci. Total Environ. 692, 1291–1303.
Christensen, J.H., Christensen, O.B., 2003. Severe summertime flooding in Europe. Nature 421 (6925), 805–806.
Chessman, B.C., 1995. Rapid assessment of rivers using macroinvertebrates: a procedure based on habitat-specific sampling, family level identification and a biotic index. Austral. J. Ecol. 20, 122–129.
Christensen, M.R., et al., 2006. Multiple anthropogenic stressors cause ecological surprises in boreal lakes. Glob. Chang. Biol. 12 (12), 2316–2322.
Chua, A.Q., Verma, M., Hsu, L.Y., Legido-Quigley, H., 2021. An analysis of national action plans on antimicrobial resistance in Southeast Asia using a governance framework approach. Lancet Reg. Health West. Pac. 7, 100084. https://doi.org/10.1016/j.lanwpc.2020.100084.
Clark, D.R., Ferguson, R.M.W., Harris, D.N., Matthews Nicholass, K.J., Prentice, H.J., Randall, K.C., et al., 2018. Streams of data from drops of water: 21st century molecular microbial ecology. WIREs Water 5, e1280.
Clark, D.R., McKew, B.A., Binley, A., Heppell, C.M., Whitby, C., Trimmer, M., 2022. Hydrological properties predict the composition of microbial communities cycling methane and nitrogen in rivers. ISME Commun. 2 (1), 1–9.
Clarke, A., Azulai, D., Dueker, M.E., Vos, M., Perron, G.G., 2019. Triclosan alters microbial communities in freshwater microcosms. Water 11, 961.
Cole, J.J., Prairie, Y.T., Caraco, N.F., McDowell, W.H., Tranvik, L.J., Striegl, R.G., et al., 2007. Plumbing the global carbon cycle: integrating inland waters into the terrestrial carbon budget. Ecosystems 10, 172–185.

Crain, C.M., Kroeker, K., Halpern, B.S., 2008. Interactive and cumulative effects of multiple human stressors in marine systems. Ecol. Lett. 11 (12), 1304–1315.
Crookes, M.J., Fisk, P., 2018. Evaluation of using mobility of chemicals in the environment to fulfil bioaccumulation criteria of the Stockholm convention. Peter Fisk Associates, pp. 1–85. 5758319(5758319).
Culp, J.M., et al., 2011. Incorporating traits in aquatic biomonitoring to enhance causal diagnosis and prediction. Integr. Environ. Assess. Manag. 7 (2), 187–197.
da Costa Naíla, B., et al., 2021. Resistance, resilience, and functional redundancy of freshwater bacterioplankton communities facing a gradient of agricultural stressors in a mesocosm experiment. Mol. Ecol. 30 (19), 4771–4788.
da Costa Naíla, B., et al., 2022. A glyphosate-based herbicide cross-selects for antibiotic resistance genes in bacterioplankton communities. mSystems 7 (2), e0148221.
Danger, M., et al., 2013. Benthic algae stimulate leaf litter decomposition in detritus-based headwater streams: a case of aquatic priming effect? Ecology 94 (7), 1604–1613.
Deng, Y., Debognies, A., Zhang, Q., Zhang, Z., Zhou, Z., Zhang, J., et al., 2022. Effects of ofloxacin on the structure and function of freshwater microbial communities. Aquat. Toxicol. 244, 106084.
Derocles, S.A.P., et al., 2018. Biomonitoring for the 21st century: integrating next-generation sequencing into ecological network analysis. In: Advances in Ecological Research. vol. 58. Academic Press, pp. 1–62.
Diaz, J.M.C., et al., 2017. Fate and effect of imidacloprid on vermicompost-amended soils under dissimilar conditions: risk for soil functions, structure, and bacterial abundance. Sci. Total Environ. 579, 1111–1119.
Dimitrov, M.R., et al., 2014. Assessing effects of the fungicide tebuconazole to heterotrophic microbes in aquatic microcosms. Sci. Total Environ. 490, 1002–1011.
Dinh, Q., Moreau-Guigon, E., Labadie, P., Alliot, F., Teil, M.J., Blanchard, M., Eurin, J., Chevreuil, M., 2017. Fate of antibiotics from hospital and domestic sources in a sewage network. Sci. Total Environ. 575, 758–766. https://doi.org/10.1016/j.scitotenv.2016.09.118.
Djemiel, C., Maron, P.-A., Terrat, S., Dequiedt, S., Cottin, A., Ranjard, L., 2022. Inferring microbiota functions from taxonomic genes: a review. Gigascience 11, 1–30. https://doi.org/10.1093/gigascience/giab090.
Dolédec, S., Statzner, B., 2008. Invertebrate traits for the biomonitoring of large European rivers: an assessment of specific types of human impact. Freshw. Biol. 53 (3), 617–634.
Donnadieu, F., et al., 2016. Influence of streambed substratum composition on stream microbial communities exposed to the fungicide tebuconazole. Freshw. Biol. 61 (12), 2026–2036.
Drzyzga, D., Lipok, J., 2018. Glyphosate dose modulates the uptake of inorganic phosphate by freshwater cyanobacteria. J. Appl. Phycol. 30 (1), 299–309.
ECHA, 2006. *C&L Inventory—ECHA., Echa.europa.eu*. Available at: https://echa.europa.eu/information-on-chemicals/cl-inventory-database/ (Accessed: 25 July 2022).
ECHA, 2007. *Registered Substances—ECHA., Echa.europa.eu*. Available at: https://echa.europa.eu/information-on-chemicals/registered-substances/ (Accessed: 25 July 2022).
ECHA, 2017. Chapter R.11: PBT/vPvB assessment. Guidance on Information Requirements and Chemical Safety Assessment, June. 3.0 European Chemicals Agency, Helsinki, Finland, pp. 66–68.
Escher, B.I., Stapleton, H.M., Schymanski, E.L., 2020. Tracking complex mixtures of chemicals in our changing environment. Science 367 (6476), 388–392.
European Environment Agency (EEA), 2022. Waterbase—Water Quality ICM. Available at: https://www.eea.europa.eu/data-and-maps/data/waterbase-water-quality-icm-2, (last access: 27 June 2022).

Falkowski, P.G., Fenchel, T., Delong, E.F., 2008. The microbial engines that drive Earth's biogeochemical cycles. Science 320 (5879), 1034–1039.

Fera, 2020. Pesticide Usage Survey Report 295, Arable Crops in the United Kingdom. https://pusstats.fera.co.uk/upload/fERxvRulVrwRsvVQVIgL3nDvXfW3ox vLeiISwzGA.pdf.

Ferguson, R.M.W., et al., 2017. The variable influence of dispersant on degradation of oil hydrocarbons in subarctic deep-sea sediments at low temperatures (0–5 C). Sci. Rep. 7 (1), 1–13.

Ferguson, R.M.W., et al., 2021. The ecological impacts of multiple environmental stressors on coastal biofilm bacteria. Glob. Chang. Biol. 27 (13), 3166–3178.

Fernández, M.L., Granados-Chinchilla, F., Rodríguez, C., 2015. A single exposure of sediment sulphate-reducing bacteria to oxytetracycline concentrations relevant to aquaculture enduringly disturbed their activity, abundance and community structure. J. Appl. Microbiol. 119 (2), 354–364.

Fick, J., Söderström, H., Lindberg, R.H., Phan, C., Tysklind, M., Larsson, D.J., 2009. Contamination of surface, ground, and drinking water from pharmaceutical production. Environ. Toxicol. Chem. 28 (12), 2522–2527. https://doi.org/10.1897/09-073.1.

Fisher, M.C., et al., 2018. Worldwide emergence of resistance to antifungal drugs challenges human health and food security. Science 360 (6390), 739–742.

Fleeger, J.W., Carman, K.R., Nisbet, R.M., 2003. Indirect effects of contaminants in aquatic ecosystems. Sci. Total Environ. 317 (1–3), 207–233.

Fordham, D.A., 2015. Mesocosms reveal ecological surprises from climate change. PLoS Biol. 13 (12), e1002323.

Galloway, J.N., Dentener, F.J., Capone, D.G., Boyer, E.W., Howarth, R.W., Seitzinger, S.P., et al., 2004. Nitrogen cycles: past, present, and future. Biogeochemistry 70, 153–226.

Garber, Kristina, US Environmental Protection Agency, Washington, DC, 2009. User's Guide and Technical Documentation KABAM Version 1.0 (KOW (based) Aquatic Bioaccumulation Model). Report of Environmental Fate and Effects, Division Office of pesticide programs. https://www.epa.gov/pesticide-science-and-assessing-pesticide-risks/models-pesticide-risk-assessment#kabam.

Gisi, U., 2014. Assessment of selection and resistance risk for demethylation inhibitor fungicides in Aspergillus fumigatus in agriculture and medicine: a critical review. Pest Manag. Sci. 70 (3), 352–364.

Government of Canada, 2012. Canadian Aquatic Biomonitoring Network (CABIN) wadeable streams protocol. https://publications.gc.ca/collections/collection_2012/ec/En84-87-2012-eng.pdf.

Graves, C.J., et al., 2016. Functional responses of salt marsh microbial communities to long-term nutrient enrichment. Appl. Environ. Microbiol. 82 (9), 2862–2871.

Gray, C., et al., 2014. Ecological networks: the missing links in biomonitoring science. J. Appl. Ecol. 51 (5), 1444–1449.

Grenni, P., 2022. Antimicrobial resistance in rivers: a review of the genes detected and new challenges. Environ. Toxicol. Chem. 41 (3), 687–714.

Grossart, H., Massana, R., McMahon, K.D., Walsh, D.A., 2020. Linking metagenomics to aquatic microbial ecology and biogeochemical cycles. Limnol. Oceanogr. 65, S2–S20.

Guan, Y., et al., 2018. Analysis of bacterial community characteristics, abundance of antibiotics and antibiotic resistance genes along a pollution gradient of Ba River in Xi'an, China. Front. Microbiol. 9, 3191.

Gustafson, D., 1989. Groundwater ubiquity score: a simple method for assessing pesticide leachability. Environ. Toxicol. Chem. 8 (4), 339–357.

Haldane, J., 1895. The action of carbonic oxide on man. J. Physiol. 18 (5–6), 430.

Hale, S.E., Arp, H.P.H., Schliebner, I., Neumann, M., 2020. Persistent, mobile and toxic (PMT) and very persistent and very mobile (vPvM) substances pose an equivalent level of concern to persistent, bioaccumulative and toxic (PBT) and very persistent and very bioaccumulative (vPvB) substances under REACH. Environ. Sci. Europe 32 (1). Springer, Berlin Heidelberg.

Harnisz, M., Korzeniewska, E., Gołaś, I., 2015. The impact of a freshwater fish farm on the community of tetracycline-resistant bacteria and the structure of tetracycline resistance genes in river water. Chemosphere 128, 134–141.

Henríquez-Núñez, H., et al., 2012. Antimicrobial susceptibility and plasmid profiles of Flavobacterium psychrophilum strains isolated in Chile. Aquaculture 354, 38–44.

Hernández-Alomia, F., Ballesteros, I., Castillejo, P., 2022. Bioremediation potential of glyphosate-degrading microorganisms in eutrophicated Ecuadorian water bodies. Saudi J. Biol. Sci. 29 (3), 1550–1558.

Hilsenhoff, W.L., 1987. An improved biotic index of organic stream pollution. Gt. Lakes Entomol. 20, 31–39.

Holmstrup, M., et al., 2010. Interactions between effects of environmental chemicals and natural stressors: a review. Sci. Total Environ. 408 (18), 3746–3762.

Hugenholtz, P., Goebel, B.M., Pace, N.R., 1998. Impact of culture-independent studies on the emerging phylogenetic view of bacterial diversity. J. Bacteriol. 180, 4765–4774.

INCHEM, 2002. *International Programme on Chemical Safety (IPCS)., Inchem.org*. Available at: https://inchem.org/#/ (Accessed: 25 July 2022).

ISO 7899-2:2000 Water Quality—Detection and Enumeration of Intestinal Enterococci—Part 2: Membrane Filtration Method.

ISO 9308-1:2014, Water Quality—Enumeration of Escherichia coli and Coliform Bacteria—Part 1: Membrane Filtration Method for Waters with Low Bacterial Background Flora.

Jackson, M.C., et al., 2016. Net effects of multiple stressors in freshwater ecosystems: a meta-analysis. Glob. Chang. Biol. 22 (1), 180–189.

Jiang, J.-G., 2006. Development of a new biotic index to assess freshwater pollution. Environ. Pollut. 139, 306–317.

Jochum, M., et al., 2012. Climate-induced changes in bottom-up and top-down processes independently alter a marine ecosystem. Philos. Trans. R Soc B Biol Sci 367 (1605), 2962–2970.

Johnson, J.S., Spakowicz, D.J., Hong, B.-Y., Petersen, L.M., Demkowicz, P., Chen, L., et al., 2019. Evaluation of 16S rRNA gene sequencing for species and strain-level microbiome analysis. Nat. Commun. 10, 5029.

Johnson, A.C., et al., 2020. Learning from the past and considering the future of chemicals in the environment. Science 367 (6476), 384–387.

Karr, J.R., 1981. Assessment of biotic integrity using fish communities. Fisheries 6, 21–27.

Kearns, P.J., et al., 2016. Nutrient enrichment induces dormancy and decreases diversity of active bacteria in salt marsh sediments. Nat. Commun. 7 (1), 1–9.

Kelly, M.G., Whitton, B.A., 1995. The trophic diatom index: a new index for monitoring eutrophication in rivers. J. Appl. Phycol. 7, 433–444.

Kidd, K.A., et al., 2014. Direct and indirect responses of a freshwater food web to a potent synthetic oestrogen. Philos. Trans. R Soc B Biol Sci 369 (1656), 20130578.

Kinney, C.A., et al., 2006. Survey of organic wastewater contaminants in biosolids destined for land application. Environ. Sci. Technol. 40 (23), 7207–7215.

Kolkwitz, R., Marsson, K., 1909. Ökologie der tierischen saprobien. Beiträge zur lehre von des biologischen gewasserbeurteilung. Int. Rev. gesamten Hydrobiol. 2, 126–152.

Krausfeldt, L.E., Farmer, A.T., Castro Gonzalez, H.F., Zepernick, B.N., Campagna, S.R., Wilhelm, S.W., 2019. Urea is both a carbon and nitrogen source for Microcystis aeruginosa: tracking 13C incorporation at bloom pH conditions. Front. Microbiol. 10, 1064.

Kuehn, K.A., et al., 2014. Priming in the microbial landscape: periphytic algal stimulation of litter-associated microbial decomposers. Ecology 95 (3), 749–762.

Kümmerer, K., Henninger, A., 2003. Promoting resistance by the emission of antibiotics from hospitals and households into effluent. Clin. Microbiol. Infect. 9 (12), 1203–1214. https://doi.org/10.1111/j.1469-0691.2003.00739.x.

Kurenbach, B., Gibson, P.S., Hill, A.M., Bitzer, A.S., Silby, M.W., Godsoe, W., Heinemann, J.A., 2017. Herbicide ingredients change Salmonella enterica sv. Typhimurium and Escherichia coli antibiotic responses. Microbiology 163 (12), 1791. https://doi.org/10.1099/mic.0.000573.

Larsson, D.J., 2014. Antibiotics in the environment. Ups. J. Med. Sci. 119 (2), 108–112. https://doi.org/10.3109/03009734.2014.896438.

Larsson, D.J., de Pedro, C., Paxeus, N., 2007. Effluent from drug manufactures contains extremely high levels of pharmaceuticals. J. Hazard. Mater. 148 (3), 751–755. https://doi.org/10.1016/j.jhazmat.2007.07.008.

Laskowski, R., et al., 2010. Interactions between toxic chemicals and natural environmental factors—A meta-analysis and case studies. Sci. Total Environ. 408 (18), 3763–3774.

Laudadio, I., Fulci, V., Palone, F., Stronati, L., Cucchiara, S., Carissimi, C., 2018. Quantitative assessment of shotgun metagenomics and 16S rDNA amplicon sequencing in the study of human gut microbiome. OMICS **22**, 248–254.

Li, D., et al., 2011. Bacterial community characteristics under long-term antibiotic selection pressures. Water Res. 45 (18), 6063–6073.

Liess, M., Von Der Ohe, P.C., 2005. Analyzing effects of pesticides on invertebrate communities in streams. Environ. Toxicol. Chem. 24 (4), 954–965.

Linz, A.M., et al., 2018. Freshwater carbon and nutrient cycles revealed through reconstructed population genomes. PeerJ 6, e6075.

Liu, C., et al., 2012. Past and future trends in grey water footprints of anthropogenic nitrogen and phosphorus inputs to major world rivers. Ecol. Indic. 18, 42–49.

Løkke, H., Ragas, Ad M.J., Holmstrup, M., 2013. Tools and perspectives for assessing chemical mixtures and multiple stressors. Toxicology 313.2-3, 73–82. https://doi.org/10.1016/j.tox.2012.11.009.

Lorenzo, P., Adriana, A., Jessica, S., Carles, B., Marinella, F., Marta, L., Luis, B.J., Pierre, S., 2018. Antibiotic resistance in urban and hospital wastewaters and their impact on a receiving freshwater ecosystem. Chemosphere 206, 70–82. https://doi.org/10.1016/j.chemosphere.2018.04.163.

Lu, T., et al., 2020. Understanding the influence of glyphosate on the structure and function of freshwater microbial community in a microcosm. Environ. Pollut. 260, 114012.

Maruyama, A., et al., 2014. The release rate of environmental DNA from juvenile and adult fish. PLoS One 9 (12), e114639.

McElroy, D.J., et al., 2015. Size-balanced community reorganization in response to nutrients and warming. Glob. Chang. Biol. 21 (11), 3971–3981.

McGinley, J., Harmon O'Driscoll, J., Healy, M.G., Ryan, P.C., Mellander, P.E., Morrison, L., Callery, O., Siggins, A., 2022. An assessment of potential pesticide transmission, considering the combined impact of soil texture and pesticide properties: a meta- analysis. Soil Use Manag. 38 (2), 1162–1171. https://doi.org/10.1111/sum.12794.

Menezes, S., Baird, D.J., Soares, A.M.V.M., 2010. Beyond taxonomy: a review of macroinvertebrate trait-based community descriptors as tools for freshwater biomonitoring. J. Appl. Ecol. 47 (4), 711–719.

Miller, T.H., et al., 2019. Biomonitoring of pesticides, pharmaceuticals and illicit drugs in a freshwater invertebrate to estimate toxic or effect pressure. Environ. Int. 129, 595–606.

Miller, T.H., et al., 2021. Multicompartment and cross-species monitoring of contaminants of emerging concern in an estuarine habitat. Environ. Pollut. 270, 116300.

Min, X.J., Hickey, D.A., 2007. Assessing the effect of varying sequence length on DNA barcoding of fungi. Molecular Ecology Notes 7 (3), 365–373.

Moe, S.J., et al., 2013. Combined and interactive effects of global climate change and toxicants on populations and communities. Environ. Toxicol. Chem. 32 (1), 49–61.
Munita, J.M., Arias, C.A., 2016. Mechanisms of antibiotic resistance. Microbiol. Spectr. 4 (2), 1–24. https://doi.org/10.1128/microbiolspec.VMBF-0016-2015.
Murphy, A., Barich, D., Fennessy, M.S., Slonczewski, J.L., 2021. An Ohio State scenic river shows elevated antibiotic resistance genes, including acinetobacter tetracycline and macrolide resistance, downstream of wastewater treatment plant effluent. Microbiol. Spectr. 9 (2), e0094121.
Murray, C.J.L., et al., 2022. Global burden of bacterial antimicrobial resistance in 2019: a systematic analysis. Lancet 399 (10325), 629–655.
O'Neil, J., 2014. Review on antibiotic resistance. Antimicrobial resistance: Tackling a crisis for the health and wealth of nations. In: Health and Wealth Nations, (December), pp. 1–16. Available at: https://amr-review.org/sites/default/files/AMR Review Paper—Tackling a crisis for the health and wealth of nations_1.pdf.
Ormerod, S.J., et al., 2010. Multiple stressors in freshwater ecosystems. Freshw. Biol. 55, 1–4.
Orr, J.A., et al., 2020. Towards a unified study of multiple stressors: divisions and common goals across research disciplines. Proc. R. Soc. B 287 (1926), 20200421.
Pascault, N., et al., 2014. A high-throughput sequencing ecotoxicology study of freshwater bacterial communities and their responses to tebuconazole. FEMS Microbiol. Ecol. 90 (3), 563–574.
Patrick, R., 1949. A proposed biological measure of stream conditions based on a survey of Conestoga Basin, Lancaster County, Pennsylvania. Proc. Acad. Nat. Sci. Philadelphia 101, 277–341.
Peng, F.-J., Diepens, N.J., Pan, C.-G., Ying, G.-G., Salvito, D., Selck, H., Van den Brink, P.J., 2019. Response of sediment bacterial community to triclosan in subtropical freshwater benthic microcosms. Environ. Pollut. 248, 676–683.
Pérez, G.L., et al., 2007. Effects of the herbicide Roundup on freshwater microbial communities: a mesocosm study. Ecol. Appl. 17 (8), 2310–2322.
Perkins, R., et al., 2021. Potential role of veterinary flea products in widespread pesticide contamination of English rivers. Sci. Total Environ. 755, 143560.
Pesce, S., Batisson, I., Bardot, C., Fajon, C., Portelli, C., Montuelle, B., Bohatier, J., 2009. Response of spring and summer riverine microbial communities following glyphosate exposure. Ecotoxicol. Environ. Saf. 72 (7), 1905–1912. https://doi.org/10.1016/j.ecoenv.2009.07.004.
Pesce, S., et al., 2016. Combined effects of drought and the fungicide tebuconazole on aquatic leaf litter decomposition. Aquat. Toxicol. 173, 120–131.
Pimentão, A.R., et al., 2020. Fungistatic effect of agrochemical and pharmaceutical fungicides on non-target aquatic decomposers does not translate into decreased fungi-or invertebrate-mediated decomposition. Sci. Total Environ. 712, 135676.
Pizarro, H., et al., 2016. Glyphosate input modifies microbial community structure in clear and turbid freshwater systems. Environ. Sci. Pollut. Res. 23 (6), 5143–5153.
Poquet, J.M., et al., 2009. The MEDiterranean Prediction And Classification System (MEDPACS): an implementation of the RIVPACS/AUSRIVAS predictive approach for assessing Mediterranean aquatic macroinvertebrate communities. Hydrobiologia 623 (1), 153–171.
Poretsky, R., Rodriguez-R, L.M., Luo, C., Tsementzi, D., Konstantinidis, K.T., 2014. Strengths and limitations of 16S rRNA gene amplicon sequencing in revealing temporal microbial community dynamics. PLoS One 9, e93827.
Quince, C., Walker, A.W., Simpson, J.T., Loman, N.J., Segata, N., 2017. Shotgun metagenomics, from sampling to analysis. Nat. Biotechnol. 35, 833–844.
Ranjan, R., Rani, A., Metwally, A., McGee, H.S., Perkins, D.L., 2016. Analysis of the microbiome: advantages of whole genome shotgun versus 16S amplicon sequencing. Biochem. Biophys. Res. Commun. 469, 967–977.

Richardson, A.K., et al., 2022. A miniaturized passive sampling-based workflow for monitoring chemicals of emerging concern in water. Sci. Total Environ. 839, 156260.

Rolph, C.A., et al., 2019. From full-scale biofilters to bioreactors: engineering biological metaldehyde removal. Sci. Total Environ. 685, 410–418.

Rosentreter, J.A., Borges, A.V., Deemer, B.R., Holgerson, M.A., Liu, S., Song, C., Eyre, B.D., 2021. Half of global methane emissions come from highly variable aquatic ecosystem sources. Nat. Geosci. 14 (4), 225–230.

Rüdel, H., Körner, W., Letzel, T., Neumann, M., Nödler, K., Reemtsma, T., 2020. Persistent, mobile and toxic substances in the environment: a spotlight on current research and regulatory activities. Environ. Sci. Eur. 32 (5). https://doi.org/10.1186/s12302-019-0286-x.

Sagova-Mareckova, M., et al., 2021. Expanding ecological assessment by integrating microorganisms into routine freshwater biomonitoring. Water Res. 191, 116767.

Sakkas, V.A., Lambropoulou, D.A., Albanis, T.A., 2002. Study of chlorothalonil photodegradation in natural waters and in the presence of humic substances. Chemosphere 48 (9), 939–945.

Saxton, M.A., et al., 2011. Glyphosate influence on phytoplankton community structure in Lake Erie. J. Great Lakes Res. 37 (4), 683–690.

Seitzinger, S.P., et al., 2010. Global river nutrient export: a scenario analysis of past and future trends. Global Biogeochem. Cycles 24, 4.

Shakya, M., Quince, C., Campbell, J.H., Yang, Z.K., Schadt, C.W., Podar, M., 2013. Comparative metagenomic and rRNA microbial diversity characterization using archaeal and bacterial synthetic communities. Environ. Microbiol. 15 (6), 1882–1899.

Sharma, K.V., Sarvalingam, B.K., Marigoudar, S.R., 2021. A review of mesocosm experiments on heavy metals in marine environment and related issues of emerging concerns. Environ. Sci. Pollut. Res. 28 (2), 1304–1316.

Simpson, J.C., Norris, R.H., 2000. Biological assessment of river quality: Development of AUSRIVAS models and outputs. In: Wright, J.F., Sutcliffe, D.W., Furse, M.T. (Eds.), Assessing the Biological Quality of Fresh Waters: RIVPACS and Other Techniques. Cumbria (UK), Freshwater Biological Association, pp. 143–163.

Singer, A.C., et al., 2016. Review of antimicrobial resistance in the environment and its relevance to environmental regulators. Front. Microbiol. 7, 1728.

Singer, A.C., Qiuying, X., Keller, V.D.J., 2019. Translating antibiotic prescribing into antibiotic resistance in the environment: a hazard characterisation case study. PLoS One 14 (9), e0221568.

Singh, R., et al., 2019. Antibiotic resistance in major rivers in the world: a systematic review on occurrence, emergence, and management strategies. J. Clean. Prod. 234, 1484–1505.

Smith, V.H., Schindler, D.W., 2009. Eutrophication science: where do we go from here? Trends Ecol. Evol. 24 (4), 201–207.

Stewart, R.I.A., et al., 2013. Mesocosm experiments as a tool for ecological climate-change research. Adv. Ecol. Res. 48, 71–181.

Stockholm Convention, United Nations Environment Programme (UNEP), Switzerland, 2019. Stockholm Convention on persistent organic pollutants (POPs). Text and Annexes. http://chm.pops.int/theconvention/overview/textoftheconvention/tabid/2232/default.aspx.

Taberlet, P., et al., 2012. Towards next-generation biodiversity assessment using DNA metabarcoding. Mol. Ecol. 21 (8), 2045–2050.

Tan, B., Ng, C., Nshimyimana, J.P., Loh, L.L., Gin, K.Y.-H., Thompson, J.R., 2015. Next-generation sequencing (NGS) for assessment of microbial water quality: current progress, challenges, and future opportunities. Front. Microbiol. 6, 1027.

Tedersoo, L., Anslan, S., 2019. Towards PacBio-based pan-eukaryote metabarcoding using full-length ITS sequences. Environ. Microbiol. Rep. 11, 659–668.

The World Health Organisation, 2020. Global Health Estimates: Leading Causes of Death. Available at: https://www.who.int/data/gho/data/themes/mortality-and-global-health-estimates/ghe-leading-causes-of-death.
Thomas, G.E., et al., 2020. Bacterial community legacy effects following the Agia Zoni II oil-spill, Greece. Front. Microbiol. 11, 1706.
Thompson, M.S.A., et al., 2016. Gene-to-ecosystem impacts of a catastrophic pesticide spill: testing a multilevel bioassessment approach in a river ecosystem. Freshw. Biol. 61 (12), 2037–2050.
Tong, T., Xie, S., 2019. Impacts of sulfanilamide and oxytetracycline on methane oxidation and methanotrophic community in freshwater sediment. Ecotoxicology 28 (4), 392–398.
Tran, P.Q., et al., 2021. Depth-discrete metagenomics reveals the roles of microbes in biogeochemical cycling in the tropical freshwater Lake Tanganyika. ISME J. 15 (7), 1971–1986.
Tyler, S., et al., 2019. Owner-reported flea treatment measures and skin disease in cats. J. Feline Med. Surg. 21 (4), 282–285.
Ulén, B., Fölster, J., 2007. Recent trends in nutrient concentrations in Swedish agricultural rivers. Sci. Total Environ. 373 (2–3), 473–487.
United Nations Environment Programme, 2019. Stockholm Convention Stockholm.
Anon., 2007. *Pesticide Properties Database., Sitem.herts.ac.uk.* University Hertfordshire. Available at: http://sitem.herts.ac.uk/aeru/ppdb/en/index.htm (Accessed: 25 July 2022).
van der Struijk, L.F., Kroeze, C., 2010. Future trends in nutrient export to the coastal waters of South America: implications for occurrence of eutrophication. Global Biogeochem. Cycles 24, 4.
Vera, M.S., et al., 2012. Direct and indirect effects of the glyphosate formulation Glifosato Atanor® on freshwater microbial communities. Ecotoxicology 21 (7), 1805–1816.
Verweij, P.E., et al., 2020. The one health problem of azole resistance in Aspergillus fumigatus: current insights and future research agenda. Fungal Biol. Rev. 34 (4), 202–214.
Veterinary Medicines Directorate (VMD), 2022. Product Information Database. https://www.vmd.defra.gov.uk/ProductInformationDatabase/Default.aspx.
Voulvoulis, N., Scrimshaw, M.D., Lester, J.N., 2000. Occurrence of four biocides utilized in antifouling paints, as alternatives to organotin compounds, in waters and sediments of a commercial estuary in the UK. Mar. Pollut. Bull. 40 (11), 938–946.
Wall, G., Lopez-Ribot, J.L., 2020. Current antimycotics, new prospects, and future approaches to antifungal therapy. Antibiotics 9 (8), 445.
Wang, H., et al., 2019. Experimental warming reduced topsoil carbon content and increased soil bacterial diversity in a subtropical planted forest. Soil Biol. Biochem. 133, 155–164.
Wang, Y., et al., 2021. Algal blooms modulate organic matter remineralization in freshwater sediments: a new insight on priming effect. Sci. Total Environ. 784, 147087.
Wauchope, R.D., Buttler, T.M., Hornsby, A.G., Augustijn-Beckers, P.W., Burt, J.P., 1992. The SCS/ARS/CES pesticide properties database for environmental decision-making. Rev. Environ. Contam. Toxicol. 123, 1–155. PMID: 1732992.
Wilkinson, J.L., et al., 2022. Pharmaceutical pollution of the world's rivers. Proc. Natl. Acad. Sci. U.S.A. 119 (8), e2113947119.
Williams, A.J., Grulke, C.M., Edwards, J., McEachran, A.D., Mansouri, K., Baker, N.C., Patlewicz, G., Shah, I., Wambaugh, J.F., Judson, R.S., Richard, A.M., 2017. The CompTox Chemistry Dashboard: a community data resource for environmental chemistry. J. Chem. 9 (1), 1–27. Springer International Publishing.
Woodiwiss, F., 1964. The biological system of stream classification used by the Trent River Board. Chem. Ind. 11, 443–447.
Wright, J.F., Furse, M.T., Armitage, P.D., 1993. RIVPACS: a technique for evaluating the biological quality of rivers in the UK. Eur. Water Pollut. Control 3, 15–25.

Yasin, J.A., Kroeze, C., Mayorga, E., 2010. Nutrients export by rivers to the coastal waters of Africa: past and future trends. Global Biogeochem. Cycles 24, 4.

Yates, M.C., Fraser, D.J., Derry, A.M., 2019. Meta-analysis supports further refinement of eDNA for monitoring aquatic species-specific abundance in nature. Environ. DNA 1 (1), 5–13.

Zhang, M., Zhou, Z., Zhang, J., Yu, Y., Sun, L., Lu, T., Qian, H., 2022. Metagenomic ecotoxicity assessment of trace difenoconazole on freshwater microbial community. Chemosphere 294, 133742.

Zhou, Z., et al., 2022. METABOLIC: high-throughput profiling of microbial genomes for functional traits, metabolism, biogeochemistry, and community-scale functional networks. Microbiome 10 (1), 1–22.

Zhu, Y., et al., 2020. Disproportionate increase in freshwater methane emissions induced by experimental warming. Nat. Clim. Change 10 (7), 685–690.

Zolkefli, N., Sharuddin, S.S., Yusoff, M.Z.M., Hassan, M.A., Maeda, T., Ramli, N., 2020. A review of current and emerging approaches for water pollution monitoring. Water 12, 3417.

Wurtsbaugh, W.A., Paerl, H.W., Dodds, W.K., 2019. Nutrients, eutrophication and harmful algal blooms along the freshwater to marine continuum. Wiley Interdiscip. Rev. Water 6 (5), e1373.

y García, C.A.S., et al., 2022. Rethinking the term "glyphosate effect" through the evaluation of different glyphosate-based herbicide effects over aquatic microbial communities. Environ. Pollut. 292, 118382.